Solution NMR of Paramagnetic Molecules

Applications to Metallobiomolecules and Models

Current Methods in Inorganic Chemistry

A book series devoted to theoretical and experimental techniques in inorganic and organometallic chemistry

Current Methods in Inorganic Chemistry Volume 2

Solution NMR of Paramagnetic Molecules

Applications to Metallobiomolecules and Models

Ivano Bertini
Magnetic Resonance Center, Department of Chemistry
University of Florence
Florence, Italy

Claudio Luchinat
Magnetic Resonance Center, Department of Agricultural Biotechnology
University of Florence
Florence, Italy

and

Giacomo Parigi
Magnetic Resonance Center, Department of Agricultural Biotechnology
University of Florence
Florence, Italy

2001

ELSEVIER
Amsterdam – London – New York – Oxford – Paris – Shannon – Tokyo

ELSEVIER SCIENCE B.V.
Sara Burgerhartstraat 25
P.O. Box 211, 1000 AE Amsterdam, The Netherlands

First edition 2001

Library of Congress Cataloging in Publication Data
A catalog record from the Library of Congress has been applied for.

ISBN: 0 444 20529 2

∞ The paper used in this publication meets the requirements of ANSI/NISO Z39.48-1992 (Permanence of Paper).
Printed in The Netherlands.

Preface

Applications of NMR to paramagnetic molecules and biomolecules in solution have grown in both number and sophistication as the technology of the instruments has improved. They now represent a respectable share of all NMR activity.

For these NMR experiments, the general theory of NMR must be understood and, on top of this, the theory of the electron–nucleus interaction and its consequences for the NMR parameters. Therefore, the field of NMR of paramagnetic molecules has its own niche in the entire scientific panorama.

The authors aim to provide an up-to-date report on the state of the field. Our main scientific activity involves structural and dynamic studies of paramagnetic metalloproteins, in a Research center at the University of Florence. The laboratory is a NMR Research Infrastructure resource supported by the European Union to perform a European service. We are thus exposed to the needs of the scientific community, and have responded to them in several ways, from the development of new instruments or part of them to the description of new phenomena and development of new software. Since 1985, together with colleagues from the Universities of Pisa and Siena, we have organized nine Chianti Workshops on Electron and Nuclear Relaxation, a series of conferences well known to the scientific community in the field.

This book is based on our scientific experience, and largely capitalizes on our previous books, of which we here maintain (and try to improve) the pictorial way of presenting theoretical aspects:

I. Bertini, C. Luchinat (1986) *NMR of Paramagnetic Molecules in Biological Systems.* Benjamin/Cummings, Menlo Park, CA.

L. Banci, I. Bertini, C. Luchinat (1991) *Nuclear and Electron Relaxation. The Magnetic Nucleus–Unpaired Electron Coupling in Solution.* VCH, Weinheim.

I. Bertini, C. Luchinat (1996) *NMR of Paramagnetic Substances, 1 edn.* Coord. Chem. Rev. 150, Elsevier, Amsterdam.

With respect to the previous books, there is an attempt here to project the field of paramagnetic molecules into the domains of cross correlations and partially oriented systems.

Chapter 1 describes the interactions between a spin, electronic or nuclear, and a magnetic field: just some basic physics which cannot be avoided. Chapter 2 deals with contact and dipolar shifts. The aim here is to be clearer and more rigorous than ever. Chapter 3 deals with relaxation: a complex subject that we have tried to make simple and pictorial, but also exhaustive and rigorous. The relaxation properties of different metal ions are discussed in the context of their suitability for NMR experiments. Theory and experiments of Nuclear Magnetic Relaxation Dispersions are described. Chapter 4 covers chemical exchange, the effect of diffusion on relaxation, and the effect of bulk magnetic susceptibility on chemical shifts.

Chapters 5 and 6 present some applications. In the former, the electron relaxation properties of various metal ions are reviewed and the consequent nuclear relaxation properties discussed in more detail. An analysis of the shifts is presented for some cases in which a connection with structural features of the metal–ligand moiety has been established. In Chapter 6, the effects of magnetic coupling on the shifts and relaxation are presented theoretically and examples are given.

In Chapters 7 and 8, one-dimensional NOE experiments and a few two-dimensional experiments are presented. Strategies to minimize adverse paramagnetic effects are discussed, as well as ways to exploit such effects to extract structural and dynamic properties. Partial orientation and cross correlation between the Curie magnetic moment relaxation and nuclear dipolar relaxation are also discussed. Chapter 9 deals with the experimental strategies necessary to achieve the highest level of performance in NMR of paramagnetic compounds in solution.

We thank our colleagues, doctors and students who have collaborated with us at our Center of Magnetic Resonance (CERM) for reading and discussing various parts of the books. Interactions with Lucia Banci, Mario Piccioli, Isabella C. Felli, Roberta Pierattelli, Antonio Rosato are acknowledged in particular. We take the opportunity to pay tribute to the late Luigi Sacconi and to William DeW. Horrocks Jr. for their influence on the early scientific career of Ivano Bertini. The help of Seymour H. Koenig at a later stage was important for both Ivano Bertini and Claudio Luchinat. A special role in our understanding of the intricacies of paramagnetic shifts and relaxation has been played over the years by Bruce R. McGarvey and Jozef Kowalewski. More recently we have enjoyed discussing the perspectives of the field with David A. Case and Martin Blackledge.

Contents

Chapter 1

Introduction

This chapter is intended to recall the principles of magnetism, the definition of magnetic induction and of magnetic induction in a vacuum which is referred to as magnetic field. Readers may not recollect that the molar magnetic susceptibility is expressed in cubic meters per mol! Some properties of electron and nuclear spins are reviewed and finally some basic concepts of the magnetic resonance experiments are refreshed. In summary, this chapter should introduce the readers into the language used by the authors.

1.1 MAGNETIC MOMENTS AND MAGNETIC FIELDS

This book will deal with NMR experiments on systems which contain unpaired electrons. Unpaired electrons disturb the experiment to such an extent that quite different conditions are needed. However, since we have to live with molecules bearing unpaired electrons, we do our best to take advantage from these properly designed NMR experiments in order to learn as much as possible regarding the properties of the unpaired electrons and the structure and dynamics or the substance. To be more precise, we are going to exploit NMR in order to learn how the unpaired electron(s) interacts with the resonating nucleus and how these perturbed nuclei provide information typical of NMR experiments.

The nucleus under investigation must have a magnetic moment in order to make the NMR experiment possible. An unpaired electron also has a magnetic moment. A magnetic moment $\boldsymbol{\mu}$ (J T^{-1}) can be visualized as a magnetic dipole (Fig. 1.1). Such magnetic moment causes a magnetic dipolar field. In electromagnetism, this vector is provided by a continuous current in a coil (Fig. 1.2). If a second magnetic moment $\boldsymbol{\mu}_2$ (which we take to be of smaller vector intensity without loss of generality) is within the dipolar field created by the former magnetic moment $\boldsymbol{\mu}_1$ anchored at a distance r, it will orient accordingly, as represented in Fig. 1.3. We can refer to the electronic magnetic moment as the large magnetic moment $\boldsymbol{\mu}_1$ and to the nuclear magnetic moment as the small magnetic moment $\boldsymbol{\mu}_2$. The absolute value of the magnetic moment associated with the electron is 658 times (see Section 1.2) that for a proton, which has the largest magnetic moment among the magnetic nuclei (except tritium). The orientation of the small magnetic moment along the dipolar field of the large magnetic moment shown in

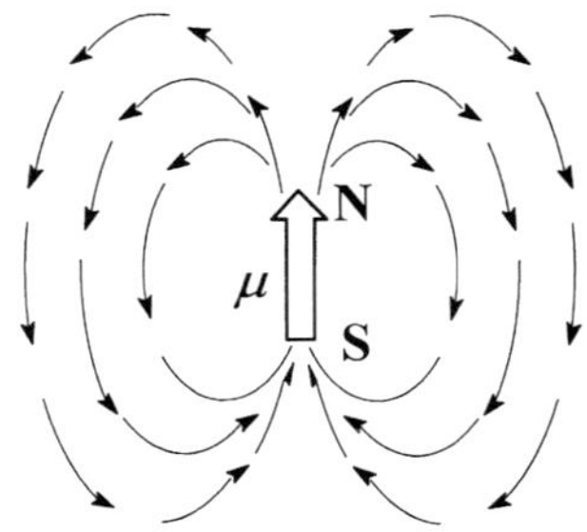

Fig. 1.1. A magnetic moment can be seen as a magnetic dipole $\boldsymbol{\mu}$ characterized by north (N) and south (S) polarities. It gives rise to a magnetic field which is indicated by force lines. The dipolar nature provides the vectorial nature of this moment, whose intensity is indicated by μ.

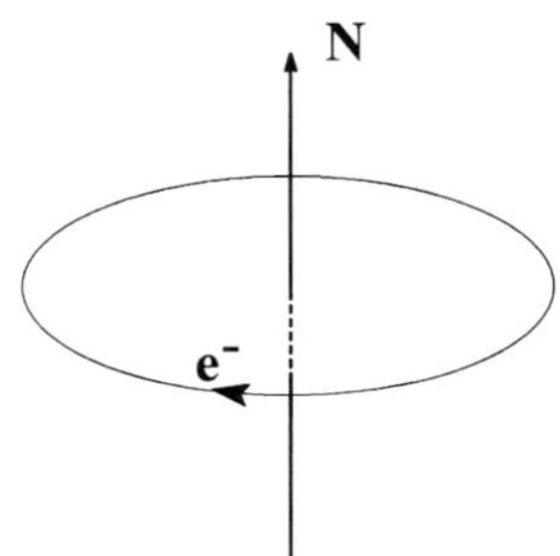

Fig. 1.2. A continuous flow of electricity in a coil provides a magnetic moment which is proportional to the intensity of the current and depends on the coil size.

Fig. 1.3 represents the minimum energy situation. In general, the energy of the interaction between the two magnetic bars depends on the relative orientation of the two vectors, if at equilibrium or fixed by external forces, according to Eq. (1.1):

$$E^{\mathrm{dip}} = -\frac{\mu_0}{4\pi}\left[\frac{3(\boldsymbol{\mu}_1 \cdot \boldsymbol{r})(\boldsymbol{\mu}_2 \cdot \boldsymbol{r})}{r^5} - \frac{\boldsymbol{\mu}_1 \cdot \boldsymbol{\mu}_2}{r^3}\right] \tag{1.1}$$

where μ_0 is the magnetic permeability of a vacuum ($\mathrm{J}^{-1}\ \mathrm{T}^2\ \mathrm{m}^3$), $\boldsymbol{r}$ is the vector connecting the two point dipoles and r is its magnitude. The energy can be negative (stabilization) or positive (destabilization) according to the relative magnitude of the two terms in parentheses.

In all our experiments the two magnetic bars are immersed in an external magnetic field. The intensity of the magnetic field is proportional to the density of force lines (Fig. 1.4). Later, we will be interested in the effective field in a given region of space, which is referred to as magnetic induction $\boldsymbol{B}$ (expressed in tesla):

$$\boldsymbol{B} = \mu_0(\boldsymbol{H} + \boldsymbol{M}) = \boldsymbol{B}_0 + \mu_0\boldsymbol{M} \tag{1.2}$$

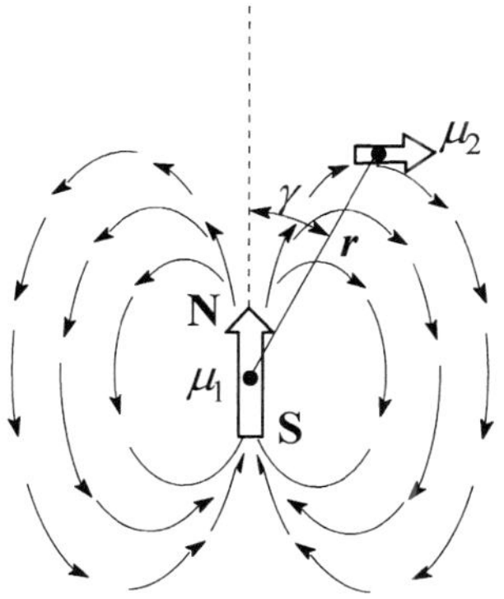

Fig. 1.3. Orientation of a small magnetic moment $\boldsymbol{\mu}_2$ (e.g. that of the nucleus) within a magnetic field generated by a larger magnetic moment $\boldsymbol{\mu}_1$ (e.g. that of the electron) at distance r. Here γ is the angle between $\boldsymbol{\mu}_1$ and $\boldsymbol{r}$.

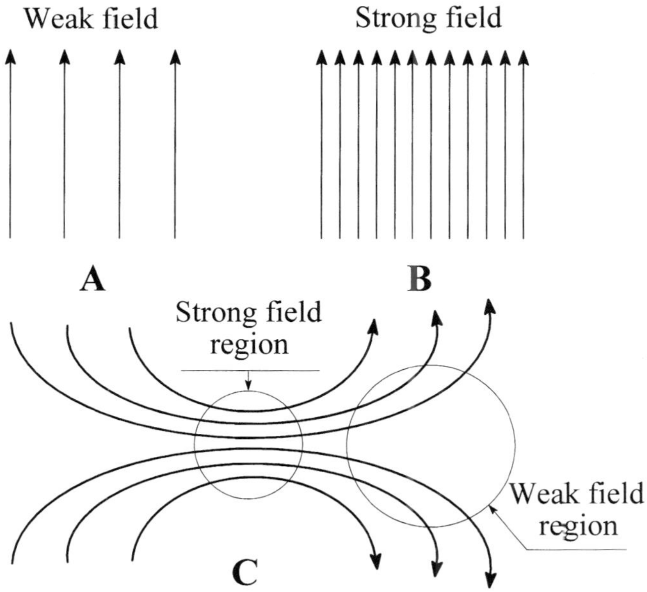

Fig. 1.4. The force lines of a magnetic field $\boldsymbol{B}_0$. (A) A weak, homogeneous field. (B) A strong, homogeneous field. (C) An inhomogeneous field with weak and strong field regions.

where $\boldsymbol{H}$ and $\boldsymbol{M}$ (J T^{-1} m^{-3}) are the magnetic field strength and the magnetization of the medium referred to unit volume respectively, and μ_0 is already defined (Eq. (1.1)). The magnetic induction is thus given by the magnetic induction in a vacuum ($\mu_0\boldsymbol{H} = \boldsymbol{B}_0$) plus a contribution ($\mu_0\boldsymbol{M}$) depending on the kind of substance constituting the medium. In this book, the magnetic induction in a vacuum $\boldsymbol{B}_0$ will always be referred to as the *external magnetic field*.

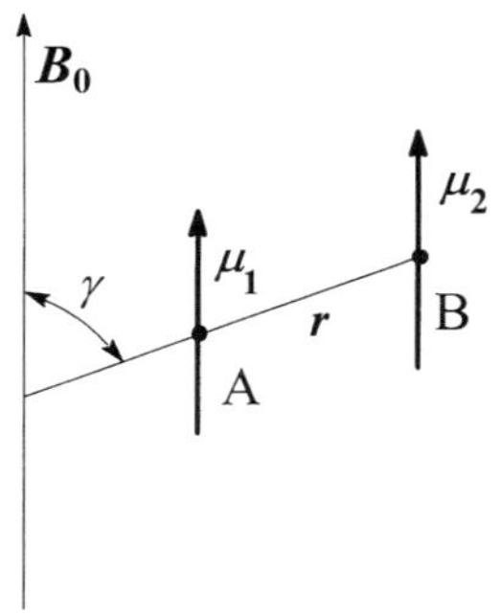

Fig. 1.5. Two magnetic bars, $\boldsymbol{\mu}_1$ and $\boldsymbol{\mu}_2$, anchored at generic points A and B at distance r in a magnetic field $\boldsymbol{B_0}$. γ is the angle between the magnetic field and the AB vector.

The energy E of a magnetic moment μ immersed in a magnetic field $\boldsymbol{B_0}$ is given by

$$\boxed{E = -\boldsymbol{\mu} \cdot \boldsymbol{B_0}} \tag{1.3}$$

Eq. (1.3) shows that the energy is at a minimum when $\boldsymbol{\mu}$ is aligned along $\boldsymbol{B_0}$.

In the absence of further limit conditions which may hold in the case of the electron (see later), we can now think of an electron spin and a nuclear spin anchored at points A and B, both aligned along the external magnetic field $\boldsymbol{B_0}$, as shown in Fig. 1.5. Since the two magnetic moments are forced to be parallel by the strong external field, the energy of the interaction between them, given by Eq. (1.1), simplifies to

$$E^{\mathrm{dip}} = -\frac{\mu_0}{4\pi}\frac{\mu_1\mu_2}{r^3}\left(3\cos^2\gamma - 1\right) \tag{1.4}$$

where γ is the angle between the direction of $\boldsymbol{B_0}$ and that of the AB vector (see Appendix II).

1.2 ABOUT THE SPIN MOMENTS

Electron and nuclear magnetic moments can be regarded as arising from a property of the particles, i.e. that they possess an intrinsic angular momentum as if they were spinning. Such angular momenta are given by

$$\boldsymbol{J}_S = \hbar\boldsymbol{S} \qquad \boldsymbol{J}_I = \hbar\boldsymbol{I} \tag{1.5}$$

for the electron and nucleus respectively, where $\boldsymbol{S}$ and $\boldsymbol{I}$ are dimensionless *spin* angular momentum vectors and $\hbar = h/2\pi$ is the Planck constant (J s rad^{-1}). The moduli of the vectors are given by

$$J_S = \hbar\sqrt{S(S+1)} \qquad J_I = \hbar\sqrt{I(I+1)} \tag{1.6}$$

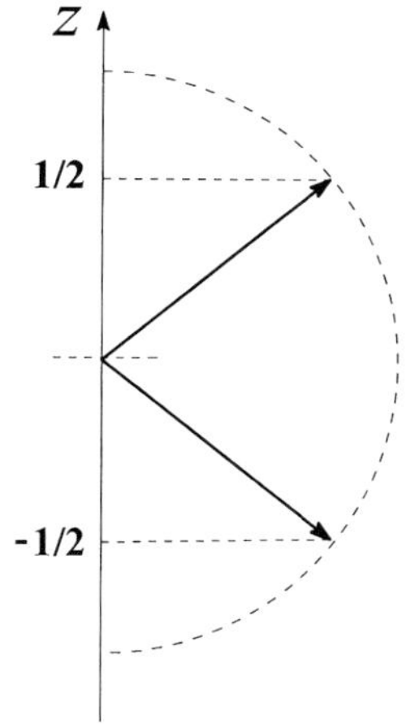

Fig. 1.6. Allowed orientations of an $I = \frac{1}{2}$ angular momentum relative to the z direction defined by the external magnetic field. The vector has modulus $\sqrt{3}/2$, and its projections on the z axis are $\frac{1}{2}$ and $-\frac{1}{2}$.

where S and I are quantum numbers associated with the spinning particle. For a single electron or for a single nucleon (proton or neutron), $S = I = \frac{1}{2}$. S or I identify sets of spin wavefunctions for the above particles. Note that the values of the angular momenta are not related to the nature of the particles.

The projection of $\mathbf{S}$ and $\mathbf{I}$ along a z direction (defined by an external magnetic field or otherwise) are $+\frac{1}{2}$ or $-\frac{1}{2}$ (Fig. 1.6). Thus we have two wavefunctions, one with S (or I) $= \frac{1}{2}$ and with a component along $z = \frac{1}{2}$ and another with S (or I) $= \frac{1}{2}$ and with a component along $z = -\frac{1}{2}$.

The component is indicated in quantum mechanics as M_S or M_I. The notation to indicate the wavefunction is thus

$$|S, M_S\rangle \quad \text{or} \quad |I, M_I\rangle$$

where $|\ \rangle$ is the 'ket' notation for wavefunctions.

These wavefunctions are eigenfunctions of the operators S^2 (I^2) and S_z (I_z):

$$\begin{aligned} S^2|S, M_S\rangle &= S(S+1)|S, M_S\rangle \\ I^2|I, M_I\rangle &= I(I+1)|I, M_I\rangle \\ S_z|S, M_S\rangle &= M_S|S, M_S\rangle \\ I_z|I, M_I\rangle &= M_I|I, M_I\rangle. \end{aligned} \tag{1.7}$$

Physically, it means that it is possible to know simultaneously the square of the intensity of the spin angular momentum and its component along z. Since the spin wavefunctions are not eigenfunctions of the operators S or I, it is impossible to know intensity and orientation of the angular momentum vector simultaneously. We will learn how to live with it!

Since the electron and the proton are charged particles, there is a magnetic moment associated with the angular momenta. The latter is related to a motion, and a motion of a charged particle produces a magnetic moment. The neutron is not charged as a result of balancing of charges of different sign. However, since the charges are not homogeneously distributed from the center of the particle, the neutron also has a magnetic moment associated with the angular momentum.

The intrinsic angular momentum $\boldsymbol{S}$ is related to the intrinsic magnetic moment $\boldsymbol{\mu}_S$ through the relation

$$\boxed{\boldsymbol{\mu}_S = -g_e \mu_B \boldsymbol{S}}$$

and therefore the moduli of $\boldsymbol{\mu}_S$ and $\boldsymbol{\mu}_I$ are given by

$$|\mu_S| = |g_e| \frac{|e|\hbar}{2m_e}\sqrt{S(S+1)} = |g_e|\,\mu_B\sqrt{S(S+1)} \tag{1.8}$$

$$|\mu_I| = |g_I| \frac{|e|\hbar}{2m_p}\sqrt{I(I+1)} = |g_I|\,\mu_N\sqrt{I(I+1)} \tag{1.9}$$

where e is the elementary charge of the electron, g_e is the so-called free electron g value, which is 2.0023 [1]. μ_B and μ_N are the electron Bohr magneton and nuclear magneton, m_e and m_p are the electron and proton masses, and g_I depends on the nucleus under consideration (see later). The ratio of $|\mu_S|$ and $|\mu_I|$ for the proton is 658.2107 [1]. Analogously to the angular momenta, only the projection along z of the magnetic moment and its modulus are known, but not its direction.

Sometimes the magnetogyric ratio γ is used to indicate the ratio between magnetic moments and angular momenta

$$\gamma_S = -\frac{g_e \mu_B}{\hbar} \qquad \gamma_I = \frac{g_I \mu_N}{\hbar}, \tag{1.10}$$

where γ_S and γ_I for the proton have opposite signs, their ratio thus being -658.2107.

If reference is made to Fig. 1.7, it appears that the angle φ is known because the modulus of vector $\boldsymbol{\mu}$ is known, as well as its projection along the z axis, but the orientation of $\boldsymbol{\mu}$ cannot be known

$$\mu_z = \mu \cos\varphi. \tag{1.11}$$

---

[1] In this book g_e is taken positive, and the equations containing g_e are explicitly written in such a way as to contain a positive g_e.

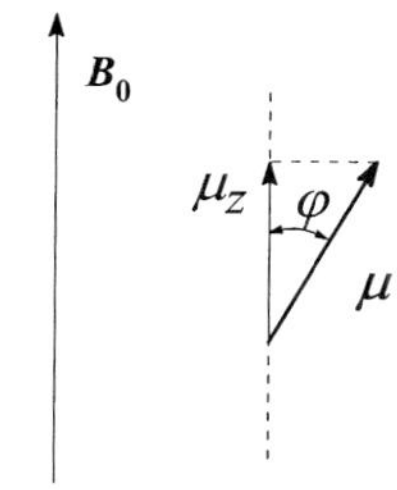

Fig. 1.7. A magnetic moment μ in a magnetic field forming an angle φ with the magnetic field direction.

This nicely reconciles the quanto-mechanical picture with classical physics, which shows that a magnetic moment which must form an angle φ with the direction of an external magnetic field *precesses* about it with an angular frequency

$$\boxed{\omega = -\gamma B_0.} \tag{1.12}$$

The resulting picture of a spin moment (magnetic or angular) in a magnetic field $\boldsymbol{B_0}$ is that it precesses about the $\boldsymbol{B_0}$ direction with an angular frequency proportional to the intensity of $\boldsymbol{B_0}$ and to its own magnetic moment, and with the φ angles such that the projection of $\boldsymbol{\mu}$ along $\boldsymbol{B_0}$ assumes the quanto-mechanically allowed values $\pm\ \mu_z$ (Fig. 1.8). The sign of ω, which is related to the sign of g (Eqs. (1.10) and (1.12)), gives the direction of precession.

In the case of more than one unpaired electron the total spin value S is ½ the number of unpaired electrons. Commonly, we will deal with one to seven unpaired electrons and S can thus take values from ½ to 7/2.

In the case of odd numbers of protons and/or neutrons, a total spin value I varying from ½ to 7 occurs. Owing to the complex intranuclear forces, the g_I values also vary from 5.96 for ^{3}H to 0.097 for ^{191}Ir. The g_I values for magnetically active nuclei are summarized in Appendix I.

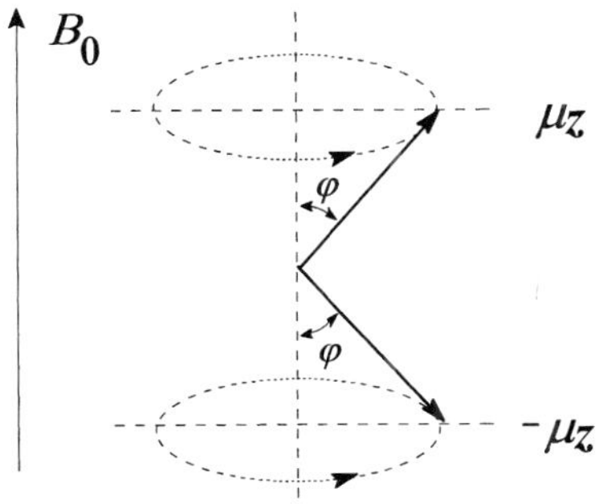

Fig. 1.8. The allowed precessions of a spin I = ½ with negative γ (positive ω) in a magnetic field.

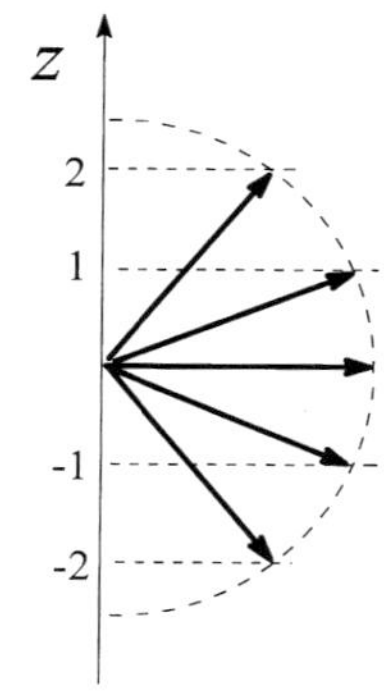

Fig. 1.9. Allowed orientations and z projections of a spin $S = 2$ (or $I = 2$) in a magnetic field.

The number of allowed values of M_S (or M_I) is $2S+1$ (or $2I+1$), and the values range from S to $-S$ (or from I to $-I$), differing by one unit. Fig. 1.9 shows the allowed orientations for a spin $S = 2$ (or $I = 2$). It is just an extension of the $I = S = 1/2$ case.

1.3 SOMETHING MORE ABOUT THE NUCLEAR SPIN

Nuclear spin vectors are localized on the nucleus, at least for the purposes discussed here. Therefore they can be treated as point dipoles. We have already shown that they are described by $2I + 1$ wavefunctions, each characterized by the value I and a value of M_I.

We have already seen that in a magnetic field there are different allowed spin orientations. We want now to point out that the different spin orientations in a magnetic field correspond to different energies. This is quite intuitive by looking at Fig. 1.9. If the orientation of the magnetic spin dipole is different in a magnetic field from case to case, then the interaction energies will be different. Actually, according to Eq. (1.3), the energy will be given by the product of the projection along z of the spin magnetic moment and the external magnetic field

$$E = -g_I \mu_N M_I B_0 \tag{1.13}$$

where $g_I \mu_N M_I$ is the projection of the spin magnetic moment along $\boldsymbol{B_0}$.

In quantum mechanical terms the energy is given by the Hamiltonian operator, which in this case is called the nuclear Zeeman Hamiltonian

$$H = -g_I \mu_N \boldsymbol{I} \cdot \boldsymbol{B_0} \tag{1.14}$$

where $\boldsymbol{I}$ is the spin operator. Now, if we define the z axis along $\boldsymbol{B_0}$, the nuclear spin vectors have a non zero projection along z, whereas they have generally zero

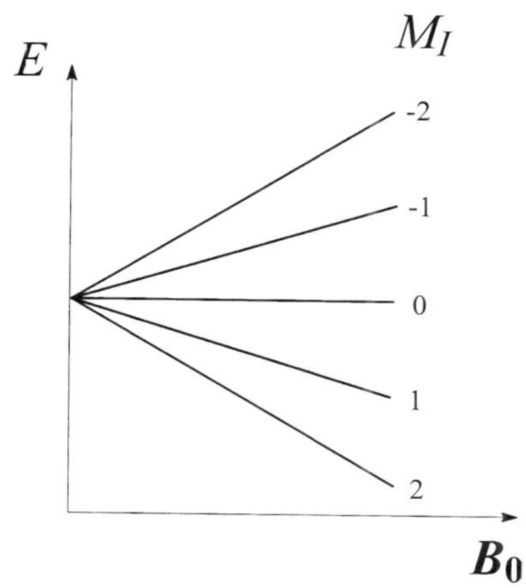

Fig. 1.10. The Zeeman energies of a nuclear spin ($I = 2$) as a function of the external magnetic field $\boldsymbol{B}_0$.

time average in the xy plane. Therefore, we can write

$$H = -g_I \mu_N B_0 I_z. \tag{1.15}$$

Since the application of I_z on a wavefunction $|I, M_I\rangle$ gives M_I (Eq. (1.7)), the energies of interaction between the spin and the magnetic field given in Eq. (1.13) are obtained. Such energies are dependent on the magnitude of the external magnetic field (see Fig. 1.10) and the energy separation ΔE between two adjacent levels is

$$\Delta E = g_I \mu_N B_0 \big(M_I - (M_I - 1)\big) = g_I \mu_N B_0. \tag{1.16}$$

In the absence of an external magnetic field the Zeeman Hamiltonian provides zero energy and all the $|I, M_I\rangle$ levels (termed as I manifold) have the same energy. However, this may not be true for nuclei with $I > ½$. In this case, the non-spherical distribution of the charge causes the presence of a quadrupole moment. Whereas a dipole can be described by a vector with two polarities, a quadrupole can be visualized by two dipoles as in Fig. 1.11.

The presence of a quadrupole moment can make the $|I, M_I\rangle$ levels inequivalent even in the absence of an external magnetic field, provided there is an electric field gradient. Only the wavefunctions with the same absolute value of M_I are pairwise degenerate in axial symmetry, i.e. $M_I = \pm 1, \pm 2$, etc. An example is reported in Fig. 1.12 for $I = 3/2$.

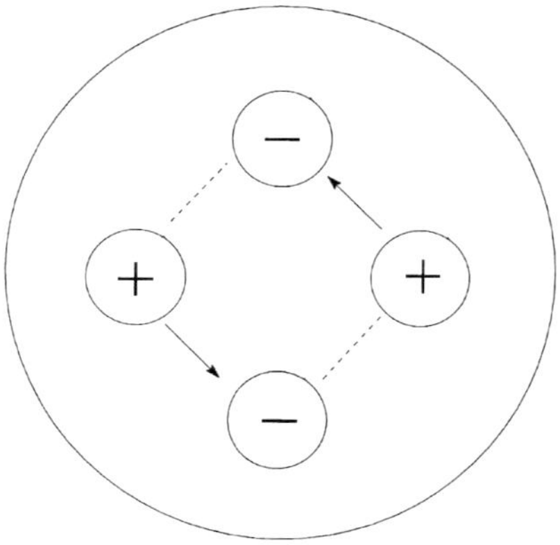

Fig. 1.11. Schematic drawing of a quadrupole moment.

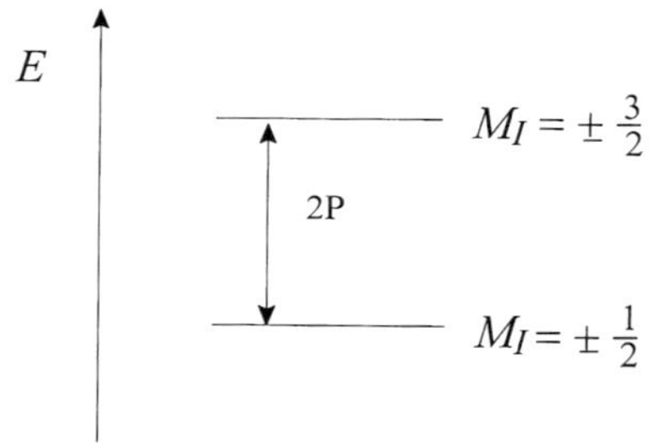

Fig. 1.12. The energy levels of a spin $I = 3/2$ at zero magnetic field in axial symmetry, with P being the product of the quadrupole moment with the electric field gradient.

1.4 A LOT MORE ABOUT THE ELECTRON SPIN

At variance with the nucleus, the electron is associated with an orbital, i.e. a wavefunction which is related to the distribution in space of the electron cloud, and which displays an angular momentum $\boldsymbol{L}$ and a magnetic moment $\boldsymbol{\mu}_L$. In analogy with the spin operators (see Eqs. (1.7)), the following relations hold

$$L^2|l, m_l\rangle = l(l+1)|l, m_l\rangle$$

$$L_z|l, m_l\rangle = m_l|l, m_l\rangle$$

where n, l, m_l are the quantum numbers describing the electron orbital, with $l = 0, \ldots, n$ and $m_l = -l, \ldots, l$.

In a naive and incorrect way we can say that the electron with $S = 1/2$ senses the orbital magnetic moment. Actually, a charged particle cannot sense the orbital magnetic moment due to its own movement. However, the electron moves in the electric potential of the charged nucleus. If we change the system of reference,

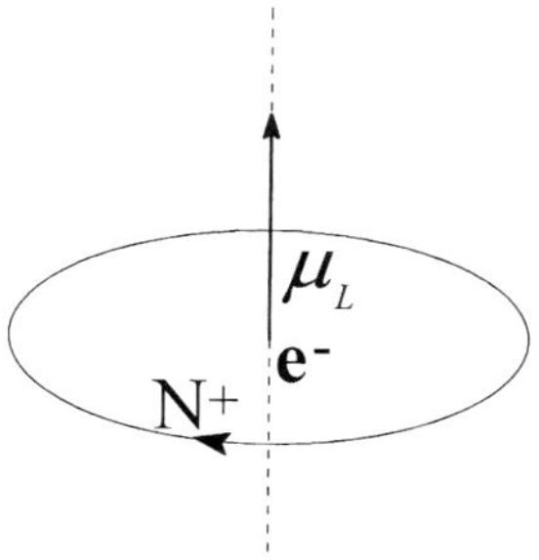

Fig. 1.13. The electron 'senses' the orbital magnetic moment.

the movement of the electron around the nucleus can be seen as a movement of the nucleus around the electron (Fig. 1.13). The 'motion' of the charged nucleus then generates a magnetic field which is sensed by the electron. A convenient way to describe the relative movement of the nucleus with respect to the electron is that of using the same n, l and m_l quantum numbers describing the electron. The resulting angular and magnetic properties will depend on the values of m_s for the spin, m_l for the orbital, and on their interaction. The latter phenomenon is called *spin–orbit coupling* and is of paramount importance in understanding the electronic properties. Traditionally, two different formalisms are used for transition metal ions of the first series on one side and lanthanides on the other. In the latter case spin–orbit coupling is strong, and l and m_l are not good quantum numbers. This case will be treated later (see Chapter 2). In the former case, spin–orbit coupling is small enough to be considered a perturbation. For more than one unpaired electron, total L and M_L can be defined. In a molecule, the ligand field defines internal direction(s) along which the orbital angular momentum is preferentially aligned (quantized). Other orientations have higher energies.

We now let the molecule interact with an external magnetic field $\boldsymbol{B_0}$. The interaction energy, as far as the orbital is concerned, is given by the orbital Zeeman operator

$$H = -\mu \cdot \boldsymbol{B_0} = \mu_B \boldsymbol{L} \cdot \boldsymbol{B_0}. \tag{1.17}$$

This interaction will tend to disalign $\boldsymbol{L}$ from its internal axes (Fig. 1.14A). As a result, when the molecule rotates with respect to $\boldsymbol{B_0}$, the interaction energy of Eq. (1.17) is orientation dependent.

In coordination chemistry, reference is often made to limiting cases in which the orbital contribution tends to zero. In this case, the treatment is equal to the nuclear case and the same Hamiltonian is used (the opposite sign with respect to Eq. (1.14) is justified by the positive g_e).

$$\boxed{H = g_e \mu_B \boldsymbol{S} \cdot \boldsymbol{B_0}} \tag{1.18}$$

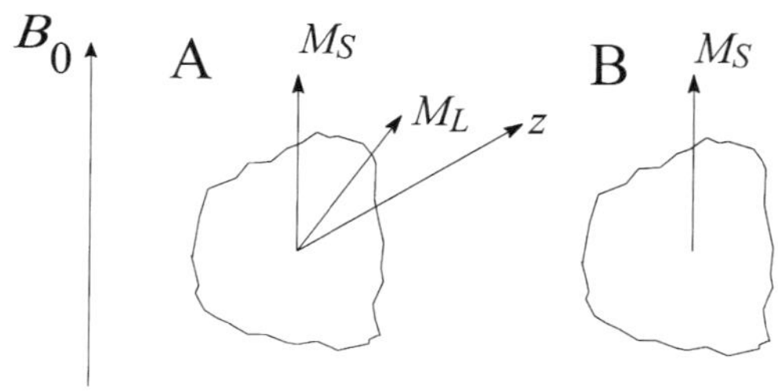

Fig. 1.14. (A) Orientation of M_S and M_L in the presence of internal molecular axes. (B) A case in which the external magnetic field determines the quantization axes.

and

$$E = g_e \mu_B M_S B_0. \tag{1.19}$$

In such a system, the external magnetic field defines the molecular z axis. If we rotate the molecule with respect to $\boldsymbol{B_0}$, the spin and its magnetic moment are not affected (Fig. 1.14B). However, in the molecule of Fig. 1.14A, a molecular z axis can be defined. When rotating the molecule, the orbital contribution to the overall magnetic moment changes, whereas the spin contribution is constant. The total Zeeman Hamiltonian is

$$\boxed{H = \mu_B \left(\boldsymbol{L} + g_e \boldsymbol{S}\right) \cdot \boldsymbol{B_0}.} \tag{1.20}$$

A convenient way to handle Eq. (1.20) is that of defining a tensor $\boldsymbol{g}$ which couples the magnetic moment $\boldsymbol{S}$ with the external magnetic field. Such a tensor defines the coupling between $\boldsymbol{S}$ and $\boldsymbol{B_0}$ for all molecular directions. We can represent the tensor as a solid ellipsoid (Fig. 1.15) with three principal directions defining the axes of the ellipsoid and of the molecule. In any kk direction we have a value of g_{kk} such that

$$g_{kk}^2 = g_{xx}^2 \cos^2 \alpha + g_{yy}^2 \cos^2 \beta + g_{zz}^2 \cos^2 \gamma \tag{1.21}$$

where $\cos\alpha$, $\cos\beta$ and $\cos\gamma$ are the direction cosines of the kk vector. The projections of the total electron magnetic moment along any kk direction defined by $\boldsymbol{B_0}$ are given by $\mu_B g_{kk} M_S$. The energy of the $|S, M_S\rangle$ function, when the magnetic field is along the kk direction is

$$E = \mu_B g_{kk} M_S B_0. \tag{1.22}$$

As we can see, the expression of the energy does not contain L. The Hamiltonian has the form

$$\boxed{H = \mu_B \boldsymbol{S} \cdot \boldsymbol{g} \cdot \boldsymbol{B_0}} \tag{1.23}$$

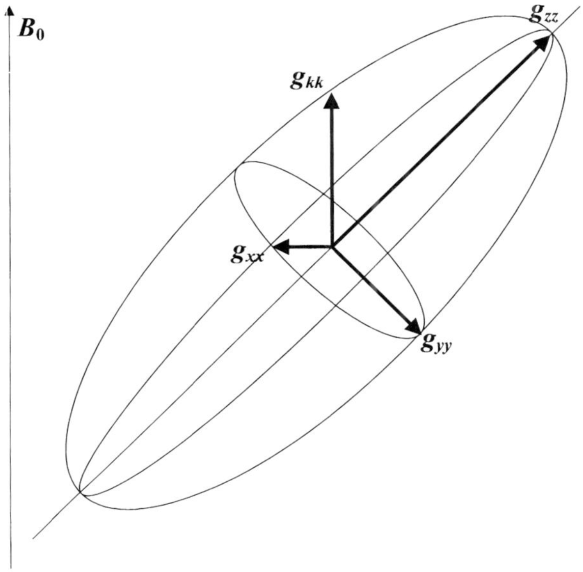

Fig. 1.15. The ellipsoid representing the components of the $\boldsymbol{g}$ tensor in every direction. The molecule to which the tensor is associated has a generic orientation in the magnetic field $\boldsymbol{B_0}$.

which is the scalar product of the $\boldsymbol{S}$ vector (defined in Section 1.2), the $\boldsymbol{g}$ tensor and the $\boldsymbol{B_0}$ vector.

This new formalism, known as spin-Hamiltonian formalism, does not contain the $\boldsymbol{L}$ operator, which would require more laborious calculations. Its effects are parametrically included in the $\boldsymbol{g}$ tensor, which would pass from ellipsoidal to spherical in the absence of orbital angular momentum.

When the molecule under investigation rotates fast with respect to the g anisotropy (i.e. the reorientation rate τ_r^{-1} (see also Section 3.2) is larger than the spreading of the different orientation-dependent energies of the spin ($\tau_r^{-1} > \Delta E/\hbar$)) we measure an average g value $\overline{g}$, which is also different from g_e. The two limit situations of isotropic and anisotropic g are illustrated in Fig. 1.16. When $g = 2.0023$, and therefore the orbital contribution is zero, the splitting of any S manifold is as in Fig. 1.16A and independent of the orientation of the molecule with respect to the external magnetic field; when there is an orbital contribution, a different splitting of the S manifold in any direction occurs (Figs. 1.16B and C), and upon rapid rotation there is an average splitting of the levels.

Besides providing a different effective magnetic moment for each orientation, spin–orbit coupling is also able to cause a splitting of an S manifold with $S > 1/2$ at zero magnetic field. When S is, let us say, $3/2$, spin–orbit coupling and low symmetry effects split the quartet in a way similar to that depicted in Fig. 1.12. When S is half integer, at least two-fold degeneracies remain (so-called Kramers

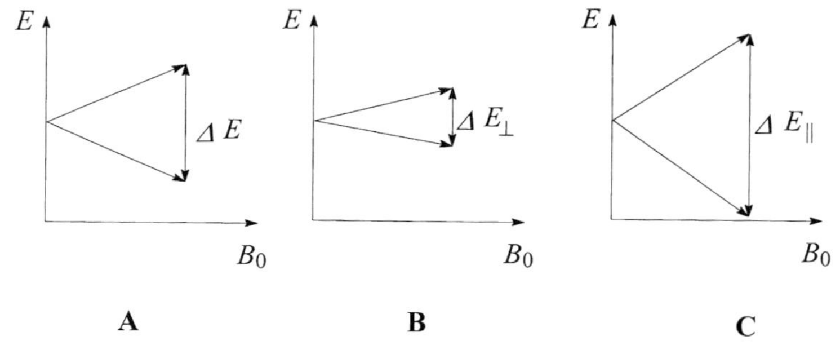

Fig. 1.16. The splitting of the $S = \frac{1}{2}$ manifold in a magnetic field B_0 when (A) g is isotropic and there are only two energy values independent of the orientation of the molecule in the magnetic field and (B,C) the energies depend on the orientation of the molecule in the magnetic field ($\Delta E_{\parallel} > \Delta E_{\perp}$).

doublets, $M_S = \pm n/2$, n integer), whereas when S is integer the splitting can remove any degeneracy (Fig. 1.17).

Such splitting is called zero field splitting and indicated as ZFS. It adds up to the Zeeman energy. In the spin-Hamiltonian formalism, i.e. when the effects of the orbital angular momentum are parameterized, it is indicated as

$$H = \boldsymbol{S} \cdot \boldsymbol{D} \cdot \boldsymbol{S} \tag{1.24}$$

where $\boldsymbol{D}$ is the ZFS tensor. It is traceless, in the sense that its effect upon rapid rotation is zero (rapid means that the rotation rate (s^{-1}) is larger than the maximum energy splitting ($\Delta E/\hbar$ (s^{-1})). However, its appearance is of paramount

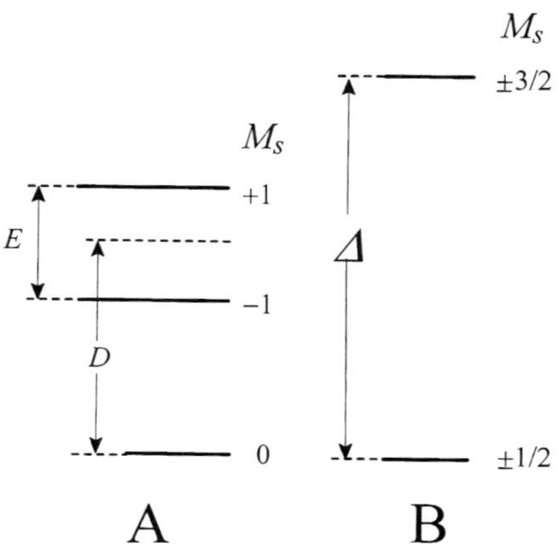

Fig. 1.17. The splitting of an $S = 1$ (A) and $S = \frac{3}{2}$ (B) manifold in the presence of spin orbit coupling and low symmetry components. D is the axial and E the rhombic ZFS parameter (the latter only shown in case A). The wavefunctions are labeled as high field eigenfunctions.

importance in electron relaxation and in determining the magnetic properties of metal complexes. The comparison of Fig. 1.12 with Fig. 1.17B shows that the nuclear quadrupole splitting and the ZFS are formally similar. In general, the ZFS is defined by two parameters, D (axial anisotropy) and E (rhombic anisotropy), that characterize the $\mathbf{D}$ tensor (Fig. 1.17).

Hamiltonian Eq. (1.24) is formally equivalent to that describing the dipolar interaction between two spins s_1 and s_2 whose sum is S. Actually, in organic radicals where spin–orbit interactions are negligibly small, the dipolar interaction between the two electron spins in an $S = 1$ system causes ZFS.

1.5 ABOUT THE ENERGIES

Up to now we have seen that S or I manifolds split in an external magnetic field according to their M_S or M_I values. The latter are the allowed components of the S or I vectors along the external magnetic field. When we said that the spins orient in a magnetic field as in Fig. 1.5, we actually referred to the projection along z relative to the low energy orientation, which is the only populated at $T = 0$ K. The excited levels are separated by the Zeeman energy (see Eqs. (1.15) and (1.16)). Such energies are about 0.3 cm^{-1} at 0.3 T for the electron, and 658 times smaller for the proton. The thermal energy kT is about 200 cm^{-1} at 300 K and about 0.7 cm^{-1} at 1 K. So, the population of the two levels is almost the same at every temperature above a few Kelvin. The Boltzmann population P_i of each M_I level is

$$P_i = \frac{\exp(-E_i/kT)}{\sum_i \exp(-E_i/kT)} \tag{1.25}$$

where E_i is the energy of the ith level with respect to the ground level and the sum is extended to all levels. When $kT \gg E_i$, as happens at room temperature, E_i/kT tends to zero, the exponential tends to unity and each level is almost equally populated. The magnetic resonance experiments are based on the small population differences. The energy of the system (for instance, an ensemble of N_A spins) is given by the sum of the energies for each level weighted by the population of the level.

1.6 MAGNETIZATION AND MAGNETIC SUSCEPTIBILITY

The effect of the external magnetic field is that of splitting the energies of the S or I manifolds (see, for example, Fig. 1.10) and, therefore, of making different the populations of the levels. The difference in population according to the Boltzmann law (Eq. (1.25)) tells us that the magnetic field has indeed changed the energy of the system. By making reference for simplicity to Fig. 1.6 (two orientations), the spins with the lower energy orientation are more than those with the higher energy orientation. As a consequence, an induced magnetic moment μ_{ind} is established. The net interaction energy of the whole system with the magnetic field is the product of the induced magnetic moment and the magnetic field. The magnetization per unit volume M (Eq. (1.2)) corresponds to the induced magnetic moment per unit volume and, for many substances, is found to be proportional to the applied magnetic field B_0:

$$M = \frac{\mu_{\text{ind}}}{V} = \chi_V H = \frac{1}{\mu_0} \chi_V B_0 \tag{1.26}$$

where χ_V, the dimensionless proportionality constant between M and H, is the magnetic susceptibility per unit volume.

Classically, this effect can be seen as an ensemble of magnetic moments randomly oriented in the absence of a magnetic field with resultant equal to zero. When an external magnetic field is applied, it tends to orient the magnetic moments and to provide a resultant different from zero (Fig. 1.18). The larger the magnetic field, the larger the resultant induced magnetic moment. From Eq. (1.26), $\chi_V = \mu_0 M/B_0 = \mu_0 \mu_{\text{ind}}/(B_0 V)$: for N_A particles, $\mu_{\text{ind}} = N_A \langle \mu \rangle$, where $\langle \mu \rangle$ is the average induced magnetic moment per particle (see later), and

$$\chi_M = V_M \chi_V = V_M \frac{\mu_0 M}{B_0} = \frac{\mu_0 N_A \langle \mu \rangle}{B_0} \tag{1.27}$$

where χ_M (m^3 mol^{-1}) is the magnetic susceptibility per mole, and V_M is the molar volume. χ_M is magnetic field independent, just like χ_V.

Let us now take an S manifold, unsplit at zero magnetic field, with no orbital angular momentum. The sum of the energies in a magnetic field would be zero

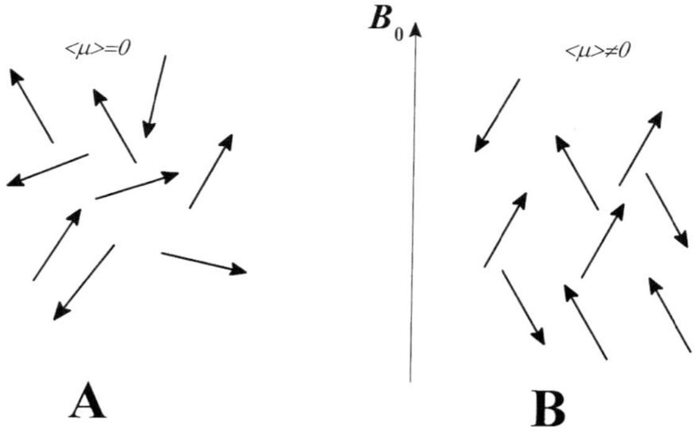

Fig. 1.18. An ensemble of magnetic moments (A) orient themselves along the applied magnetic field $\boldsymbol{B_0}$ (B). The partial orientation determines a resultant non-zero magnetic moment.

(Eq. 1.19) if the levels were equally populated:

$$E = g_e \mu_B B_0 \sum_{M_S=-S}^{S} M_S = 0 \tag{1.28}$$

However, if the populations are considered, in an ensemble of N_A particles, and by recalling that $M_S = \langle S, M_S | S_z | S, M_S \rangle$ (Eq. (1.7)), the energy is

$$E = N_A g_e \mu_B B_0 \frac{\sum \langle S, M_S | S_z | S, M_S \rangle \exp(-g_e \mu_B B_0 M_S / kT)}{\sum \exp(-g_e \mu_B B_0 M_S / kT)}. \tag{1.29}$$

If we consider that $g_e \mu_B B_0 M_S \ll kT$, the exponential is well approximated by $(1 - g_e \mu_B B_0 M_S / kT)$. It follows from Eq. (1.28) that the denominator is just $2S + 1$, and the energy is given by (see Appendix VI)

$$E = -N_A g_e^2 \mu_B^2 \frac{S(S+1) B_0^2}{3kT} = N_A g_e \mu_B B_0 \langle S_z \rangle. \tag{1.30}$$

The quantity

$$\boxed{\langle S_z \rangle = \frac{\sum \langle S, M_S | S_z | S, M_S \rangle \exp(-g_e \mu_B B_0 M_S / kT)}{\sum \exp(-g_e \mu_B B_0 M_S / kT)} = -\frac{g_e \mu_B S(S+1) B_0}{3kT}} \tag{1.31}$$

is called the expectation value of S_z. By operating with S_z on each $|S, M_S\rangle$ level, considering the population, and summing up over all the levels, we obtain an expectation value different from zero.

From the classical treatment, the energy of the system is also given by the product of the induced magnetic moment along the field, μ_{ind} ($\mu_{\text{ind}} = N_A \langle \mu \rangle$, see Eqs. (1.26) and (1.27)) and the external magnetic field B_0 (cf. Eq. (1.3)):

$$E = -\mu_{\text{ind}} B_0 \tag{1.32}$$

Therefore, by combining Eqs. (1.30) and (1.32):

$$\langle \mu \rangle = \frac{\mu_{\text{ind}}}{N_A} = \mu_B^2 g_e^2 \frac{S(S+1) B_0}{3kT} = -\mu_B g_e \langle S_z \rangle. \tag{1.33}$$

In other words, the induced magnetic moment per particle is just proportional to the expectation value $\langle S_z \rangle$. Note that the value of $\langle S_z \rangle$ is referred to a single spin S and to its fractional occupancy of the energy level ladder. In the case of a spin ensemble, the value of $\langle S_z \rangle$ provides the average value of S_z of the ensemble. $\langle S_z \rangle$ is a dimensionless number. From Eqs. (1.27) and (1.33), the magnetic susceptibility per mole is given by

$$\boxed{\chi_M = \mu_0 N_A \mu_B^2 g_e^2 \frac{S(S+1)}{3kT}.} \tag{1.34}$$

By using Eq. (1.8) we obtain

$$\chi_M = \mu_0 N_A \frac{\mu_S^2}{3kT} \tag{1.35}$$

This result of the quantum mechanical treatment coincides with the derivation of magnetic susceptibility in terms of classical magnetic moments if their moduli are taken equal to those associated to the individual spins.

All the above treatment holds for a single S manifold. If there is some orbital contribution to the magnetic moment, it cannot be neglected. All the calculations should be repeated by using Hamiltonian (1.20) for a generic direction k and, by keeping in mind (Eq. (1.27)) that $\chi_M = \mu_0 N_A \langle \mu \rangle / B_0$

$$\chi_{M_{kk}} = -\frac{\mu_0 N_A \mu_B}{B_0} \frac{\sum_i \langle \phi_i | L_{kk} + g_e S_{kk} | \phi_i \rangle \exp(-E_i/kT)}{\sum_i \exp(-E_i/kT)} \tag{1.36}$$

where the sum is over all the levels of the S manifold now containing the orbital part. E_i is the Zeeman energy and may also contain ZFS effects. In the spin-Hamiltonian formalism, neglecting ZFS effects

$$\langle \phi | L_{kk} + g_e S_{kk} | \phi \rangle = g_{kk} M_S \tag{1.37}$$

and

$$\chi_{M_{kk}} = \mu_0 N_A \mu_B^2 g_{kk}^2 \frac{S(S+1)}{3kT} \tag{1.38}$$

where g_{kk} is now different from g_e (Eq. (1.21)). χ_M is now a tensorial quantity.

Eq. (1.36) is called the Van Vleck equation. In that form the Zeeman operator operates only to first order. Indeed, we should include the second order Zeeman term, which allows the interaction between the ground S multiplet (ϕ_i functions) with all the excited ones (ϕ_j functions):

$$\sum_{j \neq i} \frac{\mu_B^2 \left| \langle \phi_i | L_{kk} + g_e S_{kk} | \phi_j \rangle \right|^2}{E_i^0 - E_j^0} \tag{1.39}$$

where E_i^0 and E_j^0 are the energies at zero magnetic field $\big(E = E^0 + \mu_B(\boldsymbol{L} +$

$g_e\boldsymbol{S}) \cdot \boldsymbol{B_0})$. The complete Van Vleck equation, including the population, is

$$\chi_{M_{kk}} = \mu_0 N_A \mu_B^2 \times \frac{\sum_i \left[\frac{\left|\langle\phi_i|L_{kk} + g_e S_{kk}|\phi_i\rangle\right|^2}{kT} - 2\sum_{j\neq i} \frac{\left|\langle\phi_i|L_{kk} + g_e S_{kk}|\phi_j\rangle\right|^2}{E_i^0 - E_j^0} \right] \exp(-E_i^0/kT)}{\sum_i \exp(-E_i^0/kT)}. \quad (1.40)$$

1.7 THE NUCLEAR MAGNETIC RESONANCE EXPERIMENT

The small difference in population among the M_I or M_S levels allows the magnetic resonance experiments. From now on we will focus on the nuclear magnetic resonance experiment, but little would be changed if we dealt with EPR.

1.7.1 The continuous wave experiment and definition of T_1 and T_2

The nuclear magnetic resonance phenomenon was first observed through a continuous wave experiment [2–6]. Resonance between the two M_I levels of an $I = 1/2$ nucleus (without loss of generality) in a magnetic field $\boldsymbol{B_0}$ can be achieved by applying a magnetic field rotating with a frequency (see Eqs. (1.12) and (1.16)) such that

$$h\nu = \hbar\omega = \hbar\,|\gamma_I|\,B_0 = \mu_N\,|g_I|\,B_0. \quad (1.41)$$

The frequency ν is in the radiofrequency (r.f.) range; for the proton it is 100 MHz at 2.3 T. In order to excite the transition, the radiation should be polarized orthogonally to the external magnetic field. The radiation stimulates upward and downward transitions equally. The difference in population accounts for a net absorption of energy. The r.f. therefore tends to bring the two levels to equal populations. When this situation is reached, we have saturation of the signal. No net absorption can be observed. The non-radiative processes which tend to bring back the system to equilibrium tend to re-establish the difference in population. Such processes provide exchange of energy between the spin system and the surrounding, which is called lattice. As described in more detail in Section 1.7.4, the achievement of the equilibrium from non-equilibrium conditions is often assumed to occur through exponential processes of the type

$$\begin{aligned} &\exp(-t/T_1) \quad \text{and} \quad \exp(-t/T_2) \quad \text{or} \\ &\exp(-R_1 t) \quad \text{and} \quad \exp(-R_2 t) \end{aligned} \quad (1.42)$$

where the time constants T_1 and T_2 are called longitudinal and transverse relaxation times. T_1^{-1} and T_2^{-1} are the corresponding rate constants, indicated as R_1 and R_2. Traditionally, NMR spectroscopists are bound to the relaxation times, although the use of relaxation rates is more convenient.

The difference in population between the two levels ensures the establishment of a magnetic moment along the z direction defined by the external magnetic field. This is the magnetic moment that we have described in detail for the electron in Section 1.6. Such a magnetic moment can be thought to be the component of a single macroscopic magnetic moment precessing about z. In other words, an ensemble of spins ½ will be a little more than 50% precessing about z with $M_I = ½$ and a little less than 50% precessing about $-z$ with $M_I = -½$. The difference will provide the magnetic moment along z, which could be thought of as being due to a number of spins equal to the excess spins, all rotating in phase (i.e. represented by coincident vectors). Such a resulting macroscopic rotating vector has the following components:

$$M_z = \text{constant} \tag{1.43}$$

$$M_y = M_0 \cos \omega t \tag{1.44}$$

$$M_x = M_0 \sin \omega t. \tag{1.45}$$

The time average component in the xy plane is zero. The frequency ω is the frequency of precession about z and is the resonance frequency ($\omega = 2\pi\nu$). Indeed, if we send r.f. with exactly that frequency, the vector senses the external magnetic field as well as the r.f. field. The magnetization vector undergoes a

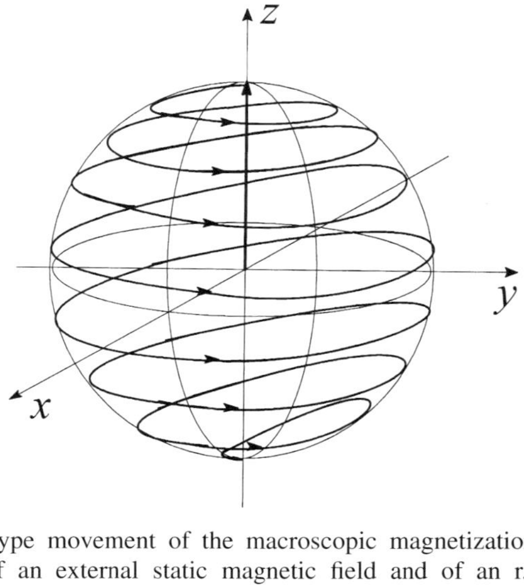

Fig. 1.19. The spiral-type movement of the macroscopic magnetization vector subject to the simultaneous action of an external static magnetic field and of an r.f. field rotating at the resonance frequency.

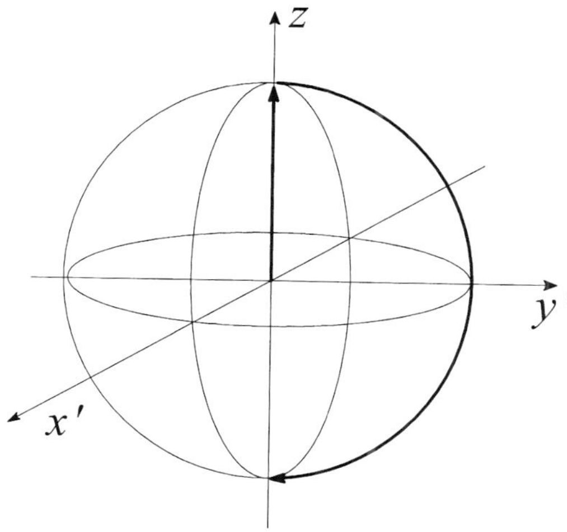

Fig. 1.20. The spiral movement of the macroscopic magnetization of Fig. 1.19 becomes a rotation about the x' (or y') axis of the rotating frame if the sphere rotates about z with the proper frequency.

spiral-type movement which brings its z component from M_z to $-M_z$ (Fig. 1.19). If we use as reference frame a rotating coordinate system with z' coinciding with z and x' and y' rotating at the ω frequency, the magnetization vector will be still in the new coordinate frame. The r.f. field rotating at frequency ω will also be a constant vector. In the rotating frame the magnetization vector only precesses about the r.f. field (Fig. 1.20).

Before the appropriate r.f. is turned on, the magnetization vector has an equilibrium component M_z along z and zero value in the xy plane. The r.f. decreases the M_z value and establishes an M_{xy} different from zero. The return to equilibrium can be followed along z, thus defining T_1, and in the xy plane, thus defining T_2. The two processes are different. To the longitudinal process, only energy exchanges with the lattice through switches of energy levels contribute. The longitudinal relaxation time is also called spin lattice relaxation time. Spin–spin flip–flop transitions not involving energy exchange do contribute to R_2. For this reason the transverse relaxation time is also called spin–spin relaxation. Because all processes contributing to R_1 also contribute to R_2, but not vice versa, the relation

$$R_2 \geq R_1 \tag{1.46}$$

always holds.

1.7.2 The pulse experiment

If we send a radiation for a time short with respect to T_1 or T_2, but with enough power to affect the spin system, we say that we send a pulse. Let us say that the pulse is at the frequency ω of the rotating frame. The magnetization vector starts precessing about the r.f. field. The precession will continue with time, at least for times shorter than T_1 and T_2, with an angular velocity proportional to the r.f. field, i.e. to the power of the pulse. The precession angle or, better, rotation angle after a given time will thus depend on the product of power and time, i.e. on the energy of the pulse. If the r.f. field is along x' in the rotating frame, the magnetization vector will rotate about x' towards y'. After a given time, let us say when the rotation is 90°, we stop the pulse and we let the system return to equilibrium. The coil which transmitted the pulse reveals now the disappearance of the magnetization in the xy plane. We say that the coil reveals the free induction decay (FID). The process is exponential with time constant T_2 (Fig. 1.21). The cosine Fourier transform of this exponential decay is

$$\int_0^\infty \exp(-t/T_2)\cos(\omega t)\,\mathrm{d}t = \frac{T_2}{1+\omega^2 T_2^2}. \tag{1.47}$$

The frequency function is a Lorentzian with linewidth at half height of $(\pi T_2)^{-1}(= R_2/\pi)$. The same, of course, holds in the continuous wave experiment. R_2 is a measure of the uncertainty of the energy levels, which gives the linewidth in every spectroscopy. The uncertainty principle, according to which the uncertainty in energy of a level is inversely proportional to the lifetime, tells us that T_2 is a measure of the lifetime of the energy levels.

If the frequency of the pulse is different from the resonating frequency, nothing changes as long as the pulse contains the latter frequency as a component. It

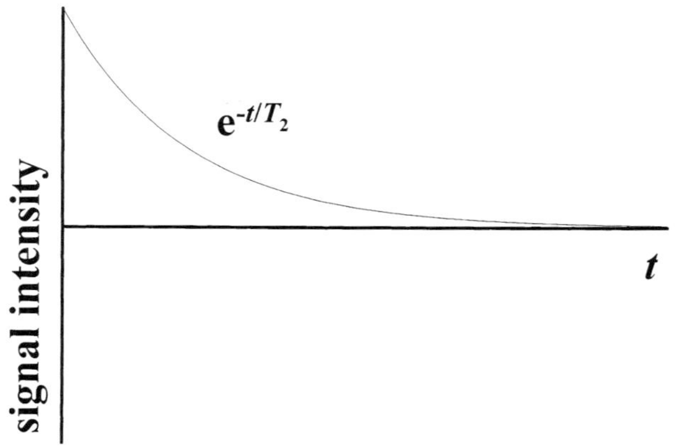

Fig. 1.21. The FID detected on resonance with the precession frequency of the signal of interest.

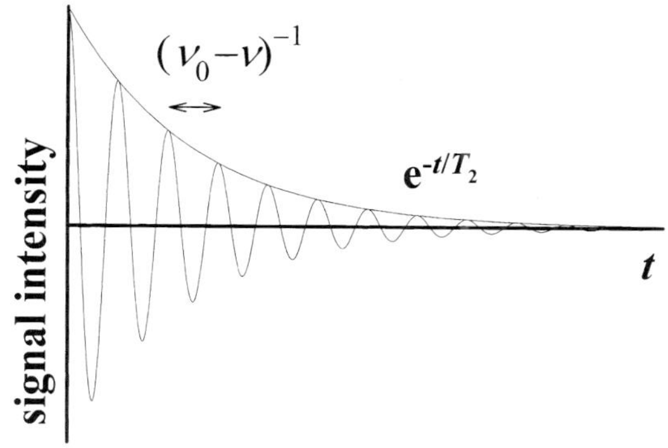

Fig. 1.22. The FID detected off resonance with respect to the precession frequency of the signal of interest. The FID has the shape of a damped oscillation, where the time separation between adjacent maxima equals the reciprocal of the difference $\nu_0 - \nu$.

can be shown that a pulse effectively covers an interval of frequencies of the order of the reciprocal of the pulse length, centered at its own frequency (carrier frequency). The FID now has the shape shown in Fig. 1.22; besides intensity and linewidth, it contains the information about the nuclear resonance frequency.

1.7.3 The chemical shift

The effective magnetic field which a nuclear spin senses when placed in an external magnetic field is the sum of several contributions, whose nature is not discussed here except for that due to the interaction with unpaired electron(s). For example, the paired electrons will decrease the external magnetic field, since they experience an induced magnetic field which opposes to the external magnetic field. This is the same phenomenon which gives rise to bulk diamagnetism. Therefore, all nuclei in a diamagnetic molecule will experience a magnetic field smaller than free nuclei. Since free nuclei are not a practical standard, nuclei in convenient chemical compounds are used as standards. Typically, tetramethylsilane (TMS) is used in proton NMR because its protons are surrounded by a relatively large amount of paired electrons and essentially are at an extreme of the range. The effective magnetic field sensed by such protons is

$$B = B_0(1 - \sigma_{\mathrm{TMS}}) \tag{1.48}$$

where σ is the so-called shielding constant and $B_0\sigma$ is the induced magnetic field which opposes to B_0 (equal to $-\mu_0 M$ of Eq. (1.2)).

Actually, the nucleus senses a different field in different directions around it. Therefore, the shift is a tensorial quantity. Upon rotation in solution an average value σ is obtained. The proton of $CHCl_3$ will experience a smaller shielding constant because chlorine, being quite electronegative, will attract the electrons

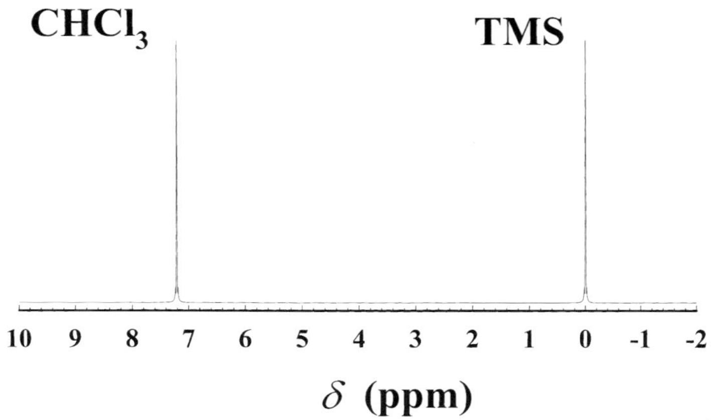

Fig. 1.23. NMR spectrum of a mixture of $CHCl_3$ and TMS taken at 600 MHz proton Larmor frequency. By taking the chemical shift of TMS as zero, the chemical shift of $CHCl_3$ is 7.28 ppm. Shifts to the left of TMS are taken as positive and called downfield. At the chosen magnetic field, the chemical shift of $CHCl_3$ is 4326 Hz.

and remove them from the proton. So, the proton of $CHCl_3$ will sense a larger effective magnetic field that the protons of TMS when placed in the same magnetic field (Fig. 1.23).

For fixed B_0, the frequency needed to have the transition will be larger for $CHCl_3$ than for TMS. The difference in frequency is called chemical shift with respect to TMS. In this case the shift is positive:

$$\Delta\nu = \nu_{\mathrm{CHCl_3}} - \nu_{\mathrm{TMS}}. \tag{1.49}$$

We sometimes say that the $CHCl_3$ proton resonates downfield (at a smaller external magnetic field) with respect to TMS, which in turn resonates upfield (at a larger external magnetic field) if we imagine to keep the frequency fixed and to vary the magnetic field.

Under the above definition the chemical shift is expressed in frequency units (Hz) and depends on the external magnetic field, which then needs to be specified. It may be convenient to express the chemical shift as a pure number, i.e. divided by the frequency of the standard:

$$\frac{\Delta\nu}{\nu_0} = \delta = \frac{\nu_{\mathrm{CHCl_3}} - \nu_{\mathrm{TMS}}}{\nu_{\mathrm{TMS}}} = 7.28 \times 10^{-6} = 7.28 \text{ ppm} \tag{1.50}$$

where ν_0 is the frequency of the spectrometer magnetic field referred to the standard and ppm means parts per million. The symbol δ will be used for chemical shift throughout the book, and all chemical shift equations will be expressed in δ.

Unpaired electrons will have a preference for being aligned along the external magnetic field, in the absence of other restrictions. Therefore, their magnetic moment will sum up and increase the effective magnetic field sensed by a nucleus. The interaction between a magnetic nucleus and unpaired electrons is called hyperfine coupling, after Fermi's account of the hyperfine splitting of the lines in the atomic spectra due to the coupling between the proton and the unpaired electron [7]. The shift in this case is positive (i.e. downfield: since there is the contribution of the electron dipole, a smaller external field is needed to bring the nucleus to resonance). The effect of the presence of unpaired electrons on the nuclear properties will be the subject of this book and of course will be discussed in detail, hopefully in a clear, concise and exhaustive way.

In terms of the Hamiltonian, such hyperfine coupling is represented as

$$\boxed{H = \mathbf{I} \cdot \mathbf{A} \cdot \mathbf{S}} \tag{1.51}$$

where $\mathbf{A}$ is the hyperfine coupling tensor. It is a tensor because it has different values depending on the orientation of the molecular frame within an external magnetic field, which defines the laboratory z axis (see Chapter 2).

1.7.4 Something more about relaxation rates

T_1 and T_2 have already been defined in Section 1.7.1. Since their understanding is fundamental to any approach to NMR experiments and theory, we repeat here the definitions by using a different wording. When an ensemble of independent and equivalent spins at equilibrium in a magnetic field is perturbed, for example by irradiation at the right frequency or by suddenly changing the magnetic field, the system is not at equilibrium any longer. For simplicity we like to consider the spins independent, i.e. not interacting one another. This assumption is convenient to define T_1 but unrealistic, as we will see in the following chapters. The spins are also assumed to experience the same chemical shift. If we refer to the Boltzmann law, which accounts for the levels' population, we can say that, after a perturbation, the spin temperature is different from the lattice temperature. For lattice we mean the environment of the nuclei, which is assumed to have an infinite heat capacity.

After the perturbation, the system tends to reach equilibrium again. We assume that the return to equilibrium is a stochastic process, i.e. each step towards equilibrium is random and uncorrelated to other steps. This implies that the spins are not related one to the other. Under these conditions the return to equilibrium is a first order kinetic process. The decay, as in Eq. (1.42), is then given by

$$F(t) = \exp(-Rt) \tag{1.52}$$

where t is the time and R is a rate constant, as anticipated in Section 1.7.1.

The magnetic field has cylindrical symmetry, i.e. the physical properties of a system do not change upon rotation about the field direction, and there is no difference between x and y directions. If the measurement of the return to equilibrium is performed along the B_0 direction, i.e. the return to equilibrium of M_z is followed, the rate constant is said to be longitudinal and is indicated as R_1. The magnetization reaches equilibrium through transitions between pairs of levels differing by ΔM_I or $\Delta M_S = \pm 1$. Such transitions are induced, like in the magnetic resonance experiments, by oscillating magnetic fields available within the lattice. Matter, even in the condensed phase, experiences continuous movements whose average kinetic energy is proportional to $kT/2$ for each degree of freedom. Matter contains magnetic dipoles, electric charges, etc. which are capable of originating magnetic fields upon sudden reorientations or sudden movements which can be seen as short pulses. Longitudinal relaxation occurs through energy exchange with the lattice until the equilibrium population of the levels is obtained and the spin temperature equals the lattice temperature. We recall that the lattice is assumed to have an infinite heat capacity.

In a typical experiment, called inversion recovery, M_z is initially inverted by applying a 180º pulse. Its recovery along the z axis is given by

$$M_z(t) = M_z(\infty) - 2M_z(\infty)\exp(-R_1 t) \tag{1.53}$$

where $M_z(\infty)$ is the equilibrium value of M_z. The value of $M_z(t)$ at various times t is sampled by applying a 90º pulse and measuring the intensity of the detected signal. R_1 can then be extracted from a fitting of the data to an exponential recovery.

If the rate of reaching equilibrium is measured orthogonally to the magnetic field, i.e. by sampling the projection of the magnetization in the xy plane obtained, for instance, after a 90º pulse, a different rate constant is obtained, which is called the transverse relaxation rate and indicated as R_2. In a pulsed experiment, the rate of reaching equilibrium is equal to the rate of disappearance of magnetization in the xy plane, and is proportional to the linewidth at half height:

$$\pi\Delta\nu_{1/2} = R_2 \tag{1.54}$$

All the mechanisms which contribute to R_1 contribute also to R_2, because the re-establishment of the equilibrium population brings zero magnetization in the xy plane. The dephasing, or fanning out, of the single spin components in the xy plane contributes only to R_2. Therefore

$$R_2 > R_1 \quad \text{or} \quad R_2 = R_1 + c \tag{1.55}$$

where c is the cumulative rate of all processes which lead to the disappearance of the magnetization in the xy plane besides those which lead to the re-establishment of Boltzmann equilibrium. Typically, local magnetic fields oscillating along B_0 at any frequency close to zero produce oscillations in the actual magnetic field and increase the linewidths; in other words, such fields $\boldsymbol{b}_z$ spread the spin moments (Fig. 1.24) through the vectorial interaction with the xy magnetization:

$$\boldsymbol{b}_z \times \boldsymbol{M}_{xy}.$$

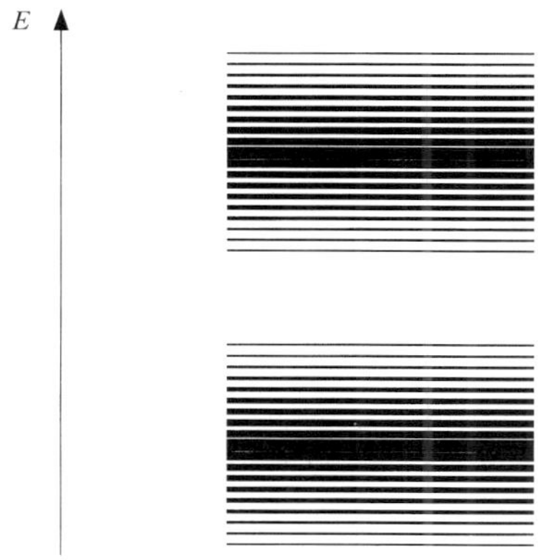

Fig. 1.24. The spreading of energy levels as due to an effective fluctuating magnetic field along z.

The dephasing of the single spins leads to the equilibrium value of the magnetization in the xy plane, which is zero. We have already seen that the simultaneous change of orientation of two spins contributes only to R_2. Besides measuring the linewidth, typical experiments for measuring R_2 are the so-called spin–echo experiments. In their simplest form, a 90° pulse is applied first, followed by a 180° pulse after a variable time t. The intensity of the signal is measured after another time t following the 180° pulse.

In another type of experiment, the magnetization is kept locked in the xy plane. A 90° pulse with B_1 along y (in the rotating frame) sends the magnetization along x. Then B_1 is aligned along x (Fig. 1.25). The field B_1 originated by the transmitter is the only field sensed by the magnetization in the rotating frame. Then the magnetization precesses along B_1. At equilibrium the magnetization will return along z. The time constant for the kinetic process of returning to equilibrium, i.e. for the disappearance of the magnetization along y when a spin-locking field B_1 is operating, is called $T_{1\rho}$ and the corresponding rate constant is called $R_{1\rho}$.

The presence of unpaired electrons will activate new relaxation pathways, and

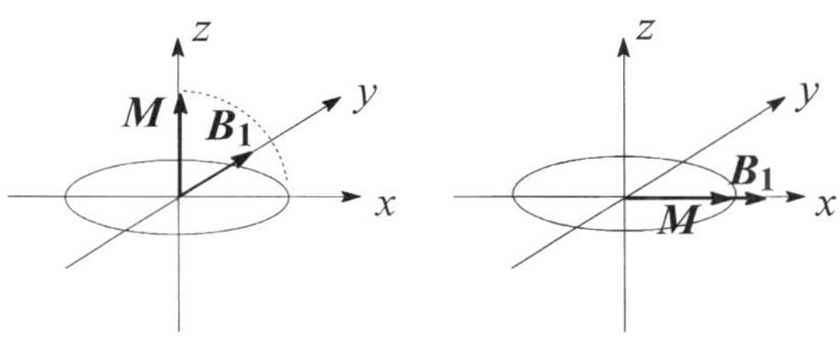

Fig. 1.25. In a spin-locking experiment the magnetization is initially brought along the x axis of the rotating frame (left), and then *locked* along that axis (right), by proper reorientation of the r.f. field $\boldsymbol{B}_1$.

we will have a nuclear R_i ($i = 1$ or 2) enhancement which is indicated as R_{iM}. Similarly, $R_{1\rho}$ also increases. This will be treated in some detail in Chapter 3.

We should end this section by recalling that, while the concept of relaxation is very general, the concept of relaxation rates is much less general, as it implies exponential return to equilibrium of magnetization. Exponential relaxation can only occur when the spin system is weakly coupled to the lattice, and the latter can be assumed to have infinite heat capacity. However, in the case of nuclei, very often the lattice is also constituted by other nuclear spin systems that have a limited heat capacity, of the same order as that of the spin system under investigation. *In the case of electron spins, sometimes the coupling with the lattice is too strong, and non-exponential relaxation may also occur*. These concepts will be further developed in Chapters 3 and 7. Despite these drawbacks, the concept of relaxation rates is still very useful and is thoroughly used, especially when dealing with nuclear relaxation rate enhancements due to unpaired electrons which, as we will see, are often truly exponential contributions.

1.8 GENERAL REFERENCES

For the physical concepts we have mainly referred to B.I. Bleaney and B. Bleaney (1976) *Electricity and Magnetism.* Oxford University Press, 3rd edition.

For the CW NMR experiments one can refer, among others, to: J.A. Pople, W.G. Schneider and H.J. Bernstein (1959) *High-Resolution Nuclear Magnetic Resonance.* McGraw-Hill, New York.

There are numerous books on FT-NMR. The reference book is: R.R. Ernst, G. Bodenhausen and A. Wokaun (1987) *Principles of Nuclear Magnetic Resonance in One and Two Dimensions.* Oxford University Press, Oxford. We may also indicate: T.C. Farrar and E.D. Becker (1971) *Introduction to Pulse and Fourier Transform NMR Methods.* Academic Press, New York.

Previous books on NMR of paramagnetic molecules are: G.N. La Mar, W.DeW. Horrocks, Jr. and R.H. Holm (Eds.) (1973) *NMR of Paramagnetic Molecules.* Academic Press, New York. I. Bertini and C. Luchinat (1986) *NMR of Paramagnetic Molecules in Biological Systems.* Benjamin-Cummings, Menlo Park.

1.9 REFERENCES

[1] S.H. Koenig, A.G. Prodell, P. Kusch (1952) Phys. Rev. 88, 191.
[2] F. Bloch, W.W. Hansen, M. Packard (1946) Phys. Rev. 69, 127.
[3] F. Bloch (1946) Phys. Rev. 70, 460.
[4] F. Bloch, W.W. Hansen, M. Packard (1946) Phys. Rev. 70, 474.
[5] E.M. Purcell, H.C. Torrey, R.V. Pound (1946) Phys. Rev. 69, 37.
[6] E.M. Purcell (1946) Phys. Rev. 69, 681.
[7] E. Fermi (1930) Z. Phys. 60, 320.

Chapter 2

The Hyperfine Shift

This chapter recalls the principles of the hyperfine coupling between electrons and nuclei in terms of energy and deals with its consequences on chemical shift. The equations for contact and pseudocontact shifts are derived and illustrated in a pictorial way. Their physical/chemical backgrounds are discussed as well as their limits of validity. The mechanisms of spin delocalization are illustrated. The perspectives when high field magnets are used are highlighted.

2.1 NUCLEAR HYPERFINE SHIFT AND RELAXATION

How does the nucleus sense the electron? As mentioned in Section 1.7.4, nuclear and electron spins continuously change their levels in a magnetic field but the population distribution remains constant at a given temperature. The change of level occurs through mechanisms which are different or have different degrees of efficiency for the two sets of spins, with the result that electrons relax orders of magnitude faster than nuclei. As a consequence, the nucleus in each of its M_I energy levels sees one electron rapidly changing its orientation among the M_S levels. The nucleus therefore senses an average magnetic moment of the electron which would be zero if the populations of the M_S levels were equal. As it is not so in a magnetic field, the nucleus senses the average magnetic moment $\langle \mu \rangle$ which is just proportional to $\langle S_z \rangle$ (see Section 1.6). It should be recalled that $\langle S_z \rangle$ is negative (Eq. (1.31)) due to the slight excess of population in the $M_S = -\frac{1}{2}$ state.

The average electronic induced magnetic moment affects the energy of the nuclear $|I, M_I\rangle$ levels, i.e. affects the energy of the nucleus, because it represents a contribution to the overall magnetic field sensed by the nucleus. Such electron–nucleus interaction is represented by Eq. (1.51) and follows the treatment of the interaction between magnetic moments discussed in the previous chapter. In particular, there will be a dipolar interaction energy which averages zero in solution (see Eq. (2.19)) or averages a value different from zero when the electron induced magnetic moment changes intensity with orientation as described in Fig. 1.16B and C (see also Eq. (2.20)). A fraction of electron can sit on the resonating nucleus and so provides a direct shielding constant (see Section 2.1.1). This contribution is called contact contribution.

If the interaction energy fluctuates, then nuclear relaxation enhancements occur. Such relaxation enhancements are proportional to the average squared interaction energy

$$R_{1,2} \propto \langle E^2 \rangle.$$

Upon rotation the average electron induced magnetic moment causes nuclear relaxation. This is called Curie relaxation (see Section 3.6). However, the full electron magnetic moment causes nuclear relaxation through dipolar and contact mechanisms by jumping over the various Zeeman levels. The electron jumps over the Zeeman levels with rates equal to the electronic R_1 or R_2 values and such jumps give also rise to nuclear relaxation.

The present Chapter deals with the hyperfine shifts which are only due to the average electron induced magnetic moment and therefore are related to $\langle S_z \rangle$. Chapter 3 will deal with nuclear hyperfine relaxation which, as discussed above, depends on both average electron induced magnetic moment (Curie relaxation) and on the full electron magnetic moment (dipolar and contact relaxation).

2.1.1 The spin density

The unpaired electron has the complication that it is not localized on a single point but, in general, is delocalized on the entire molecule. So, in every point of space where the molecular orbital (MO) containing the unpaired electron has a non-zero value, the average electron magnetic moment sensed by the nucleus is different from zero and is proportional to $\langle S_z \rangle$ times the fraction of unpaired electron present at that point. Such a fraction is called spin density ρ, which for a single electron is given by the square of its wavefunction at that point.

The situation is further complicated by the fact that the presence of an unpaired electron in a MO polarizes the paired electrons in the core. This is called spin polarization. The MO containing the paired electrons is modified by the presence of the unpaired electron in another MO in such a way that the electron with the spin aligned with the unpaired electron will have a slight preference to occupy the region of space of its MO which is closer to the unpaired electron itself (this is a manifestation of Hund's rule). Conversely, the other electron with spin antiparallel to the unpaired electron will have a slight preference to occupy regions of its MO far from the unpaired electron. This is illustrated in Fig. 2.1 for the case of two atomic orbitals.

At this point we recall that the unpaired electron may have some probability to sit just on the nucleus. Type s orbitals have maximal electron density on the nucleus, as their wavefunction is of the type $\exp(-r)$, where r is the electron–nucleus distance. Therefore, if the unpaired electron occupies an s orbital, or an MO containing an s orbital, there will be a finite probability that the electron resides on that nucleus. The amount of unpaired electrons residing on the nucleus is the spin density ρ at the nucleus. The spin density at the nucleus or in

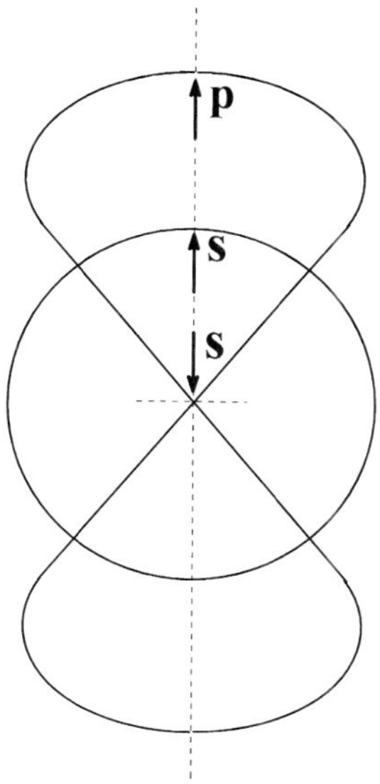

Fig. 2.1. The unpaired electron in a 2p orbital affects the distribution of the two paired electrons in a 1s orbital.

any other point is, in general, not just the positive quantity arising from the direct delocalization explained above, but also contains contributions due to spin polarization of paired electrons. The latter mechanism accounts for the presence of spin density on nuclei when the unpaired electron occupies a d orbital, which, like the p orbital, has a node at the nucleus. The unpaired electron on a p (or d) orbital can spin-polarize the electron pair occupying any s orbital. As shown in Fig. 2.1, the electron of a 1s pair with the same spin as the unpaired electron in a p or d orbital will preferentially be in the outer part of the 1s orbital, whereas the electron with reverse spin will preferentially be in the inner part of the 1s orbital. The two electrons of the pair will now have different probabilities to be at the nucleus. The difference provides the spin density at the nucleus. This contribution has opposite sign with respect to the contribution from the spin density in the orbital containing the unpaired electron. Since the unpaired electron has excess spin value $M_S = -1/2$ (*positive spin density*), the spin density at the nucleus arises from slight excess spin value $M_S = 1/2$, and is thus negative. Note, therefore, that although the unpaired electron always preferentially aligns with its spin along the external magnetic field, spin polarization may induce on the nucleus negative spin density as if the electron were polarized opposite to the magnetic field. Spin polarization is a local phenomenon and the g value of the spin density in an s orbital is rigorously 2.0023.

2.2 THE MAGNETIC NUCLEUS–UNPAIRED ELECTRON INTERACTION: THE HYPERFINE SHIFT

The average induced electron magnetic moment gives rise to a contribution to the chemical shift which is called hyperfine shift. Experimentally it is defined as the difference between the chemical shift in a paramagnetic system and that in an analogous diamagnetic system. It is customary to separately consider the effect of the spin density on the resonating nucleus from that outside it. The former part of the interaction is called contact or Fermi contact interaction. The latter is called dipolar interaction because it is a through space interaction which can be described by the dipolar interaction of two magnetic dipoles. The evaluation of the dipolar interaction requires, in principle, the evaluation of the integral all over the space of the interaction energy between the resonating nucleus and the spin density on each point of the space. In practice the problem will be simplified.

2.2.1 The Fermi contact coupling

The contact shift is given by an additional magnetic field generated at the nucleus by the electron magnetic moment located at the nucleus itself. The latter magnetic moment originates from the spin density at the nucleus, weighted by $\langle S_z \rangle$. For an isolated electron spin, $\langle S_z \rangle$ is always negative as long as the spin–orbit coupling is a small perturbation. In contrast, spin density can be either positive or negative because of spin polarization. For each s orbital the spin density at the nucleus, ρ, is given by

$$\rho = \Psi_{-1/2}(0)^2 - \Psi_{1/2}(0)^2 \tag{2.1}$$

where $\Psi(0)$ is the value of the MO wavefunction at zero distance from the nucleus for $-1/2$ and $1/2$ spins. Therefore, if the excess spin value is $-1/2$, the contribution to the magnetic field will be positive (Fig. 2.2) as for a free electron and the chemical shift in frequency will be positive. The reverse holds for $+1/2$ spin. The spin density (when normalized to one electron) is proportional to what is called the contact coupling constant A, which expresses how much the nucleus and the electron sense each other:

$$\boxed{A = \frac{2\mu_0}{3}\hbar\gamma_I g_e \mu_B \rho.} \tag{2.2}$$

Since the s orbitals do not have orbital magnetic moment, $g = g_e = 2.0023$. A is an energy, which in SI units is expressed in joule. Sometimes it is convenient to have it expressed in frequency units, i.e. in hertz or in radians per second, and then we have to divide A by h or by $\hbar$ ($\hbar = h/2\pi$). In magnetic resonance there often is a factor 2π which complicates life. The frequency which we refer to is the angular frequency or the Larmor precession frequency. Such frequency ω is

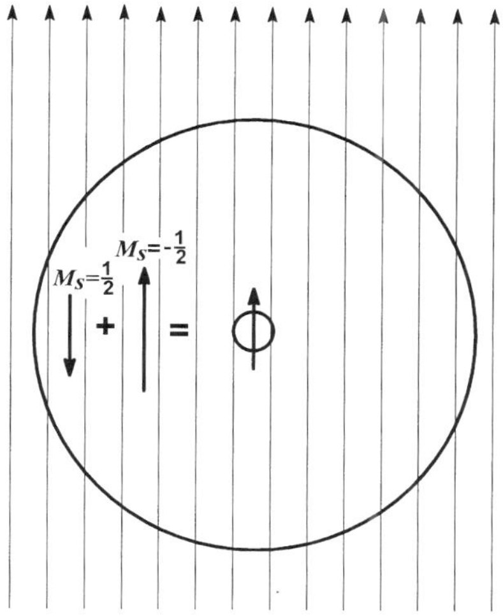

Fig. 2.2. Excess −½ electron spin density contributes a magnetic field at the nucleus that adds to the external magnetic field.

such that $\omega = 2\pi\nu$ (Eq. (1.41)), where ν is the frequency in hertz or s^{-1} and ω is expressed in radians per second. The general relationship between energy in joule and frequency or angular frequency is

$$E = h\nu = h\frac{\omega}{2\pi} = \hbar\omega. \tag{2.3}$$

In order to express the energy in wavenumbers, we should remember that $E = hc/\lambda$, and then E is proportional to the inverse of a length.

In quantum mechanical terms, the Hamiltonian for the contact interaction is

$$H = A\boldsymbol{I} \cdot \boldsymbol{S} \tag{2.4}$$

and the energy is obtained by evaluating the matrix elements $\langle \Psi_{S,M_S,I,M_I} | H | \Psi_{S,M_S,I,M_I} \rangle$. In the high field approximation ($g_e\mu_B B_0 \gg A$), which always holds for high field NMR experiments, the contact contribution to the chemical shift is (Appendix III) [1]

$$\boxed{\delta^{\text{con}} = \frac{A}{\hbar}\frac{g_e\mu_B S(S+1)}{3\gamma_I kT}.} \tag{2.5}$$

This equation holds for a single electron in an orbital well separated from the excited ones. The nuclear I value of the resonating nucleus does not appear in Eq. (2.5) because the frequency is referred to any $\Delta M_I = 1$ transition. By using the definition of $\langle S_z \rangle$ of Eq. (1.31), the following relation is obtained (Appendix III):

$$\delta^{\text{con}} = -\frac{A}{\hbar\gamma_I B_0}\langle S_z \rangle. \tag{2.6}$$

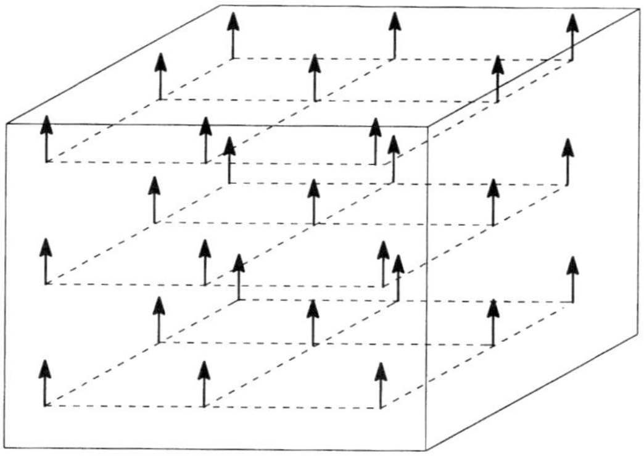

Fig. 2.3. Ideal crystal with all molecules iso-oriented.

Eq. (2.6) tells us that the fractional variation in frequency is due to the presence of unpaired spin density, which determines A. No nuclear parameter is contained in the expression of the contact shift. The γ_I appearing in both Eqs. (2.5) and (2.6) cancel the γ_I contained in A (Eq. (2.2)).

We should note that if $g = g_e$, the contact shift is isotropic (independent of orientation). If g is different from g_e and anisotropic (see Section 1.4), then the contact shift is also anisotropic. The anisotropy of the shift is due to the fact that: (1) the energy spreading of the Zeeman levels is different for each orientation (see Fig. 1.16), and therefore the value of $\langle S_z \rangle$ will be orientation dependent; and (2) the values of $\langle S, M_S | S_z | S, M_S \rangle$ of Eq. (1.31) are orientation dependent as the result of efficient spin–orbit coupling. On the contrary, the contact coupling constant A is a constant whose value does not depend on the molecular orientation.

Let us suppose now that we have a solid with all molecules aligned with one another (Fig. 2.3) and that we perform the NMR experiment on a single crystal. If g of the S manifold equals g_e, the contact shift contribution will be independent of the crystal orientation in the magnetic field. If, however, g has a different value in any k direction, then spin–orbit coupling is not negligible and the contact shift will be orientation dependent. Specific calculations are needed. If, however, we arbitrarily neglect the anisotropy of $\langle \psi | S_z | \psi \rangle$, the following equation can be written

$$\delta_{kk}^{\text{con}} = \frac{A}{\hbar} \frac{g_{kk} \mu_B S(S+1)}{3\gamma_I kT}. \tag{2.7}$$

When the solid is dissolved in a liquid, the rotational rate of the molecules is fast with respect to the difference in hyperfine shift due to the electron Zeeman anisotropy, and the contact shift will be an average:

$$\delta_{\text{sol}}^{\text{con}} = \frac{A}{\hbar} \frac{\overline{g} \mu_B S(S+1)}{3\gamma_I kT}. \tag{2.8}$$

Typical values for proton hyperfine shift anisotropy are hardly larger than $10^5\ s^{-1}$, i.e. much smaller than τ_r^{-1} for all practical cases.

A correct equation for a singly populated S manifold is

$$\delta_{\text{sol}}^{\text{con}} = \frac{1}{\mu_0}\frac{A}{\hbar}\frac{1}{3\gamma_I \mu_B}\left(\frac{\chi_{xx}}{g_{xx}} + \frac{\chi_{yy}}{g_{yy}} + \frac{\chi_{zz}}{g_{zz}}\right) \tag{2.9}$$

where χ_{ii} and g_{ii} are the principal components of the molecular susceptibility (χ_M/N_A, cf. Eq. (1.27)) and $\boldsymbol{g}$ tensors. The reader can find a justification of it by comparing Eq. (2.9) with Eq. (2.8) through Eq. (1.38). Of course Eq. (2.9) is hardly used, as the principal χ values (and g) are generally unknown.

We note here that Eq. (2.8) holds for a single electron in an orbital which is well separated by any other excited level. In the case of multiple unpaired electrons in different molecular orbitals, Eq. (2.8) still may hold in the absence of strong spin–orbit coupling effects but the interpretation of the hyperfine constant becomes complicated: the hyperfine coupling is the sum of that for each molecular orbital. Indeed, each metal orbital which contains an unpaired electron is involved in a molecular orbital and provides a contribution to the total ρ for the various nuclei. The experimental data, however, provides through Eq. (2.8) the sum of the A values and therefore the sum of ρ. In order to make the spin density or contact constants comparable for different systems independent of the value of S, i.e. independent of the number of electrons, the value of ρ is normalized to one electron, i.e. it is divided by the number of electrons which is just $2S$ (in such a way that $\sum_i \rho_i/2S = 1$). Eq. (2.2) becomes

$$A = \frac{\mu_0}{3S}\hbar\gamma_I g_e \mu_B \sum_i \rho_i. \tag{2.10}$$

A is now the average contact coupling per electron.

When there are multiple electrons, ZFS occurs, and the Boltzmann population of the levels may be taken into account. Then Eq. (2.9) should be used. As an example we develop a formula expressing the Fermi contact shift for a simple system; that is, $S = 5/2$ with tetragonal symmetry and no orbital angular momentum. The zero field split energy levels, for $\boldsymbol{B_0}$ along z, can be represented as a function of the M_S values as represented in Fig. 2.4, where D is the ZFS parameter. The principal susceptibility values can then be calculated through the Van Vleck equation, shown in Equation (1.36), and, as long as $2D$ is larger than the perturbation due to an external magnetic field, they take the form:

$$\chi_{zz} = \mu_0 \frac{g_\parallel^2 \mu_B^2}{4kT}\left(\frac{1 + 9e^{-2X} + 25e^{-6X}}{1 + e^{-2X} + e^{-6X}}\right), \qquad X = \frac{D}{kT} \tag{2.11}$$

$$\chi_{xx} = \chi_{yy} = \mu_0 \frac{g_\perp^2 \mu_B^2}{4kT}\left(\frac{9 + \frac{8}{X} - \frac{11}{2X}e^{-2X} - \frac{5}{2X}e^{-6X}}{1 + e^{-2X} + e^{-6X}}\right). \tag{2.12}$$

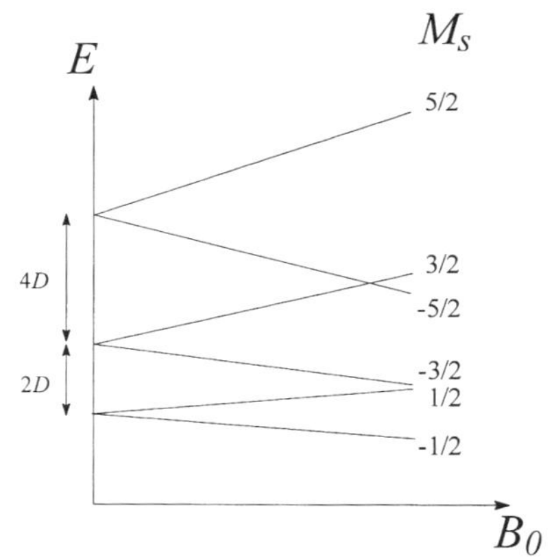

Fig. 2.4. Electron energy levels in the presence of axial ZFS, for $\boldsymbol{B_0}$ along z, in the $S = 5/2$ case.

By substituting into Equation (2.9), and taking $e^{-X} = 1 - X$ $(D \ll kT)$, the following expression is obtained [2]:

$$\delta^{\text{con}} = \left(\frac{A}{\hbar}\right) \frac{35\overline{g}\mu_B}{12\gamma_I kT} \left[1 - \frac{32(g_{\parallel} - g_{\perp})D}{45\overline{g}kT}\right]. \tag{2.13}$$

This expression contains one term dependent on T^{-1} and another dependent on T^{-2}, the latter term being a consequence of the Boltzmann population of the zero field split levels. The g values are not the experimental values obtained by EPR but the molecular g values, which can be obtained through a proper analysis of the experimental values. For $S = 5/2$ and excited states far elevated in energy, the molecular g values are close to the free electron, g_e, value and D is relatively small; therefore, the term in T^{-2} can be neglected.

Concluding, it is probable that the use of Eq. (2.8) instead of Eq. (2.9) does not introduce serious errors for all metal ions provided that empty orbitals are far in energy.

Another case is that of one unpaired electron in a metal orbital in the presence of 1–2 empty metal orbitals close in energy. Spin–orbit coupling mixes the various orbitals so that the unpaired electron resides in an orbital which is a linear combination of all orbitals available. A typical case is that of low spin octahedral Fe(III), where there is a single unpaired electron but d_{xy}, d_{zz} and d_{yz} are close in energy. Here the g values are largely anisotropic (from $g_{\max} \approx 3$ to $g_{\min} \approx 1$) and the excited states may be partially populated. Eq. (2.8) may be an inadequate mean of describing the real case especially as far as the temperature dependence is concerned. In fact, each excited level has a different hyperfine coupling constant and contributes according to its population which can be estimated from the Boltzmann distribution [3].

It is now obvious that the case of $S > 1/2$ and low lying excited states can hardly be analyzed by means of Eq. (2.8) but requires a specific treatment which takes into full account the effects of spin–orbit and Zeeman interactions. It should be said that these treatments are rare in the literature, as they require a lot of skill and wisdom in choosing the right parameters.

2.2.2 The dipolar coupling

This contribution to the shift is quite difficult to evaluate, because in general the spin density distribution all over the space is not known. The approach to this problem should be stepwise. We first consider that the unpaired electron is localized on the metal nucleus in a paramagnetic metal complex. We refer to this as to the metal-centered point–dipole approximation. Surely this contribution will always be present and often dominant. Then we will discuss the consequences of relaxing this condition. Even in the metal-centered approximation several cases should be discussed.

2.2.2.1 Metal-centered point–dipole approximation

Let us now refer to a set of molecules with their x, y and z axes iso-oriented in an idealized solid state (Fig. 2.5). If the external magnetic field is aligned with the z axis, the dipolar interaction energy between the nuclear magnetic moment and the electron magnetic moments, according to Eq. (1.4), is

$$E^{\mathrm{dip}} = -\frac{\mu_0}{4\pi}\hbar\gamma_I I_z \langle\mu\rangle \frac{1}{r^3}(3\cos^2\gamma - 1). \tag{2.14}$$

From Eqs. (1.27), (1.33) and (1.34) we can write E^{dip} in terms of the z

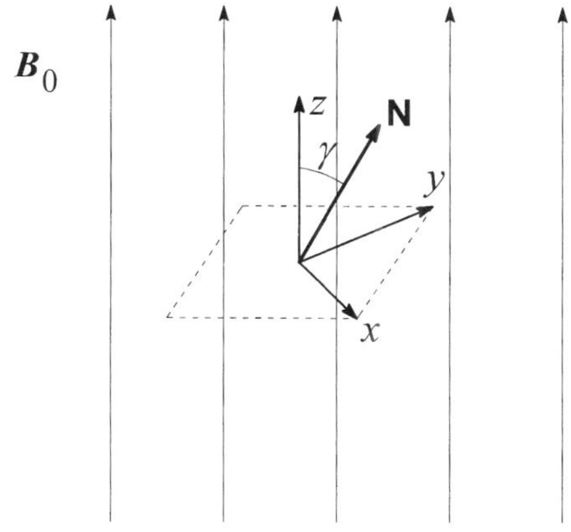

Fig. 2.5. A molecule possessing magnetic anisotropy, with the z axis oriented along the external magnetic field and a nucleus having a metal–nucleus vector at an angle γ with the external magnetic field.

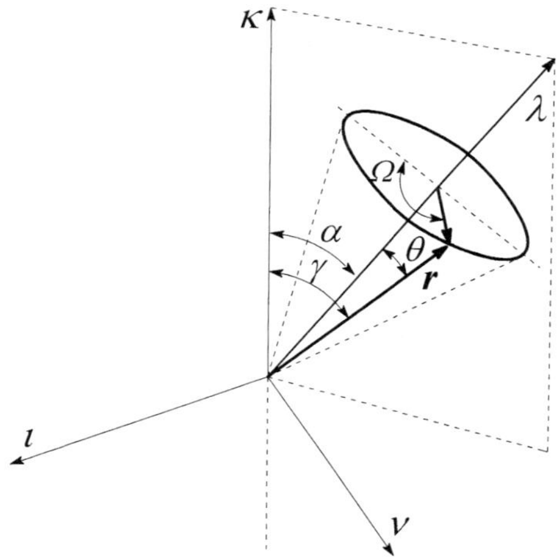

Fig. 2.6. A molecule possessing magnetic anisotropy, with the z axis oriented along a generic γ direction. α is the angle of the z axis with the external field direction κ; θ is the angle between the metal–nucleus vector $\boldsymbol{r}$ and the z-axis; γ is the angle between the metal–nucleus vector and the external field; Ω defines the position of $\boldsymbol{r}$ on the surface of the cone about λ.

component of the molecular susceptibility tensor $\boldsymbol{\chi} = \chi_M/N_A$

$$E^{\text{dip}} = -\frac{B_0}{4\pi r^3}\hbar\gamma_I I_z \chi_{zz}(3\cos^2\gamma - 1) \tag{2.15}$$

where γ is the angle between the metal nucleus vector and the external magnetic field, which is coincident with the molecular χ_{zz} direction. In a generic k direction, reference will be made to Fig. 2.6 and the energy, for an axial $\boldsymbol{\chi}$ tensor, will be given by (Appendix IV)

$$\begin{aligned} E^{\text{dip}} = -\frac{B_0}{4\pi r^3}\hbar\gamma_I I_z\Big[&\chi_\parallel \cos^2\alpha(3\cos^2\theta - 1) \\ &+ \chi_\perp \sin^2\alpha(3\sin^2\theta\cos^2\Omega - 1) \\ &+ \tfrac{3}{4}(\chi_\parallel + \chi_\perp)\sin 2\alpha \sin 2\theta \cos\Omega\Big] \end{aligned} \tag{2.16}$$

where α is the angle between the molecular z axis and the external magnetic field, θ is the angle between the metal–nucleus vector $\boldsymbol{r}$ and the molecular z axis, and Ω is an angle related to the projection of $\boldsymbol{r}$ on the molecular xy plane. Ω is zero when $\boldsymbol{r}$ lies in the plane defined by the external magnetic field and the molecular z axis.

From the dipolar interaction energy, the dipolar shift can be obtained by evaluating from Eq. (2.16) ΔE^{dip} between two states differing by $\Delta M_I = \pm 1$ and dividing it by the nuclear Zeeman energy $\hbar\gamma_I B_0$ (Appendix IV):

$$\delta^{\mathrm{dip}} = \frac{1}{4\pi r^3}\Big[\chi_{\parallel}\cos^2\alpha(3\cos^2\theta - 1) + \chi_{\perp}\sin^2\alpha(3\sin^2\theta\cos^2\Omega - 1) + \tfrac{3}{4}(\chi_{\parallel} + \chi_{\perp})\sin 2\alpha\sin 2\theta\cos\Omega\Big]. \tag{2.17}$$

This is indeed what is expected in ^{1}H ENDOR spectroscopy in single crystals of isooriented molecules. By integration of Eq. (2.17) over all molecular orientations, the following equation is obtained (Appendix IV):

$$\delta^{\mathrm{pc}} = \frac{1}{12\pi r^3}(\chi_{\parallel} - \chi_{\perp})(3\cos^2\theta - 1). \tag{2.18}$$

This contribution to the shift is isotropic because it is already averaged out over all the orientations. Then it is similar to the contact shift and is called pseudocontact shift δ^{pc}. In the literature it is also referred to as dipolar shift or isotropic dipolar shift.

Note that, when $\boldsymbol{\chi}$ is isotropic, the pseudocontact shift is zero (Eq. (2.18)), but a dipolar shift is observed in the solid state (Eq. (2.17) and Eq. (IV.9)). That the dipolar interaction energy averages zero for isotropic $\boldsymbol{\chi}$ can be easily verified by averaging Eq. (2.15) over all orientations, as

$$\int_{-1}^{1}(3\cos^2\gamma - 1)\,\mathrm{d}\cos\gamma = 0. \tag{2.19}$$

There is also an intuitive mode to see the effect of an isotropic electron magnetic moment on a nucleus. Fig. 2.7 illustrates that, indeed, the dipolar magnetic field experienced by the nucleus for different orientations of the metal–nucleus vector with the magnetic field changes sign, being positive when the

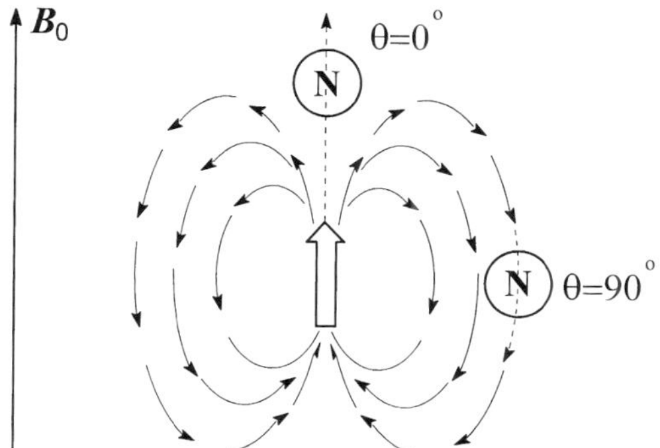

Fig. 2.7. A nucleus N immersed in a dipolar magnetic field of an electron spin, the latter aligned along the external magnetic field, experiences an extra field that adds ($\theta = 0°$) or subtracts ($\theta = 90°$) to the external field.

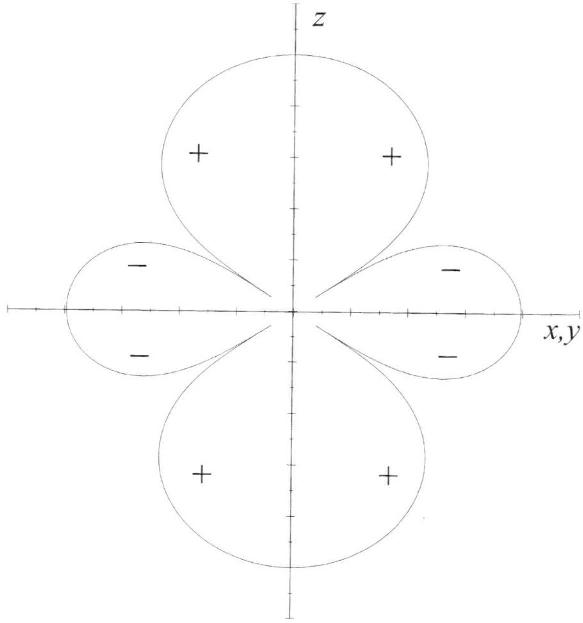

Fig. 2.8. Angular dependence of the pseudocontact shift for an axial system, shown as a surface of constant absolute value of δ^{pc}. In the example δ^{pc} is positive along the z axis and negative in the xy plane. The three-dimensional shape of the surface is similar to the representation of a d_{z^2} orbital.

vector is along the field and negative when the vector is perpendicular to the field. Therefore, one can accept that the average over all the orientations is zero. However, if the induced magnetic field changes intensity with the molecular orientation because $\boldsymbol{\chi}$ is anisotropic, complete cancellation upon rotation does not occur any longer. It should be remembered that the anisotropy of $\boldsymbol{\chi}$ (see Sections 1.4 and 1.6) is due to orbital contributions to the magnetic moments.

The angular dependence of the pseudocontact shift is shown in Fig. 2.8. For $\theta = 54.74°$, the pseudocontact shift is zero. If $\chi_\parallel > \chi_\perp$, then the pseudocontact shift is positive for θ values less than 54.74°. In the case in which χ_{xx} and χ_{yy} differ, then the angle Ω between the projection of the metal–nucleus vector on the xy plane and the y axis should also be considered (Fig. 2.9). The result is [4]

$$\boxed{\delta^{pc} = \frac{1}{12\pi}\frac{1}{r^3}\left[\Delta\chi_{ax}(3\cos^2\theta - 1) + \tfrac{3}{2}\Delta\chi_{rh}\sin^2\theta\cos 2\Omega\right]} \qquad (2.20)$$

where $\Delta\chi_{ax} = \chi_{zz} - \dfrac{\chi_{xx} + \chi_{yy}}{2}$

$$\Delta\chi_{rh} = \chi_{xx} - \chi_{yy}.$$

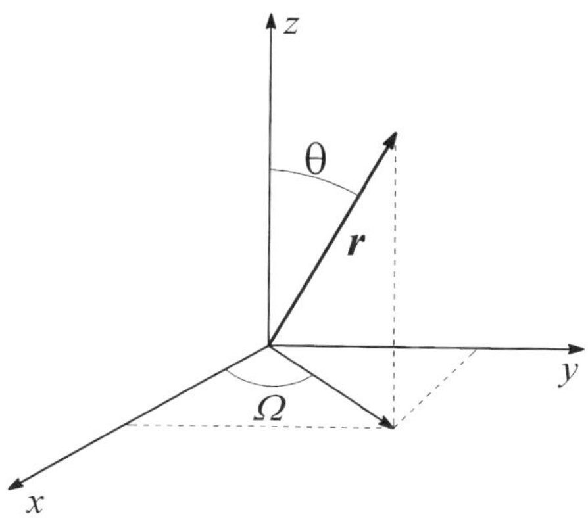

Fig. 2.9. Definition of the in-plane angle Ω for non-axial systems.

The contribution of the in-plane anisotropy of χ varies with $\cos 2\Omega$.

Eq. (2.20) can also be expressed in terms of the direction cosines l, m and n of the metal-nucleus vector with respect to the principal directions of the χ-tensor (Fig. 2.9):

$$\delta^{\mathrm{pc}} = \frac{1}{24\pi}\frac{1}{r^3}\Big\{\big[2\chi_{zz} - (\chi_{xx} + \chi_{yy})\big](3n^2 - 1) + 3(\chi_{xx} - \chi_{yy})(l^2 - m^2)\Big\}. \tag{2.21}$$

Eq. (2.20), or its simplified version in the axial case, Eq. (2.18), are of general validity. However, the principal directions and components of the molecular χ tensor are seldom available. Pseudocontact shifts can be still evaluated by expressing the principal molecular magnetic susceptibility values as a function of the principal g values, in analogy with Eq. (1.38):

$$\boxed{\chi_{kk} = \mu_0\mu_B^2 g_{kk}^2 \frac{S(S+1)}{3kT}.} \tag{2.22}$$

As seen in Section 1.6, this approximation holds when the spin multiplet ground state is well isolated from excited electronic states, and ZFS is negligible. If $S = 1/2$, the g values can easily be measured through EPR spectroscopy and the g directions can be determined by single-crystal EPR measurements. When the latter measurements are not available, sometimes the principal g directions can be guessed from the symmetry of the molecule, and an independent estimate of the pseudocontact shift can still be attempted.

By substituting the molecular χ values with the expression in Eq. (2.22), Eq. (2.18) and (2.20) respectively become

$$\delta^{\mathrm{pc}} = \frac{\mu_0}{4\pi}\frac{\mu_B^2 S(S+1)}{9kT}(g_\parallel^2 - g_\perp^2)\frac{1}{r^3}(3\cos^2\theta - 1) \tag{2.23}$$

$$\delta^{\mathrm{pc}} = \frac{\mu_0}{4\pi}\frac{\mu_B^2 S(S+1)}{18kT}\frac{1}{r^3}\Big\{\big[2g_{zz}^2 - (g_{xx}^2 + g_{yy}^2)\big](3\cos^2\theta - 1) + 3(g_{xx}^2 - g_{yy}^2)\sin^2\theta\cos 2\Omega\Big\} \tag{2.24}$$

As in the case of contact shift, the pseudocontact shift does not contain parameters of the resonating nucleus. At variance with contact shifts, the pseudocontact shifts can be evaluated even if more than one S multiplet is populated, provided that $\Delta\chi$ values are experimentally available.

Note that Eqs. (2.23) and (2.24) (and, likewise, Eqs. (2.18) and (2.20)) can be written in a way analogous to Eq. (2.8):

$$\delta^{\mathrm{pc}} = \frac{A^{\mathrm{pc}}}{\hbar}\frac{\overline{g}\mu_B S(S+1)}{3\gamma_I kT} \tag{2.25}$$

where A^{pc} from e.g. Eq. (2.23) is given by

$$A^{\mathrm{pc}} = \frac{\mu_0}{12\pi}\hbar\gamma_I\overline{g}\mu_B\left(\frac{g_\parallel^2}{\overline{g}^2} - \frac{g_\perp^2}{\overline{g}^2}\right)\frac{1}{r^3}(3\cos^2\theta - 1). \tag{2.26}$$

A of Eqs. (2.2) or (2.4), together with A^{pc} of Eq. (2.26), provide orientational average values of the hyperfine coupling tensor of Eq. (1.51).

2.2.2.2 Ligand-centered contributions

The first improvement of the dipole–dipole approximation is introduced by the consideration that the unpaired electron(s) is at least spread within the cage formed by the donor atoms. In the case of 4f electrons this does not produce a significant effect, whereas a 3d electron appears to the nucleus as a magnetic point–dipole only at large distances. An estimate of the lower limit of this distance is 700 pm [4]. In order to properly evaluate the pseudocontact term at shorter distances it is necessary to consider the appropriate MOs and to apply the dipolar interaction Hamiltonian to these wavefunctions. Attempts at this evaluation assuming a series of point multipoles instead of a simple point–dipole, with the aim of mimicking the electron distribution within the metal atomic orbitals, have been partially successful [5–9]. These considerations provide considerable corrections on nuclei close to the paramagnetic center. Nuclei far from the paramagnetic center still see the electron localized at the metal.

Finally, if the electron is considered to be delocalized on a large part of the molecule as the result of π delocalization, the problem is further complicated. Pseudocontact shift is also provided by spin density in a p orbital which has an orbital contribution to the magnetic moment different from zero. This is very relevant for heteronuclei like ^{15}N or ^{13}C which may bear unpaired spin density on p_z orbitals. The picture is of the type of Fig. 2.1 where, besides spin polarization, dipolar interactions occur. For example, fractions of unpaired electrons delocalized onto nitrogen p orbitals in $Fe(CN)_6^{3-}$ are predicted to affect seriously the overall ^{14}N hyperfine shifts [10]. Therefore, the analysis of hyperfine shifts for heteronuclei requires ligand-centered pseudocontact shifts to be taken into account. Hydrogen nuclei have no accessible p orbitals and hence any ligand-centered effect can only arise from unpaired spin density on p_π orbitals of neighboring atoms. Thus, the effects are expected to be smaller, although not necessarily negligible.

2.3 SHIFT AND SPIN PATTERNS FOR PROTONS AND DEUTERONS IN SOLUTION

From the equations given so far, it is difficult to predict any behavior of chemical compounds from the hyperfine shift point of view. Still, any effort is welcome in providing rules and expectations under certain conditions. We will see here that this is somewhat possible.

Chemical bonds involving protons and deuterons are very similar to one another. Therefore, since the contact and pseudocontact shifts are independent of the nuclear parameters, the shifts are essentially the same. Both nuclei have the characteristic that only 1s orbitals are involved. Therefore, protons and deuterons are discussed together. Sometimes, the use of ^2H NMR can be appropriate because it has a smaller linewidth than ^1H (see Section 3.9).

Let us first try to discuss the contact contribution to the shift for protons and deuterons. As we shall see, there are conditions under which the pseudocontact shifts are small and negligible, or can be determined separately and subtracted from the hyperfine shifts. Furthermore, in the case of a monodentate ligand which exchanges rapidly from bound to free, the average pseudocontact shift in the absence of geometric constraints is zero.

The contact coupling constant is proportional to the spin density normalized to one electron. The proportionality constant depends on the magnetic properties of the two interacting particles. For the hydrogen atom such a constant K is experimentally found to be equal to 1.42×10^9 Hz [11][1]. In hydrogen and hydrogen isotopes (deuterium, tritium) there is only one pathway to generate contact spin density at the nucleus, i.e. through the 1s orbital, and no other orbital contributes to the contact or pseudocontact shift. The following equation then relates the contact hyperfine coupling with spin density

$$\frac{A}{h} = \frac{K\rho}{2S} \tag{2.27}$$

where K for the other hydrogen isotopes (by recalling that A contains γ_I (Eq.

---

[1] The constant K is also theoretically calculated to be 1.42×10^9 Hz by using Eqs. (2.27) and (2.10): $K = \frac{2\mu_0}{6\pi}\gamma_I g_e \mu_B \psi(0)^2$, where $\psi(0)^2 = \frac{4}{4\pi}\left(\frac{1}{a_0}\right)^3$ and $a_0 = 53$ pm.

(2.2)) and that $\sum_k \rho_k$ is the same for all isotopes) is given by ($k = 2, 3$)

$$K(^k\mathrm{H}) = \frac{\gamma(^k\mathrm{H})}{\gamma(^1\mathrm{H})} K(^1\mathrm{H}).$$

Let us now consider an organic radical, e.g. an alkane. The unpaired electron will be in a MO. If the alkane radical is negative, the electron will be in the *lowest unoccupied* MO (LUMO), if positive or neutral in the *highest occupied* MO (HOMO). The amount of unpaired electron directly delocalized at the hydrogen nucleus will be proportional to the coefficient representing its 1s orbital in the MO containing the unpaired electron. Furthermore, we may have spin polarization from the fraction of an unpaired electron in a p orbital of a carbon or nitrogen atom to the hydrogen 1s orbital. Typical examples of spin polarization will be discussed later.

When the unpaired electron in the molecule resides on a metal ion, spin delocalization on the ligand can occur through different mechanisms, and different signs of spin density can arise. Positive contributions to the spin density on the ligand arise from the MO(s) involving partially filled metal orbitals. The HOMO of the ligand which is involved in the MO together with the metal orbital bears a fraction of unpaired electron. Negative contributions occur through spin polarization from unpaired electrons in the atomic orbitals of the metal and doubly occupied MOs of the metal ligand complex.

The variety of spin density distributions on the same ligand by changing the metal ions depends on the different weights of the two above contributions. Spin density distributions can also change if σ and π mechanisms contribute to different extents.

2.3.1 Metal ion–water interactions

Let us now consider the mechanism of spin delocalization in the water molecule when bound to a paramagnetic metal ion. Table 2.1 reports the proton A/h values for some hexaaqua metal complexes and the vanadyl pentaaqua complex. They have been determined from hyperfine shifts, although sometimes they have been determined as absolute values from relaxation studies (see Chapter 3). Spin density has been evaluated through Eq. (2.27). The proton hyperfine shifts, assumed to be mainly contact in origin as a result of the averaged high symmetry, are downfield and the hyperfine coupling constants are positive. With the possible exceptions of Ti^{3+}, Fe^{3+} and Mn^{2+}, it has been suggested that the hydrogen spin density, which is also listed in Table 2.1, tends to correlate with the number of unpaired t_{2g} electrons. The t_{2g} orbitals have the correct symmetry for π bonding. Unpaired electrons in the e_g orbitals, which have the correct symmetry for σ bonding, do not affect the value of ρ, and indeed the experimental proton hyperfine coupling constants in Cu^{2+} (d^9) and Ni^{2+} (d^8) complexes, having no unpaired electrons in t_{2g} orbitals, are very small (Fig. 2.10). It appears, therefore, that unpaired π

TABLE 2.1
Hyperfine coupling constants [a] of water nuclei in aqua complexes of common metal ions

Metal ion	Electronic configuration	$2S$	n [b]	A/h (MHz)		Reference
				^{1}H [c]	^{17}O	
Ti^{3+}	d^1	1	1	4.5 (3.2×10^{-3}) [d]	4.4 [d]	[114,115]
VO^{2+}	d^1	1	1	2.1 (1.4×10^{-3}) [d]	3.8 [d]	[116,117]
V^{2+}	d^3	3	3	2.0 (4.2×10^{-3})		[118]
Cr^{3+}	d^3	3	3	2.0 (4.2×10^{-3})		[12,118]
Mn^{2+}	d^5	5	3	0.6–1.0 ((2.2–3.5) $\times 10^{-3}$)	5.3	[118–120]
Fe^{3+}	d^5	5	3	0.4–1.3 ((1.4–4.6) $\times 10^{-3}$)		[118,121]
Fe^{2+}	d^6	4	2		9.4	[119]
Co^{2+}	d^7	3	1	0.4 (1.1×10^{-3})	14.8–15.7	[12,119,122]
Ni^{2+}	d^8	2	0	0.2 (3×10^{-4})	20.4	[12,123]
Cu^{2+}	d^9	1	0	0.06–0.19 (4.2×10^{-5}–1.3×10^{-4})	50.0	[12,124]
Ln^{3+}	f^1–f^6, f^8–f^{13}				0.6–0.8	[125]

[a] Contact for protons; contact + ligand centered pseudocontact for oxygen. Positive signs indicate downfield shifts for ^{1}H and upfield shifts for ^{17}O signals.
[b] Number of unpaired electrons in t_{2g} orbitals for high-spin complexes with O_h symmetry.
[c] ρ_H values in parentheses, estimated from Eq. (2.27).
[d] Absolute values (obtained from relaxation data).

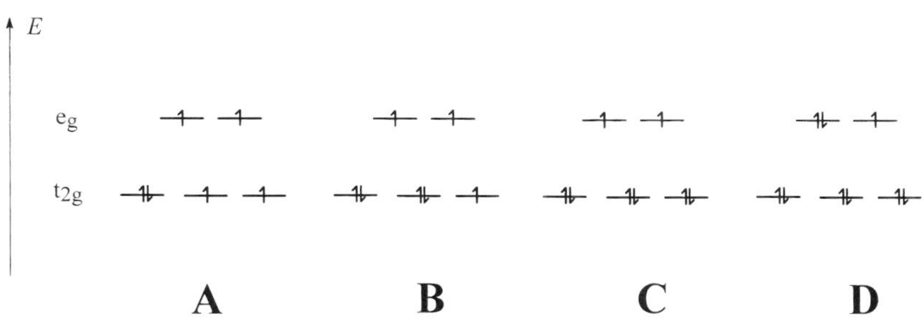

Fig. 2.10. Schematic representation of the energy of the five 3d orbitals in a iron(II) (A), cobalt(II) (B), nickel(II) (C) or copper(II) (D) complex of octahedral symmetry. The two sets of levels are identified by the group-theoretical labels t_{2g} and e_g. The latter two complexes have no unpaired electrons in t_{2g} orbitals.

electrons of the metal ions delocalize onto the H_2O moiety. One fully occupied water MO of σ type is capable of directly overlapping with the t_{2g} 3d orbital set of the metal. Thus the unpaired t_{2g} electron is capable of directly producing spin density on the protons. A further π interaction with the metal t_{2g} set arises from a non-bonding p orbital of oxygen, which would place spin density on the protons only through spin polarization. The sign of the observed shifts shows that the former mechanism is dominant [12]. The ^{17}O data (Table 2.1) show large hyperfine coupling constants when unpaired electrons are located in the e_g orbitals, which have the correct symmetry for σ bonding. The data for $Cu(H_2O)_6^{2+}$ and $Ni(H_2O)_6^{2+}$ demonstrate this clearly. It is obvious that direct delocalization through metal–oxygen σ bonding involving oxygen s orbitals gives rise to quite sizable spin density on the oxygen nucleus.

2.3.2 Other cases

Let us now consider the complexes between amines (am) and nickel(II). The stoichiometry is $Ni(am)_6^{2+}$ and the geometry octahedral. The splitting of the d orbitals with the electron occupancy is analogous to that discussed for the hexaaqua complex (Fig. 2.10 and Table 2.1) The two e_g unpaired electrons will have an excess of $M_S = -\frac{1}{2}$ occupancy (positive spin density). Such $-\frac{1}{2}$ spin will be transmitted on the ligand through a MO which is a linear combination of the d orbitals and the ligand HOMO. The latter is mainly constituted by the atomic orbitals of the nitrogen atoms in accordance with the lone pair being localized on the nitrogen. The contribution to the ligand HOMO of atomic orbitals of the other ligand atoms decreases dramatically with the number of bonds. It follows that the spin density transmitted through this mechanism will rapidly decrease to zero as the number of chemical bonds between the metal and the resonating proton increases. The proton shifts fall rapidly to zero as we move away from the α-carbon [13,14]. The positive shift (in frequency) is due to $-\frac{1}{2}$ spin density (see Eqs. (2.6) and (2.27)), which in turn is the consequence of direct transfer with the same sign from d orbitals to ligand orbitals.

A second mechanism involving spin polarization may also be operative, in which spin polarization yields alternate positive and negative unpaired spin density along the ligand backbone. The protons bound to nitrogen experience upfield shift, as is also observed in coordinated ammonia [15]. It is believed that it is due to prevailing spin polarization from a nitrogen p orbital to the hydrogen 1s orbital [14,15]. ^{1}H, ^{13}C and ^{14}N shift values are available for ethylamine coordinated to a bis(acetylacetonate) nickel(II) complex [14]. It is reasonable to believe that the hyperfine shift is mainly contact in origin. The ^{14}N, $^{13}C_\beta$, and α-proton hyperfine couplings are positive (see Table 2.2), whereas α-carbons and amine protons have negative hyperfine couplings probably due to spin polarization. The spin polarization mechanisms which account for upfield shifts do provide a downfield mechanism for α-CH protons (see Section 2.4). It is possible that spin polarization and direct delocalization mechanisms are simultaneously operative. The contact shifts are upfield for protons of ammonia coordinated to Cu^{2+}, Ni^{2+} and Co^{2+} [15] for the same mechanism as discussed above; the contact shifts are downfield for Mn^{2+} owing to a dominant mechanism of direct transfer of spin density from t_{2g} orbitals to 1s hydrogen orbitals [15]. Table 2.2 also shows the hyperfine shifts (essentially contact, see later) in a nickel(II) bis(alkylxantate) complex with

TABLE 2.2

Estimated A/h values (MHz) for ^{1}H, ^{13}C and ^{14}N nuclei of nitrogen- and sulfur-containing ligands of nickel(II)

Ligand								Reference
Ammonia			N^1H_3					[15]
			−1.5					
Ethylamine		^{14}N	N^1H_2	$^{13}CH_2$	C^1H_2	$^{13}CH_3$	C^1H_3	[14]
		14.4	−2.64	−0.70	1.02	1.80	0.22	
$(bpy)Ni(xan\text{-}R)_2$ [a]	R	α-C^{1}H	β-C^{1}H	γ-C^{1}H	δ-C^{1}H	ε-C^{1}H	ζ-C^{1}H	[16]
	ethyl	−0.03	0.03					
	propyl	−0.03	0.04	0.02				
	butyl	−0.03	0.04	0.01	0.003			
	pentyl	−0.03	0.04	0.01	0	0.001		
	hexyl	−0.03	0.04	0.01	0	0	0	
	cyclo-	−0.23	0.09	0.005	0.001			
	hexyl		0.01	0.001				
	R	α-^{13}CH	β-^{13}CH	γ-^{13}CH	δ-^{13}CH	ε-^{13}CH	ζ-^{13}CH	
	ethyl	0.20	0.18					
	propyl	0.18	0.17	0.02				
	butyl	0.18	0.17	0.02	0.02			
	pentyl	0.18	0.17	0.02	0.02	0.004		
	hexyl	0.18	0.16	0.03	0.02	0.003	0.002	
	cyclohexyl	0.15	0.04	0.02	0.01			

[a] bpy = bipyridyl; xan-R = $R-O-C(=S)S^-$.

aliphatic chains of variable length [16]. The upfield shift of α-CH protons is due to spin polarization by π spin density in the delocalized xanthate ring [16].

The case of the hexa(pyridine) nickel(II) complex has been extensively debated in the early literature of NMR of paramagnetic complexes [17–21]. The shift pattern with α-H > γ-H > β-H (Fig. 2.11 and Table 2.3) was soon recognized to be predominantly of σ-type. The ligand has a σ MO system which has the correct symmetry to overlap with the $d_{x^2-y^2}$ and d_{z^2} orbitals. However, spin polarization can induce ½ spin density in the π system. Once some unpaired spin density is in a p_π orbital, it spin-polarizes the electrons of the C−H σ bond, thus producing a further mechanism for transferring spin density on the proton. The proton A/h value from this mechanism is proportional to the spin density on the carbon p_z orbital, ρ_C^π, through a proportionality constant Q_{CH}^H:

$$\frac{A}{h} = Q_{CH}^H \frac{\rho_C^\pi}{2S}. \tag{2.28}$$

The constant Q_{CH}^H is negative and of the order of −70 MHz (see Section 2.4

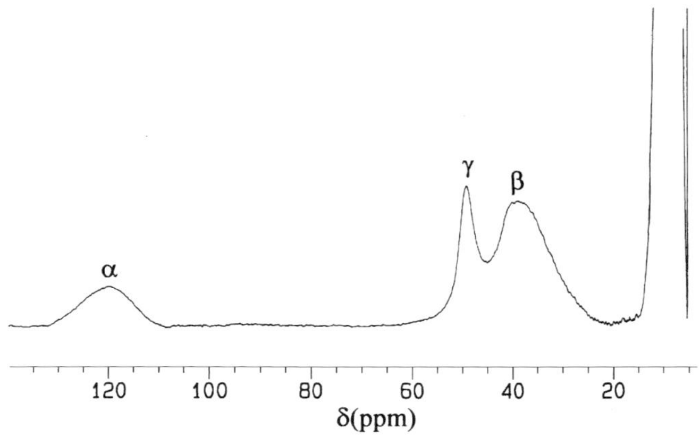

Fig. 2.11. ^{1}H NMR spectrum of hexa(pyridine)nickel(II). The shift pattern is α-H ≫ γ-H > β-H, indicating predominance of a σ-type mechanism with a π contribution.

TABLE 2.3

Estimated A/h values (MHz) for ^{1}H, ^{13}C and ^{14}N nuclei of pyridine and γ-picoline axially bis-coordinated to nickel(II) bis-acetylacetonate

Ligand		A/h (MHz)	Reference
Pyridine	^{14}N	+18	[126]
Pyridine	α-^{13}C	−0.3	[22,23]
Pyridine	β-^{13}C	+0.7	[22,23]
Pyridine	γ-^{13}C	−0.1	[22,23]
γ-Picoline	γ-$^{13}CH_3$	+0.1	[23]
Pyridine	α-^{1}H	+0.9	[36]
Pyridine	β-^{1}H	+0.3	[36]
Pyridine	γ-^{1}H	+0.35	[36]
γ-Picoline	γ-$C^{1}H_3$	−0.1	[23]

and Table 2.6). Therefore, we can say that the spin density on the proton has opposite sign of that on the attached sp^2 carbon (Fig. 2.12). Note that Eq. (2.28) is analogous to Eq. (2.27) for the free hydrogen atom, but the unpaired spin density now resides on the neighboring atom and the proportionality constant is much smaller.

If the spin delocalization mechanism on the pyridine ring were σ solely, substitution of γ-H with γ-CH_3 would produce almost zero spin density on the γ-CH_3 protons. On the contrary (Table 2.3 and Fig. 2.11), some upfield (negative) shift is observed [22,23]. Spin polarization, from e.g. positive spin density on the p_z orbital of an sp^2 carbon, produces negative spin density on the attached proton (Fig. 2.12), and positive spin density again on the protons of an attached CH_3 moiety (see also Section 2.4). Therefore, if the γ-CH_3 protons experience upfield

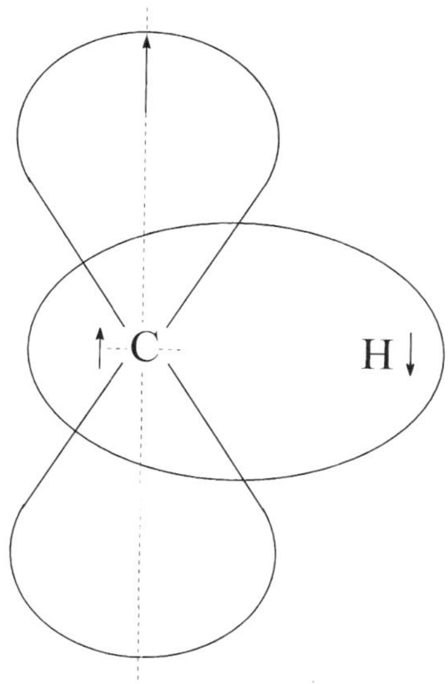

Fig. 2.12. The spin density on protons has the opposite sign of that on the attached sp^2 carbon.

shift, they experience ½ spin density which comes from ½ spin density of the p_z of the γ-carbon, which in turn comes from −½ spin density in the σ MO. Several calculations are available which try to separate the various contributions [21,23–32].

In the case of pyridine bound to chromium(III), the three unpaired electrons sit in d_{xy}, d_{xz} and d_{yz} orbitals which do not overlap with ligand σ orbitals. Therefore, there is no mechanism for direct unpaired electron transfer onto the ligand through σ bonds. Spin density is produced in the doubly occupied ligand HOMO through ligand π bonding to the d_{xy}, d_{xz} and d_{yz} orbitals. The spin and shift patterns in bis malonato bis pyridine chromium(III) are reversed with respect to the nickel case, with more tendency to alternation. Therefore, the spin pattern is mainly due to the σ bonding in the Ni complex (since all the shifts of the pyridine protons are downfield), with a contribution due to the π spin density between the carbons of the pyridine ring, whereas it is mainly due to the π spin density in Cr complex (since the shifts of the meta protons are downfield and those of the *ortho* and *para* protons are upfield), with a minor contribution from the σ bonding [33].

Imidazole ligands have been studied with iron, either 2+ ($S = 2$) or 3+ (S = ½, 5/2), Co^{2+}, Ni^{2+}, Cu^{2+} [34–36]. The spectrum of the high spin complex $Fe(NEtIm)_6^{2+}$ [35] is reported in Fig. 2.13. The shifts are downfield and the order, if all substituents are taken into account, is N(1)H > 5H > 4H > 2H > NCH_3 ≥ $5CH_3$. Probably this is the result of σ spin delocalization, π spin delocalization, and spin polarization effects. In the case of low spin iron(III), 2H is sizably upfield in $Fe(CN)_5Im^{2-}$ (about 20 ppm), NCH_3 or $5CH_3$ are downfield, and the other

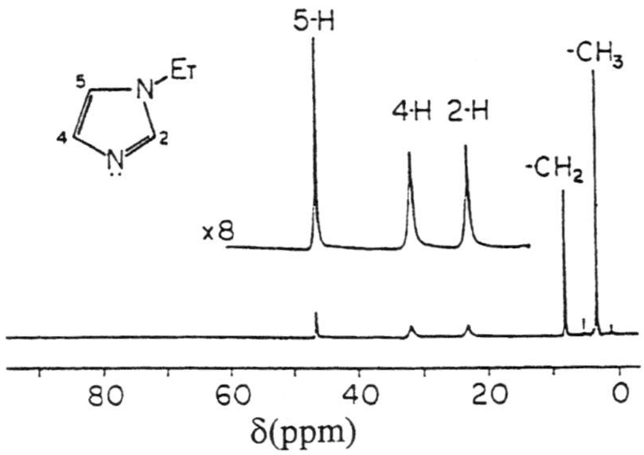

Fig. 2.13. ^{1}H NMR spectra of $Fe(NetIm)_6^{2+}$ [35].

TABLE 2.4
Hyperfine shifts (ppm) of ring nuclei in octahedral complexes of Ni^{2+} with aniline and pyridine-*N*-oxide

Ligand		*Ortho*	*Meta*	*Para*	Reference
Aniline	^{1}H	−6.9	+3.8	−7.6	[38]
Py-NO	^{1}H	−13	+11	−17	[36]
	^{13}C	+166	−65	+90	[37]

ring protons are slightly upfield. Alternation of proton and methyl shifts has been interpreted as due to a predominance of direct π spin density onto the imidazole ring [35].

Completely different is the shift pattern for pyridine-*N*-oxide in hexa(pyridine-*N*-oxide) nickel(II) [36–38] (Table 2.4). This is also a pertinent example of how spin delocalization occurs in a six-membered π system. Here the *ortho* and *para* protons experience upfield shift, whereas the *meta* protons experience downfield shift. The absolute values of the hyperfine shifts are not sensitive to the distance from the paramagnetic center. This behavior could be accounted for by the non-orthogonality of the p_π of nitrogen with the M−O coordination bond (see also Section 2.4). The M−O coordination bond involves the e_g orbitals which have unpaired electrons. The unpaired electron in the p_z of nitrogen delocalizes into the ring through a π orbital of either bonding or antibonding character. If ½ spin is present in a π MO orbital, it polarizes the doubly occupied π MOs. So, in any p_π of each carbon or nitrogen atom of the ring there is ½ spin of the given MO and a tendency to have the same spin from the other π MOs through spin polarization. However, the total spin on every doubly occupied MO has to be zero. Therefore, if spin polarization causes ½ spin contribution on a p_z orbital it has to cause −½ spin polarization of the same amount in absolute value on a p_z of another atom, in order to make zero spin density all over the six atoms per doubly occupied MO. As a result, total $-\frac{1}{2}$ spin density occurs at p_z of *ortho* and *para* carbons, whereas $\frac{1}{2}$ spin density occurs at p_z of *meta* carbons. Note that from the shifts the amounts of $-\frac{1}{2}$ and $\frac{1}{2}$ electron spins appear to be within a factor of two in absolute values.

A similar pattern is shown in d^5–d^9 ions [38]. However, the ratios of the shifts varies with the metal ion. In the case of d^1 ions the shifts are very small, while in the case of Cr^{3+} the shifts are all downfield [36]. This possibly indicates that more mechanisms contribute to the overall shifts.

Phenylamine types of ligand behave like pyridine-*N*-oxide (Table 2.4) [38]. The same holds for phosphines [39]. It is also likely that phenolates are similar to the above systems.

When the number of atoms between the metal and the aromatic ring increases, like in Ph−C=N−R [26], $Ph-CONH_2$ [38], Ph_3PO [40], $Ph-CH_2-NH_2$ [41], the patterns are similar to those described above but the shifts on the aromatic ring are about one order of magnitude smaller.

Other ligands like phenanthroline [42–44] and bipyridyl [42,45] are expected to behave similar to pyridine. Salicylaldiminates (Table 2.5) display two spin delocalization mechanisms: one through the oxygen atom and the other through the −C=NR group, which is of minor relevance. The sum of the two mechanisms qualitatively accounts for the observed values (Table 2.5), although often a bias of negative shifts is observed [46,47].

Historically, bis(aminotroponeiminato) nickel(II) complexes have been very instructive. The compounds are either pseudotetrahedral or display a tetrahedral–planar equilibrium. The ligands contain seven-membered rings showing alternation of proton shifts and spin densities (Table 2.5). The interest lies in the variety of R derivatives which show how spin density can be transmitted through π bonds, whereas it cannot be transmitted through sp^3 carbons or through ethereal oxygen atoms [48,49].

Different chemical shift values are observed for the various carboxylate groups with monodentate, bidentate or bridging bidentate metal coordination geometry, as a result of different spin delocalization mechanisms. For monodentate groups, delocalization of unpaired spin density can occur for: (1) a σ-symmetry metal orbital through the σ-bond framework, which contributes to downfield NMR shifts for all atoms in the carboxyl ligand; (2) spin polarization, yielding alternate positive and negative unpaired spin density along the ligand backbone and thus providing an upfield or downfield contribution; (3) spin delocalization occurring through π-spin delocalization and causing downfield shifts, whose magnitude depends on the dihedral angle between the COO^- plane and the CH_2 group attached to it. For bidentate groups, the latter effect is expected to be larger as the metal ion is forced to be coplanar with the carboxylate [50]. αCH_2 signals for a bidentate carboxylate ion have been observed in the 100–140 ppm region in high spin iron(III) complexes [50].

TABLE 2.5
Estimated A/h values for protons in nickel(II) bis-salicylaldiminato and bis-aminotroponeiminato systems

Structure	Proton	A/h (MHz)	Reference
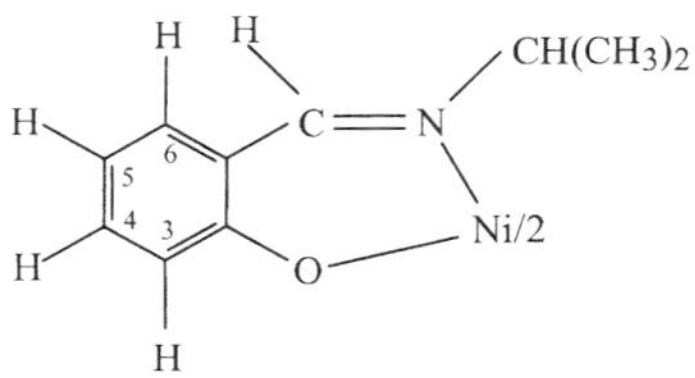	3	-0.30 [a]	
	4	$+0.30$ [a]	
	5	-0.29 [a]	[46]
	6	$+0.14$ [a]	
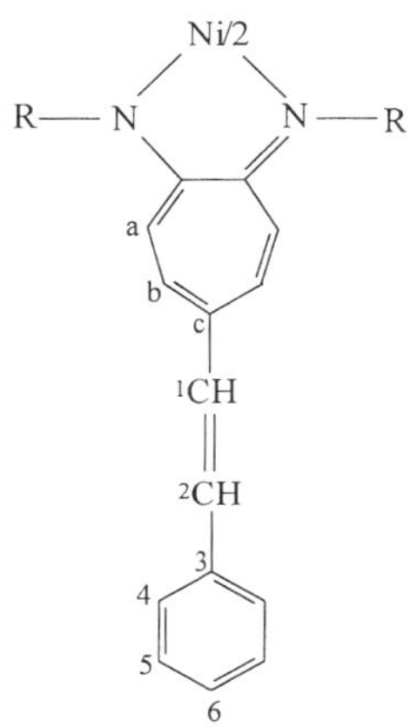	a	-1.29 [b]	
	b	$+0.66$ [b]	[48]
	(c)	(-1.80 [b])	
	1	$+0.60$ [b]	
	2	-0.87 [b]	
	4	-0.19 [b]	[49]
	5	$+0.05$ [b]	
	6	-0.21 [b]	

[a] Pseudocontact contributions to the hyperfine shifts have been subtracted.
[b] Pseudocontact contributions to the hyperfine shifts have been neglected.

2.4 PROTON HYPERFINE COUPLING AND CONFORMATION

The contact hyperfine shift contains information on conformational arrangements because it is somehow related to chemical bonds. The information, however, is generally hidden, with the exception of few cases. For example, if we have a CH moiety of an sp^3 carbon bound to an sp^2 carbon which bears spin density in the p_z orbital, we have already seen that a spin density of the same sign is transferred to the methyl protons (Section 2.3.2). A mechanism to originate spin density at the methyl proton is the direct overlap between the sp^2 carbon p_z and the 1s hydrogen orbital [51–54] (Fig. 2.14). Such overlap depends on the dihedral angle θ between the p_z(axis)–C–C plane and the C–C–H plane (Fig. 2.14). The relationship is described by the general Karplus equation [55]

$$\frac{A}{h} = (a\cos^2\theta + b\cos\theta + c)\frac{\rho_C^\pi}{2S} \tag{2.29}$$

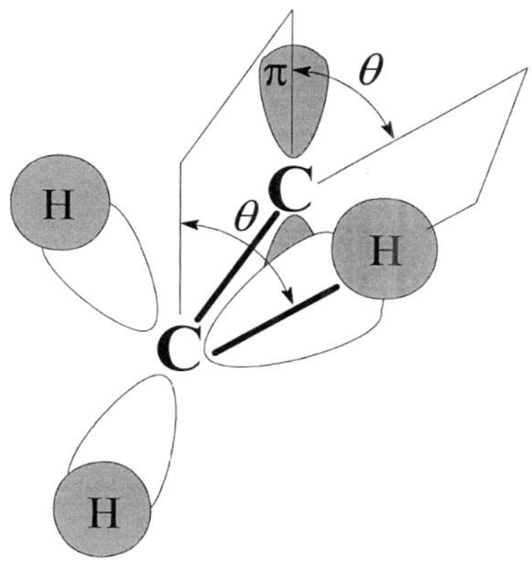

Fig. 2.14. CH moiety of an sp^3 carbon bound to an sp^2 carbon, illustrating the definition of the dihedral angle θ.

where ρ_C^π is the spin density on the p_z orbital and a, b and c are constants. For the present moiety, b equals zero. The constant a represents the maximal direct overlap of the 1s hydrogen orbital with the carbon p_z orbital, which occurs when $\theta = 0$ (Fig. 2.14). The constant c represents the θ-independent transfer of spin density through the C−C−H σ bonds, which is the only mechanism available when $\theta = 90^\circ$. The latter mechanism is less efficient than the former, and c is much smaller than a [51]. The constant a is about 140 MHz [51,55]. When the CH moiety belongs to a CH_3 group and the group is free to rotate, an average over all positions should be calculated. The average value of $\cos^2\theta$ is ½, so that $a\langle\cos^2\theta\rangle \cong 70$ MHz. Experimentally, a value of $a\langle\cos^2\theta\rangle + c$ of ≅75 MHz is found [51]. In other words, if there is a positive (−½) spin density on the p_z orbital of the sp^2 carbon atom, a positive spin density is present on average on a proton of a CH_3 group, whereas it would be negative if the proton were directly attached to the sp^2 carbon atom (see Fig. 2.12 and Section 2.3.2). In general, we can say that A/h is always given by an equation such as Eq. (2.27), where the K value (which equals 1420 MHz for the isolated hydrogen atom) is scaled down to a constant Q, called McConnell constant, which depends on the geometry of the system. As we have seen, the constant Q is ≈ -70 MHz for a proton directly attached to an sp^2 carbon (Eq. (2.28)) or $\approx +75$ MHz for a freely rotating CH_3 group. Some other Q values are collected in Table 2.6 [56–59].

This reasoning induces one to think that, even in a moiety of the type M–D–C–H, where D is the donor atom, the CH proton experiences a contact coupling depending on the orientation of the D–C–H plane (Fig. 2.15). If the spin density is along the M−D bond (Fig. 2.15A), then the coupling depends on the dihedral angle between the M–D–C and D–C–H plane in a fashion similar to Eq. (2.29), where b may not be negligible [13], θ being now defined with respect to the M−D bond rather than with respect to a non-bonding p_z orbital as in Fig. 2.14. This is

TABLE 2.6

Q values (MHz) for fragments involved in π electron spin delocalization systems [a]

Q^{C}_{CH}	Q^{H}_{CH}	$Q^{C}_{CC'}$	$Q^{C}_{C'C}$	$\langle Q^{H}_{CCH_3} \rangle$	Q^{F}_{CF}	Q^{F}_{FC}	Reference
+54.6	−65.8	+40.3	−39.0				[56]
	−75.6	+53.2					[57]
				+75			[58]
					−410	+2370	[59]

[a] The notation Q^{Y}_{XY} means the McConnell constant for a nucleus Y attached to an atom X in sp^2 configuration and bearing unpaired electron spin density on its p_z orbital. The notation Q^{X}_{XY} means the McConnell constant for the nucleus of atom X in the same moiety.

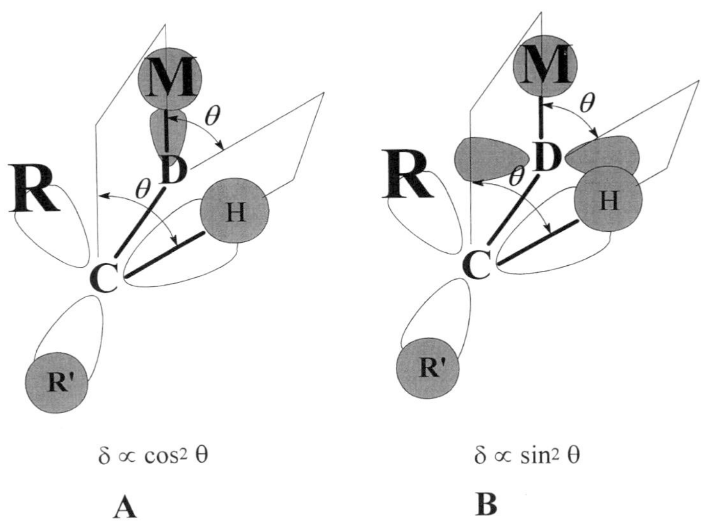

Fig. 2.15. Dihedral angle θ between the M–D–C plane and the D–C–H plane of M–D–C–H moieties. The case of the spin density being along the M−D bond is shown in A and the case of the spin density being in a p orbital of the D atom orthogonal to the M−D bond is shown in B.

the case when D is an amino nitrogen [13–15]. If the spin density is in a p orbital of the D atom orthogonal to the M−D bond (Fig. 2.15B), a dependence on $\sin^2\theta$ is more appropriate. This seems to be the case for *S* as donor atom [60].

In the case of $Fe_4S_4^{2+}$ clusters it has indeed been found that the contact shifts of the βCH_2 protons of coordinated cysteines depend on θ as

$$\delta = a\sin^2\theta + b\cos\theta + c \qquad a \cong 10.3 \quad b \cong -2.2 \quad c \cong 3.9 \tag{2.30}$$

The relative values of the c and a constants give an idea of the importance of the two processes of electron delocalization, through M−D σ and D π bonds, the latter being dominant.

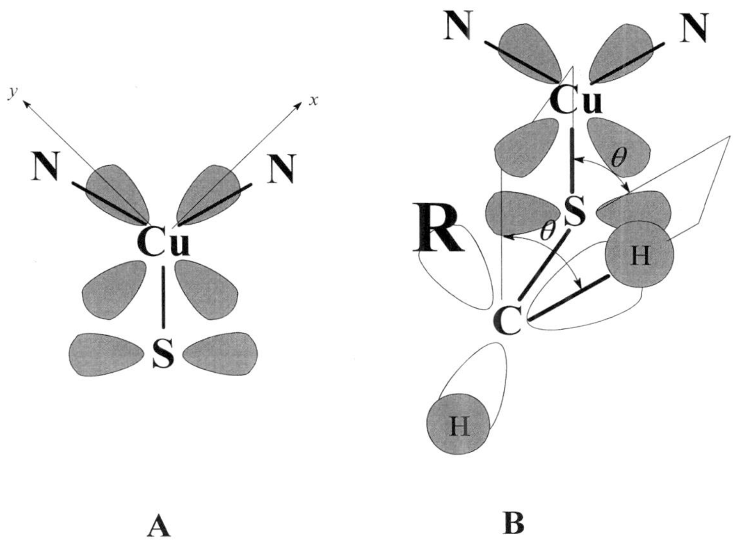

Fig. 2.16. Arrangement of the donor atoms in the equatorial plane of copper and of the copper $d_{x^2-y^2}$ and sulfur p orbitals (A). The dihedral angle θ between the Cu–S–C plane and one of the S–C–H planes in depicted in B [113].

In the case of blue proteins there is a plane with Cu(II), two histidine nitrogens and one cysteine sulfur, with the x and y directions defined as in Fig. 2.16A. The $d_{x^2-y^2}$ orbital contains the unpaired electron. The sulfur donor atom has a π orbital which overlaps the $d_{x^2-y^2}$ orbital (Fig. 2.16(B)). Large downfield shifts are observed (see also Chapter 5), and again the angular dependence of the shifts of the CH_2 protons is expected to be mainly of $\sin^2$ type [61].

2.5 THE ORIGIN OF THE SHIFTS IN HETERONUCLEI

The predictions of the proton shifts is difficult; the prediction is even more difficult in the case of heteroatoms. In the case of σ spin density like in a aliphatic amine (Table 2.2A), the ^{13}C Fermi contact shift for the α-carbon of the aliphatic amine is upfield (A/h is negative) because of predominant spin polarization effects, whereas that for other carbon atoms is downfield, and rapidly attenuates with the number of bonds. A sizable downfield shift is experienced by the ^{14}N nucleus when nitrogen is a donor atom.

More hints are available for aromatic moieties. If we consider three sp^2 carbon atoms C′−C−C″ in a delocalized moiety (e.g. in an aromatic ring), and we assume that unpaired spin density resides in the p_z orbitals, we have the following contributions to the contact interaction at the central carbon atom C [62]:

(1) spin polarization of the 1s orbital from the p_z orbital of the same atom C;

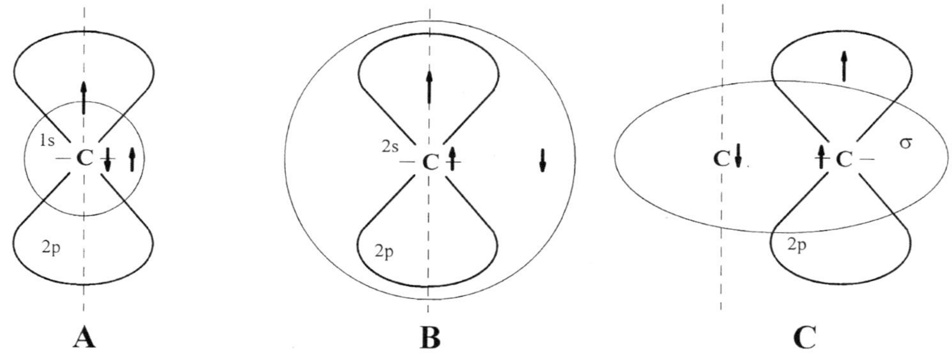

Fig. 2.17. Spin polarization mechanisms arising from unpaired spin density on a $2p_z$ orbital, illustrated for a ^{13}C nucleus. (A) Polarization through a 1s orbital; (B) polarization through a 2s orbital; (C) polarization on a nearby carbon through a σ bond.

(2) spin polarization of the 2s orbital from the p_z orbital of the same atom C, through the three σ bonds in which the 2s orbital is involved;
(3) spin polarization of the 2s orbital from the p_z orbitals of the neighboring carbon atoms C′ and C″, through their σ bonds to atom C.

The first contribution is proportional to the spin density on the p_z orbital, ρ_C^π, through a negative proportionality constant, S^C, which has been estimated to be −35.5 MHz. Therefore, the resulting spin density at the C nucleus is opposite in sign to that of the p_z orbital (Fig. 2.17A).

The second contribution is again proportional to ρ_C^π and arises from the spin polarization of the σ bonds involving C′, C″ and the H atom bound to C. The proportionality constants, $Q_{CC'}^C$, $Q_{CC''}^C$ and Q_{CH}^C are of the same type of Q_{CH}^H in Eq. (2.28). They are all positive (Table 2.6), and therefore the resulting spin density at the C nucleus through the involvement of its 2s orbital in the three σ bonds is of the same sign as that of the p_z orbital (Fig. 2.17B).

At variance with the first two, the third contribution is proportional to the p_z spin density on the neighboring carbon atoms, $\rho_{C'}^\pi$ and $\rho_{C''}^\pi$. The proportionality constants, $Q_{C'C}^C$ and $Q_{C''C}^C$ are negative (Table 2.6) and the sign of the resulting spin density at the C nucleus is opposite to that on the p_z orbitals of the neighboring carbons (Fig. 2.17C).

The hyperfine coupling constant on the carbon atom C is therefore given by:

$$\frac{A}{h} = \left(S^C + Q_{CC'}^C + Q_{CC''}^C + Q_{CH}^C\right)\rho_C^\pi + Q_{C'C}^C\rho_{C'}^\pi + Q_{C''C}^C\rho_{C''}^\pi. \tag{2.31}$$

From the foregoing analysis, it appears that there are several terms making a significant contribution to the hyperfine coupling constant. As they are both positive and negative and of similar magnitude (Table 2.6), the final value of the hyperfine coupling constant is therefore the result of considerable cancellation among the various contributions. In Eq. (2.31) the three spin densities ρ_C^π, $\rho_{C'}^\pi$

and $\rho^{\pi}_{C''}$ are unknown. It appears that, even considering the shift of the proton attached to the carbon, which provides an independent estimate of ρ^{π}_{C} through Eq. (2.28), the number of parameters defining the shift of the heteronucleus is larger than the two experimental shifts, unless the shifts of the protons attached to the neighboring carbons are also considered. Similar considerations hold if the central atom is an sp^2 nitrogen. A deeper insight can be achieved by using experimental data on nuclear relaxation (see Section 3.9).

^{13}C data are available on several ligands. In Tables 2.3 and 2.4 the relative ^{13}C hyperfine shifts of pyridine and pyridine-*N*-oxide are reported [22,37]. In both cases it appears that there is alternation in the shifts, which would indicate π spin delocalization. However, the ratio between the shifts of protons and carbons at each position are different one from the other, and differ from what would be expected on the basis of a simple spin polarization mechanism. It is concluded that more than one mechanism is operative.

Data are available on ^{13}C and ^{15}N of cyanide bound to a heme moiety. Cyanide is the most common reactant of oxidized heme moieties [63]. All the monocyano and dicyano derivatives of isolated porphyrins and hemoproteins are low spin and contain either diamagnetic iron(II) or paramagnetic (S = ½) iron(III). ^{13}C and ^{15}N NMR spectra have been reported for the latter systems and, as shown in Tables 2.7 and 2.8, dramatic paramagnetic effects have been measured for both nuclei: the ^{13}C hyperfine shifts range from −2000 to −2400 ppm upfield [64] whereas the ^{15}N shifts range from 600 to 1200 ppm downfield [65–68]. The spin delocalization would be dominantly spin polarization through an Fe−C π bond, which may also account for the opposite sign of carbon and nitrogen shifts. Possible metal-centered pseudocontact contributions estimated from the known metal–nucleus distance and magnetic anisotropy values are of the order of 100–300 ppm downfield. Theoretical studies [10] have estimated sizable ligand-centered pseudocontact contributions to the shifts of nitrogen in $Fe(CN)_6^{3-}$ (Section 2.2.2).

Although the magnitude of the shifts is only slightly sensitive to variations

TABLE 2.7
^{13}C chemical shifts (ppm from TMS) for coordinated cyanide in low spin iron(III) porphyrins and hemoproteins [64]

Iron species	Porphyrin 3,8-R group	^{13}C chemical shift
Fe protoporphyrin IX	$-CH{=}CH_2$	−2393
Hemin c	$-CH(CH_3)SCH_2-CH(NH_2)COO^-$	−2300
Fe 3,8-disulfonate-deuteroporphyrin IX	$-SO_3^-$	−2167
Fe 5,10,15,20 tetra(4-carboxyphenyl)-porphyrin		−1968
(C_2H_5NC)-myoglobin (diamagnetic)		171.3

TABLE 2.8
^{15}N chemical shifts (ppm from $^{15}NO_3^-$) for coordinated cyanide in low spin iron(III) porphyrins and hemoproteins

Iron species	Solvent	^{15}N chemical shift	Reference
Hemin	Me_2SO	732	
	Py	696	
	CH_3OD	506	[65]
	$CH_3OD + D_2O$ (1 : 1)	480	
	H_2O	448	
(4-AcPy)hemin	4-AcPy + D_2O	945	[65]
(Py)hemin	Py + D_2O	989	[65]
(3,5-Lu)hemin	3,5-Lu + D_2O	1070	[65]
(1-CH_3Im)hemin	1-CH_3Im + D_2O	926	[65]
Myoglobin	D_2O	931–948 [a]	[65]
Cytochrome *c*	D_2O	841	[66]
Hemoglobin	D_2O	975 (α)	[67]
		1047 (β)	
Horseradish peroxidase	D_2O	576	[68]
Cytochrome *c* peroxidase	D_2O	587	[68]
Lignin peroxidase	D_2O	608	[68]
Manganese peroxidase	D_2O	639	[68]

[a] Depending on pH.
Abbreviations: AcPy: acetylpyridine; Py: pyridine; Lu: lutidine; Im: imidazole.

in the nature of the heme and to its contacts with the surrounding protein, it is strongly dependent on the identity of the trans ligand. For example, the ^{15}N signal in protohemin cyanide shifts from 945 to 989 to 1070 ppm downfield on passing from 4-acetylpyridine to pyridine to 3,5-lutidine as the *trans* ligand [66]. Both ^{13}C and ^{15}N hyperfine shifts are also very sensitive to changes in the solvent. Aprotic solvents usually give the largest absolute values, which decrease by as much as 200 ppm on passing from DMSO to methanol to water [65,66]. In general, the axial ligand field strength of cyanide, and hence the extent of unpaired spin delocalization, decreases with the increasing hydrogen bonding capabilities of the solvent.

^{15}N NMR data are available for cyanometmyoglobin, cyanocytochrome *c*, cyanomethemoglobin, and some peroxidases (Table 2.8) [65–68]. The large observable differences may be related to the different coordination strength of the trans histidine and/or to the different hydrophobicity of the CN binding site. The difference in the ^{15}N chemical shift (about 70 ppm) of the α and β subunits in hemoglobin (Table 2.8) [67] illustrates the sensitivity of the cyanide nuclei to the protein environment around the axial positions.

For ^{19}F Fermi contact shift of fluorine bound to sp^2 carbon atoms, spin density on the nucleus arises from spin polarization by ρ_C^π, as for analogous CC′ moieties, and from spin polarization by ρ_F^π, which occurs via direct delocalization through C−F π bonding (Fig. 2.18). The hyperfine coupling is therefore

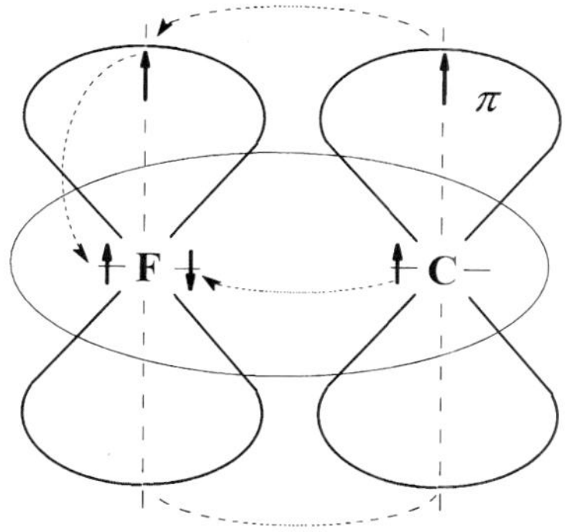

Fig. 2.18. Spin polarization (through σ bonding) and direct delocalization (through π bonding) mechanisms for unpaired spin density transfer from an sp^2 carbon to a fluorine nucleus.

$$\frac{A}{h} = Q^{\mathrm{F}}_{\mathrm{CF}}\rho^{\pi}_{\mathrm{C}} + Q^{\mathrm{F}}_{\mathrm{FC}}\rho^{\pi}_{\mathrm{F}} \tag{2.32}$$

where $Q^{\mathrm{F}}_{\mathrm{CF}}$ is analogous to $Q^{\mathrm{H}}_{\mathrm{CH}}$ or $Q^{\mathrm{C}}_{\mathrm{C'C}}$ and is negative, and $Q^{\mathrm{F}}_{\mathrm{FC}}$, which is the polarization effect of ρ^{π}_{F} on the C−F σ bond, is positive by analogy to $Q^{\mathrm{C}}_{\mathrm{CC'}}$ (see Table 2.6).

^{31}P contact shifts in ligands coordinating through P–O moieties in oxovanadium(IV), cobalt(II) and nickel(II) complexes are downfield and of the order of 10^3 ppm [69–71]. Smaller downfield shifts are observed for nickel-coordinated P–S moieties [72]. Direct P–M coordination gives rise to even larger hyperfine coupling than in P–O moieties. In low spin cobalt(II) complexes A/h values of the order of 10^2 MHz are found [73].

2.6 WHEN IS METAL-CENTERED PSEUDOCONTACT SHIFT EXPECTED?

Pseudocontact shift is expected every time there are energy levels close to the ground state. This causes orbital contributions to the ground state, and such contributions are orientation dependent. Therefore, the magnetic susceptibility tensor is anisotropic. Anisotropy of the magnetic susceptibility tensors arises also from sizable ZFS of the S manifold (Section 1.4).

Six-coordinated high spin cobalt(II) in pure octahedral symmetry has a $^4\mathrm{T}_{2g}$ ground state which is triply degenerate (see Section 5.2). Under spin–orbit coupling and low symmetry components up to six Kramers doublets can be obtained (Fig. 2.19) within 1000 cm^{-1}. The orbital contribution to the magnetic susceptibility is sizable, and the orbital contribution is intrinsically anisotropic. The pseudocontact shifts estimated [74] for a tetragonal complex are reported in Table 2.9. Octahedral nickel(II) complexes have a $^3\mathrm{A}_{2g}$ ground state which is orbitally non-degenerate, with the first excited state at about 10,000 cm^{-1}. So,

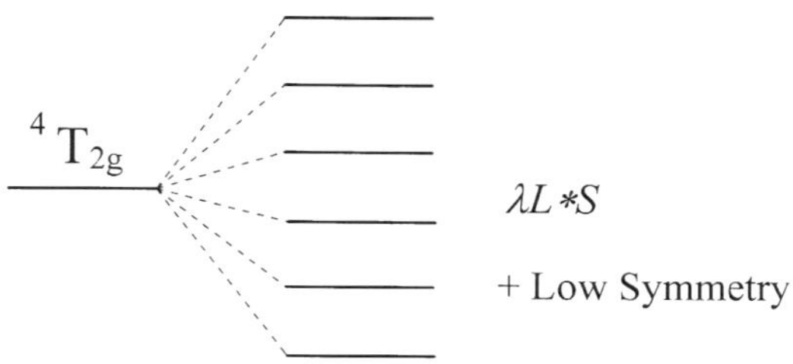

Fig. 2.19. A ^{4}T state is split into six Kramers' doublets by spin orbit coupling and low symmetry components.

TABLE 2.9
Hyperfine (δ^{hyp}) and pseudocontact (δ^{pc}) shifts [a] of the ^{1}H and ^{13}C nuclei of pyridine bis-coordinated to cobalt(II) bis-acetylacetonate [74]

Atom	δ^{hyp} (ppm)	δ^{pc} (ppm)
α-H	+32.9	−39.5
β-H	+5.0	−18.1
γ-H	−9.4	−15.6
α-C	−199	−92.5
β-C	+229	−35.7
γ-C	−73.8	−28.3

[a] Pseudocontact shifts are estimated using single crystal magnetic susceptibility data.

to a first approximation, its effects on the magnetic susceptibility tensor can be neglected.

It is foreseeable that pseudocontact shifts occur in low spin iron(III) porphyrins which in octahedral symmetry would have an orbitally triply degenerate ground state. Therefore, g and χ values in iron porphyrins are highly anisotropic (see Sections 2.3 and 5.1.2) [75,76].

High spin iron–porphyrin systems have $S = 5/2$ and are orbitally non-degenerate. The large tetragonal distortion induces a ZFS of the sextet as shown in Fig. 2.20, with D of the order of 10 cm^{-1}. The principal susceptibility values can then be calculated through the Van Vleck equation (Eq. (1.40)), as long as $2D$ is larger than the Zeeman energy, and are given by Eqs. (2.11) and (2.12). Note that most of the magnetic anisotropy is determined by D, as $g_\parallel$ and $g_\perp$ are similar. By substituting the magnetic susceptibility values in Eq. (2.18) with expressions (2.11) and (2.12), and taking $e^X = 1 + X (D \ll kT)$, the following equation is obtained [2]:

$$\delta^{pc} = \frac{\mu_0}{4\pi}\frac{35\mu_B^2(g_\parallel^2 - g_\perp^2)}{36kT}\frac{3\cos^2\theta - 1}{r^3}\left[1 - \frac{32(g_\parallel^2 + \frac{1}{2}g_\perp^2)}{15(g_\parallel^2 - g_\perp^2)}\frac{D}{kT}\right]. \quad (2.33)$$

This equation gives the sum of the magnetic susceptibility contributions of

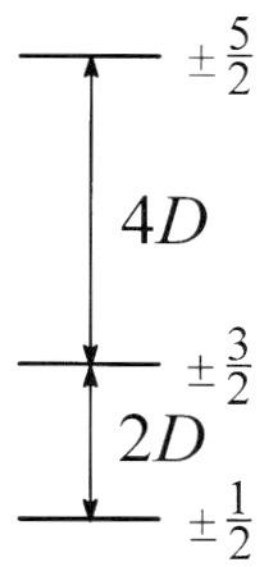

Fig. 2.20. ZFS of the spin sextet levels of high spin iron(III).

each state i in Van Vleck equation, weighted by the population according to the Boltzmann distribution. In contrast to the case of the Fermi contact shift (Eq. (2.13)), the term depending on T^{-2} is already sizable for relatively small values of D/kT, since $g_{\parallel}^2 - g_{\perp}^2$ is small [2]. The term in T^{-1}, which depends on g anisotropy, is always small.

2.7 ATTEMPTS TO SEPARATE CONTACT AND PSEUDOCONTACT SHIFTS

As a result of the nature of the pseudocontact shifts, a knowledge of the metal magnetic susceptibility tensor provides a method for their independent evaluation, if it is assumed that the shifts are metal-centered dipolar in origin. In principle, therefore, by performing magnetic susceptibility measurements on single crystals with several orientations of the magnetic field, it is possible to determine the magnetic anisotropy and the principal axes of the metal magnetic tensor[2]. By knowing the geometrical coordinates, the pseudocontact shifts can be easily predicted using Eq. (2.20). Such a procedure assumes that the magnetic anisotropy and the molecular axes are the same in the solid state and in solution. This is somewhat unjustified. An example of this procedure is shown in Table 2.9.

The procedure based on the direct use of the **g** tensor anisotropy and Eq. (2.24) is quite common for $S = 1/2$ systems, since g values from frozen solutions are easily obtainable. In this case, both the second order Zeeman contributions and possibly the effects of temperature on the g values are neglected. Furthermore, the directions of the molecular axes are arbitrarily assumed unless single-crystal data are available. Attempts are available in the literature regarding low spin cobalt(II) [77] and copper(II) [61].

Other semiquantitative methods for evaluating the metal-centered pseudocontact contributions to the shifts are based on the pattern predicted for a series of metal ions of the pseudocontact shifts assuming axial symmetry. Eq. (2.18), which

[2] The diamagnetic contribution, if not negligible, should be appropriately subtracted.

is valid for the axial case, can be rewritten as

$$\delta^{pc} = G_{r,\theta} \cdot D_{\chi} \tag{2.34}$$

where $G_{r,\theta}$ is the geometric factor which can be calculated from the structure and D_{χ} is the term containing the magnetic susceptibility. Since the latter term is constant for a given molecule, the ratios between the pseudocontact shifts should be the same as the ratios between the $G_{r,\theta}$ values. When this equation is obeyed, it can be concluded that the observed hyperfine shifts are mainly pseudocontact in origin. In other cases this same property allows the pseudocontact contribution on all nuclei to be estimated by assuming that at a certain nucleus position the contact contribution is negligible. Section 2.9 will be devoted to this case in paramagnetic metalloproteins. Horrocks has developed a variant [78] which is based on a comparison between the proton hyperfine shift pattern for a molecule where the shifts are only contact in origin, and an analogous molecule whose proton shifts have to be separated into their contributions. The spin delocalization mechanisms have to be the same, as it is possibly the case for nickel(II) and cobalt(II) pairs.

Particular cases applicable to low spin and high spin Fe(III) will be discussed in Section 5.2. The case of lanthanides will be discussed in Section 2.8.4.

2.8 THE CASE OF LANTHANIDES AND ACTINIDES

2.8.1 Electronic properties of lanthanides

The electronic properties of lanthanides are peculiar, in that spin–orbit interactions are very large, larger in fact than ligand field effects. Therefore, a somewhat different formalism is used.

Spin–orbit interactions couple the orbital L and spin S angular momenta. The resulting free ion terms are characterized by J values which are vector combinations of L and S. Again the level of interest is the ground level, which is that with largest S, largest L with equally large S, smallest J for f^1 to f^6 lanthanides, and largest J for f^8 to f^{13} lanthanides (Table 2.10). For example, Ce(III) has a ground state with $J = 5/2$, since one electron in an f orbital gives rise to a free ion term ^2F with $S = 1/2$ and $L = 3$. Two terms arise through spin–orbit coupling: one with $J = 3 - 1/2 = 5/2$ and the other with $J = 3 + 1/2 = 7/2$. The former term is the ground state, as shown in Table 2.10. The J values of the ground state vary from $5/2$ to 8, including $J = 0$ for Eu(III) and Sm(II). In general, it can be assumed that the ground term is the only one of interest; however, this is not true for Sm(III), which has an excited J manifold at about 1000 cm^{-1}, and Eu(III) and Sm(II), whose ground terms have $J = 0$ and would be diamagnetic if excited terms were not taken into consideration.

In actinides, the J terms are closer in energy and more heavily admixed by spin–orbit coupling.

TABLE 2.10
Some electronic properties of lanthanide ions

Ion	Configuration	$^{2S+1}L_J$ of ground state (multiplicity in parentheses)	g_J [a]	$\langle S_z \rangle_J$ [b]	$\langle S_z \rangle_J$ [c]	δ^{pc} [d] (ppm)
Ce^{3+}	$4f^1$	$^2F_{5/2}$ (6)	6/7	1.07	0.98	+1.6
Pr^{3+}	$4f^2$	$^3H_4$ (9)	4/5	3.20	2.97	+2.7
Nd^{3+}	$4f^3$	$^4I_{9/2}$ (10)	8/11	4.91	4.49	+1.05
Pm^{3+}	$4f^4$	$^5I_4$ (9)	3/5	4.80	4.01	−0.6
Sm^{3+}	$4f^5$	$^6H_{5/2}$ (6)	2/7	1.79	−0.06	+0.17 [e]
Eu^{3+}(Sm^{2+})	$4f^6$	$^7F_0$ (1)	–	–	−10.68	∼ −1.0 [e]
Gd^{3+}(Eu^{2+})	$4f^7$	$^8S_{7/2}$ (8)	2	−31.5	−31.50	∼0
Tb^{3+}	$4f^8$	$^7F_6$ (13)	3/2	−31.5	−31.82	+20.7
Dy^{3+}	$4f^9$	$^6H_{15/2}$ (16)	4/3	−28.30	−28.54	+23.8
Ho^{3+}	$4f^{10}$	$^5I_8$ (17)	5/4	−22.50	−22.63	+9.4
Er^{3+}	$4f^{11}$	$^4I_{15/2}$ (16)	6/5	−15.30	−15.37	−7.7
Tm^{3+}	$4f^{12}$	$^3H_6$ (13)	7/6	−8.17	−8.21	−12.7
Yb^{3+}	$4f^{13}$	$^2F_{7/2}$ (8)	8/7	−2.57	−2.59	−5.2

[a] Calculated from Eq. (2.35). The equation does not hold for f^6 ions.
[b] Calculated according to Eq. (2.38).
[c] Calculated by inclusion of the excited states [81].
[d] Pseudocontact shift predicted from Eq. (2.37) for $r = 300$ pm, $3\cos^2\theta - 1 = 1$, $T = 300$ K, and D_z values for each lanthanide estimated from Ref. [79].
[e] Including contributions from excited J manifolds [79]. For f^6 ions, the contribution of the ground state manifold is zero.

Ligand field effects split the J manifold in a way that is not easily predicted without specific calculations. However, the overall splitting is such that many of the levels are appreciably populated at room temperature. An elegant procedure that takes such effects into account in a general way with respect to pseudocontact shifts of metal-centered origin has been provided by Bleaney [79].

2.8.2 The pseudocontact contribution to the hyperfine shifts

Ligand field effects remove the spherical symmetry around the metal ion and cause magnetic susceptibility anisotropy. The pseudocontact shifts are then given by Eq. (2.20). As already mentioned in Section 2.6 for high spin iron(III), the susceptibility can be viewed as arising from two contributions: one is given by a Zeeman term of the type $g\mu_B B_0 S_z$, and the other is given by the splitting of the magnetic manifold, which provides anisotropy. As the orbital and spin angular moments are strongly coupled, the g factor in lanthanide free ions is a function of both L and S and of their vector combination J; it is given by

$$g_J = 1 + \frac{J(J+1) - L(L+1) + S(S+1)}{2J(J+1)} \tag{2.35}$$

and the Zeeman term then becomes $g_J \mu_B B_0 J_z$. Note that when $L = 0$, $g_J = 2$, and when $S = 0$, $g_J = 1$, as expected from general theory. In analogy with Eq. (2.23), the Zeeman contribution provides a contribution to pseudocontact shift in T^{-1}. If D is much larger than the Zeeman energy and smaller than kT, the ligand field splitting provides a contribution in T^{-2} (see Eq. (2.33)). The general expression is [79]

$$\boxed{\begin{aligned}\delta^{\mathrm{pc}} &= -\frac{\mu_0}{4\pi}\frac{g_J^2\mu_B^2 J(J+1)(2J-1)(2J+3)}{60(kT)^2}\\ &\quad\times\frac{D_z(3\cos^2\theta-1)+(D_x-D_y)\sin^2\theta\cos 2\Omega}{r^3}\end{aligned}} \tag{2.36}$$

where r, θ, and Ω are the polar coordinates of the nucleus with respect to the principal directions of the $\boldsymbol{D}$ tensor (Section 1.4), and g_J is given by Eq. (2.35) (Table 2.10). Eq. (2.36) differs from Eq. (2.33) in that only the term in T^{-2} has remained. In fact, in the present approximation, g_J is isotropic.

For axial symmetry, Eq. (2.36) becomes

$$\delta^{\mathrm{pc}} = -\frac{\mu_0}{4\pi}\frac{g_J^2\mu_B^2 J(J+1)(2J-1)(2J+3)}{60(kT)^2}\frac{D_z(3\cos^2\theta-1)}{r^3}. \tag{2.37}$$

The parameter D_z accounts for crystal field effects, assuming that the splitting of the $2J+1$ levels does not exceed kT in energy. The shifts at 300 K for a nucleus at 300 pm from the metal and $3\cos^2\theta - 1 = 1$, calculated following Bleaney's treatment [79] and using a plausible value for the ligand field parameter to calculate D_z for each lanthanide, are reported in the last column of Table 2.10 and in Fig. 2.21. For Sm(III), the shift value has been obtained by also considering the first excited $J = 7/2$ manifold. For Eu(III), the ground state $J = 0$ does not contribute to the shift; instead, the excited levels with $J = 1$ and $J = 2$, which lie at about 400 and 1200 cm^{-1} above it, provide the pseudocontact shift, which has a rather complex temperature dependence.

For f^7 ions, the ligand field splitting is zero under first-order conditions, whereas higher order effects account for the splitting of the $J = 7/2$ level and for a small pseudocontact shift.

The accuracy of the present point–dipole approximation with respect to a model with delocalized unpaired electrons (Section 2.2.2) has also been evaluated for f^n systems [80]. The deviations have been found to be substantially smaller than those estimated for 3d metal ions.

2.8.3 The contact contribution to the hyperfine shifts

Although the contact contribution can be sizable and sometimes dominant for ^{17}O and ^{14}N nuclei directly coordinated to the metal [81], it is often negligible for

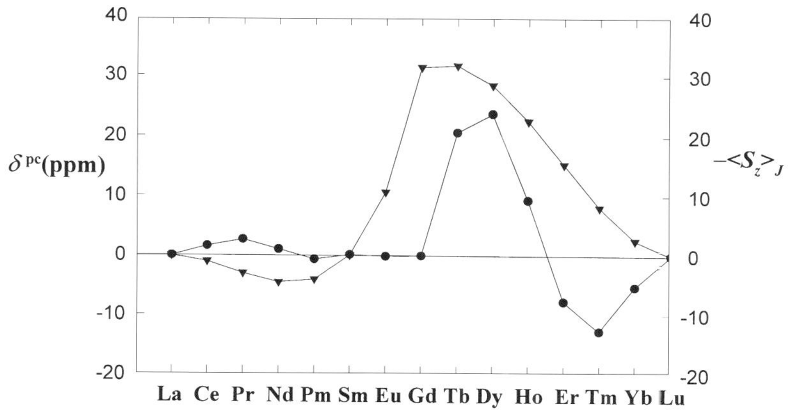

Fig. 2.21. Calculated patterns of: dipolar shifts (•) [79] and $-\langle S_z \rangle$ (▼) [81] values (proportional to the contact shifts) induced by lanthanide ions.

nuclei a few bonds apart from the metal ion, unless efficient spin delocalization mechanisms are operative. The contact shift can be evaluated (see Eq. (2.5)) through the expectation value $\langle S_z \rangle_J$ on the ground J manifold, the Zeeman splitting of which is less than kT [82]:

$$\langle S_z \rangle_J = -g_J(g_J - 1)J(J+1)\frac{\mu_B B_0}{3kT}. \tag{2.38}$$

In this approximation, g_J is isotropic. We need to consider $\langle S_z \rangle_J$ and not $\langle J_z \rangle$ because only the spin is delocalized onto the molecule [81]. The contact contribution is therefore

$$\boxed{\delta^{\text{con}} = \frac{A}{\hbar}\frac{g_J(g_J - 1)\mu_B J(J+1)}{3\gamma_I kT}} \tag{2.39}$$

or

$$\delta^{\text{con}} = \frac{1}{\mu_0}\frac{A}{\hbar}\frac{g_J - 1}{\gamma_I \mu_B g_J}\chi \tag{2.40}$$

where χ is the magnetic susceptibility without ligand field splitting. Although this simple treatment does not take into account the ligand field splitting of the J manifold, it provides a useful criterion for predicting the general trend of contact shifts along the series. Note that this method does not apply to Sm(III), Eu(III) and Sm(II) salts for the reason already outlined. The values of $\langle S_z \rangle_J$, with and without inclusion of the excited states, are reported in Table 2.10 and in Fig. 2.21. It is noteworthy that $\langle S_z \rangle_J$ in lanthanides can be either negative or positive owing to the large orbital contribution to the determination of the energy levels. Since orbital contributions are larger than spin contributions, $\langle S_z \rangle_J$ is different in sign with respect to the spin-only value when the spin orbit coupling constant is negative, as it occurs in the first half of the series.

It is also often assumed that A is constant along a series of lanthanide complexes with the same ligand. The mechanism providing spin density delocalization from the metal to the ligands is due to a weak covalent bonding involving the 6s metal orbital, which, in turn, can transfer unpaired spin density on nearby nuclei through spin polarization from 4f orbitals [82].

2.8.4 Separation of pseudocontact and contact contributions

The following method for separation of pseudocontact and contact shifts, proposed by Dobson et al. [83], is valid for axial systems assuming that: (1) there is a nucleus in such a position in the molecule that the contact contribution is zero because it is far from the paramagnetic center; and (2) the contact hyperfine coupling is constant ($\pm 10\%$) for Nd–Tm.

For nuclei i and j:

$$\frac{\delta_i^{\mathrm{pc}}}{\delta_j^{\mathrm{pc}}} = R_{ij}. \tag{2.41}$$

If $\delta_j^{\mathrm{con}} = 0$, then the ratio of the total hyperfine shifts δ^{hyp} for nuclei i and j is

$$\frac{\delta_i^{\mathrm{hyp}}}{\delta_j^{\mathrm{hyp}}} = R_{ij} + \frac{\delta_i^{\mathrm{con}}}{\delta_j^{\mathrm{iso}}} = R_{ij} + \frac{A}{\hbar}\frac{g_J(g_J-1)\mu_B J(J+1)}{3\gamma_N kT\delta_j^{\mathrm{iso}}}\nu_0. \tag{2.42}$$

By plotting $\delta_i^{\mathrm{iso}}/\delta_j^{\mathrm{iso}}$ vs. $g_J(g_J-1)J(J+1)/\delta_j^{\mathrm{iso}}$ for all the lanthanides, the intercept gives R_{ij} and hence the ratio between pairs of pseudocontact shifts (Fig. 2.22). Within this approximation, the requirement for axial symmetry was found to hold for many systems, even with low symmetry, possibly as a result of ligand rearrangement which might be fast on the NMR time scale.

Separation of pseudocontact and contact shifts in axial systems can also be achieved, again by using a variety of lanthanides, without assuming that for a given signal the contact shift is zero, as long as the expected patterns of both pseudocontact and contact shifts (Fig. 2.21) hold. Here, the hyperfine shift can be expressed as the sum of contact and pseudocontact shifts (Eqs. (2.39) and (2.37)):

$$\delta_{ij}^{\mathrm{hyp}} = \delta_{ij}^{\mathrm{con}} + \delta_{ij}^{\mathrm{pc}} = KA_i\langle S_z\rangle_j + G_i D_j \tag{2.43}$$

where the subscript i refers to the observed nucleus and j to the lanthanide employed (cf. Eq. (2.34)). $\langle S_z\rangle_j$ and D_j depend only on the lanthanide, whereas A_i and G_i are assumed to depend only on the nucleus and not on the lanthanide. G_i is given by $(3\cos^2\theta_i - 1)/r_i^3$ of Eq. (2.37). The unknowns for a single shift are A_i and G_i. If n signals are observed, the unknowns are $2n$; that is, A_i and G_i for the n nuclei. For a second lanthanide derivative, A_i and G_i are the same, and the

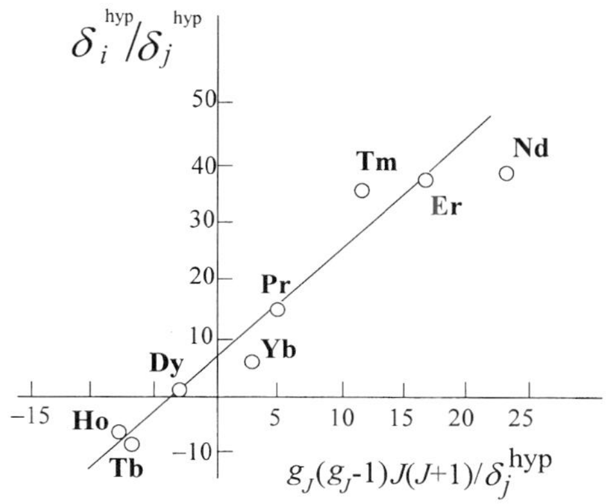

Fig. 2.22. Observed ratios between the hyperfine shifts of the ^{31}P (δ_i^{hyp}) and ^{1}H (δ_j^{hyp}) resonances of cytidine monophosphate interacting with lanthanide ions, as a function of $g_J(g_J-1)J(J+1)/\delta_i^{\mathrm{hyp}}$ for each lanthanide [83].

experimental data are now $2n$, that is as many as the unknowns. By using a variety of lanthanides, it is possible, in general, to set up and solve the appropriate number of simultaneous equations and thus evaluate the terms A_i and G_i for each nucleus [84–86]. This procedure is again valid as long as A_i and G_i depend only on the nucleus, and $\langle S_z\rangle_j$ and D_j depend only on the lanthanide in a predictable way and not on the nucleus under observation. The accuracy of the results depends on how precisely the patterns of Fig. 2.21 are capable of reproducing the contact and pseudocontact shifts within the lanthanide series. As we have seen, precise estimates of $\langle S_z\rangle$ in the equation for δ^{con} (Eqs. (2.38) and (2.39)) require evaluation of the splitting of the J manifold for the particular complex examined; the equation for δ^{pc} (Eq. 2.37), besides being valid only for axial symmetry, assumes that the splitting of the $2J+1$ levels does not exceed kT. It has been pointed out that such an assumption is not correct [87], although a thorough treatment [88] has shown that Eqs. (2.36) and (2.37) are capable of accounting for pseudocontact shifts at or near room temperature with an accuracy of 10–20%.

When contact contributions can be neglected, the pseudocontact shifts for ligand protons in series of lanthanide complexes can be reproduced by using the complete equation for non-axial cases (Eq. (2.36)) and the geometric factors calculated from the X-ray structure [89–95]. The values for the D_z and $D_x - D_y$ parameters obtained through best fitting procedures are in fair to good agreement with the single-crystal anisotropy data [91].

2.9 THE PSEUDOCONTACT SHIFTS IN PARAMAGNETIC METALLOPROTEINS

In paramagnetic metalloproteins there are many nuclei close to the paramagnetic center as a result of the folding of the protein chain. Such nuclei, if separated from the metal ion by several chemical bonds, can experience only pseudocontact shifts. We are in the lucky situation of having many of these nuclei and therefore many pseudocontact shift values. If a structure or a structural model is available, the r', θ' and φ' values for each nucleus is known in any arbitrary coordinated system. At this point the unknowns of Eq. (2.20) are $\Delta\chi_{ax}$, $\Delta\chi_{rh}$ and three angles which define the molecular axes with respect to the arbitrary coordinated system. With a minimum of five experimental pseudocontact shifts, the above parameters are found. A program (Fantasian) is available at http://www.cerm.unifi.it. Besides the structural model, the assignment of the signals must be available through the usual 2D/3D experiments and the chemical shifts of the analogous compound without unpaired electrons should be known. This latter condition can be fulfilled by analyzing the protein with La^{3+} or Ca^{3+} in the case of lanthanides or by using zinc(II) for cobalt(II) systems or low spin iron(II) for low spin iron(III). Sometimes the analysis of the blank diamagnetic compound can be substituted with an estimation of the diamagnetic chemical shifts. Programs are available for that [96] (http://www.amber.ucsf.edu/amber/amber.html). In this case some errors are introduced which may affect the accuracy of the determined parameters.

Only some metal ions, as discussed in Section 2.6, can provide significant $\Delta\chi$ values so that significant pseudocontact shifts are observed. The procedure outlined above is the only procedure which provides information on the magnetic susceptibility of the metal ion, as single crystal measurements are not possible in metalloproteins. A limitation of the procedure is that the structure in solution may differ from the structural model. The latter in general is that obtained from X-ray analysis in the solid state. Eq. (2.20) has also been used to refine such model, allowing movements of the atoms which minimize the discrepancy between calculated and experimental data [97,98]. Recently a program was proposed [99] which allows us to calculate the solution structure by using pseudocontact shifts as constraints together with NOEs (see Chapter 7) and other classical constraints. In order to avoid that ligand centered contributions undermine the treatment, an indetermination proportional to the pseudocontact shift values is assigned to each experimental pseudocontact shift. In fact ligand centered contributions may be effective for nuclei close to the metal ion, which experience the largest pseudocontact shifts.

Once the $\Delta\chi$ values and the molecular directions are obtained, the metal centered pseudocontact shifts can be calculated of nuclei which experience also contact and ligand centered pseudocontact shifts. With this procedure the contact plus ligand centered pseudocontact shift (which is small [100]) have been calculated for several systems. In Table 2.11 the data relative to the low spin iron(III) containing heme in cytochrome b_5 are reported [101].

Besides low spin ferric heme containing proteins [102,103], four- [104] and five-coordinated [105] high spin cobalt(II)-containing proteins have been studied. Lanthanides have been used from the early times of NMR of paramagnetic molecules [106–108].

TABLE 2.11
Separation of pseudocontact and contact (including any ligand centered pseudocontact) contributions to the hyperfine shifts of heme and axial ligand protons for oxidized rat microsomal cytochrome b_5 at 313 K (heme numbering as in Fig. 5.7B) [127]

Atom name	Residue name	pcs (ppm)	cs + lcpcs (ppm)
8-CH_3	Heme	−2.82	2.01
1-CH_3	Heme	−1.54	8.60
2-Hα	Heme	−4.47	23.27
2-Hβ (*trans*)	Heme	−1.71	−10.01
2-Hβ (*cis*)	Heme	−1.72	−9.99
meso-Hα	Heme	−11.84	−0.60
3-CH_3	Heme	−4.67	15.97
4-Hα	Heme	−2.66	−1.29
4-Hβ (*trans*)	Heme	−1.58	−1.61
4-Hβ (*cis*)	Heme	−2.66	−0.58
meso-βH	Heme	−3.82	3.28
5-CH_3	Heme	−1.50	17.67
meso-γH	Heme	−12.30	2.31
7-Hα	Heme	−5.88	20.98
7-Hα′	Heme	−4.01	−1.42
6-Hα	Heme	−4.72	16.30
6-Hα′	Heme	−2.79	13.07
HN	His 39	2.39	0.63
Hα	His 39	3.82	0.55
Hβ1	His 39	5.89	9.81
Hβ2	His 39	5.85	1.32
HN	His 63	2.94	2.19
Hα	His 63	4.18	0.76
Hε 1	His 63	13.26	−26.94

2.10 THE EFFECT OF HIGH MAGNETIC FIELDS

Pseudocontact shifts occur when the metal magnetic susceptibility is anisotropic and they are derived (see Appendix IV) under the assumption that every orientation of the molecule is equally probable. However, the magnetic susceptibility tensor axes are fixed within the molecule and the magnetic anisotropy introduces a dependence on orientation of the interaction energy between the paramagnetic ion and the magnetic field. If the energy of interaction between the molecule and the magnetic field is orientation dependent, it follows that not all molecular orientations are equally probable. As a consequence, changes in the observed pseudocontact shifts should be expected. Actually, it is not the metal magnetic susceptibility the only cause of the orientation dependence of the molecular energy, but the overall molecular magnetic anisotropy, which includes the diamagnetic part. The latter is likely to increase with the size of the molecule, and it has been indeed found relevant in biological macromolcules.

An equation for the effect of an axial magnetic anisotropy of a paramagnetic center on the pseudocontact shift is [3,109]:

$$\delta^{pc} = \frac{(3\cos^2\theta - 1)(\chi_{\parallel} - \chi_{\perp})}{12\pi r^3}\left(1 + \frac{B_0^2}{6\mu_0 kT}(2\chi_{\parallel} + \chi_{\perp})\right) \tag{2.44}$$

where θ is the angle between the electron–proton direction $\boldsymbol{r}$ and $\boldsymbol{\chi}_{\parallel}$.

The effect is predicted to increase with the square of the magnetic field. At high magnetic fields (i.e. $\boldsymbol{B_0} > 10$ T) the effect is expected to be measurable. However, it is hard to be observed in practice, as it is masked by a larger effect, which could be predicted [109] on the ground of the known field dependence of the magnetic susceptibility of a paramagnetic center [110,111], as shown below. On the other hand, when a paramagnetic molecule is oriented mechanically by using orienting devices such as liquid crystals, a large and measurable change in hyperfine shift can be observed [112].

In deriving the contact shift (Eq. (2.9) and Appendix III) and the pseudocontact shift (Eq. (2.20) and Appendix IV) the shift in Hertz was assumed to be linearly dependent on $\boldsymbol{B_0}$ because of the assumption that the magnetic susceptibility is linearly dependent on $\boldsymbol{B_0}$. In other words, the difference in population among the electron Zeeman levels was considered to be proportional to $\boldsymbol{B_0}$. This is true as long as the electronic Zeeman is small with respect to kT. However, this is not true any more at high magnetic fields. The phenomenon is described in magnetochemistry by the Brillouin equation [110,111]. The effect is enhanced by the presence of ZFS effects. The consequent alteration of the pseudocontact shifts can be even larger than that due to Eq. (2.44), but at the same time is a small fraction of the pseudocontact shift. Therefore, quantitative estimates have not been attempted.

2.11 REFERENCES

[1] H.M. McConnell, D.B. Chesnut (1958) J. Chem. Phys. 28, 107.
[2] R.J. Kurland, B.R. McGarvey (1970) J. Magn. Reson. 2, 286.
[3] I. Bertini, C. Luchinat, G. Parigi (2000) Eur. J. Inorg. Chem. 2473.
[4] R.M. Golding, L.C. Stubbs (1979) J. Magn. Reson. 33, 627.
[5] A.D. Buckingham, P.J. Stiles (1972) Mol. Phys. 24, 99.
[6] P.J. Stiles (1974) Mol. Phys. 27, 501.
[7] P.J. Stiles (1975) Mol. Phys. 29, 1271.
[8] R.M. Golding, R.O. Pascual, L.C. Stubbs (1976) Mol. Phys. 31, 1933.
[9] J.P. Riley, W.T. Raynes (1977) Mol. Phys. 33, 619.
[10] R.M. Golding, R.O. Pascual, J. Vrbancich (1976) J. Mol. Phys. 31, 731.
[11] R. Beringer, M.A. Heald (1954) Phys. Rev. 95, 1474.
[12] Z. Luz, R.G. Shulman (1965) J. Chem. Phys. 43, 3750.
[13] R.J. Fitzgerald, R.S. Drago (1967) J. Am. Chem. Soc. 89, 2879.
[14] J.P. Quaegebeur, C. Chachaty, T. Yasukawa (1979) Mol. Phys. 37, 409.
[15] B.B. Wayland, W.L. Rice (1967) Inorg. Chem. 6, 2270.

[16] S.K. Sheppard, G.R. Eaton, S.S. Eaton (1989) Inorg. Chem. 28, 4496.
[17] J.A. Happe, R.L. Ward (1963) J. Chem. Phys. 39, 1211.
[18] R.W. Kluiber, W.D. Horrocks, Jr. (1967) Inorg. Chem. 6, 166.
[19] R.H. Holm, G.W. Everett, W.D. Horrocks, Jr. (1966) J. Am. Chem. Soc. 88, 1071.
[20] G.M. Zhidomirov, P.V. Schastnev, N.D. Chuvylkin (1970) J. Struct. Chem. 11, 458.
[21] R.E. Cramer, R.S. Drago (1970) J. Am. Chem. Soc. 92, 66.
[22] K. Hayamizu, M. Murata, O. Yamamoto (1975) Bull. Chem. Soc. Jpn. 48, 1842.
[23] D.M. Doddrell, J.D. Roberts (1970) J. Am. Chem. Soc. 92, 6839.
[24] A.D. McLachlan (1960) Mol. Phys. 3, 233.
[25] R.W. Kluiber, W.D. Horrocks, Jr. (1967) Inorg. Chem. 6, 430.
[26] I. Bertini, D.L. Johnston, W.D. Horrocks, Jr. (1970) Inorg. Chem. 9, 693.
[27] W.D. Perry, R.S. Drago, D.W. Herlocker, G.K. Pagenkopf, K. Czworniak (1971) Inorg. Chem. 10, 1087.
[28] M.J. Scarlett, A.T. Casey, R.A. Craig (1970) Aust. J. Chem. 23, 1333.
[29] W.D. Horrocks, Jr., D.L. Johnston (1971) Inorg. Chem. 10, 1835.
[30] P. Fantucci, V. Valenti (1976) J. Am. Chem. Soc. 98, 3832.
[31] P. Fantucci, P. Balzarini, V. Valenti (1977) Inorg. Chim. Acta 25, 113.
[32] S.F. Sontum, D.A. Case, M. Karplus (1983) J. Chem. Phys. 79, 2881.
[33] W.D. Wheeler, S. Kaizaki, J.I. Legg (1982) Inorg. Chem. 21, 3248.
[34] I. Bertini, G. Canti, C. Luchinat, F. Mani (1981) J. Am. Chem. Soc. 103, 7784.
[35] F.-J. Wu, D.M. Kurtz, Jr. (1989) J. Am. Chem. Soc. 111, 6563.
[36] Anonymous (1995) Unpublished results from our laboratory.
[37] I. Bertini, D. Gatteschi, A. Scozzafava (1972) Inorg. Chim. Acta 6, 185.
[38] I. Bertini, C. Luchinat, A. Scozzafava (1976) Inorg. Chim. Acta 19, 201.
[39] G.N. La Mar, W.D. Horrocks, Jr., L.C. Allen (1964) J. Chem. Phys. 41, 2126.
[40] I. Bertini, C. Luchinat, A. Scozzafava (1979) Inorg. Nucl. Chem. Lett. 15, 89.
[41] N.M. Karayannis, C.M. Mikulski, L.L. Pytlewski (1971) Inorg. Chim. Acta 5, 69.
[42] R.W. Kluiber, W.D. Horrocks, Jr. (1965) J. Am. Chem. Soc. 87, 5350.
[43] G.N. La Mar, G.R. Van Hecke (1970) Inorg. Chem. 9, 1546.
[44] I. Bertini, L.J. Wilson (1971) J. Chem. Soc. Inorg. Phys. Theor. 489
[45] M.L. Wicholas, R.S. Drago (1968) J. Am. Chem. Soc. 90, 2196.
[46] C. Benelli, I. Bertini, D. Gatteschi (1972) J. Chem. Soc. Dalton Trans. 661.
[47] R.H. Holm, C.J. Hawkins (1973) In: La Mar G.N., Horrocks W.D. Jr., Holm R.H. (eds) NMR of Paramagnetic Molecules Principles and Applications. Academic Press, New York: pp. 243.
[48] D.R. Eaton, A.D. Josey, W.D. Phillips, R.E. Benson (1962) J. Chem Phys. 37, 347.
[49] D.R. Eaton, A.D. Josey, R.E. Benson, W.D. Phillips, T.L. Cairns (1962) J. Am. Chem. Soc. 84, 4100.
[50] I.M. Arafa, H.M. Goff, S.S. David, B.P. Murch, L. Que Jr. (1987) Inorg. Chem. 26, 2779.
[51] E.W. Stone, A.H. Maki (1962) J. Chem. Phys. 37, 1326.
[52] C. Heller, H.M. McConnell (1960) J. Chem. Phys. 32, 1535.
[53] A.D. McLachlan (1958) Mol. Phys. 1, 233.
[54] P.G. Lykos (1960) J. Chem. Phys. 32, 625.
[55] M. Karplus (1963) J. Am. Chem. Soc. 85, 2870.
[56] M. Karplus, G.K. Fraenkel (1961) J. Chem. Phys. 35, 1312.
[57] E.T. Strom, G.R. Underwood, D. Jurkowitz (1972) Mol. Phys. 24, 901.
[58] W. Derbishire (1962) Mol. Phys. 5, 225.
[59] D.R. Eaton, A.D. Josey, W.D. Phillips, R.E. Benson (1962) Mol. Phys. 5, 407.
[60] I. Bertini, F. Capozzi, C. Luchinat, M. Piccioli, A.J. Vila (1994) J. Am. Chem. Soc. 116, 651.

[61] I. Bertini, S. Ciurli, A. Dikiy, R. Gasanov, C. Luchinat, G. Martini, N. Safarov (1999) J. Am. Chem. Soc. 121, 2037.
[62] H.M. McConnell (1972) Proc. Natl. Acad. Sci. USA 69, 335.
[63] H.M. Goff (1983) In: Lever A.B.P., Gray H.B. (eds) Iron Porphyrins. Addison-Wesley, Reading, MA: pp. 237.
[64] H.M. Goff (1977) J. Am. Chem. Soc. 99, 7723.
[65] I. Morishima, T. Inubushi, S. Neya, S. Ogawa, T. Yonezawa (1977) Biochem. Biophys. Res. Commun. 78, 739.
[66] I. Morishima, T. Inubushi (1978) J. Am. Chem. Soc. 100, 3568.
[67] D.V. Behere, E. Gonzales-Vergara, H.M. Goff (1985) Biochem. Biophys. Res. Commun. 131, 607.
[68] L. Banci, I. Bertini, I.-C. Kuan, M. Tien, P. Turano, A.J. Vila (1993) Biochemistry 32, 13483.
[69] H. Pan, B.C. Gerstein (1991) J. Magn. Reson. 92, 618.
[70] A.C. McLaughlin, C. Grathwohl, R.E. Richards (1978) J. Magn. Reson. 31, 283.
[71] J.-C. Boubel, J.-J. Delpuech, A. Peguy (1978) J. Chem. Soc., Dalton Trans. 1506.
[72] C.P. Bhasin, G. Srivastava, R.C. Mehrotra (1987) Inorg. Chim. Acta 128, 69.
[73] G. Labauze, J.B. Raynor (1981) J. Chem. Soc., Dalton Trans. 590.
[74] W.D. Horrocks, Jr., D.D. Hall (1971) Inorg. Chem. 10, 2368.
[75] W.D. Horrocks, Jr., E.S. Greenberg (1973) Biochim. Biophys. Acta 322, 38.
[76] J. Trehwella, P.E. Wright (1980) Biochim. Biophys. Acta 625, 202.
[77] A. Shirazi, H.M. Goff (1982) Inorg. Chem. 21, 3420.
[78] W.D. Horrocks, Jr. (1970) Inorg. Chem. 9, 670.
[79] B. Bleaney (1972) J. Magn. Reson. 8, 91.
[80] R.M. Golding, R.O. Pascual, B.R. McGarvey (1982) J. Magn. Reson. 46, 30.
[81] R.M. Golding, M.P. Halton (1972) Aust. J. Chem. 25, 2577.
[82] W.B. Lewis, J.A. Jackson, J.F. Lemons, H. Taube (1962) J. Chem. Phys. 36, 694.
[83] C.M. Dobson, R.J.P. Williams, A.V. Xavier (1973) J. Chem. Soc. Dalton Trans. 2662.
[84] C.N. Reilley, B.W. Wood, J.F. Desreux (1975) Anal. Chem. 47, 2110.
[85] J. Reuben, G.A. Elgavish (1980) J. Magn. Reson. 39, 421.
[86] J.G. Shelling, M.E. Bjorson, R.S. Hodges, A.K. Taneja, B.D. Sykes (1984) J. Magn. Reson. 57, 99.
[87] W.D. Horrocks, Jr. (1977) J. Magn. Reson. 26, 333.
[88] B.R. McGarvey (1979) J. Magn. Reson. 33, 445.
[89] R.E. Cramer, R. Dubois, C.K. Furuike (1975) Inorg. Chem. 14, 1005.
[90] R.E. Cramer, R. Dubois, K. Seff (1974) J. Am. Chem. Soc. 96, 4125.
[91] R.E. Cramer, R.B. Maynard (1978) J. Magn. Reson. 31, 295.
[92] M.V.R. Stainer, J. Takats (1983) J. Am. Chem. Soc. 105, 410.
[93] B.G. Jenkins, R.B. Lauffer (1988) Inorg. Chem. 27, 4730.
[94] C.F.G.C. Geraldes, A.D. Sherry, G.E. Kiefer (1992) J. Magn. Reson. 97, 290.
[95] J. Lisowski, J.L. Sessler, V. Lynch, T.D. Mody (1995) J. Am. Chem. Soc. 117, 2273.
[96] K. Ösapay, D.A. Case (1991) J. Am. Chem. Soc. 113, 9436.
[97] M. Gochin, H. Roder (1995) Bull. Magn. Reson. 17, 1.
[98] M. Gochin, H. Roder (1995) Protein Sci. 4, 296.
[99] L. Banci, I. Bertini, M.A. Cremonini, G. Gori Savellini, C. Luchinat, K. Wüthrich, P. Güntert (1998) J. Biomol. NMR 12, 553.
[100] L. Banci, I. Bertini, C. Luchinat, R. Pierattelli, N.V. Shokhirev, F.A. Walker (1998) J. Am. Chem. Soc. 120, 8472.
[101] L. Banci, I. Bertini, C. Luchinat, P. Turano (2000) In: Kadish K.M., Smith K.M., Guilard R. (eds) The Porphyrin Handbook. Academic Press, San Diego, CA: pp. 323.
[102] Y. Gao, J. Boyd, G.J. Pielak, R.J.P. Williams (1991) Biochemistry 30, 1928.

[103] S.D. Emerson, G.N. La Mar (1990) Biochemistry 29, 1556.
[104] I. Bertini, C. Luchinat, M. Piccioli, M. Vicens Oliver, M.S. Viezzoli (1991) Eur. Biophys. J. 20, 269.
[105] L. Banci, L.B. Dugad, G.N. La Mar, K.A. Keating, C. Luchinat, R. Pierattelli (1992) Biophys. J. 63, 530.
[106] I.D. Campbell, C.M. Dobson, R.J.P. Williams, A.V. Xavier (1973) Ann. New York Acad. Sci. 222, 163.
[107] I.D. Campbell, C.M. Dobson, R.J.P. Williams (1975) Proc. Roy. Soc. London A345, 41.
[108] L. Lee, B.D. Sykes (1983) Biochemistry 22, 4366.
[109] I. Bertini, I.C. Felli, C. Luchinat (1998) J. Magn. Reson. 134, 360.
[110] O. Kahn (1993) Molecular magnetism. VCH Publishers, New York
[111] B.I. Bleaney, B. Bleaney (1976) Electricity and magnetism. Third edition edn. Oxford University Press, Oxford.
[112] I. Bertini, F. Castellani, C. Luchinat, G. Martini, G. Parigi, S. Ristori (2000) J. Phys. Chem. 104, 10653.
[113] R.H. Holm, P. Kennepohl, E.I. Solomon (1996) Chem. Rev. 96, 2239.
[114] I. Bertini, C. Luchinat, Z. Xia (1992) Inorg. Chem. 31, 3152.
[115] A.M. Chmelnick, D. Fiat (1969) J. Chem. Phys. 51, 4238.
[116] I. Bertini, C. Luchinat, Z. Xia (1992) J. Magn. Reson. 99, 235.
[117] J. Reuben, D. Fiat (1967) Inorg. Chem. 6, 579.
[118] N. Bloembergen, L.O. Morgan (1961) J. Chem. Phys. 34, 842.
[119] Y. Ducommun, K.E. Newman, A. Merbach (1980) Inorg. Chem. 19, 3696.
[120] I. Bertini, F. Briganti, C. Luchinat, Z. Xia (1993) J. Magn. Reson. 101, 198.
[121] I. Bertini, F. Capozzi, C. Luchinat, Z. Xia (1993) J. Phys. Chem. 97, 1134.
[122] T.J. Swift, R.E. Connick (1962) J. Chem. Phys. 37, 307.
[123] Y. Ducommun, W.L. Earl, A.E. Merbach (1979) Inorg. Chem. 18, 2754.
[124] W.B. Lewis, M. Alei, L.O. Morgan (1966) J. Chem. Phys. 44, 2409.
[125] J. Reuben, D. Fiat (1969) J. Chem. Phys. 51, 4909.
[126] S. Ito, Y. Yano (1984) Bull. Chem. Soc. Jpn. 57, 2824.
[127] F. Arnesano, L. Bauci, I. Bertini, I.C. Felli (1998) Biochemistry 37, 173.

Chapter 3

Relaxation

The sole presence of an electron spin causes nuclear relaxation. The correlation time for the electron nucleus interaction is presented as well as equations valid for dipolar and contact interaction. To do so, electron relaxation mechanisms need to be quickly reviewed. All the subtleties of nuclear relaxation enhancements are presented pictorially and quantitatively.

3.1 INTRODUCTION

In paramagnetic molecules, the magnetic nucleus does not see unpaired electrons as localized, but as spin density distributed throughout space, with an integrated intensity equal to that of the total electrons. As a result, in every unit volume, the spin density will spend some time in the low Zeeman energy level(s) (negative M_S, see Chapter 1) and some time (slightly less) in the upper energy level(s) (positive M_S). Changes in M_S values involve changes in the orientation of the electron magnetic moment. The time sharing of the levels occurs through electron relaxation. Electron relaxation thus provides fluctuating magnetic fields and causes nuclear relaxation. When reference is made to spin density at the resonating nucleus, nuclear relaxation is contact in origin. The remaining spin density on the molecule and the relative electron relaxation are sensed by the resonating nucleus through dipolar coupling. The consequent nuclear relaxation is dipolar in origin. Often, reference is made to metal-centered relaxation; in this case, the whole electron is considered localized on the metal. Dipolar relaxation due to spin density elsewhere is called ligand-centered relaxation. This will be discussed in Section 3.4.1.

Restricting ourselves to metal-centered nuclear relaxation, the fluctuation of the electron dipolar field at the nucleus due to electron relaxation can be easily visualized as depicted in Fig. 3.1A. We now want to mention other mechanisms, besides electron relaxation, which occur in solution and through which unpaired electrons cause nuclear relaxation. All of these mechanisms will be discussed in detail in this chapter.

(1) Rotation of the molecular frame causes the nucleus to see the electron in different positions (Fig. 3.1B). If the rotation is faster than the electron relaxation time, on the rotational time scale the nucleus sees the electron with the same

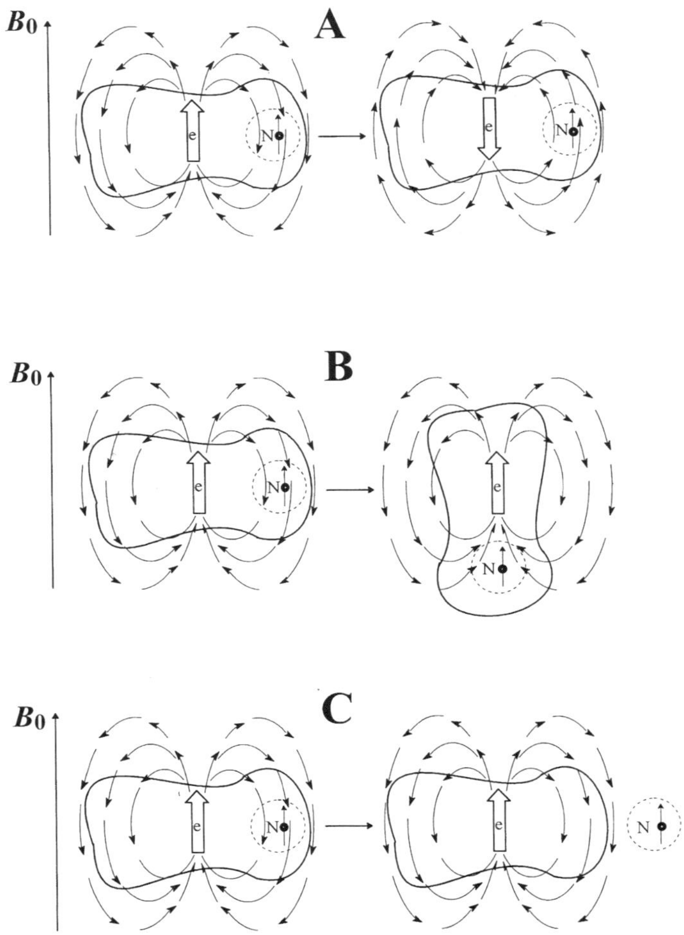

Fig. 3.1. Pictorial representation of the motions causing nuclear relaxation: electron spin relaxation (A), molecular rotation (B) and chemical exchange (C). It can be seen that the electron dipolar field at the nucleus fluctuates with time in direction (A), intensity (C) or both (B).

M_S value but on different positions in space. This random motion of the electron around the nucleus is again seen as a fluctuating magnetic field that causes nuclear relaxation through dipolar coupling.

(2) In addition, the nucleus sees the induced electronic magnetic moment $\langle \mu \rangle$ (Eq. (1.33)) which is aligned along the magnetic field as a result of the time average over the Zeeman levels. Upon rotation, the induced electronic magnetic field causes fluctuating magnetic fields sensed by the nucleus through space. This is another dipolar mechanism, that can be also visualized by referring to Fig. 3.1B, where the electronic vector is now the difference in electron population of the Zeeman levels and called Curie spin relaxation.

(3) Finally, the approach and binding of a moiety containing the resonating nucleus to a group containing the unpaired electron, and the following detachment (chemical exchange), cause fluctuating magnetic fields at the nucleus (Fig. 3.1C) through both contact and dipolar mechanisms. A limiting case of chemical exchange is the outer sphere interaction (see Section 4.5).

Relaxation measurements provide a wealth of information both on the extent of the interaction between the resonating nuclei and the unpaired electrons, and on the time dependence of the parameters associated with the interaction. Whereas the dipolar coupling depends on the electron–nucleus distance, and therefore contains structural information, the contact contribution is related to the unpaired spin density on the various resonating nuclei and therefore to the topology (through chemical bonds) and the overall electronic structure of the molecule. The time-dependent phenomena associated with electron–nucleus interactions are related to the molecular system, and to the lifetimes of different chemical situations, for the resonating nucleus. Obtaining either structural or dynamic information, however, is only possible if an in-depth analysis of a series of experimental results provides sufficient data to characterize the system within the theoretical framework discussed in this chapter.

In real systems, each unpaired electron is delocalized through chemical bonds in the neighborhood of the metal, to an extent depending on the chemical properties of the system. Therefore, the dipolar interaction with the resonating nucleus should be evaluated by integrating over all the points where there is a finite unpaired spin density. Approximate procedures will be given in Section 3.4.1. An account of the nuclear relaxation times based on this general approach requires details of the electronic structure of the system, including the unpaired spin densities on the various non-s orbitals of the molecule, which is difficult to treat in general terms.

3.2 THE CORRELATION TIME

Electrons switch between levels characterized by M_S values. Let us examine now an ensemble of n molecules, each with an unpaired electron, in a magnetic field at a given temperature. The bulk system is at constant energy but at the molecular level electrons move, molecules rotate, there are concerted atomic motions (vibrations) within the molecules and, in solution, molecular collisions. Is it possible to have information on these dynamics on a system which is at equilibrium? The answer is yes, through the *correlation function*. The correlation function is a product of the value of any time-dependent property at time zero with the value at time t, summed up to a large number n of particles. It is a function of time. In this case the property can be the M_S value of an unpaired electron and the particles are the molecules. The correlation function has its maximum value at $t = 0$; since each molecule has one unpaired electron, the product of the

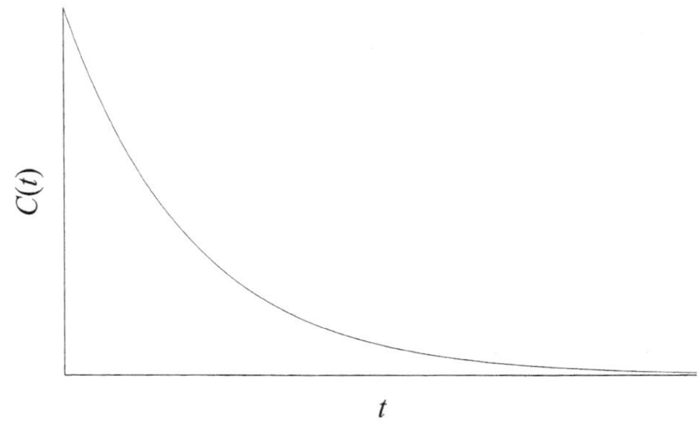

Fig. 3.2. Exponential decay with time of the correlation function.

M_S value at time zero times the M_S value at time t ($t = 0$) is either $\frac{1}{2} \cdot \frac{1}{2}$ or $-\frac{1}{2} \cdot (-\frac{1}{2})$, i.e. always equal to $+\frac{1}{4}$. With time, some spins will change their orientation (Fig. 3.1A), so some particles will contribute by $-\frac{1}{4}$. The value of the correlation function $C(t)$ decreases. It can be shown that the decay of the correlation function can be approximated by an exponential in the absence of constraints (see Fig. 3.2) [1].

$$C(t) = \sum_i M_S(0) M_S(t) \approx \left(\sum_i \frac{1}{4}\right) \exp(-t/\tau_c). \tag{3.1}$$

From Eq. (3.1), the correlation time τ_c is defined as the time constant for which the correlation function exponentially decays to zero. At a time small compared with τ_c, $\exp(-t/\tau_c) \to 1$ and we expect that essentially all spins maintain their original value. Therefore, most of the products will be $\frac{1}{4}$. At times long with respect to τ_c, we expect that all spins have changed their orientations many times, so that on the average half of the spins will result with the same M_S with respect to their original values, and the other half will have the other M_S value. Under these conditions, statistically half of the products will be $\frac{1}{4}$ and half $-\frac{1}{4}$, and the summation over a large number n of spins will yield zero.

Strictly speaking, the time dependence of $C(t)$ is not exponential for $t \ll \tau_c$; by referring to the above example, after a time much shorter than τ_c but still larger than zero, it is likely that $C(t) = C(0)$. Indeed, an expansion of the initial part of $C(t)$ would show a flat region around zero. In other words, it must be that $\mathrm{d}C(0)/\mathrm{d}t = 0$, whereas for a true exponential $\mathrm{d}C(0)/\mathrm{d}t = 1/4\tau_c$. Only for longer times does $C(t)$ become effectively exponential. The small initial deviation

[1] Here, and in the following, only positive values of time t are considered. Otherwise, all equations of the type of Eq. (3.1) should contain the absolute value of time in the exponential decay term, $\exp(-|t|/\tau_c)$.

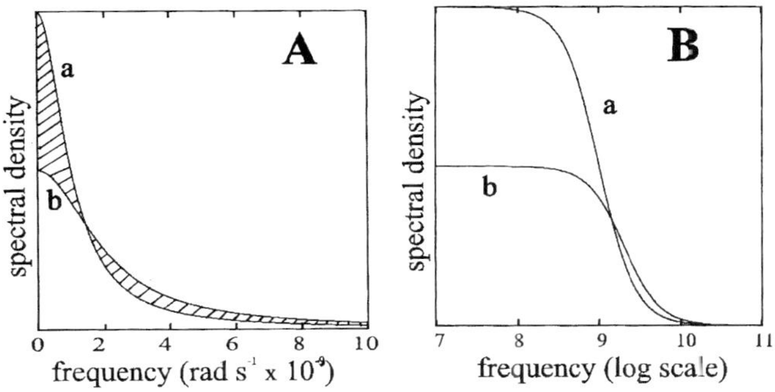

Fig. 3.3. Spectral density function $J(\omega)$ as a function of frequency, linear scale (A) and log scale (B). The profiles are obtained for $\tau_c = 1 \times 10^{-9}$ s (a) and 5×10^{-10} s (b). The two dashed areas in (A) are equivalent, showing that the area under the a and b curves is the same. The inflection points occur at $\omega\tau_c = 1$.

is, however, irrelevant for our purposes. Real cases of strong deviations from an exponential behavior do exist, but we will often restrict our interest to exponential correlation functions. In any case, the correlation time of a certain time-dependent process can be defined as the integral of the correlation function of that process, independently of the actual law for the time dependence.

The Fourier transform of an exponential decay in the time domain, as seen in Section 1.7.2, is a Lorentzian in the frequency domain:

$$J(\omega) = \frac{\tau_c}{1 + \omega^2\tau_c^2}. \tag{3.2}$$

This function (Fig. 3.3) has its inflection point at $|\omega|\tau_c = 1$. We say that the function undergoes a dispersion at $|\omega|\tau_c = 1$. This function provides the intensity of the various frequencies available in the lattice due to the fluctuations to which τ_c is related. It is also called noise, power, or spectral density function. The nucleus picks up the needed ω frequency for its relaxation. The probability of this to occur depends on the spectral density (i.e. on the value of function (3.2)) at that frequency.

Transverse nuclear relaxation can also occur when the local fields at the nucleus fluctuate slowly, i.e. with an ω frequency near zero (see also Section 3.4). Then the spectral density function will take the form

$$J(0) = \tau_c. \tag{3.3}$$

This is a so-called non-dispersive term. This effect can be viewed as transient increases or decreases of the precession frequency due to local field fluctuations along z (the $\boldsymbol{B_0}$ direction). These changes in the precession frequency do not appreciably affect the z-projection of the nuclear magnetic moment but 'dephase' its projection in the xy plane. In cases where two nuclear spins relax each other, the sum and the difference of the two frequencies (zero and double quantum transitions) are needed besides ω, and then we will also have functions of the type

$$J(\omega_1 - \omega_2) = \frac{\tau_c}{1 + (\omega_1 - \omega_2)^2 \tau_c^2}$$
$$J(\omega_1 + \omega_2) = \frac{\tau_c}{1 + (\omega_1 + \omega_2)^2 \tau_c^2}. \tag{3.4}$$

Finally, local field fluctuations along z for the first spin can also be caused by fluctuations of the xy component of the second spin. This originates a term of the type of $J(0)$ but 'offset' by the precessional frequency of the second spin, ω_2:

$$J(\omega_2) = \frac{\tau_c}{1 + \omega_2^2 \tau_c^2}. \tag{3.5}$$

Analogous spectral densities will also be found in the treatment of electron relaxation (see Eqs. (3.11) and (3.12)).

Once the correlation time is fixed, the value of the spectral density function at a given frequency can be determined. When τ_c is long, the spectral density is large in the small frequency range below τ_c^{-1}; when τ_c is short, we have a larger range of frequencies available but with a lower intensity (Fig. 3.3). Of course, τ_c refers to real physical mechanisms of fluctuations. We use the word fluctuation to indicate a movement which is random and unpredictable. We also use the words uncorrelated and stochastic for this behavior. The ω frequencies generally arise from sudden and short movements, which provide a range of frequencies just like in pulsed NMR (Section 1.7.2).

The movements capable of relaxing the nuclear spin that are of interest here are related to the presence of unpaired electrons, as has been discussed in Section 3.1. They are electron spin relaxation, molecular rotation, and chemical exchange. These correlation times are indicated as τ_s (electronic relaxation correlation time), τ_r (rotational correlation time), and τ_M (exchange correlation time). All of them can modulate the dipolar coupling energy and therefore can cause nuclear relaxation. Each of them contributes to the decay of the correlation function. If these movements are independent of one another, then the correlation function decays according to the product

$$\exp(-t/\tau_s)\exp(-t/\tau_r)\exp(-t/\tau_M) = \exp\left[-(\tau_s^{-1} + \tau_r^{-1} + \tau_M^{-1})t\right].$$

Therefore, the overall correlation time for dipolar coupling is such that

$$\boxed{(\tau_c^{-1})^{\text{dip}} = \tau_s^{-1} + \tau_r^{-1} + \tau_M^{-1}} \tag{3.6}$$

i.e. its reciprocal is the sum of the single reciprocal correlation times. It often happens that only one dominates. In the case of relaxation by contact or isotropic

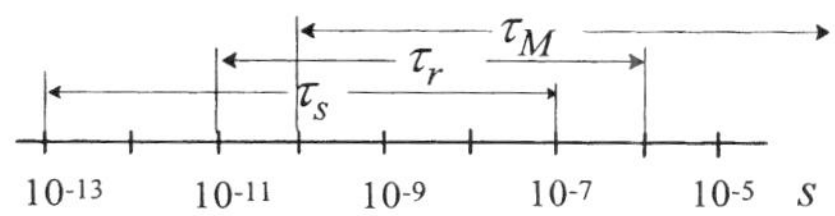

Fig. 3.4. Ranges of typical values for τ_s, τ_r and τ_M as defined in Eq. (3.6).

coupling, only chemical exchange and electron relaxation can modulate the coupling. Thus

$$\boxed{(\tau_c^{-1})^{\text{con}} = \tau_s^{-1} + \tau_M^{-1}.} \tag{3.7}$$

Sometimes distinction can be made between τ_{c1} and τ_{c2} if τ_s is specified as τ_{s1} (longitudinal) or τ_{s2} (transverse) electronic correlation time (see Section 3.4).

Electronic relaxation times fall in the range 10^{-7} to 10^{-13} s (Fig. 3.4), whereas the exchange time can be indefinitely long or as short as 10^{-10} s, depending on the chemical bond strength. As far as rotation is concerned, the rotational correlation time can be predicted for spherical rigid particles [1–3].

$$\boxed{\tau_r = \frac{4\pi\eta a^3}{3kT} = \frac{\eta \text{MW}}{dN_A kT}} \tag{3.8}$$

where η is the viscosity (kg s^{-1} m^{-1}) of the solvent, a is the radius of the molecule, assumed spherical, MW is the molecular weight, (kg mol^{-1}), d is the density (kg m^{-3}) of the molecule (usually taken equal to 10^3), and N_A is Avogadro's constant. Values for τ_r in water solutions at room temperature range from 3×10^{-11} s for hexaaqua metal complexes, to 10^{-8} s for small proteins (MW $\approx$ 10 kg mol^{-1}) to 10^{-6} s for larger macromolecules (MW $\approx 10^3$ kg mol^{-1}) (Fig. 3.4). The value of τ_r may also be estimated from T_1 measured at several magnetic fields on a diamagnetic analog of the metal complexes of interest. In fact, in diamagnetic systems the nuclear relaxation times are usually determined by dipolar coupling with other nuclei in the same molecule, the correlation time for the interaction being the rotational correlation time of the molecule itself (in the absence of chemical exchange).

3.3 ELECTRON RELAXATION

From the foregoing discussion it appears that electron relaxation may be important for NMR on paramagnetic substances because τ_s may be the correlation time. Indeed, it is always important. Let us say we have two paramagnetic ($S = 1/2$) compounds in water, one with τ_s of 10^{-8} s and the other with τ_s of 10^{-12} s. The size of each molecule is the same and such that the rotational correlation time is 10^{-10} s. For the former complex $\tau_c \approx \tau_r = 10^{-10}$ s and for the latter $\tau_c \approx \tau_s = 10^{-12}$ s. So, in one case nuclear relaxation depends on a correlation time which is due to rotation, whereas in the other it depends on the electron relaxation time. When the rotational mechanism determines the correlation time, and the latter happens to be long, nuclear relaxation rates will be invariably large and the NMR lines broad. As a consequence, we lose resolution and the capability of revealing connectivities among signals.

As will be seen soon, as a practical consequence, we can state that high resolution NMR can be fruitfully attempted for those systems for which τ_c is determined by τ_s: the shorter τ_s, the better. Detection of signals and exploitation of the NMR experiment is still convenient for diamagnetic moieties interacting with paramagnetic systems with long τ_s when in the presence of fast chemical exchange. Under these circumstances the linewidths of nuclei of the moiety which is in molar excess and exchanging with the moiety bound to the paramagnetic center are reduced by the molar fraction of the bound moiety, and still contain structural an dynamic information (see Chapter 4).

In Table 3.1 estimates of the electronic relaxation times at high NMR fields are reported for some paramagnetic metal ions in solution and at room temperature. As τ_s values may be strongly field dependent, the low field limiting values τ_{s0}, when available, are also reported. The table also reports the calculated linewidth at 800 MHz of a proton at 5 Å distance (in the presence of dipolar coupling, see later) for two values of τ_r. The underlined metal ions are those for which high resolution NMR is feasible and 2D NMR can be attempted also for nuclei experiencing hyperfine shifts. In particular, lanthanides(III) (except gadolinium(III)) and low spin iron(III) are particularly suitable for NMR. Then, tetrahedral nickel(II) and high spin six-coordinated cobalt(II) complexes have short electronic relaxation times. Manganese(III), low spin cobalt(II) and high spin heme-containing iron(III) compounds are at the borderline. Oxovanadium(IV), chromium(III), copper(II), manganese(II) and gadolinium(III) are not suitable for NMR experiments, except on nuclei in rapid exchange with diamagnetic species in excess. In Chapter 6 we will see that magnetic coupling in polymetallic systems may provide short effective electronic relaxation times independently of the electronic relaxation times of the monometallic system. We are now going to cursorily review the principles of electron relaxation in order to be able to account for the values of the electronic relaxation times or to predict them.

3.3.1 The main mechanisms for electron relaxation

Mechanisms analogous to those illustrated in Fig. 3.1 apply also to electron relaxation (see below). However, electrons have other more efficient relaxation mechanisms which overcome the former ones. They are based on the presence of spin orbit coupling. Molecular motions modulate the orbital magnetic moment and then affect the electron spin. Several possible mechanisms for electron relaxation

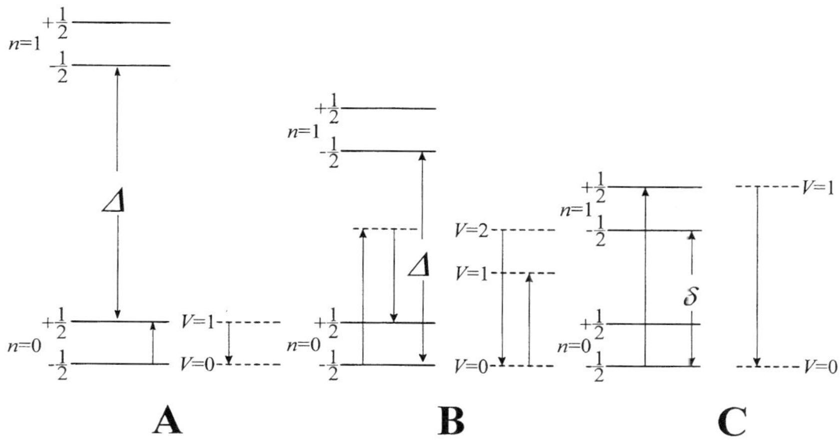

Fig. 3.5. Lattice and spin transitions are coupled by (A) direct processes, (B) Raman processes, (C) Orbach processes. The proximity of the excited electronic state favors both Orbach and Raman processes. The electronic states are labeled with n, the lattice vibrational states are labeled with V. Δ and δ indicate energy separations with excited states coupled to the ground state by spin–orbit coupling.

have been successfully identified, and equations have been derived for many of them [4,5].

Solid-state theories ascribe electron relaxation to the coupling of electronic spin transitions with transitions between lattice vibrational levels, or more generally with *phonons*. Disappearance (depopulation of a vibrational level) or creation (population of a vibrational level) of phonons modulate the orbital component of the electron magnetic moment.

The orbit–lattice interaction allows the electron to change energy level, which may imply changes in the electron spin state. It is customary to separate the coupling into direct processes [6], Orbach processes [7], and Raman processes [6,7], all of which are shown in Fig. 3.5. In the direct process, a phonon with the same energy as the electron spin transition is required. The direct process may be important only at about liquid helium temperature, where only low energy phonons are available. The Orbach process requires low-lying electronic energy levels at about the energy of the phonons. At 300 K, corresponding to a thermal energy of about 200 cm^{-1}, the most probable phonon energies range between 50 and 1000 cm^{-1}. So, the electron–phonon coupling can provide the energy for jumps from the ground to the excited states and vice versa. Such jumps may involve spin changes. The Raman process is operative when there are not enough phonons available for electronic transitions which are too small in energy. Therefore, in the Raman process two phonons of high energy may simultaneously interact with the spin system so that their energy difference equals the electron Zeeman energy. This phenomenon can also be regarded as the scattering of a single phonon upon collision with the electron. The Orbach process provides the

TABLE 3.1

Electronic relaxation times of some common paramagnetic metal ions and nuclear relaxation rates for a proton at 5 Å from the metal, at 800 MHz ^{1}H resonance frequency, due to dipolar and Curie relaxation, estimated from Eqs. (3.16), (3.17), (3.29) and (3.30), with $\tau_c = \tau_s$

Metal ion	Electronic configuration	S	τ_s (s)	Small molecules ($\tau_r = 10^{-10}$ s)	Large molecules ($\tau_r = 10^{-8}$ s)		Reference
				$R^I_{1,2M}$	R^I_{1M}	R^I_{2M}	
Ti^{3+}	d^1	1/2	10^{-10}–10^{-11}	40–300	50–400	100–500	[58,59]
VO^{2+}	d^1	1/2	$\approx 10^{-8}$ [a]	300–500	40	15,000	[60–62]
V^{3+}	d^2	1	$\approx 10^{-11}$	100–150	100	500	[63,64]
V^{2+}	d^3	3/2	$\approx 10^{-9}$	1500–2000	1000	15,000	[65–67]
Cr^{3+}	d^3	3/2	5×10^{-9}–5×10^{-10}	1500–2000	200–1500	10,000–60,000	[13,68,69]
Cr^{2+}	d^4	2	10^{-11}–10^{-12}	50–500	40–400	4000	est. from [70]
Mn^{3+}	d^4	2	10^{-10}–10^{-11} [b]	300–2000	300–3000	4000–8000	[64,69,71]
Mn^{2+}	d^5	5/2	$\approx 10^{-8}$ [c]	4000–6000	400	200,000	[13,72]
Fe^{3+} (H.S.)	d^5	5/2	10^{-9}–10^{-11} [d]	500–6000	500–3000	8000–50,000	[69,73,74]
Fe^{3+} (L.S.)	d^5	1/2	10^{-11}–10^{-13}	2–50	1–60	50–150	[75,76]
Fe^{2+} (H.S.)	d^6, 5–6 coord.	2	10^{-12}–10^{-13}	50–150	10–50	4000	[77,78], est. from [79]
	d^6, 4 coord.	2	$\approx 10^{-11}$	300–500	400	4000	est. from [80]
Co^{2+} (H.S.)	d^7, 5–6 coord.	3/2	5×10^{-12}–10^{-13}	20–200	10–200	1000–2000	[81,82]
	d^7, 4 coord.	3/2	$\approx 10^{-11}$	200–300	200	2000	[44]
Co^{2+} (L.S.)	d^7	1/2	10^{-9}–10^{-10}	200–400	200–400	500–3000	[66]
Ni^{2+}	d^8, 5–6 coord.	1	$\approx 10^{-10}$ [e]	600–700	1000	2000	[83]
	d^8, 4 coord.	1	$\approx 10^{-12}$	20–30	20	400	[84,85]
Cu^{2+}	d^9	1/2	$(1–5) \times 10^{-9}$	300–500	40–200	3000–20,000	[86]
Ru^{3+}	d^5	1/2	10^{-11}–10^{-12}	5–50	5–50	50–150	[64,87]
Re^{3+}	d^4	2	10^{-12}–10^{-13}	50–150	10–50	4000	[88]
Gd^{3+}	f^7	7/2	10^{-8}–10^{-9} [f]	5000–15,000	800–5000	100,000–400,000	[22–26]

TABLE 3.1 (continued)

Metal ion	Electronic configuration	S	τ_s (s)	Small molecules ($\tau_r = 10^{-10}$ s)	Large molecules ($\tau_r = 10^{-8}$ s)		Reference
				$R^I_{1,2M}$	R^I_{1M}	R^I_{2M}	
$\underline{Ln}^{3+}$		J					[20,21,39,89]
Ce^{3+}	f^1	5/2	4×10^{-13}–10^{-13}	6–12	3–7	200–300	
Pr^{3+}	f^2	4	3×10^{-13}–6×10^{-14}	16–30	4–13	1000	
Nd^{3+}	f^3	9/2	8×10^{-13}–2×10^{-13}	20–40	10–20	1100	
Sm^{3+}	f^5	5/2	2×10^{-13}–5×10^{-14}	0.2–0.6	0.2–0.6	3–4	
Eu^{2+}	f^7	7/2	5×10^{-14}–10^{-14}	300–400	20–30	30,000	
Tb^{3+}	f^8	6	8×10^{-13}–2×10^{-13}	800–1000	100–200	60,000	
Dy^{3+}	f^9	15/2	10^{-12}–4×10^{-13}	1100–1500	170–300	80,000	
Ho^{3+}	f^{10}	8	8×10^{-13}–2×10^{-13}	1000–1500	150–250	80,000	
Er^{3+}	f^{11}	15/2	8×10^{-13}–3×10^{-13}	700–1000	120–200	50,000	
Tm^{3+}	f^{12}	6	2×10^{-12}–5×10^{-13}	300–500	80–200	20,000	
Yb^{3+}	f^{13}	7/2	5×10^{-13}–2×10^{-13}	50–70	16–30	3000	

Metal ions more suitable for high resolution NMR are underlined.

[a] Field dependence of τ_s with a τ_{s0} value of 5×10^{-10} s has also been reported [90].

[b] τ_s values in the 10^{-11}–10^{-12} s range have also been proposed [91].

[c] τ_s is strongly field dependent. τ_{s0} values of 10^{-9}–10^{-10} s have been reported [92,93].

[d] τ_s is field dependent, τ_{s0} values being highly variable and ranging from 10^{-9} [74] to less than 10^{-11} s [94].

[e] τ_s is strongly field dependent. τ_{s0} values of 10^{-11}–10^{-12} s have been reported [95,96].

[f] τ_s is field dependent. τ_{s0} values of 10^{-8}–4×10^{-10} s have been reported [24,25].

fastest mechanism. The Raman process may successfully compete when there are low-lying excited levels.

Following the solid-state approach, equations have been derived [8,9] also for the electron spin relaxation of $S = ½$ ions in solution determined by the aforementioned processes. Instead of phonons, collisions with solvent should be taken into consideration, whose correlation time is usually in the range 10^{-11} to 10^{-12} s. However, there is no satisfactory theory that unifies relaxation in the solid state and in solution. The reason for this is that the solid state theory was developed for low temperatures, while solution theories were developed for room temperature. The phonon description is a powerful one when phonons are few. By increasing temperature, the treatment becomes cumbersome, and it is more convenient to use stochastic theory (see Section 3.2) instead of analyzing the countless vibrational transitions that become active.

Two mechanisms proposed for $S = ½$ ions are typical of solutions and are determined by the rotation of the molecule. The first is the so-called spin-rotation mechanism [10]. This mechanism arises from the fact that, upon a sudden change in the rotational motion of a molecule, for example after a collision, the molecular skeleton and the electron cloud may be slightly misplaced from one another, thus providing an instantaneous electric dipole moment. Therefore, there is a coupling between the spin angular momentum and the magnetic field generated by the instantaneous electric dipole moment, the latter related to the rotational angular momentum of the molecule; this coupling causes relaxation. The smaller the molecule, the larger the effect; for macromolecules, this mechanism is indeed negligible. The contribution to the relaxation rate of this mechanism is thus inversely proportional to that of the mechanisms modulated by τ_r, which increases with the size of the molecule. The effect is also proportional to the departure of $\overline{g}$ from g_e. The second kind of mechanism is due to fluctuations in electron Hamiltonian terms, due to Brownian motions. The electron Hamiltonian is given by

$$H = \mu_B \boldsymbol{L} \cdot \boldsymbol{B_0} + g_e \mu_B \boldsymbol{S} \cdot \boldsymbol{B_0} + \boldsymbol{S} \cdot \boldsymbol{D} \cdot \boldsymbol{S} + \boldsymbol{I} \cdot \boldsymbol{A} \cdot \boldsymbol{S} \tag{3.9}$$

where the first two terms represent the coupling between the orbital and the intrinsic magnetic moment with the external magnetic field, the third term the ZFS and the forth the hyperfine coupling between unpaired electron and metal nucleus. In the spin Hamiltonian formalism, the first two terms may be written as

$$\mu_B (\boldsymbol{L} + g_e \boldsymbol{S}) \cdot \boldsymbol{B_0} = \mu_B \boldsymbol{S} \cdot \boldsymbol{g} \cdot \boldsymbol{B_0} \tag{3.10}$$

where $\boldsymbol{g}$ is a tensor which contains the anisotropy introduced by the orbital angular operator. Fluctuations of $\boldsymbol{g}$, $\boldsymbol{A}$ and $\boldsymbol{D}$ can cause electron relaxation. If g or A are anisotropic, their values are orientation dependent, and are therefore modulated by rotation [11]. As already seen, electrons have an efficient relaxation mechanism with respect to nuclear relaxation also because of the presence of spin–orbit coupling, that arises from the coupling between electron spin magnetic moment and orbital momentum, due to its motion around the electrically charged nucleus. In $S > 1/2$ systems, the presence of spin–spin coupling must also be considered. In the spin Hamiltonian formalism all these terms are included in the ZFS Hamiltonian $\boldsymbol{S} \cdot \boldsymbol{D} \cdot \boldsymbol{S}$. Therefore, for ions with $S > 1/2$, relaxation mechanisms can arise from modulation of the quadratic ZFS in solution; these mechanisms have been discussed in detail by Bloembergen and Morgan [12] and Rubinstein et al. [13]. Deformations of the coordination polyhedron by collision with solvent molecules cause a transient ZFS which allows the coupling of rotation with spin transitions. The following equations have been derived for R_{1e} and R_{2e} for $S = 1$

$$\boxed{R_{1e} = \frac{2\Delta_t^2}{50}\left[4S(S+1) - 3\right]\left(\frac{\tau_v}{1+\omega_s^2\tau_v^2} + \frac{4\tau_v}{1+4\omega_s^2\tau_v^2}\right)} \tag{3.11}$$

$$\boxed{R_{2e} = \frac{\Delta_t^2}{50}\left[4S(S+1) - 3\right]\left(3\tau_v + \frac{5\tau_v}{1+\omega_s^2\tau_v^2} + \frac{2\tau_v}{1+4\omega_s^2\tau_v^2}\right)} \tag{3.12}$$

where Δ_t^2 is the mean squared fluctuation of the ZFS and τ_v is the correlation time for the instantaneous distortions of the metal coordination polyhedron. Analogous equations also hold for the two electronic transitions $1/2 \rightarrow -1/2$ and $1/2 \rightarrow 3/2$ of $S = 3/2$ systems [13]. For $S = 5/2$ three different relaxation times for each type should be considered [13]. Eqs. (3.11) and (3.12) are still good approximations for their effective averages. From Eqs. (3.11) and (3.12), the low-field limiting value of the electronic relaxation rate τ_{s0}^{-1} is given by $\tau_{s0}^{-1} = (\Delta_t^2/5)\,[4S(S+1) - 3]\tau_v$. Eqs. (3.11) and (3.12) have been derived in the absence of a static ZFS term in the electron Hamiltonian.

From the foregoing, the importance of the contribution of the various electronic relaxation mechanisms can be considered qualitatively for various cases. The least efficient mechanisms are the rotational mechanisms ($\boldsymbol{g}$, $\boldsymbol{A}$ modulation) and spin rotation, which account for R_{1e} values up to 10^9 s^{-1} and are only detected in the absence of other mechanisms. Contributions from the modulation of the quadratic ZFS to the relaxation rates through solvent bombardment can range from 10^8 to 10^{11} s^{-1}, depending on the extent of the instantaneous geometrical distortions and spin–orbit coupling. It should be remembered that the extent of the splitting of an S manifold depends on the spin–orbit coupling constant, which increases with the atomic number and thus from left to right and from top to bottom in the periodic table. A new mechanism based on a generalization of the spin rotation mechanism, and which is valid when the spin–orbit coupling energy is larger than the energy separation from the first excited electronic state, has been proposed which accounts for electron relaxation rates as large as 10^{11} s^{-1}, e.g. for low spin ruthenium(III) complexes [14,15]. The Orbach and Raman processes may be the only processes that can account for relaxation rates higher than 10^{11} s^{-1}. Although

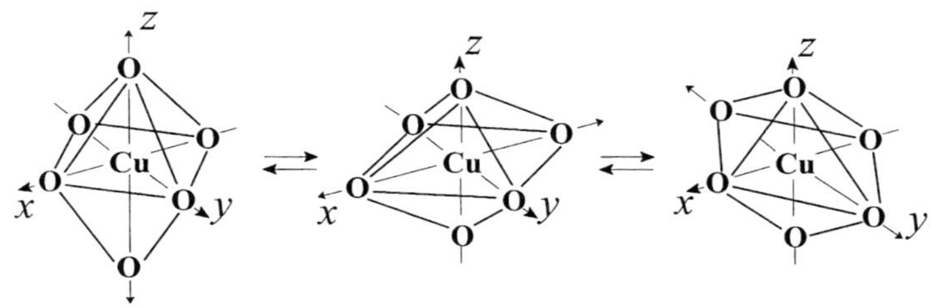

Fig. 3.6. Elongations along the three principal axes of the Cu–O bonds in the hexaaqua copper(II) complex due to dynamic Jahn–Teller effects occur randomly with a correlation time of about 5×10^{-12} s.

in solution there are no phonons, vibrational transitions within molecular clusters, and thus local, may allow for fast electron relaxation. The Orbach process at room temperature occurs when there are energy separations between ground and excited states in the range 100–1000 cm^{-1}.

$S = 1/2$ ions like Cu^{2+} have no ZFS, small magnetic anisotropy and, in general, excited states far above the ground state in energy. Electronic relaxation times are therefore long. The main relaxation mechanism operative for copper(II) complexes is probably the Raman process [16] in the solid state. In solution, besides a Raman-type relaxation mechanism [16], contributions from modulation of g and A anisotropy by molecular tumbling may also be operative for small complexes. Symmetric copper complexes are known to experience dynamic Jahn–Teller effects. In the case of six-coordinated complexes elongation occurs along the three principal axes (Fig. 3.6). The occurrence of elongation is random with a correlation time of about 5×10^{-12} s. Elongation changes the hyperfine coupling constant between the copper nucleus and the unpaired electrons. This can be a further electron relaxation mechanism operative for example for the $Cu(OH_2)_6^{2+}$ complex [17–19]. $S = 1/2$ ions like low spin iron(III), with ground levels deriving from an orbitally degenerate ground level in cubic symmetry (Fig. 3.7), may have low-lying excited states; therefore, Orbach processes are likely to be very efficient and short relaxation times are expected. The same holds for pseudooctahedral cobalt(II) and pseudotetrahedral nickel(II) chromophores. $S = 5/2$ ions, like high spin manganese(II) and high spin iron(III), have small g anisotropy and first

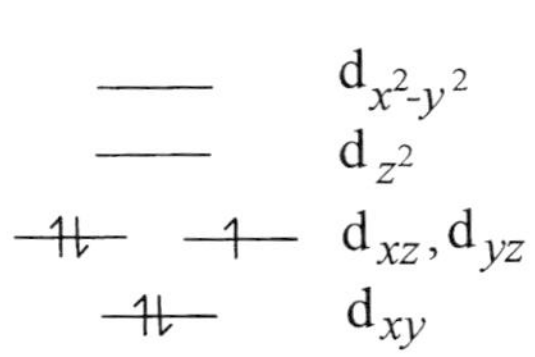

Fig. 3.7. A typical low symmetry splitting of the orbitally degenerate $^2T_2$ ground state of low spin iron(III).

excited levels much higher than the ground level, since they arise from a different free ion term. In these cases, modulation of the quadratic ZFS is probably the most efficient relaxation mechanism in solution; such a mechanism is much more efficient for iron(III) than for manganese(II), since in the former spin–orbit coupling is larger and the first excited state is closer. Pseudooctahedral nickel(II) and pseudotetrahedral cobalt(II) have excited states higher in energy than those in the pseudotetrahedral and pseudooctahedral analogs already discussed, so the Orbach and Raman relaxation mechanisms are expected to be relatively less efficient and probably comparable with the modulation of the quadratic ZFS. All lanthanide ions, with the exception of gadolinium(III), are likely to be relaxed by Orbach-type processes, although some very efficient Raman processes are also invoked on the basis of the temperature dependence [20,21]. For gadolinium(III), which has an ^8S free ion ground term, the modulation of the quadratic ZFS is again the dominant mechanism [22–26]. The electron relaxation mechanisms operative for several different metal ions will be discussed in more detail in Chapter 5. For the particular case of the aqua ions, they are summarized in Table 5.6.

3.4 NUCLEAR RELAXATION DUE TO DIPOLAR COUPLING WITH UNPAIRED ELECTRONS

From the classic description of the coupling between two point dipoles with magnetic moments $\boldsymbol{\mu}_1$ and $\boldsymbol{\mu}_2$, the energy of the interaction E^{dip} is proportional to their scalar product and to the inverse of the third power of their distance (Eq. (1.1)). If the reciprocal orientation of the two vectors changes randomly with time, the fluctuating magnetic field produced by one at the location of the other may induce spin transitions on the other center, and thus provide a relaxation pathway. The interest here is limited to nuclear spin transitions induced by fluctuating magnetic fields originating from unpaired electron spins. It should be remembered that the electronic magnetic moment is orders of magnitude larger than that of nuclei; for example, the magnetic moment of the free electron is 658 times that of the proton (Section 1.2). The interaction between protons will anyway be present, even if usually much smaller, and it contributes to the diamagnetic relaxation.

While keeping in mind the general picture of nuclear relaxation in paramagnetic systems as described in Section 3.1, it is appropriate to consider first the simple case of dipolar coupling between two point-dipoles as if the unpaired electrons were localized on the metal ion. The enhancement of the nuclear longitudinal relaxation rate R_{1M} due to dipolar coupling with unpaired electrons can be calculated starting from the Hamiltonian for the system:

$$H = g_e\mu_B \boldsymbol{S}\cdot\boldsymbol{B}_0 - \hbar\gamma_I \boldsymbol{I}\cdot\boldsymbol{B}_0 + \boldsymbol{I}\cdot\boldsymbol{A}\cdot\boldsymbol{S}. \tag{3.13}$$

This Hamiltonian is constituted by two time-independent terms, which refer to the Zeeman interactions for the electron (first term) and the nucleus ($\hbar\gamma_I = g_I\mu_N$, second term), as seen in Chapter 1, plus the interaction term between the electron and the nucleus spin (Eq. (1.51)). The latter term fluctuates with time, and relaxation is due to its fluctuation[2]. As a result[3], nuclear relaxation rates will be proportional to the average of the square of this interaction energy and to the appropriate spectral density functions[4]:

$$R_{1M} \propto < (E_{\text{int}})^2 > f(\omega, \tau_c). \tag{3.14}$$

Consider a single electron and a single nucleus separated by a distance r in a magnetic field B_0, and whose interaction is dipolar in origin. The $\boldsymbol{I} \cdot \boldsymbol{A} \cdot \boldsymbol{S}$ term is the dipolar coupling between the two (which average zero in solution in the absence of magnetic anisotropy). The energy levels and the transition frequencies ω are shown in Fig. 3.8. Each transition has associated a relative transition probability per unit time, W_0 for ω_0, W_1^S for ω_1^S, W_1^I for ω_1^I, W_2 for ω_2. The nuclear longitudinal relaxation rate, which describes the decay of the nuclear spin magnetization, will be expressed in terms of the probabilities of the transitions involving nuclear spin flipping (see Appendix V):

---

[2] Nuclear relaxation thus originates from the modulation of the hyperfine interaction

$$H_1(t) = \boldsymbol{I} \cdot \boldsymbol{A} \cdot \boldsymbol{S} = \frac{\mu_0}{4\pi} \hbar^2 \gamma_I \gamma_S \left(\frac{3(\boldsymbol{I} \cdot \boldsymbol{r})(\boldsymbol{S} \cdot \boldsymbol{r})}{r^5} - \frac{\boldsymbol{I} \cdot \boldsymbol{S}}{r^3} \right) + \frac{2}{3} \mu_0 g_e \mu_B g_I \mu_N \sum_i \delta(r)_i \boldsymbol{S} \cdot \boldsymbol{I}.$$

The first term provides the dipolar relaxation; the second, corresponding to the limit of the former for $r = 0$, provides the contact relaxation. δ indicates the Dirac δ operator and Σ_i represents the sum over all unpaired electrons.

[3] R_{1M} can be expressed in terms of the probabilities of all possible transitions involving different spin states. From the time dependent perturbation theory, if H_0 is the static Hamiltonian (first two terms in Eq. (3.13)) with eigenvalues $E_n = \hbar\omega_n$ and H_1 is the time dependent perturbing Hamiltonian (third term in Eq. (3.13)), which describes an interaction with energy $E_1(t)$ and whose explicit form will depend on the kind of interaction, the probability to have a transition from the m to the n state is proportional to

$$w_{mn} = \frac{1}{\hbar^2} \int_0^t \overline{E_{1(mn)}(t') E_{1(mn)}(t'') e^{-i(\omega_m - \omega_n)(t''-t')}} \, \mathrm{d}(t'' - t') + \text{c.c.}$$

Since $H_1(t)$ is a 'stationary random perturbation', it is invariant under a change in the origin of the time, and the time (or ensemble) average will thus be

$$\overline{E_1(t')E_1(t'+t)} = \overline{E_1(0)^2} e^{-t/\tau_c}$$

where τ_c is the correlation time constant relative to the fluctuations. From the integration over the time, the spin density functions are thus obtained.

[4] We recall here that the time average of the interaction energy E_{dip} between two point-dipoles is zero (Chapter 2 and Appendix II), while the time average of its square is different from zero. The average dipolar shift in solution (pseudocontact shift) depends on the presence of magnetic anisotropy, which is necessary to give a non-zero average of E_{dip}; dipolar relaxation, on the contrary, will occur independently of the presence of magnetic anisotropy.

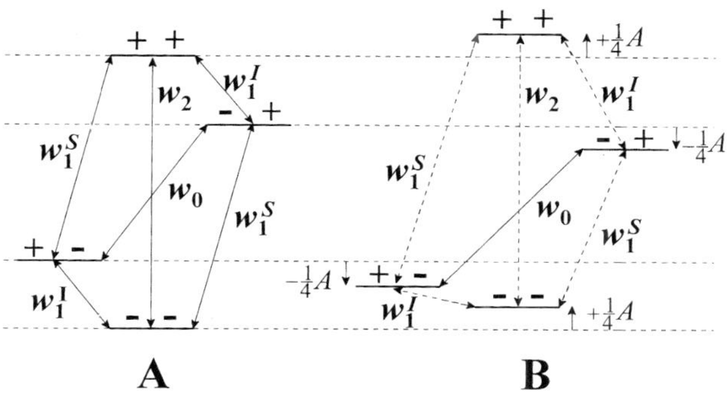

Fig. 3.8. Energy levels, transition frequencies and transition probabilities per unit time (W_0, W_1 and W_2) in a magnetically coupled I–S system ((A) dipolar coupling; (B) contact coupling). A is the contact hyperfine coupling constant. The order of the levels is for $g_I < 0$.

$$R_{1M} = \frac{1}{10}\left(\frac{\mu_0}{4\pi}\right)^2 \frac{\hbar^2\gamma_I^2\gamma_S^2}{r^6}\left[\frac{\tau_c}{1+(\omega_I-\omega_S)^2\tau_c^2} + \frac{3\tau_c}{1+\omega_I^2\tau_c^2} + \frac{6\tau_c}{1+(\omega_I+\omega_S)^2\tau_c^2}\right] \tag{3.15}$$

which is the Solomon equation [27]. The three terms in parentheses are proportional to the transition probabilities and contain the spectral density functions $f(\omega, \tau_c)$ already discussed (Eqs. (3.2) and (3.4)). The frequencies needed for the nuclear transitions are: ω_I (proton spin changes, electron spin remains the same), $\omega_I + \omega_S$ (proton and electron spins increase or decrease their energy) and $\omega_I - \omega_S$ (proton spin energy level increases and electron spin energy level decreases, or opposite). This equation can be generalized [28,29] to systems with $S > \frac{1}{2}$

$$R_{1M} = \frac{2}{15}\left(\frac{\mu_0}{4\pi}\right)^2 \frac{\gamma_I^2 g_e^2\mu_B^2 S(S+1)}{r^6}\left[\frac{\tau_c}{1+(\omega_I-\omega_S)^2\tau_c^2} + \frac{3\tau_c}{1+\omega_I^2\tau_c^2} + \frac{6\tau_c}{1+(\omega_I+\omega_S)^2\tau_c^2}\right]. \tag{3.16}$$

Note that Eq. (3.16) contains S but not I; R_{1M} is independent of I because it refers to $\Delta M_I = 1$ transitions [27,29]. J is needed instead of S for lanthanides. Analogously [27–29], an equation can be derived for R_{2M}:

$$R_{2M} = \frac{1}{15}\left(\frac{\mu_0}{4\pi}\right)^2 \frac{\gamma_I^2 g_e^2\mu_B^2 S(S+1)}{r^6}\left[4\tau_c + \frac{\tau_c}{1+(\omega_I-\omega_S)^2\tau_c^2} + \frac{3\tau_c}{1+\omega_I^2\tau_c^2} + \frac{6\tau_c}{1+(\omega_I+\omega_S)^2\tau_c^2} + \frac{6\tau_c}{1+\omega_S^2\tau_c^2}\right]. \tag{3.17}$$

R_{2M} differs from R_{1M} principally for the first and last terms. It is well known in NMR that frequencies near zero contribute to R_2 but not to R_1 (first term). This term originates from a $J(0)$ term (Eq. (3.3)), as described in Section 3.2. The last term is present because R_{2M} must also contain the probabilities of transitions connecting states with the same I_z but different S_z. When ω_S and ω_I are much smaller than τ_c^{-1} (fast motion limit) the denominators of the fractions between brackets become unity and the Solomon Eqs. (3.16) and (3.17) reduce to

$$R_{1M} = R_{2M} = \frac{4}{3}\left(\frac{\mu_0}{4\pi}\right)^2 \frac{\gamma_I^2 g_e^2 \mu_B^2 S(S+1)}{r^6}\tau_c. \tag{3.18}$$

At a magnetic field of 2.35 T the proton transition frequency (Larmor frequency) is: $\nu_I = 100$ MHz; $|\omega_I| = 6.28 \times 10^8$ rad s^{-1}; $|\omega_S| = 4.13 \times 10^{11}$ rad s^{-1}. The ratio $|\omega_S|/|\omega_I|$ is 658 if the nucleus is a proton, and higher for other nuclei (except tritium, see Appendix I). Therefore, the values $|\omega_I \pm \omega_S|$ in Eqs. (3.16) and (3.17) can be approximated by $|\omega_S|$.

It should be recalled that the longitudinal relaxation process requires energy exchange with the lattice. In this case, the lattice is represented by the electron spin system, which exchanges energy with the nuclear spin system. As was pointed out in Section 1.7.1, every process that causes R_1 relaxation also contributes to R_2 relaxation. In the fast motion limit, the coupling with the lattice has the same effect on the magnetization along the z axis and on that in the xy plane, and $R_1 = R_2$. However, outside this limit, the two functions $f_1(\omega, \tau_c)$ for R_{1M} and $f_2(\omega, \tau_c)/2$ for R_{2M}, corresponding to the expressions in parentheses in Eqs. (3.16) and (3.17), diverge, leading to $R_{1M} < R_{2M}$. The function $f_1(\omega, \tau_c)$ has its maximum value when the rate constant τ_c^{-1} equals the nuclear Larmor precession frequency $|\omega_I|$ (Fig. 3.9), which makes the coupling of the electron and nuclear spins most efficient. The function $f_2(\omega, \tau_c)/2$ increases monotonically with τ_c and differs

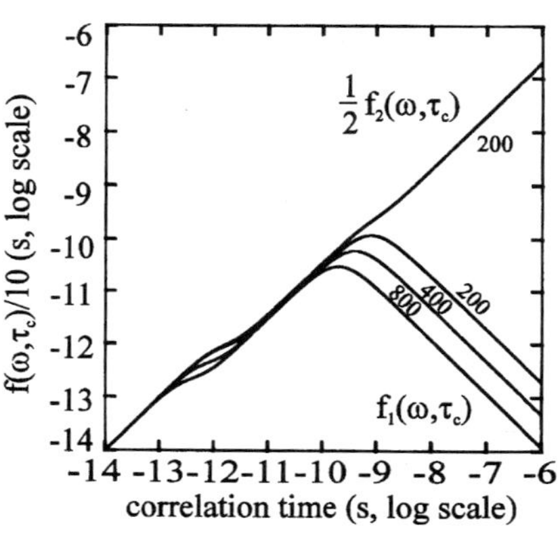

Fig. 3.9. Plot of $f_1(\omega, \tau_c)$ and $\frac{1}{2}f_2(\omega, \tau_c)$ of Eqs. (3.16) and (3.17) against τ_c for various proton Larmor frequencies (MHz).

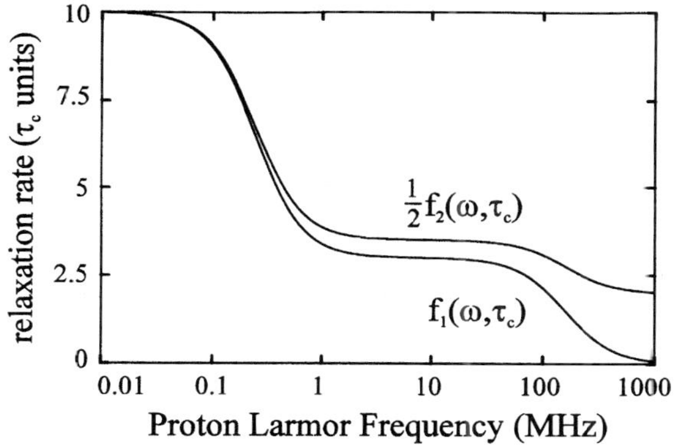

Fig. 3.10. Plot of $f_1(\omega, \tau_c)$ and $\frac{1}{2}f_2(\omega, \tau_c)$ of Eqs. (3.16) and (3.17) as a function of magnetic field (expressed as proton Larmor frequency; log scale). $\tau_c = 2 \times 10^{-9}$ s.

from $f_1(\omega, \tau_c)$ for $\tau_c > |\omega_I|^{-1}$ (Fig. 3.9) because of the frequency-independent term $4\tau_c$.

R_{1M} and R_{2M} values also depend on the applied magnetic field, as ω_I and ω_S are determined by it ($|\omega_I| = \gamma_I B_0$; $|\omega_S| = \gamma_S B_0 = 658|\omega_I|$ for protons). The magnetic fields currently used range from that of the Earth (25–65 μT) to 18.8 T ($|\omega_I|/2\pi = 800$ MHz for protons). The field dependence of $f_1(\omega, \tau_c)$ (Fig. 3.10) shows two plateaus whose heights are in a 10/3 ratio and depend on the value of τ_c. The τ_c parameter is considered constant within the magnetic field range used, but this may not always be so (see Eqs. (3.11) and (3.12), Section 3.3). The inflection point of the first dispersion is halfway between the two plateaus and corresponds to a magnetic field value at which $|\omega_S| = \tau_c^{-1}$; the inflection point of the second dispersion is halfway between the second plateau and zero, and corresponds to $|\omega_I| = \tau_c^{-1}$. Profiles of the type of that shown in Fig. 3.10 and in similar figures in the remainder of this Chapter can be obtained experimentally as described in Section 9.6, and are called Nuclear Magnetic Relaxation Dispersions (NMRD). Examples of experimental NMRD profiles will be found in Chapter 5.

The field dependence of $f_2(\omega, \tau_c)/2$ is similar to that of $f_1(\omega, \tau_c)$ (Fig. 3.10), except that the relative height of the two plateaus is 20/7 and at high field the function levels off at a finite value that is one-fifth of the low field limit, because of the field independent term $4\tau_c$.

In the foregoing discussion, the electron–nucleus spin system was assumed to be rigidly held within a molecule isotropically rotating in solution. If the molecule cannot be treated as a sphere, its motion is, in general, anisotropic and three different correlation times should be considered. Furthermore, if the molecule is not rigid, the correlation times for rotations about single bonds should also be taken into account. For the case of a metal ion rigidly held to a macromolecule, but with a coordinated water molecule freely rotating about the M–O bond, the effect of neglecting the latter rotation on the calculation of R_{1M} for the protons is estimated to be about 20% [30]. The exact magnitude of the effect is a function of the angle that the metal–proton vector makes with the M–O bond.

When τ_c is significantly affected by τ_s, then it can be magnetic field dependent, since electron relaxation can often be described by equations similar to those for nuclear relaxation (see for example Eqs. (3.11) and (3.12)). Owing to this dependence, longitudinal (T_{1e}) and transverse (T_{2e}) electronic relaxation times may not be equal and, strictly speaking, when electron relaxation is the dominant correlation time, Eqs. (3.16) and (3.17) should be written as

$$R_{1M} = \frac{2}{15}\left(\frac{\mu_0}{4\pi}\right)^2 \frac{\gamma_I^2 g_e^2 \mu_B^2 S(S+1)}{r^6}\left(\frac{7\tau_{s2}}{1+\omega_S^2\tau_{s2}^2}+\frac{3\tau_{s1}}{1+\omega_I^2\tau_{s1}^2}\right) \quad (3.19)$$

$$R_{2M} = \frac{1}{15}\left(\frac{\mu_0}{4\pi}\right)^2 \frac{\gamma_I^2 g_e^2 \mu_B^2 S(S+1)}{r^6}\left(4\tau_{s1}+\frac{13\tau_{s2}}{1+\omega_S^2\tau_{s2}^2}+\frac{3\tau_{s1}}{1+\omega_I^2\tau_{s1}^2}\right) \quad (3.20)$$

where the terms containing ω_S, $\omega_I + \omega_S$ and $\omega_I - \omega_S$ have been collected for simplicity, since $|\omega_I| \ll |\omega_S|$. The longitudinal relaxation time $T_{1\rho}$ in the rotating frame is also shortened through coupling with unpaired electrons, according to the following equation [29]:

$$R_{1\rho M} = \frac{1}{15}\left(\frac{\mu_0}{4\pi}\right)^2 \frac{\gamma_I^2 g_e^2 \mu_B^2 S(S+1)}{r^6}\left(\frac{4\tau_{s1}}{1+\omega_1^2\tau_{s1}^2}+\frac{13\tau_{s2}}{1+\omega_S^2\tau_{s2}^2}+\frac{3\tau_{s1}}{1+\omega_I^2\tau_{s1}^2}\right). \quad (3.21)$$

It differs from the equation for R_{2M} for the substitution of the non-dispersive term with an $\omega_1\tau_s$ dependent term, where ω_1 is the nuclear Larmor frequency in the B_1 field. Since the latter frequency is always such that $|\omega_1|\tau_c \ll 1$, $R_{1\rho M}$ can be safely taken equal to R_{2M}. This also holds for the other nuclear relaxation mechanisms discussed in the following sections.

In the low field (or fast motion) limit, in the absence of chemical exchange, and for $T_{2e} = T_{1e}$, Eqs. (3.19)–(3.21) reduce to

$$R_{1M} = R_{2M} = R_{1\rho M} = \frac{4}{3}\left(\frac{\mu_0}{4\pi}\right)^2 \frac{\gamma_I^2 g_e^2 \mu_B^2 S(S+1)}{r^6} T_{1e} \quad (3.22)$$

justifying the qualitative statement that the shorter the electronic relaxation times the smaller the paramagnetic effects on nuclear relaxation.

3.4.1 Generalized dipolar coupling

As anticipated in Sections 2.2.2 and 3.1, the unpaired electrons should not be considered as point-dipoles centered on the metal ion. They are at the least delocalized over the atomic orbitals of the metal ion itself. The effect of the deviation from the point–dipole approximation under these conditions is estimated to be negligible for nuclei already 3–4 Å away [31]. Electron delocalization onto the ligands, however, may heavily affect the overall relaxation phenomena. In this case the experimental R_{iM} may be higher than expected, and the ratios between the R_{iM} values of different nuclei does not follow the sixth power of the ratios between metal to nucleus distances. In the case of hexaaqua metal complexes the point–dipole approximation provides shorter distances than observed in the solid state (Table 3.2) for both ^{1}H and ^{17}O. This implies spin density delocalization on the oxygen atom. Ab initio calculations of R_{1M} have been performed for both ^{1}H and ^{17}O nuclei in a series of hexaaqua complexes (Table 3.2). The calculated metal nucleus distances in the assumption of a purely metal-centered dipolar relaxation mechanism are sizably smaller than the crystallographic values for ^{17}O, and the difference dramatically increases from 3d^5 to 3d^9 metal ions [32]. The differences for protons are quite smaller [32].

R_{1M} calculations in the presence of ligand-centered contributions are possible for metal complexes with ligands having dominant π spin density delocalization mechanisms. With certain approximations, the relaxation rates of protons and carbon atoms in sp^2 CH moieties can be expressed [33–35] as the sum of a

TABLE 3.2

Crystallographic values of metal–hydrogen and metal–oxygen distances in hexaaqua complexes of divalent 3d metal ions [a] compared with calculated effective distances

Measurements	Metal ions				
	Mn	Fe	Co	Ni	Cu
^{1}H data					
r_{cryst} (pm)	290.6	290.6	278.1	275.5	270.4
r_{eff} (pm)	290.0	289.5	276.3	273.1	268.4
$r^6_{\text{cryst}} \cdot \langle 1/r^3 \rangle^2_{\text{eff}}$ [b]	1.01	1.02	1.04	1.05	1.05
^{17}O data					
r_{cryst} (pm)	222.0	222.0	209.0	206.1	201.0
r_{eff} (pm)	219.8	211.7	190.5	174.9	172.3
$r^6_{\text{cryst}} \cdot \langle 1/r^3 \rangle^2_{\text{eff}}$ [b]	1.06	1.33	1.74	2.67	2.52

[a] From [32]. M–H distances were calculated using $r_{\text{O–H}} = 95.7$ pm and HOH = 104.52.
[b] The effective distances account for the experimental T_{1M}^{-1} by considering the electronic distribution predicted from ab initio MO calculations. $r^6_{\text{cryst}} \cdot \langle 1/r^3 \rangle^2_{\text{eff}}$ is the predicted enhancement coefficient of T_{1M}^{-1} with respect to a pure metal-centered dipolar relaxation mechanism.

metal-centered term (given by the Solomon equation), a ligand-centered term and a cross term. The ligand-centered contributions will be proportional to the spin density ρ_{2p_z} in the carbon $2p_z$ orbital, and to the square of the reciprocal third power of the average nucleus-$2p_z$ electron distance. Equations of the type of Eq. (3.23) have been proposed [35]:

$$R_{1M} = \frac{2}{15}\left(\frac{\mu_0}{4\pi}\right)^2 \gamma_I^2 g_e^2 \mu_B^2 S(S+1) \times \left[\rho_M^2 r_M^{-6} + \frac{4}{25}\rho_{2p_z}^2 \langle r_{2p_z}^{-3}\rangle^2 - \frac{2}{5}\rho_{2p_z}\langle r_{2p_z}^{-3}\rangle \rho_M r_M^{-3}\right] f_1(\omega, \tau_c) \quad (3.23)$$

where $\langle r_{2p_z}^{-3}\rangle$ is calculated to be 5.3×10^{-7} pm^{-3} for protons and 3.2×10^{-6} pm^{-3} for carbon atoms [33,34]. In general, ρ_{2p_z} can be evaluated from the contact part of the proton hyperfine shift. This can be an easy-to-follow procedure for estimating ligand-centered contributions to the relaxation rates for π delocalized spin density [35].

3.5 NUCLEAR RELAXATION DUE TO CONTACT COUPLING WITH UNPAIRED ELECTRONS

In the case of relaxation due to the contact term, the transition probabilities as defined in Fig. 3.8, where the splitting of the levels is now due to the contact contribution, are found to be

$$W_1^I = W_1^S = W_2 = 0. \quad (3.24)$$

This is because any scalar relaxation occurs through a flip-flop mechanism (Fig. 3.8)[5];

$$W_0 = \frac{1}{2}\left(\frac{A}{\hbar}\right)^2 \frac{\tau_c}{1 + (\omega_S - \omega_I)^2 \tau_c^2}. \quad (3.25)$$

Here the correlation time is τ_c^{con} of Eq. (3.7) and A is defined in Eq. (2.2). This differs from τ_c^{dip} of Eq. (3.6) because of the absence of the contribution from the rotational correlation time since the contact interaction, by its nature, is not

[5] In fact,

$$H_1^{\mathrm{cont}} = A\boldsymbol{I} \cdot \boldsymbol{S} = A(I_x S_x + I_y S_y + I_z S_z) = \frac{A}{2}(S_+ I_- + S_- I_+ + 2S_z I_z),$$

(see also Appendix III) where $S_\pm$ and $I_\pm$ are the usual flip up and down spin operators. Therefore, contact relaxation contains only the $|\omega_I - \omega_S|$ term, since the transitions $|\omega_I + \omega_S|$ (corresponding to S_+I_+, S_-I_-), ω_S (S_+I_z, S_-I_z) and ω_I (S_zI_+, S_zI_-) do not contribute, at variance with dipolar relaxation.

altered by reorientation of the molecules. In the absence of chemical exchange phenomena ($\tau_M^{-1} \ll \tau_s^{-1}$), the correlation time equals the electronic relaxation time.

The enhancement of the longitudinal contact relaxation rate, after extension to the general case $S > 1/2$, is given by the Bloembergen equation [29,36]

$$R_{1M} = \frac{2}{3} S(S+1) \left(\frac{A}{\hbar}\right)^2 \frac{\tau_{s2}}{1 + \omega_S^2 \tau_{s2}^2} \tag{3.26}$$

where τ_{s2} is the actual parameter determining τ_c. The equation for the transverse contact relaxation rate is [29,36]

$$R_{2M} = \frac{1}{3} S(S+1) \left(\frac{A}{\hbar}\right)^2 \left(\tau_{s1} + \frac{\tau_{s2}}{1 + \omega_S^2 \tau_{s2}^2}\right) \tag{3.27}$$

The term $S(S+1)(A/\hbar)^2$ is proportional to the square of the coupling energy between an electron spin vector S and the nuclear spin vector I. The same equations hold for lanthanides.

In the fast motion limit, and for $\tau_{s1} = \tau_{s2}$, Eqs. (3.26) and (3.27) reduce to

$$R_{1M} = R_{2M} = \frac{2}{3} S(S+1) \left(\frac{A}{\hbar}\right)^2 \tau_{s1}. \tag{3.28}$$

As in the case of dipolar relaxation, the presence of a field-independent term in the equation for R_{2M} causes the two nuclear relaxation values to diverge for $\omega_S^2 \tau_{s2}^2 > 1$ (Figs. 3.11 and 3.12).

3.6 CURIE NUCLEAR SPIN RELAXATION

In deriving the dipolar and contact relaxation contributions due to the presence of unpaired electrons, as shown in Eqs. (3.19), (3.20), (3.26) and (3.27), the small difference in the population of the electron spin levels according to the Boltzmann distribution has been neglected. Such a difference accounts for the time-averaged magnetic moment of the molecule $\langle \mu \rangle$ (Eq. (1.33)), which is related to the $\langle S_z \rangle$ value defined in Eq. (1.31) for a simplified case. The interaction of the nuclear spins with this static magnetic moment related to $\langle S_z \rangle$ provides a further relaxation contribution. Such interaction, of course, cannot be modulated by electron relaxation since $\langle S_z \rangle$ is already an average over the electron spin states. The correlation time for the coupling is only determined by τ_r (or possibly by τ_M). This relaxation mechanism is usually called magnetic susceptibility relaxation or Curie spin relaxation, to reflect its relationship with the magnetic susceptibility of a sample (Curie law).

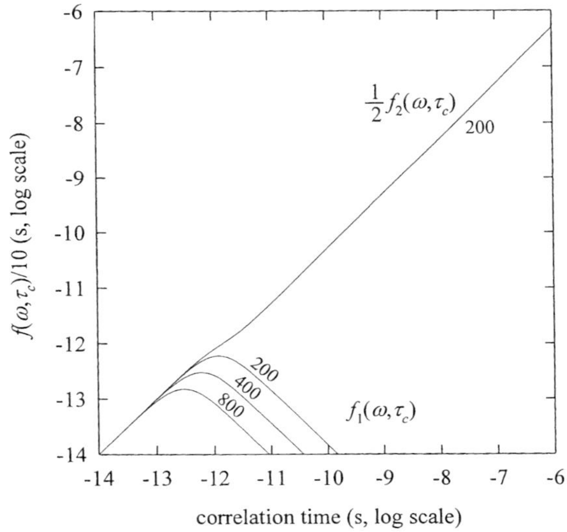

Fig. 3.11. Plot of $f_1(\omega, \tau_c)$ and $\frac{1}{2}f_2(\omega, \tau_c)$ of Eqs. (3.26) and (3.27) against τ_c for various proton Larmor frequencies (MHz).

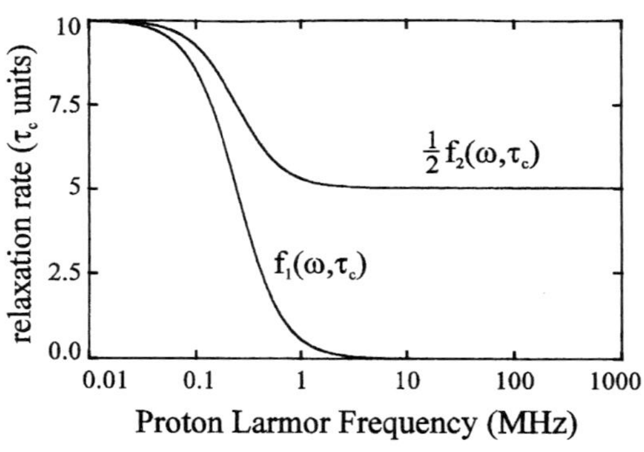

Fig. 3.12. Plot of $f_1(\omega, \tau_c)$ and $\frac{1}{2}f_2(\omega, \tau_c)$ of Eqs. (3.26) and (3.27) as a function of magnetic field (expressed as proton Larmor frequency; log scale). $\tau_c = 5 \times 10^{-11}$ s.

The dipolar contributions to R_{1M} and R_{2M} provided by this mechanism are [29,37,38]

$$R_{1M} = \frac{2}{5}\left(\frac{\mu_0}{4\pi}\right)^2 \frac{\gamma_I^2 g_e^2 \mu_B^2}{r^6} \langle S_z \rangle^2 \frac{3\tau_r}{1+\omega_I^2\tau_r^2}$$

$$= \boxed{\frac{2}{5}\left(\frac{\mu_0}{4\pi}\right)^2 \frac{\omega_I^2 g_e^4 \mu_B^4 S^2(S+1)^2}{(3kT)^2 r^6} \frac{3\tau_r}{1+\omega_I^2\tau_r^2}} \quad (3.29)$$

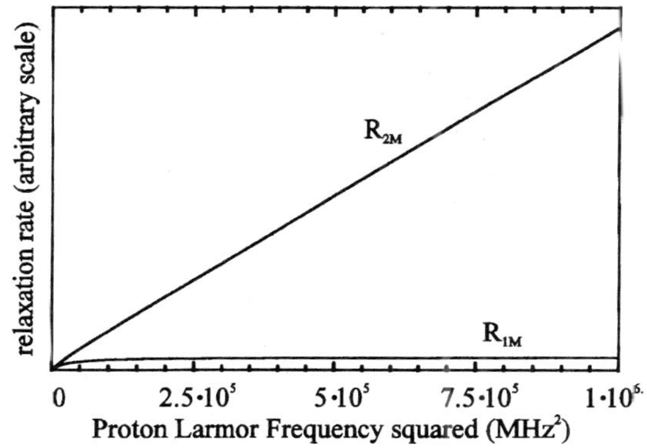

Fig. 3.13. Field dependence of the Curie spin relaxation contributions to R_{1M} and R_{2M} (arbitrary scale); $\tau_r = 2 \times 10^{-9}$ s.

$$R_{2M} = \frac{1}{5}\left(\frac{\mu_0}{4\pi}\right)^2 \frac{\gamma_I^2 g_e^2 \mu_B^2}{r^6} \langle S_z \rangle^2 \left(4\tau_r + \frac{3\tau_r}{1+\omega_I^2\tau_r^2}\right)$$

$$= \boxed{\frac{1}{5}\left(\frac{\mu_0}{4\pi}\right)^2 \frac{\omega_I^2 g_e^4 \mu_B^4 S^2(S+1)^2}{(3kT)^2 r^6} \left(4\tau_r + \frac{3\tau_r}{1+\omega_I^2\tau_r^2}\right)}. \quad (3.30)$$

In the case of lanthanides $\langle J_z \rangle^2$ should replace $\langle S_z \rangle^2$. Although the value of $\langle S_z \rangle^2$ is much smaller than the value of $S(S+1)/3$, which appears in the corresponding Solomon equations (Eqs. (3.19) and (3.20)), Curie spin relaxation may be significant when the dipolar coupling described by the Solomon equations is governed by the electronic relaxation times; that is, when the latter are much smaller than the rotational correlation times. Furthermore, since $\langle S_z \rangle^2$ depends on the square of the external magnetic field, Curie spin relaxation increases with the square of the nuclear Larmor frequency ω_I — see Eqs. (3.29) and (3.30) and Fig. 3.13. For example, when $\tau_r = 5 \times 10^{-11}$ s, $\tau_s = 5 \times 10^{-13}$ s and $B_0 = 14$ T, Curie spin relaxation can be sizably larger than the Solomon contribution to both R_1 and R_2. This is the case of aqualanthanide(III) ions except gadolinium(III) [39]. It is also quite common for Curie relaxation to dominate R_2 in macromolecules. For example, when $\tau_r = 10^{-7}$ s, $\tau_c = T_{1e} = 10^{-12}$ s, and $B_0 = 6$ T, Curie spin relaxation for R_{2M} is one order of magnitude larger than the Solomon contribution. Under these conditions Curie spin relaxation does not significantly contribute to R_{1M}; at high magnetic field ($\omega_I^2\tau_r^2 \gg 1$) the dependence on ω_I^2 in the numerator is canceled out. R_{1M} levels off at a value that is often small compared with the dipolar contribution.

The occurrence of this particular relaxation mechanism can be recognized through the field and temperature dependence of the linewidths. In the absence of other effects, the line broadening is proportional to the square of the magnetic field

(Fig. 3.13) and to η/T^3, the latter dependence arising from the $1/T^2$ dependence of $\langle S_z\rangle^2$ (Eqs. (3.29) and (3.30)) and the η/T dependence of τ_r (Eq. (3.8)).

In principle, there may also be a Curie relaxation contribution of the contact-type whenever there is chemical exchange or intramolecular rotation to modulate the coupling. The contribution to R_{2M} would then be

$$R_{2M} = \frac{4}{3}\left(\frac{A}{\hbar}\right)^2 \langle S_z\rangle^2 \tau_M = \frac{4}{3}(\Delta\omega_I)^2\tau_M \tag{3.31}$$

where $\Delta\omega_I$ is the observed contact shift ($\Delta\omega_I = 2\pi\,\Delta\nu_I$).

The relaxation equations discussed here and in Section 3.4 and 3.5 take a different form in the case of lanthanides and actinides. For these systems, in fact, the J quantum number substitutes the S quantum number and g_J substitutes g_e. In the absence of chemical exchange phenomena ($\tau_M^{-1} \ll \tau_s^{-1}$) the equations for the dipolar relaxation thus become:

$$\boxed{R_{1M} = \frac{2}{15}\left(\frac{\mu_0}{4\pi}\right)^2 \frac{\gamma_I^2 g_J^2 \mu_B^2 J(J+1)}{r^6}\left(\frac{7\tau_{s2}}{1+\omega_S^2\tau_{S2}^2} + \frac{3\tau_{s1}}{1+\omega_I^2\tau_{s1}^2}\right)} \tag{3.32}$$

$$\boxed{\begin{aligned} R_{2M} = {} & \frac{1}{15}\left(\frac{\mu_0}{4\pi}\right)^2 \frac{\gamma_I^2 g_J^2 \mu_B^2 J(J+1)}{r^6} \\ & \times\left(4\tau_{s1} + \frac{13\tau_{s2}}{1+\omega_S^2\tau_{s2}^2} + \frac{3\tau_{s1}}{1+\omega_I^2\tau_{s1}^2}\right);\end{aligned}} \tag{3.33}$$

the equations for contact relaxation become:

$$R_{1M} = \frac{2}{3}J(J+1)\left(\frac{A}{\hbar}\right)^2 \frac{\tau_{s2}}{1+\omega_S^2\tau_{s2}^2} \tag{3.34}$$

$$R_{2M} = \frac{1}{3}J(J+1)\left(\frac{A}{\hbar}\right)^2 \left(\tau_{s1} + \frac{\tau_{s2}}{1+\omega_S^2\tau_{s2}^2}\right); \tag{3.35}$$

and the equations for Curie relaxation become:

$$\boxed{R_{1M} = \frac{2}{5}\left(\frac{\mu_0}{4\pi}\right)^2 \frac{\omega_I^2 g_J^4 \mu_B^4 \left(J(J+1)\right)^2}{(3kT)^2 r^6}\frac{3\tau_r}{1+\omega_I^2\tau_r^2}} \tag{3.36}$$

$$\boxed{R_{2M} = \frac{1}{5}\left(\frac{\mu_0}{4\pi}\right)^2 \frac{\omega_I^2 g_J^4 \mu_B^4 \left(J(J+1)\right)^2}{(3kT)^2 r^6}\left(4\tau_r + \frac{3\tau_r}{1+\omega_I^2\tau_r^2}\right).} \tag{3.37}$$

The term $g_J^2\mu_B^2 J(J+1)$ in Eqs. (3.32) and (3.33), and the term $g_J^4\mu_B^4\left(J(J+1)\right)^2$ in Eqs. (3.36) and (3.37) can be more properly replaced by the experimental μ_{eff}^2 and μ_{eff}^4 values when available.

3.7 FURTHER ELECTRONIC EFFECTS ON NUCLEAR RELAXATION

3.7.1 The effect of g anisotropy and of the splitting of the S manifold at zero magnetic field

The electronic term which is the first term in the Hamiltonian written in Eq. (3.13) and used to derive the Solomon and Bloembergen equations (Eqs. (3.16), (3.17), (3.19), (3.20), (3.26), (3.27)) may be inappropriate in many cases, since the electron energy levels may be strongly affected by the presence of ZFS or hyperfine coupling with the metal nucleus. Therefore, the electron static Hamiltonian to be solved to find the ω_S values, i.e. all electron energy transitions, and their probabilities, will be, in general,

$$H_0 = \mu_B \boldsymbol{S}\cdot\boldsymbol{g}\cdot\boldsymbol{B_0} + \boldsymbol{S}\cdot\boldsymbol{D}\cdot\boldsymbol{S} + \boldsymbol{I}\cdot\boldsymbol{A}\cdot\boldsymbol{S}. \tag{3.38}$$

Let us discuss first the case in which only the first term is present. In the Solomon and Bloembergen equations for R_i ($i = 1, 2$) there is the ω_S parameter at the denominator of a Lorentzian function. Up to now ω_S has been taken equal to that of the free electron. However, in the presence of orbital contributions, the Zeeman splitting of the M_S levels changes its value and ω_S equals $|\gamma_S|B_0$ or $(g/\hbar)\mu_B B_0$. When g is anisotropic (see Fig. 1.16), the value of ω_S is different from that of the free electron and is orientation dependent. The principal consequence is that another parameter (at least) is needed, i.e. the θ angle between the metal–nucleus vector and the z direction of the g tensor (see Section 1.4). A second consequence is that the ω_S fluctuations in solution must be taken into account when integrating over all the orientations. Appropriate equations for nuclear relaxation have been derived for both the cases in which rotation is faster [40,41] or slower [42,43] than the electronic relaxation time. In practical cases, the deviations from the Solomon profile are within 10–20% (see for example Fig. 3.14).

When there is splitting of the S manifold at zero magnetic field (second term in Eq. (3.38)), we should distinguish between half integer and integer S values. In the former case we always have an $M_S = -\frac{1}{2} \rightarrow \frac{1}{2}$ transition with energy of the order of $g\mu_B B_0$. The $M_S = \frac{1}{2} \rightarrow \frac{3}{2}$ transition, for example, may contain the term D (see for example Fig. 1.17). If D is much greater than the Zeeman energy, as is often the case, $|\omega_S|$ is much larger than in the case of the $-\frac{1}{2} \rightarrow \frac{1}{2}$ transition. Then the condition $|\omega_S|\tau_c \gg 1$ holds and the Lorentzian function has a negligible value. Such electronic transitions do not contribute to nuclear relaxation. It should also be mentioned that this holds when $D \gg \hbar\tau_c^{-1}$. Neglecting the ZFS may

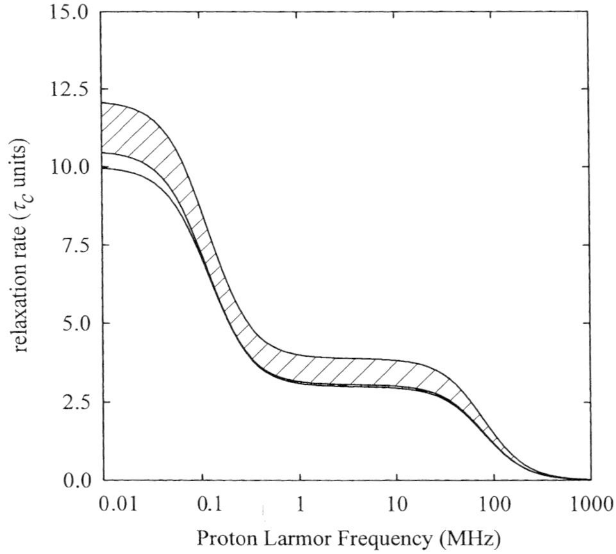

Fig. 3.14. Plot of the spectral density functions for dipolar relaxation in the presence of an axially symmetric $\boldsymbol{g}$ tensor. Conditions: $g_{\parallel} = 2.3$, $g_{\perp} = 2.0$, $\tau_c = 2 \times 10^{-9}$ s, $\theta = 0°$ (upper curve) and $\theta = 90°$ (middle curve) compared with the Solomon behavior (lower curve).

introduce an error of about a factor two in R_1 (Fig. 3.15C). In the case of integer S, all transitions contain D and it may happen that the term in ω_S never contributes to nuclear relaxation (Fig. 3.15A,B).

In any case, the presence of ZFS may cause the occurrence of a further dispersion in the plot of relaxation rate as a function of proton Larmor frequency, corresponding to the transition from the dominant ZFS limit to the dominant Zeeman limit[6] (Fig. 3.15).

Another mechanism to provide splitting of the S manifold is the hyperfine coupling between the unpaired electron and the metal nucleus. For example, at zero magnetic field an $S = 1/2$ $I = 1/2$ system gives two sets of levels of degeneracy 3 and 1, separated by A (see Appendix III) where A is the metal–nucleus-unpaired-electron hyperfine coupling. The effect of this splitting is

---

[6] The electron Hamiltonian is in fact given by

$$H = g_e \mu_B \boldsymbol{S} \cdot \boldsymbol{B_0} + \boldsymbol{S} \cdot \boldsymbol{D} \cdot \boldsymbol{S}$$

where $\boldsymbol{D}$ is the traceless ZFS tensor. The axial and rhombic components of ZFS, D and E (see Chapter 1), are defined as

$$D = D_z - \frac{D_x + D_y}{2} \quad \text{and} \quad E = \frac{D_x - D_y}{2}.$$

In the low field limit the ZFS term (the second in the above equation) is dominant, while in the high field limit the Zeeman term (the first in the above equation) is dominant.

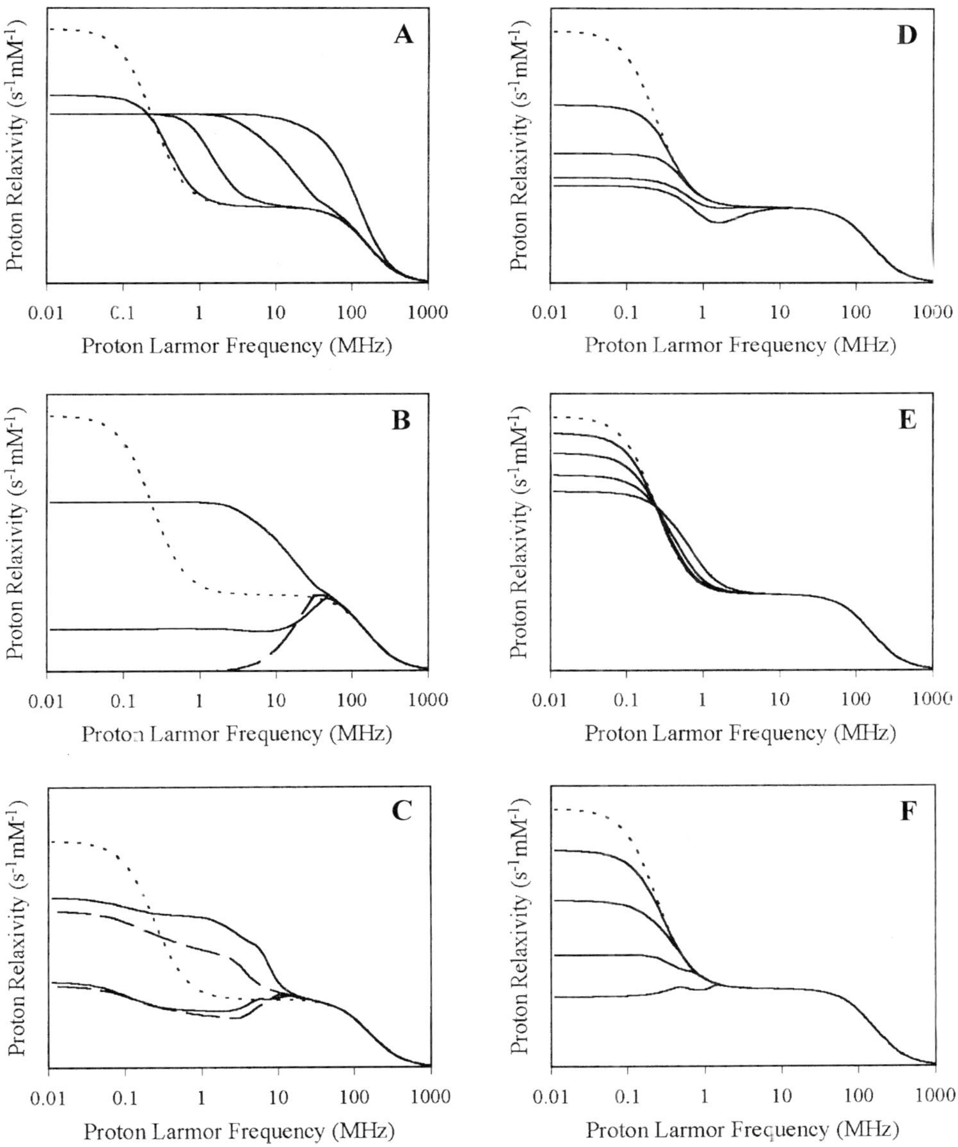

Fig. 3.15. Plot of the nuclear longitudinal dipolar relaxation rate in the presence of ZFS, for (A) $S = 1$, $\theta = 0$, $D = 0.01$, 0.1, 1, 10 cm^{-1}, $E = 0$ (axial case, solid lines); (B) $S = 1$, $D + E = 1$ cm^{-1}, $E = 0$, $\theta = 0$ (upper solid line), $\theta = 90°$ (lower solid line) and $E/D = 1/3$, $\theta = 0$ and 90° (dashed line); (C) $S = 5/2$, $D + E = 0.1$ cm^{-1}, $E = 0$, $\theta = 0$ (upper solid line), $\theta = 90°$ (lower solid line) and $E/D = 1/3$, $\theta = 0$ and 90° (dashed lines); (D) $I = 3/2$, $\theta = 0$, $A_x = A_y = A_z = 200$, 100, 50, 25 $\times 10^{-4}$ cm^{-1} (solid lines in order of increasing low field relaxivity); (E) $I = 3/2$, $\theta = 0$, $A_z = 200$, 100, 50, 25 $\times 10^{-4}$ cm^{-1}, $A_x = A_y = 0$; (F) $I = 3/2$, $\theta = 90°$, $A_z = 200$, 100, 50, 25 $\times 10^{-4}$ cm^{-1}, $A_x = A_y = 0$. The dotted line shows the Solomon profile. τ_c is 10^{-9} s in all cases.

sizable, however, when $A > \hbar\tau_c^{-1}$ and at low magnetic fields where $A > g\mu_B B_0$. In high resolution NMR, the latter condition is practically never met.

A further effect of D and A when they influence nuclear relaxation is that again an angle (if axial) or two angles (if rhombic) are needed which takes into account the location of the resonating nucleus within the D (A) tensor frame. The theoretical approach to the description of R_1 under these circumstances gives approximate analytical solutions [44]. Numerical solutions [45,46], and a computer program (http://www.cerm.unifi.it) [47] are also available.

3.7.2 Field dependence of electron relaxation

In case the electron relaxation is influencing the correlation time for nuclear relaxation, according to Eqs. (3.6) and (3.7), the field dependence of the proton relaxation rate also reflects the field dependence of τ_s, if present. In general, the latter is described by Eqs. (3.11) and (3.12). It can be seen that τ_s is constant until the $\omega_s\tau_v$ dispersion occurs. Since τ_s must be larger than τ_v to stay within the range of validity of all the mentioned equations (Redfield limit, see Section 3.14), as τ_v is the correlation time for electron relaxation, it results that the electron relaxation time starts to increase after the occurrence of the $\omega_s\tau_s$ dispersion. As far as τ_s increases, R_{1M} increases until the $\omega_I\tau_s$ dispersion occurs (see Eq. (3.19)). Therefore, the overall shape of the profile obtained by plotting the relaxation rate as a function of proton Larmor frequency results in a peak in the high field region, between the $\omega_s\tau_s$ dispersion and the $\omega_I\tau_s$ dispersion. Fig. 3.16A and B show that with increasing the value of τ_v, both the value of the maximum of the peak increases and the peak moves toward lower frequencies. The increase in relaxation rate at high fields may be even larger for R_{2M} in case of the presence of contact relaxation (Fig. 3.16C). As shown by Eq. (3.27), R_{2M} in fact increases with the electron relaxation time, that, in presence of a field dependence as described by Eq. (3.11) goes to infinity. The transverse relaxation rate at high fields would thus be limited by exchange processes only, if present, that contribute to τ_c through the exchange time τ_M (Eq. (3.7)). Exchange phenomena are described in more detail in Chapter 4. Some examples where these behaviors have been observed will be reported in Chapter 5.

3.8 A COMPARISON OF DIPOLAR, CONTACT, AND CURIE NUCLEAR SPIN RELAXATION

Once R_{1M} and R_{2M} have been measured, it is useful to try to understand the relative weight of dipolar, contact, and Curie spin contributions to the overall relaxation effect. Indeed, each of the three contributions is independently capable of providing valuable information whereas the whole value may not. The dipolar and Curie relaxation mechanisms — Eqs. (3.16), (3.17), (3.29) and (3.30) —

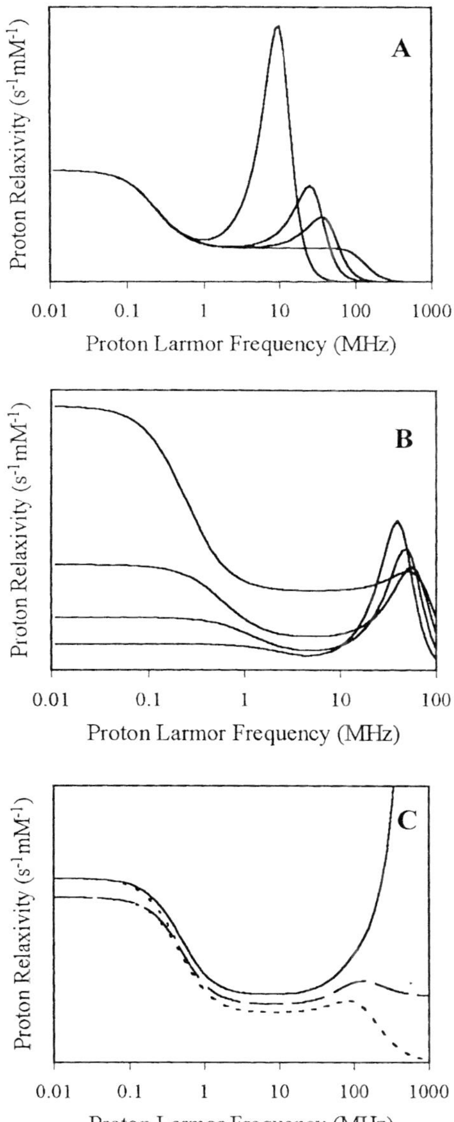

Fig. 3.16. (A) Plot of the nuclear longitudinal dipolar relaxation rate in the presence of a field dependent electron relaxation time ($S = 5/2$) for $\tau_{s0} = 10^{-9}$ s and $\tau_v = 1, 5, 10, 50 \times 10^{-12}$ s. The amplitude of the peak that appears in the high field part of the profile increases with increasing τ_v. (B) Plot of the longitudinal dipolar relaxation rate in the presence of a field dependent electron relaxation time ($S = 5/2$) for $\Delta_t = 0.047$ cm^{-1} and $\tau_v = 2, 5, 10, 20 \times 10^{-12}$ s. (C) Plot of the longitudinal relaxation rate in the presence of a field dependent electron relaxation time with $A/h = 0$ and 1 MHz (dotted lines) and of the transverse relaxation rate with $A/h = 0$ (dashed line) and 1 MHz (solid line). Conditions: $S = 5/2$, $\Delta_t = 0.047$ cm^{-1}, $\tau_v = 2 \times 10^{-12}$ s ($\tau_{s0} = 10^{-9}$ s) and $\tau_r = 10^{-9}$ s.

provide experimental values that are related to the distance of the nucleus from the paramagnetic metal and τ_c or τ_r, assuming that ligand-centered effects can be neglected. Measurements at various magnetic fields lead, in principle, to an estimation of these parameters. The field dependence of the dipolar contribution can be observed in any region of the accessible magnetic field range, that is from 0 to 800 MHz, whereas Curie relaxation can usually be observed only in the region above 100 MHz. Even at a single magnetic field value, measurements of dipolar R_{1M} contributions on various nuclei of the same moiety give information on their relative distance from the metal, as long as τ_c is the same. This is also true for R_{2M} values when they are determined by dipolar and/or Curie contributions.

The presence of contact relaxation indicates that a given moiety is covalently bound to a paramagnetic metal ion and provides an estimate of the absolute value of A (Eqs. (3.26) and (3.27)). Sometimes the contact coupling constant can be evaluated by chemical shift measurements, and it is therefore possible to predict whether the contact relaxation contributions to R_{1M}, R_{2M}, or both, are negligible or sizable.

Unlike hyperfine isotropic shifts, which often contain pseudocontact and contact contributions of the same order of magnitude, relaxation rates can often be recognized to be dominated by only one of the possible contributions. In addition, whereas contact and pseudocontact shifts may happen to have different signs, thereby making their separation more uncertain, relaxation contributions are obviously always positive and additive.

Some qualitative guidelines can be given to make an a priori estimate of the relative weight of dipolar, contact, and Curie relaxation contributions. Consider first the fast motion limit where $R_{1M} = R_{2M}$ and none of the frequency-dependent terms is dispersed. The equations take the simple form already noted:

$$R_{1M} = R_{2M} = \frac{4}{3}\left(\frac{\mu_0}{4\pi}\right)^2 \frac{\gamma_I^2 g_e^2 \mu_B^2 S(S+1)}{r^6}\tau_c^{\text{dip}} + \frac{2}{3}S(S+1)\left(\frac{A}{\hbar}\right)^2 \tau_c^{\text{con}}. \quad (3.39)$$

Under these conditions the Curie spin contribution is always negligible (see Section 3.6). If $\tau_c^{\text{dip}} = \tau_c^{\text{con}}$ it is only necessary to compare the following two expressions:

$$\frac{4}{3}\frac{\gamma_I^2}{r^6}8.6 \times 10^{-61} \quad \text{and} \quad \frac{2}{3}\left(\frac{A}{\hbar}\right)^2$$

where 8.6×10^{-61} is $(\mu_0/4\pi)^2 g_e^2 \mu_B^2$ (T^2 m^6). For hydrogen nuclei, γ_I^2 is 7.16×10^{16} rad^2 s^{-2} T^{-2}, whereas $(A/\hbar)^2$ does not usually exceed 10^{12} rad^2 s^{-2}, and is often found to be much lower. Thus, for metal–nucleus distances of 500 pm or smaller, the dipolar term is largely dominant. The value of the dipolar term drops dramatically with increasing r owing to its r^{-6} dependence, and for larger distances it could become smaller than the contact term. However, the $(A/\hbar)^2$ value is also qualitatively expected to decrease with increasing distance from the paramagnetic center. Therefore, unless some efficient unpaired spin delocalization pathway is operative, the dipolar term in proton relaxation is usually dominant for most systems of chemical interest.

The situation is different when the rotational correlation time is shorter than τ_s and therefore dominates the overall correlation time τ_c in the dipolar term. In this case, the relative importance of the dipolar term is decreased by a factor τ_s/τ_r, which can be as large as 10^2 to 10^3 for small complexes with rotational correlation times of 10^{-10} to 10^{-11} s and electronic relaxation times of 10^{-8} to 10^{-9} s – for example, Cu^{2+}, Mn^{2+}, and VO^{2+}. In macromolecular complexes, rotational correlation times are much larger, and situations of this type do not occur.

Outside the fast motion limit the relative weight of contact and dipolar interactions on R_{1M} and R_{2M} may also be different. The following considerations are particularly relevant to proton relaxation. Curie contributions to R_{2M} can be sizable. By comparing Eqs. (3.20) and (3.27) on the one hand and Eq. (3.30) on the other, it can be noted that they contain terms of the type

$$S(S+1)\tau_s \quad \text{and} \quad \omega_I^2 S^2 (S+1)^2 \tau_r$$

respectively. Therefore, it is expected that Curie contributions will be comparatively higher the higher the field, the higher S, and the higher the τ_r/τ_s ratio.

Once the Curie contribution to R_{2M} is estimated and subtracted, the contribution of contact and dipolar interactions can be estimated by examining the correlation time dependence of the paramagnetic relaxation depicted in Figs. 3.9 and 3.11. It appears that the maximum for R_{1M} occurs at $|\omega_I|\tau_c^{\text{dip}} \approx 1$ in the dipolar term and at $|\omega_S|\tau_c^{\text{con}} \approx 1$ in the contact term. Taking for simplicity $\tau_c^{\text{dip}} = \tau_c^{\text{con}}$, this means that in the intermediate situation where $|\omega_S|\tau_c^{\text{dip}} > 1 > |\omega_I|\tau_c^{\text{con}}$ the relative importance of the contact term is even smaller than that estimated in the fast motion limit. The equation for R_{2M} has non-dispersive terms in both the dipolar and contact contributions (accounting for one-fifth and one-half of the total effect measured in the fast motion limit respectively), and therefore the conclusions drawn in the fast motion limit are still qualitatively correct.

A comparison of R_{1M} and R_{2M} values may thus be useful to evaluate the occurrence of relaxation by contact interactions. Taking again $\tau_c^{\text{dip}} = \tau_c^{\text{con}} = \tau_c$, and keeping in mind the τ_c (or field) dependence of R_{1M} and R_{2M} as given by Eqs. (3.26) and (3.27), the R_{1M}/R_{2M} ratios are expected to be as reported in Table 3.3. When the estimate of τ_c is such that the intermediate situation ($|\omega_S|\tau_c > 1 > |\omega_I|\tau_c$) occurs, dominant dipolar relaxation will still give R_{1M}/R_{2M} ratios close to unity, whereas dominant contact relaxation will give $R_{2M} \gg R_{1M}$. In the latter case, no information is obtained concerning the mechanism controlling R_{1M}; however, an idea of what happens to R_{1M} can be perceived by the following procedure:

(1) use R_{2M} and a reasonable estimate for τ_c^{con} to calculate $(A/\hbar)^2$, as if R_{2M} were completely determined by contact relaxation;

TABLE 3.3
Ratios between R_{1M} and R_{2M} for dipolar and contact contributions in the various motional regimes

	$\omega_s\tau_c < 1$	$\omega_s\tau_c > 1 > \omega_I\tau_c$	$\omega_I\tau_c > 1$
R_{1M}/R_{2M} (dipolar)	1	6/7	$\ll 1$
R_{1M}/R_{2M} (contact)	1	$\ll 1$	$\ll 1$

(2) use the values of τ_c^{con} and $(A/\hbar)^2$ in the contact equation for R_{1M} to estimate the upper limit of the contact contribution of the longitudinal relaxation;
(3) compare this value with the experimental value of R_{1M} to check whether the contact contribution is negligible.

When electron relaxation has a field dependence as described by Eq. (3.11), contact relaxation can determine very large values of R_{2M}, as shown by Eq. (3.27), while the contribution from dipolar relaxation can be negligible if $\tau_c^{\text{dip}} = \tau_r$.

3.9 NUCLEAR PARAMETERS AND RELAXATION

For nuclei other than protons, the magnetic moment is smaller (with the exception of ^{3}H). The nuclei with $I > ½$ have a nuclear electric quadrupole moment which is an efficient source of nuclear relaxation by itself. Neglecting this contribution to relaxation, we may say that any dipolar coupling with the unpaired electron is smaller than in the case of the proton because the magnetic moment is smaller. Since the latter is proportional to γ_I and relaxation depends on the square of the coupling energy, R_i ($i = 1, 2$) depends on γ_I^2. As a consequence, the linewidth increase for, for example, ^{15}N nuclei due to dipolar interaction with unpaired electrons is 1/100 that of ^{1}H when the two nuclei are at the same distance from the paramagnetic center. On the one hand, it is a pity that we soon lose such a wealth of information on the nucleus–electron coupling! On the other hand, this property can be exploited in the case of metal ions with slow electron relaxation, that cause broad ^{1}H lines. However, ^{2}H lines would be much sharper. ^{2}H NMR constitutes a common practice, for example, in small copper(II)-containing ligands. Of course, when the compounds are large in molecular size, quadrupolar relaxation becomes dominant and ^{2}H spectroscopy may not be convenient.

As far as contact contributions are concerned, the nuclear γ_I parameter is contained in A (Eq. (2.2)) and therefore the same γ_I^2 dependence as in dipolar relaxation is introduced in contact relaxation. Again, heteronuclei are expected to be less relaxed owing to their smaller γ_I. However, heteronuclei can be directly coordinated to the paramagnetic metal ion. In this case the spin density on the nucleus can be very large and thus A can be very large compared with the proton case. Values of $(A/\hbar)^2$ as large as 10^{17} rad^2 s^{-2} can be obtained for directly coordinated nuclei like ^{17}O, ^{19}F, ^{14}N, and ^{15}N. In such cases, contact interaction may easily be the dominant mechanism for nuclear relaxation, especially for R_{2M}. In the case of imidazole complexes, even the non-coordinated nitrogen is quite broad. In this case, contact and/or ligand-centered effects provide efficient nuclear relaxation even for a nucleus at 4–4.5 Å from the metal [48].

3.10 THE EFFECT OF TEMPERATURE ON THE ELECTRON–NUCLEUS SPIN INTERACTION

No explicit temperature dependence is included in the equations for R_{1M} and R_{2M}, except for cases where Curie spin relaxation is the dominant term (Section 3.6). In the latter case, Curie paramagnetism has a T^{-1} dependence and therefore relaxation depends on T^{-2}. The effect of temperature on linewidths determined by Curie relaxation is dramatic also because of the τ_r dependence on temperature, as shown in Eq. (3.8). All the correlation times modulating the electron–nucleus coupling, either contact or dipolar, are generally temperature dependent, although in different ways, and their variation will therefore be reflected in the values of R_{1M} and R_{2M}.

In the limits of validity for the Solomon and Bloembergen equations, the correlation times for contact and dipolar relaxation are given by Eqs. (3.6) and (3.7) respectively. The exchange time τ_M is rarely short enough to dominate in Eqs. (3.6) and (3.7); however, when this is the case, a strong temperature dependence of the nuclear relaxation times is expected, because the variation of the exchange rate with temperature is generally exponential [49]. When the rotational correlation time is the dominant term, the effect of temperature would be anticipated to be less, since, as already seen, τ_r may be approximated by the Stokes–Einstein Eq. (3.8). The effect is enhanced, however, by the change in viscosity η of the solvent; in the case of water, η decreases as much as 2.7 times from 0 to 40°C [7]. Qualitatively, both τ_r and τ_M decrease with increasing temperature; this causes a decrease in the nuclear relaxation rates in the non-dispersive regions and an increase in the second half of the ω_I dispersive region (Fig. 3.17).

The temperature dependence of τ_s, including its sign, may be a matter of extensive speculation, mainly because the origin of the electron–spin relaxation mechanism in each particular case may be different, and in general is not known in detail. When equations of the type such as (3.11) and (3.12) hold, and by assuming that the correlation time for electron relaxation τ_v has an exponential temperature dependence, τ_s would be expected to increase with increasing temperature in the non-dispersive region ($|\omega_S|\tau_v \ll 1$) and to decrease in the dispersive region (high magnetic field). In fact, from Eqs. (3.11) and (3.12), τ_s^{-1} is proportional to

[7] The transport properties are also well modeled by exponential laws, but the energy is generally smaller than for chemical exchange.

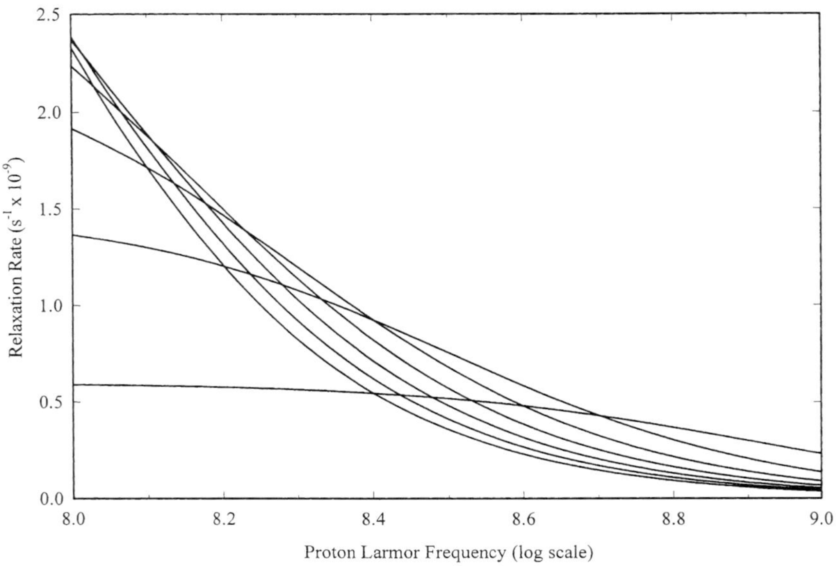

Fig. 3.17. Plot of $f_1(\omega, \tau_c)$ of Eq. (3.16) as a function of the proton Larmor frequency in the high frequency region. The curves are calculated for τ_c values decreasing from 2×10^{-9} s (upper curve on the left) in steps of 3×10^{-10} s.

$\tau_v/(1 + \omega_s^2\tau_v^2)$, i.e. proportional to τ_v as long as $|\omega_s|\tau_v \ll 1$, and to τ_v^{-1} when $|\omega_s|\tau_v \gg 1$. However, this picture may be further complicated by a possible temperature dependence of the Δ_t^2 term in Eqs. (3.11) and (3.12), making it difficult to predict the overall behavior of τ_s. In addition, different electronic relaxation mechanisms may be operative at different temperatures, although the possibility that the switching from one mechanism to another occurs in a few tens of degrees around room temperature is rather unlikely.

In the coupled metal systems that will be discussed in Chapter 6, the effect of temperature is generally more complicated. Besides the effects described above, the overall temperature dependence will obviously also depend on the sign and magnitude of the exchange coupling constant J. The reader is referred to Chapter 6 for more details.

The effects of chemical exchange on the NMR parameters, and hence the influence of temperature through modulation of the exchange process, will be considered in Section 4.3.

3.11 STABLE FREE RADICALS

Free radicals have electronic relaxation times of about 10^{-7} s, which are the longest among paramagnetic compounds [4,5]. The correlation time for the NMR experiment when the radical is free in solution is the rotational correlation time.

Small molecules are already at the limit of detectability as far as proton NMR is concerned, whereas they are easily detected through ^2H NMR. If other factors concur favorably (interactions between radicals at high concentration, tendency to aggregate, availability of low-lying excited states), the proton NMR signals can also be detected for radicals. Nevertheless, it is, in general, a hard task to perform high resolution studies on radicals.

When the radical interacts rigidly with a macromolecule its rotational correlation time increases. For very large macromolecules, the correlation time for the NMR experiment eventually becomes the electronic relaxation time. If the bound radical still has motional freedom, this motion may be faster than the rotational time of the whole molecule and thus determine the correlation time. Of course, if the system is in chemical exchange, the exchange time may also be shorter than the electronic relaxation time and so represents the limiting NMR correlation time. An immobilized radical produces a very large broadening effect on the NMR line of nearby nuclei, which is probably the largest effect that a paramagnetic species can cause.

3.12 NUCLEAR RELAXATION PARAMETERS AND STRUCTURAL INFORMATION

When a molecule is interacting with a metal ion without direct covalent bond there are no contact contributions nor ligand-centered effects to relaxation. This simplified case may occur in metalloproteins when a molecule occupies a protein pocket nearby the metal ion. The nuclear R_{1M} and R_{2M} can provide distances of the nuclei from the metal ion if the correlation time is known. However, in general this is an unknown. Recently a protocol has been suggested to use nuclear relaxation rates as structural constraints for the determination of solution structures of paramagnetic metalloproteins [50]. Measurements at variable magnetic fields could be of help if we are lucky enough to be in one of the dispersion regions. Alternatively, substitution of protons with deuterons provides additional experimental data which permits the determination of the correlation time. If the pseudocontact shifts can be determined, we have a further hint to map the nuclei within the molecular frame (see Section 2.9). Finally, NOE or NOESY experiments (see Chapter 8) may provide further constraints to locate protons close to one another. In the early times of NMR, mapping procedures have been applied to small molecules binding to the metal in metalloproteins. In this case the results can be dramatically hampered by ligand-centered effects. These tend to make the experimental R_{1M} values close to one another, and they may then completely mislocate the molecule.

Several attempts to obtain structural information on molecules in solution have been made by using shift reagents (see Sections 2.9 and 5.7.4). In flexible molecules, many conformers may exist, the site of binding at the metal may be unknown, and the donor atoms may be more than one. As a result, the structural information should be analyzed with caution.

A safer procedure has been designed by using lanthanides at the calcium binding sites of proteins, that have essentially rigid structures. Lanthanides provide pseudocontact shifts which contain structural information. However, it is generally difficult to assign the signals without additional hints, for example from bidimensional spectroscopies. By using gadolinium, further information on the metal–proton distances can be obtained from nuclear relaxation.

Some attempts to obtain structural information by using spin labels as relaxation agents are possible. Again, it is difficult to proceed with the assignment, but if this is somehow obtained (see also Chapter 8), then R_{1M} and R_{2M} measurements provide distances, as the nuclear relaxation mechanism is dipolar in origin. Owing to the long correlation times, the effects on transverse relaxation are detectable on nuclei that are as far as 1.5 and 2 nm away from the paramagnetic probe. However, nuclei that are much closer may not be detectable because of the excessive line broadening. Therefore, radicals are best used as long-distance relaxation probes. This property has recently been used to obtain structural information on a radical-Pt-DNA moiety [51]. As another example, long-range intermolecular distance information in protein–RNA and protein–DNA complexes has been extracted from intermolecular electron–proton dipolar relaxation induced by nitroxide spin-labeled RNA [52].

3.13 EXPERIMENTAL ACCESSIBILITY OF NUCLEAR RELAXATION PARAMETERS

As anticipated in Section 3.2, a nuclear longitudinal relaxation rate R_1 can be defined only when relaxation is an exponential process. This is at variance with nuclear transverse relaxation, which is always exponential and always defined by the transverse relaxation rate R_2. As far as longitudinal relaxation is concerned, when the return to the equilibrium value $M_z(\infty)$ of longitudinal nuclear magnetization after a 180° pulse is exponential, we can write (see Section 1.7.4)

$$M_z(t) = M_z(\infty) - 2M_z(\infty)\exp(-R_1 t). \tag{3.40}$$

If we consider that relaxation is further enhanced by interaction with a paramagnetic center, Eq. (3.40) becomes

$$\begin{aligned} M_z(t) &= M_z(\infty) - 2M_z(\infty)\exp(-R_1 t)\exp(-R_{1M} t) \\ &= M_z(\infty) - 2M_z(\infty)\exp[-(R_1 + R_{1M})t]. \end{aligned} \tag{3.41}$$

The meaning of Eq. (3.41) is that any further decay function is multiplied by the original decay function if they are independent (see also Section 3.2). If both are exponential, then the whole process remains exponential and the rate constant is $R_{1tot} = R_1 + R_{1M}$. As already mentioned in Chapter 1, and as will be further explained in Chapter 7, longitudinal nuclear relaxation is often a non-exponential process. In our definition of R_1 (Section 1.7.4), it was assumed that all nuclei were independent of one another. In diamagnetic systems nuclei are often coupled to one another, influencing each other in such a way that one cannot be considered a lattice with infinite heat capacity for the other, and R_1 cannot be defined. Paramagnetic relaxation, however, is an exponential process within the broad range of validity of the equations given in this chapter. This is due to the fact that electrons have a much larger magnetic moment and relax so much faster than nuclei that they behave effectively as a lattice with infinite heat capacity. So, Eq. (3.41) takes the more general form

$$M_z(t) = M_z(\infty) - 2M_z(\infty) f(t) \exp(-R_{1M} t) \tag{3.42}$$

where $f(t)$ is a generic non-exponential decay accounting for diamagnetic interactions. In such a case, even if paramagnetic relaxation is exponential, R_{1M} cannot be easily extracted from the analysis of the overall decay. If paramagnetic relaxation is dominant, exponentiality is effectively imposed to the overall nuclear relaxation. If not, the detailed dependence of longitudinal nuclear magnetization as a function of time must be analyzed [53].

Experimental techniques are available to measure magnetization recovery under different conditions. These are, for instance, the selective and non-selective variants of the inversion recovery experiment described in Section 1.7.4, and will be discussed in more detail in Chapter 9. We anticipate here, from Chapters 7 and 9, in a qualitative way, the kind of information contained in these experiments.

In a non-selective inversion recovery experiment, the 180° pulse inverts the magnetization of the nuclear spin under consideration as well as that of all other nuclear spins coupled to it. All spins will return to equilibrium and, simultaneously, will influence each other by, for example, dipolar coupling. In a selective experiment, only the nuclear spin under consideration is inverted by the pulse. During its return to equilibrium, it will also be influenced by dipolar coupling with the other, initially unperturbed spins. In both cases, diamagnetic relaxation mechanisms will superimpose to the exponential recovery due to the coupling with the paramagnetic center, and deviations from exponentiality will occur. These deviations will be, however, less severe for a non-selective experiment. Indeed, mutual influence between two nuclear spins I and J occurs through W_0 and W_2 terms (Chapter 7) analogous to the mutual influence between nuclear and electron spins I and S illustrated in Section 3.2 and Fig. 3.8. This influence will be maximal when $2\omega_I \tau_c \gg 1$, i.e. in the slow motion limit, where the W_2 terms become negligible. If we refer to the later conditions, magnetization recovery of the inverted I spin in a selective experiment occurs by partially *decreasing* M_z of the neighbor spin J through magnetization transfer (Fig. 3.18). The presence of the neighbor thus 'helps' the inverted spin to

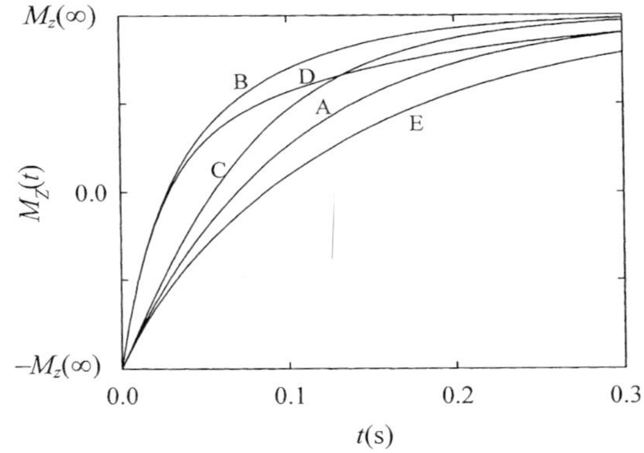

Fig. 3.18. Exponential recovery (A) of $M_z(t)$ of a nuclear spin I dipole coupled to a paramagnetic metal ion. When I is also coupled to another nuclear spin J, the latter also coupled to the metal ion, non-exponentiality occurs. If J relaxes slower than I, curves B and C are obtained for a selective and a non-selective experiment respectively. If J relaxes slower than I, curves D (selective) and E (non-selective) are obtained. If J relaxes at the same rate as I, a selective experiment gives an intermediate behavior between curves B and D (not shown), while a non-selective experiment gives pure exponential recovery (A). It is apparent that in all cases non-selective experiments perform better than selective experiments, as they are less sensitive to the non-exponentiality introduced by I–J coupling. Conditions: $R^I_{1M} = 10$ s^{-1}; $R^J_{1M} = 20$ s^{-1} (B,C), 5 s^{-1} (D,E) and 10 s^{-1} (A). The I–J cross-relaxation rate σ_{IJ} (Chapter 7) is -20 s^{-1}.

relax faster (curves B and D). This transfer of magnetization is maximal at $t = 0$, and becomes less effective with time (hence non-exponentiality arises). On the contrary, in a non-selective experiment (curves C and E), the transfer of magnetization will be zero at $t = 0$, because both spins I and J are inverted. With time, transfer occurs from the slower relaxing to the faster relaxing spin: the slower will then relax slightly faster and the faster will relax slightly slower (again non-exponentiality occurs). If the two spins accidentally relax with the same rate, no net magnetization transfer occurs and recovery is again exponential. In any case, it is apparent that *R_{1M} values can always be better evaluated from non-selective rather than from selective experiments* because of this partial compensation effect, and particularly from the initial points of the decay when magnetization transfer is negligible. The stronger the I–J coupling, the worse the selective experiments perform with respect to non-selective ones. Although the underlying theory has been well known for many years, this simple rule has never been plainly formulated, even by specialists in the field. See also Section 7.2.2.

3.14 REDFIELD LIMIT AND BEYOND

The relaxation equation derived so far for electrons and nuclei share a common assumption usually called the perturbation regime or Redfield limit [54]. The assumption allows the use of the linear response theory; it states that the energy of the coupling between the spin and the lattice, $E_{\text{int}}/\hbar$ (in frequency units), whose modulation is responsible for the spin relaxation, must be smaller than the inverse of the correlation time, τ_c, for the modulation of the coupling itself, $E_{\text{int}}/\hbar \ll \tau_c^{-1}$. The relaxation equations that hold in the Redfield limit have the general form (in the low field or extreme narrowing limit)

$$T_1^{-1} = \left(\frac{E_{\text{int}}}{\hbar}\right)^2 \tau_c \tag{3.43}$$

(see Eq. (3.14)) in addition to a proportionality coefficient which is of the order of a few units. As $E_{\text{int}}/\hbar \ll \tau_c^{-1}$ must hold, we can substitute τ_c^{-1} for $E_{\text{int}}/\hbar$ in Eq. (3.43) and obtain

$$T_1^{-1} \ll \tau_c^{-1} \tag{3.44}$$

i.e. the T_1 arising from the interaction can never be as short as the correlation time with which the interaction is modulated. This is another way of expressing the Redfield limit. Furthermore, from Eq. (3.43) we obtain

$$T_1^{-1} \ll \frac{E_{\text{int}}}{\hbar} \tag{3.45}$$

which is another way of defining the Redfield limit, i.e. that the energy of the spin–lattice coupling (in frequency units) is always larger than the relaxation rate enhancement induced on the spin by the coupling.

The fact that equations are not available when this limit does not hold (strong coupling regime) should not automatically mean that a real case cannot exist in which the interaction energy is larger than τ_c^{-1}. It only means that the consequences of this fact according to the standard relaxation equations (i.e. that the correlation time is longer than T_1) do not make physical sense, because it is not conceivable that a modulation can induce a change in a physical property (the orientation of a spin) at a rate higher than the modulation rate itself. Therefore, what happens must be that, if we imagine that in a system initially within the Redfield limit we can increase the spin–lattice interaction energy continuously, we should have an increase in T_1^{-1} that initially is proportional to the square of $E_{\text{int}}/\hbar$, and then becomes slower and slower as T_1^{-1} approaches $E_{\text{int}}/\hbar$ (and simultaneously τ_c^{-1}). The limiting value should probably be of the same order as $E_{\text{int}}/\hbar$ and τ_c^{-1}.

In practice nuclear spin–lattice relaxation is always within the Redfield limit, i.e. the interaction energy with the lattice is always much smaller than τ_c^{-1}. This is true even with paramagnetic systems, where the nuclear spin–lattice interaction energy is often much larger than usual. On the other hand, it is not obvious that electrons are always in the Redfield limit. When electrons are outside the Redfield limit, although nuclear relaxation is in the Redfield limit, it is not easy to define a correlation time because the electron relaxation time may not be known and might even be undefined. Outside the Redfield limit, in fact, electron relaxation may not be described by an exponential and therefore a relaxation time may not be defined. This problem was theoretically solved by treating the electron spin and the lattice to which it is so strongly coupled as a whole, using the Liouville superoperator formalism [55–57]. The nuclear spin is coupled to this *generalized lattice* containing both the electron and its lattice. No electron relaxation time is defined, or any nucleus–lattice correlation time. The only motion of the generalized lattice considered is rotation and it is described by a rotational diffusion equation. This approach has been applied to Ni(II) systems, where the electron relaxation, outside the Redfield limit, is mainly caused by transient ZFS of the $S = 1$ manifold (see Section 3.3.1).

3.15 REFERENCES

[1] G. Stokes (1956) Trans. Cambridge Philos. Soc 9, 5.
[2] A. Einstein (1956) Investigations on the Theory of the Brownian Movement. Dover, New York.
[3] P. Debye (1929) Polar Molecules. Dover, New York.
[4] I. Bertini, C. Luchinat, G. Martini (1994) In: Poole CP (ed) Handbook of electron spin resonance. American Institute of Physics, New York: pp. 51.
[5] I. Bertini, C. Luchinat, G. Martini (1994) In: Poole CP (ed) Handbook of electron spin resonance. American Institute of Physics, New York: pp. 79.
[6] J.H. Van Vleck (1940) Phys. Rev. 57, 426.
[7] R. Orbach (1961) Proc. R. Soc. London, Ser. A264, 458.
[8] D. Kivelson (1960) J. Chem. Phys. 33, 1094.
[9] S.A. Al'tshuler, K.A. Valiev (1959) Soviet Physics JETP 35, 661.
[10] P.W. Atkins, D. Kivelson (1966) J. Chem. Phys. 44, 169.
[11] G.V. Bruno, J.K. Harrington, M.P. Eastman (1977) J. Phys. Chem. 81, 1111.
[12] N. Bloembergen, L.O. Morgan (1961) J. Chem. Phys. 34, 842.
[13] M. Rubinstein, A. Baram, Z. Luz (1971) Mol. Phys. 20, 67.
[14] Yu.A. Serebrennikov, U.E. Steiner (1994) J. Chem. Phys. 100, 7508.
[15] U.E. Steiner, Yu.A. Serebrennikov (1994) J. Chem. Phys. 100, 7503.
[16] I. Bertini, C. Luchinat, R.D. Brown III, S.H. Koenig (1989) J. Am. Chem. Soc. 111, 3532.
[17] Powell DH, Helm L, Merbach AE (1991) Inorg Chem (in press).
[18] J.H. Freed, R.G. Kooser (1968) J. Chem. Phys. 49, 4715.
[19] R.G. Kooser, W.V. Volland, J.H. Freed (1969) J. Chem. Phys. 50, 5243.
[20] S.K. Misra, U. Orhun (1987) Solid State Commun. 63, 867.
[21] V.M. Malhotra, H.A. Buckmaster, J.M. Dixon (1980) J. Phys. C: Solid State Phys. 13, 3921.
[22] G.P. Vishnevskaya, B.M. Kozyrev (1966) J. Struct. Chem. 7, 20.
[23] A. Hudson, J.W.E. Lewis (1970) Trans. Faraday Soc. 66, 1297.
[24] S.H. Koenig (1991) Magn. Reson. Med. 22, 183.
[25] G. Hernandez, M. Tweedle, R.G. Bryant (1990) Inorg. Chem. 29, 5110.
[26] S. Aime, L. Barbero, M. Botta (1991) Magn. Res. Imaging 9, 843.
[27] I. Solomon (1955) Phys. Rev. 99, 559.

[28] A. Abragam (1961) The Principles of Nuclear Magnetism. Oxford University Press, Oxford.
[29] S.H. Koenig (1982) J. Magn. Reson. 47, 441.
[30] D.E. Woessner (1962) J. Chem. Phys. 3, 647.
[31] R.M. Golding, R.O. Pascual, B.R. McGarvey (1982) J. Magn. Reson. 46, 30.
[32] L. Nordenskiöld, L. Laaksonen, J. Kowalewski (1982) J. Am. Chem. Soc. 104, 379.
[33] H.P.W. Gottlieb, M. Barfield, D.M. Doddrell (1977) J. Chem. Phys. 67, 3785.
[34] D.M. Doddrell, P.C. Healy, M.R. Bendall (1978) J. Magn. Reson. 29, 163.
[35] J. Mispelter, M. Momenteau, J.-M. Lhoste (1981) J. Chem. Soc., Dalton Trans. 1729.
[36] N. Bloembergen (1957) J. Chem. Phys. 27, 575.
[37] M. Guéron (1975) J. Magn. Reson. 19, 58.
[38] A.J. Vega, D. Fiat (1976) Mol. Phys. 31, 347.
[39] I. Bertini, F. Capozzi, C. Luchinat, G. Nicastro, Z. Xia (1993) J. Phys. Chem. 101, 198.
[40] H. Sternlicht (1965) J. Chem. Phys. 42, 2250.
[41] K.V. Vasavada, B.D. Nageswara Rao (1989) J. Magn. Reson. 81, 275.
[42] I. Bertini, C. Luchinat, K.V. Vasavada (1990) J. Magn. Reson. 89, 243.
[43] I. Bertini, F. Briganti, C. Luchinat, M. Mancini, G. Spina (1985) J. Magn. Reson. 63, 41.
[44] I. Bertini, C. Luchinat, M. Mancini, G. Spina (1984) J. Magn. Reson. 59, 213.
[45] L. Banci, I. Bertini, F. Briganti, C. Luchinat (1986) J. Magn. Reson. 66, 58.
[46] R.R. Sharp (1993) J. Chem. Phys. 98, 2507.
[47] I. Bertini, O. Galas, C. Luchinat, G. Parigi (1995) J. Magn. Reson. Ser. A 113, 151.
[48] Y. Yamamoto, N. Nanai, Y. Inoue, R. Chujo (1989) J. Chem. Soc. Chem. Commun. 1419.
[49] F. Basolo, R.G. Pearson (1963) Mechanisms of inorganic reactions. A study of metal complexes in solutions, 1 edn. New York.
[50] I. Bertini, C. Luchinat, A. Rosato (1996) Progr. Biophys. Mol. Biol. 66, 43.
[51] S.U. Dunham, S.J. Lippard (1995) J. Am. Chem. Soc. 117, 10702.
[52] A. Ramos, G. Varani (1998) J. Am. Chem. Soc. 120, 10992.
[53] I. Bertini, A. Donaire, C. Luchinat, A. Rosato (1997) Proteins Struct. Funct. Genet. 29, 348.
[54] A.G. Redfield (1957) Res. Dev. 1, 19.
[55] N. Benetis, J. Kowalewski, L. Nordenskiöld, Wennerström, P.-O. Westlund (1983) Mol. Phys. 48, 2.
[56] N. Benetis, J. Kowalewski, L. Nordenskiöld, H. Wennerström, P.-O. Westlund (1983) Mol. Phys. 50, 515.
[57] I. Bertini, J. Kowalewski, C. Luchinat, T. Nilsson, G. Parigi (1999) J. Chem. Phys. 111, 5795.
[58] R.C. Wilson, R.J. Myers (1976) J. Chem. Phys. 64, 2208.
[59] I. Bertini, C. Luchinat, Z. Xia (1992) Inorg. Chem. 31, 3152.
[60] D. Kivelson (1966) J. Chem. Phys. 45, 1324.
[61] D.C. McCain, R.J. Myers (1967) J. Phys. Chem. 71, 192.
[62] R.N. Rogers, G.E. Pake (1960) J. Phys. Chem. 33, 1107.
[63] A.M. Chmelnick, D. Fiat (1972) J. Magn. Reson. 8, 325.
[64] D.R. Eaton (1965) J. Am. Chem. Soc. 87, 3097.
[65] G.L. McPherson, R.C. Koch, G.D. Stucky (1974) J. Chem. Phys. 60, 1424.
[66] A. Bencini, D. Gatteschi (1982) In: Melson G.A., Figgis B.N. (eds) Transition Metal Chemistry. Marcel Dekker, New York, pp. 1.
[67] B.A. Goodman, J.B. Raynor (1970) In: Emeleus H.J., Sharpe A.G. (eds) Advances in Inorganic Chemistry and Radiochemistry. Academic Press, New York.
[68] B.R. McGarvey (1957) J. Phys. Chem. 61, 1232.
[69] G.N. La Mar, F.A. Walker (1973) J. Am. Chem. Soc. 95, 1782.
[70] G.N. La Mar, G.R. Van Hecke (1969) J. Am. Chem. Soc. 91, 3442.

[71] S.H. Koenig, R.D. Brown III, M. Spiller (1987) Magn. Reson. Med. 4, 252.
[72] M. Grant, H.W. Dodgen, J.P. Hunt (1971) Inorg. Chem. 10, 71.
[73] I. Bertini, F. Capozzi, C. Luchinat, Z. Xia (1993) J. Phys. Chem. 97, 1134.
[74] I. Bertini, O. Galas, C. Luchinat, L. Messori, G. Parigi (1995) J. Phys. Chem. 99, 14217.
[75] L.P. Yu, G.N. La Mar, K. Rajarathnam (1990) J. Am. Chem. Soc. 112, 9527.
[76] L. Banci, I. Bertini, S. Marconi, R. Pierattelli, S.G. Sligar (1994) J. Am. Chem. Soc. 116, 4866.
[77] Y. Ducommun, K.E. Newman, A. Merbach (1980) Inorg. Chem. 19, 3696.
[78] Anonymous (1995) Unpublished results from our laboratory.
[79] L. Banci, I. Bertini, S. Marconi, R. Pierattelli (1993) Eur. J. Biochem. 215, 431.
[80] M.T. Werth, D.M. Kurtz, Jr., I. Moura, J. LeGall (1987) J. Am. Chem. Soc. 109, 273.
[81] A.M. Chmelnick, D. Fiat (1967) J. Chem. Phys. 47, 3986.
[82] L. Banci, I. Bertini, C. Luchinat (1985) Inorg. Chim. Acta 100, 173.
[83] C. Chachaty, A. Forchioni, J. Verlet, J.C. Ronfard-Haret (1974) Chem. Phys. Lett. 29, 436.
[84] L.-J. Ming, L. Banci, C. Luchinat, I. Bertini, J.S. Valentine (1988) Inorg. Chem. 27, 4458.
[85] G.N. La Mar (1965) J. Am. Chem. Soc. 87, 3567.
[86] L. Banci, I. Bertini, C. Luchinat (1986) Magn. Reson. Rev. 11, 1.
[87] U.E. Steiner, D. Bürßner (1990) Z. Phys. Chem. N.F. 169, 159.
[88] A. Paulo, A. Domingos, A. Pires di Matos, I. Santos, M.F.N.N. Carvalho, A.J.L. Pombiero (1994) Inorg. Chem. 33, 4729.
[89] S. Aime, M. Botta, G. Ermondi (1992) Inorg. Chem. 31, 4291.
[90] I. Bertini, C. Luchinat, Z. Xia (1992) J. Magn. Reson. 99, 235.
[91] Z. Luz, D. Fiat (1967) J. Chem. Phys. 46, 469.
[92] S.H. Koenig, R.D. Brown III (1985) J. Magn. Reson. 61, 426.
[93] I. Bertini, F. Briganti, C. Luchinat, Z. Xia (1993) J. Magn. Reson. 101, 198.
[94] J.B. Bloch, G. Navon (1980) J. Inorg. Nucl. Chem. 42, 693.
[95] H.L. Friedman, M. Holz, H.G. Hertz (1979) J. Chem. Phys. 70, 3369.
[96] J. Kowalewski, T. Larsson, P.-O. Westlund (1987) J. Magn. Reson. 74, 56.

Chapter 4

Chemical Exchange, Chemical Equilibria and Dynamics

Chemical exchange is capable of affecting the NMR parameters both in terms of shifts and relaxation. The matter is complex and the various detailed treatments may be laborious. Here the exchange of a nucleus between two sites is presented both qualitatively and quantitatively. The exchange time can be measured through saturation transfer experiments. Some thermodynamic parameters are obtained from the NMR parameters by varying the concentration of the species at equilibrium. In this chapter the relaxation due to outer-sphere interactions is introduced as well as the chemical shifts due to bulk paramagnetic effects, which are used to measure the paramagnetism of molecules in solution.

4.1 INTRODUCTION

If a chemical species containing the investigated nucleus undergoes chemical exchange among different chemical environments, the nuclear NMR parameters are affected in several ways. A full treatment of chemical exchange is rather cumbersome, both from the points of view of shift and of relaxation properties, and is beyond the scope of this book. It is worth noting that even in the simple case of exchange between two sites, the general solution implies non-exponentiality of relaxation. Fortunately, there are some limiting cases which are much easier to treat and for which simple equations can be given. These limiting cases also often hold when one of the chemical environments is paramagnetic.

In general, in the presence of chemical exchange, the nuclei of interest will spend a fraction of time residing in each of n different chemical environments. The probability P_i of finding a chosen nucleus in a particular site i, will be equal to the molar fraction f_i of nuclei in that environment:

$$P_i = f_i = \frac{N_i}{\sum_{j=1}^{n} N_j} \qquad (4.1)$$

where N_i is the number of nuclei in the ith site, $\sum_j N_j$ is the total number of nuclei in the sample, and $\sum_i P_i = 1$ by definition.

Here, the only approximation is to neglect the time spent by a nucleus *in moving* from one environment to another. For example, consider a 0.1 mM water solution of a $Ni(CH_3NH_2)_6^{2+}$ complex, in the presence of 0.1 M excess CH_3NH_2, assuming that the complex is fully formed under these conditions. If there is chemical exchange between bound and free methylamine, the methyl protons will spend

$$P_{\text{bound}} = \frac{6 \times 10^{-4}}{10^{-1} + 6 \times 10^{-4}} \approx 6 \times 10^{-3}$$

of the time in the nickel-bound state and

$$P_{\text{free}} = \frac{10^{-1}}{10^{-1} + 6 \times 10^{-4}} \approx 1$$

of the time in the bulk solution.

The NH_2 protons can also exchange with water molecules. The probability of finding such a proton in each of the three environments will be

$$P_{\text{bound}} = \frac{6 \times 10^{-4}}{55.5 + 10^{-1} + 6 \times 10^{-4}} \approx 1.1 \times 10^{-5}$$

$$P_{\text{free}} = \frac{10^{-1}}{55.5 + 10^{-1} + 6 \times 10^{-4}} \approx 1.8 \times 10^{-3}$$

$$P_{\text{H}_2\text{O}} = \frac{55.5}{55.5 + 10^{-1} + 6 \times 10^{-4}} \approx 1$$

where 55.5 is the approximate molarity of water. Therefore, the exchangeable protons will only spend about 1/100,000 (i.e. P_{bound}) of their time in a bound position.

The probabilities, or molar fractions, are equal to the fraction of time a nucleus will spend in a particular environment when observed for a suitably long period of time. They do not reveal *how long* the nucleus will stay, on average, in each particular position; in other words, the average lifetime of the nucleus in each environment cannot be calculated from the molar fractions. As will be seen, the chemical lifetime or its reciprocal, the chemical dissociation rate, are crucial parameters governing the influence of chemical exchange on NMR parameters.

4.2 A PICTORIAL VIEW OF CHEMICAL EXCHANGE

Consider, for simplicity, the exchange of a nucleus with spin I between two sites, A and B:

$$I^{\text{A}} \underset{\tau_{\text{B}}^{-1}}{\overset{\tau_{\text{A}}^{-1}}{\rightleftarrows}} I^{\text{B}}$$

Although the dissociation rate constants τ_A^{-1} and τ_B^{-1} are not known, it is known that, at equilibrium, $\tau_A/\tau_B = P_A/P_B$. We will give a full description of the effect of variation of τ_A (and τ_B) on the nuclear NMR parameters and describe the relevant equations involved [1,2].

The nuclei in the A site will resonate at a Larmor precession frequency ω_A, and those in the B site will resonate at a frequency ω_B. Under very slow chemical exchange conditions, two signals will be observed, centered at frequency ω_A and ω_B, with intensities proportional to P_A and P_B respectively. When $P_A = P_B$ the two signals will be of equal intensity and $\tau_A^{-1} = \tau_B^{-1} = \tau_M^{-1}$ (Fig. 4.1A(1)). If $R_{2A} = R_{2B}$ the two signals have the same linewidth, given by $R_{2A}/\pi = R_{2B}/\pi$. We will also assume that $R_{1A} = R_{1B}$. When the exchange rate increases, i.e. the lifetime in each chemical environment decreases, the uncertainty principle states that the uncertainty in the energy of the nucleus in each of the two environments must increase. The energy uncertainty is reflected in an increase of signal linewidth. The effect becomes appreciable when τ_M becomes comparable with, or shorter than T_{2A}, T_{2B}. Then the effective T_2 for the signals becomes close to τ_M (Fig. 4.1A(2–4)). In contrast, T_1 is only affected by the efficiency of coupling of the spin system with the lattice in order to undergo spin transitions. Therefore, if these mechanisms are equal in each environment ($R_{1A} = R_{1B}$), longitudinal relaxation will not be affected by exchange between the two sites.

When the exchange rate is further increased to the point where it is about equal to the difference $\Delta\omega$ in resonance frequency of the two environments, the linewidth is also of the same magnitude. Therefore, the two signals are as broad as their separation and a single broad resonance is observed, extending from ω_A to ω_B (Fig. 4.1A(5)).

When the exchange rate becomes faster than $\Delta\omega$, then the single resonance, centered at $(\omega_A + \omega_B)/2$, becomes sharper and sharper. From the NMR point of view, the nuclei are now experiencing an average environment and, as the exchange time decreases further, the linewidth eventually returns to the value of the separate signals in the absence of exchange (Fig. 4.1A(6–9)).

It has been shown that by increasing τ_M^{-1} the resonance frequency of each signal passes from ω_A or ω_B to the average value $(\omega_A + \omega_B)/2$. This transition is not smooth, and the two signals begin to collapse only for τ_M^{-1} values very close to $\Delta\omega$.

The situation $\tau_M^{-1} = \Delta\omega/\sqrt{2}$ (Fig. 4.1A(5)) represents signal coalescence, i.e. the borderline between the so-called slow- and fast-exchange regions. Thus, slow- or fast-exchange rates can only be defined with respect to the difference in Larmor frequency $\Delta\omega$, and therefore are dependent on the strength of the external magnetic field and on the nuclear species involved. For instance, exchange between two sites for a given nuclear species can be fast at 200 MHz and slow at 800 MHz; likewise, for a given field, exchange may be fast for deuterons and slow for protons in the same chemical environment.

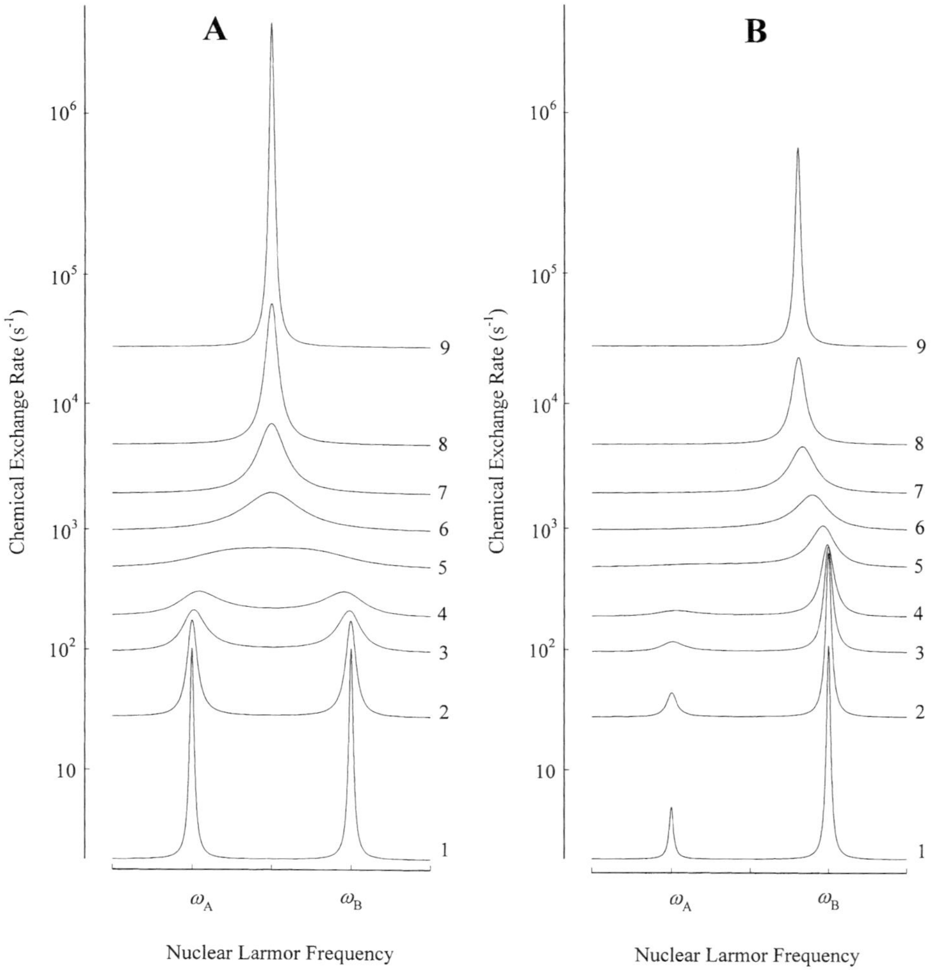

Fig. 4.1. Calculated NMR spectra of a system constituted by a nuclear species in chemical exchange between two sites, A and B, as a function of the chemical exchange rate, τ_M^{-1}. The chemical shift separation between A and B signals is $\Delta\omega = 1000$ rad s^{-1}. The various situations are: (A) $R_{2A} = R_{2B}$, $P_A = P_B$; (B) $R_{2A} = R_{2B}$, $P_A < P_B$.

4.3 NMR PARAMETERS IN THE PRESENCE OF EXCHANGE

4.3.1 Exact solutions for two-site exchange

Having introduced the concept of exchange in a pictorial way, we treat now the more general case of exchange between two sites of different population and different intrinsic relaxation times. Under such conditions relaxation is not a single exponential but rather a double exponential process, where the individual

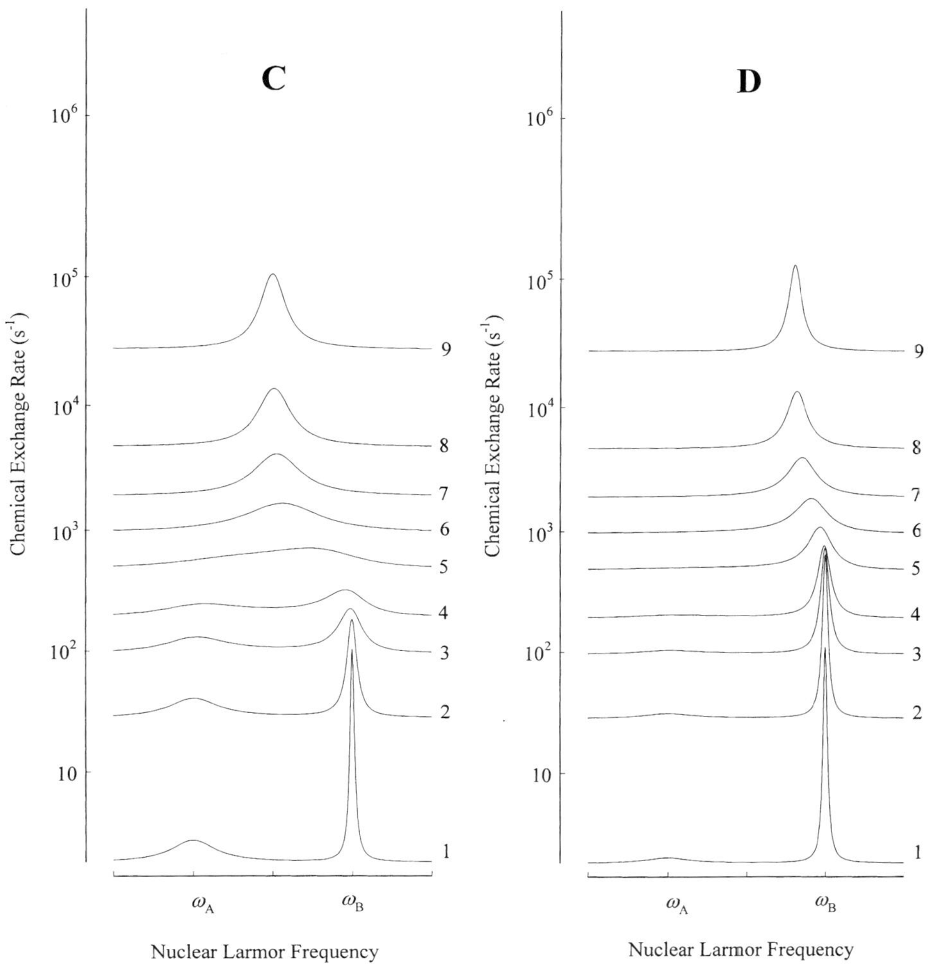

Fig. 4.1 (continued). (C) $R_{2A} > R_{2B}$, $P_A = P_B$; (D) $R_{2A} > R_{2B}$, $P_A < P_B$; and no exchange (1); slow exchange (2–4); coalescence (5); fast exchange (6–9).

rate constants are given by Eqs. (4.2) and (4.3). The chemical shifts are given by Eq. (4.4) [3,4]:

$$R_{1\text{exch}} = A_1 \pm \left[A_1^2 - \left(\tau_{1A}^{-1}\tau_{1B}^{-1} - \tau_A^{-1}\tau_B^{-1}\right)\right]^{1/2} \tag{4.2}$$

$$R_{2\text{exch}} = A_2 \pm \left[\frac{G + (G^2 + H^2)^{1/2}}{2}\right]^{1/2} \tag{4.3}$$

$$\omega_{\text{exch}} = \frac{\omega_A + \omega_B}{2} \pm \left[\frac{-G + (G^2 + H^2)^{1/2}}{2}\right]^{1/2} \tag{4.4}$$

where

$$A_1 = \frac{\tau_{1A}^{-1} + \tau_{1B}^{-1}}{2} \qquad A_2 = \frac{\tau_{2A}^{-1} + \tau_{2B}^{-1}}{2}$$

$$G = \frac{\left(\tau_{2A}^{-1} - \tau_{2B}^{-1}\right)^2}{4} + \tau_A^{-1}\tau_B^{-1} - \frac{\Delta\omega^2}{4}$$

$$H = \frac{\tau_{2A}^{-1} - \tau_{2B}^{-1}}{2}\Delta\omega$$

$$\tau_{1A}^{-1} = R_{1A} + \tau_A^{-1} \qquad \tau_{1B}^{-1} = R_{1B} + \tau_B^{-1}$$

$$\tau_{2A}^{-1} = R_{2A} + \tau_A^{-1} \qquad \tau_{2B}^{-1} = R_{2B} + \tau_B^{-1}$$

$$\tau_{AB}^{-1} = \tau_A^{-1} + \tau_B^{-1}$$

Fig. 4.1B–D illustrates the situation where the populations are unequal (B and D) and the relaxation times are unequal (C and D). It can be seen that when the populations are different the average signal sits at the weighted average position and has a linewidth which is the weighted average of the two linewidths. The linewidths of the various intermediate exchange situations can be appreciated in the plots of Fig. 4.1(B–D).

In the special case of $\tau_A = \tau_B = \tau$, $R_{1A} = R_{1B} = R_1$, and $R_{2A} = R_{2B} = R_2$, the above quantities reduce to:

$$A_1 = R_1 + \tau^{-1} \qquad A_2 = R_2 + \tau^{-1}$$

$$G = \tau^{-2} - \Delta\omega^2/4 \qquad H = 0$$

and Eqs. (4.2)–(4.4) become

$$R_{1\text{exch}} = R_1 = \text{constant} \tag{4.5}$$

$$R_{2\text{exch}} = R_2 + \tau^{-1} \pm \left(\frac{G + |G|}{2}\right)^{1/2} \tag{4.6}$$

$$\omega_{\text{exch}} = \frac{\omega_A + \omega_B}{2} \pm \left(\frac{-G + |G|}{2}\right)^{1/2} \tag{4.7}$$

Eq. (4.6) accounts for the linewidths of the signals reported in Fig. 4.1A.

4.3.2 *Exchange of excess metal ligands*

When the paramagnetic site is the least populated, and the difference in population is very large, Eqs. (4.2)–(4.4) can be simplified [3–6]. For example, the chemical shift for the signal in the diamagnetic site evolves as follows as a

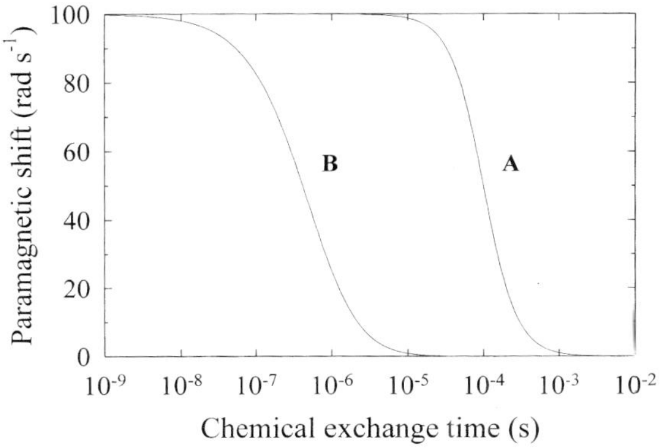

Fig. 4.2. Paramagnetic contribution to the chemical shift $\Delta\omega_p$ as a function of the exchange rate of the nucleus from the paramagnetic site τ_M. Conditions: $\Delta\omega_M = 10^4$ rad s^{-1}, $f_M = 10^{-2}$, $R_{2M} = 10^2$ (A) or 10^6 (B) s^{-1}.

function of τ_M

$$\boxed{\Delta\omega_p = \omega_p - \omega_d = \frac{f_M}{\tau_M^2} \frac{\Delta\omega_M}{\left(R_{2M} + \tau_M^{-1}\right)^2 + \left(\Delta\omega_M\right)^2}} \tag{4.8}$$

where ω_p is the experimental shift, ω_d is the chemical shift of the diamagnetic site, and $\Delta\omega_M$ is the difference in chemical shift between the metal site and the diamagnetic site. This paramagnetic contribution to the shift in the metal site is given by the equations described in Section 2.2 [1]. R_{2M} is the transverse nuclear relaxation rate enhancement in the paramagnetic site (Sections 3.4–3.6). The behavior is illustrated in Fig. 4.2. For $\tau_M^{-1} \gg R_{2M}, \Delta\omega_M$, Eq. (4.8) becomes

$$\Delta\omega_p = f_M \Delta\omega_M \tag{4.9}$$

i.e. the signal sits at the weighted average position between ω_d and ω_M.

In analogy with the chemical shift, the relaxation rates of the bulk nuclei will be termed R_{1d} and R_{2d}. In the presence of chemical exchange with nuclei in the paramagnetic site, such rates will be enhanced by an amount R_{1p} or R_{2p}, and the measured values will be

$$R_{1\,\mathrm{meas}} = R_{1d} + R_{1p} \qquad R_{2\,\mathrm{meas}} = R_{2d} + R_{2p}$$

[1] Note that the equations in Section 2.2 report the paramagnetic shifts (δ^{con}, δ^{pc}, etc.) in ppm. To obtain $\Delta\omega_M$ (in units of rad s^{-1}) one should multiply the shift in ppm by the nuclear Larmor frequency (in MHz), and further multiply by 2π. For example, a proton experiencing $\delta^{\mathrm{con}} = 3.75$ ppm yields $\Delta\omega_M = 3.75 \times 500 \times 2\pi = 11,775$ rad s^{-1} when measured in a 500 MHz spectrometer, whereas a deuteron in the same chemical environment yields $\Delta\omega_M = 3.75 \times 76.75 \times 2\pi = 1807.5$ rad s^{-1}, where 76.75 MHz is the deuterium Larmor frequency in a magnetic field corresponding to a 500 MHz proton Larmor frequency.

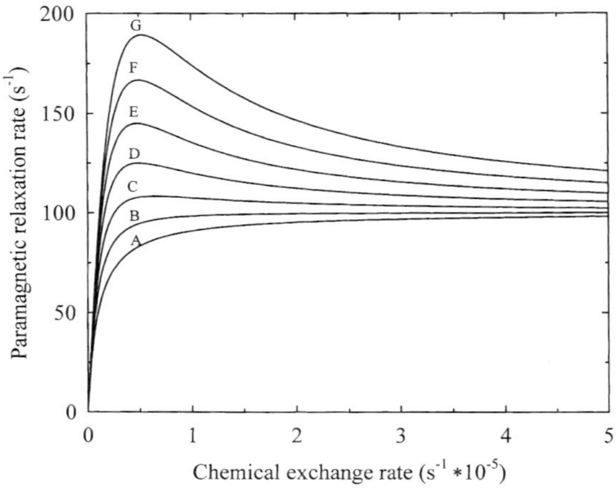

Fig. 4.3. Paramagnetic contributions to the nuclear relaxation rates R_{1p} (A) and R_{2p} (A–G) as a function of the exchange rate of the nucleus τ_M^{-1} from the paramagnetic site. Conditions: $R_{1M} = R_{2M} = 10^4$ s^{-1}, $f_M = 10^{-2}$, $\Delta\omega_M = 0$ ((A) coincident with R_{1p}), 1×10^4 (B), 1.5×10^4 (C), 2×10^4 (D), 2.5×10^4 (E), 3×10^4 (F), 3.5×10^4 (G) rad s^{-1}.

The longitudinal relaxation rate enhancement R_{1p} goes asymptotically from zero to $f_M R_{1M}$ with increasing τ_M^{-1} (Fig. 4.3A), according to the following equation:

$$\boxed{R_{1p} = f_M R_{1M} \frac{\tau_M^{-1}}{\left(R_{1M} + \tau_M^{-1}\right)^2} = f_M \left(T_{1M} + \tau_M\right)^{-1}} \tag{4.10}$$

where R_{1M} is the nuclear longitudinal relaxation enhancement in the paramagnetic site (Sections 3.4–3.6). τ_M thus acts as a limiting factor for the propagation of the effect from the paramagnetic site to the bulk nuclei. In other words, in the absence of exchange there is no paramagnetic contribution to the relaxation rate of the bulk nuclei, whereas in the fast-exchange limit ($\tau_M^{-1} \ll R_{1M}$) the measured relaxation rate is the weighted average between the diamagnetic R_{1d} and paramagnetic R_{1M} rates:

$$R_{1\ \text{meas}} = R_{1d} + f_M R_{1M}$$

where the molar fraction of bound nuclei f_M is the weighting factor. It should be remembered that, as in the case of the paramagnetic shift, R_{1M} can be very large compared with R_{1d}, so that even for small f_M values a noticeable relaxation rate enhancement may be measured.

The transverse relaxation rate enhancement R_{2p} could be treated in the same way if it were not for the difference in chemical shift $\Delta\omega_M$ between the paramagnetic and the diamagnetic species. As has already been shown, the difference in chemical shift causes a line broadening — an increasing in R_{2M} — even when the exchange is between two diamagnetic sites. Furthermore, the effect is expected to be much larger in this case since much larger chemical shift differences are involved. The general formula relating R_{2p} to $\Delta\omega_M$ and τ_M^{-1} of the bulk nuclei is [5,6]

$$R_{2p} = \frac{f_M}{\tau_M} \frac{R_{2M}^2 + R_{2M}\tau_M^{-1} + (\Delta\omega_M)^2}{\left(R_{2M} + \tau_M^{-1}\right)^2 + (\Delta\omega_M)^2} \tag{4.11}$$

Fig. 4.3 shows the variation of R_{2p} as a function of τ_M^{-1} for different values of $\Delta\omega_M$.

Note that in the absence of exchange ($\tau_M^{-1} = 0$), $R_{2p} = 0$, and for very fast exchange ($\tau_M^{-1} \gg R_{2M}, (\Delta\omega_M)^2/R_{2M}$), $R_{2p} = f_M R_{2M}$, as found for the longitudinal relaxation rate.

In the intermediate region, where τ_M^{-1} is comparable with $\Delta\omega_M$, R_{2p} may be larger than R_{1p} even if $R_{2M} = R_{1M}$ because of exchange broadening effects.

If little or no isotropic shift is present ($\Delta\omega_M \ll R_{2M}$), Eq. (4.11) reduces to

$$R_{2p} = f_M R_{2M} \frac{\tau_M^{-1}}{R_{2M} + \tau_M^{-1}} = f_M (T_{2M} + \tau_M)^{-1} \tag{4.12}$$

which has the same form as Eq. (4.10) for R_{1p}.

In the very fast exchange regions, the observed paramagnetic effects on the chemical shift $\Delta\omega_p$ and the relaxation rates R_{1p}, R_{2p}, are simply proportional to the full paramagnetic effects in the metal binding site, as summarized below:

$$\Delta\omega_p = f_M \Delta\omega_M \qquad R_{1p} = f_M R_{1M} \qquad R_{2p} = f_M R_{2M} \tag{4.13}$$

Therefore, if the molar fraction of bound ligand f_M is known, the full paramagnetic effects can be calculated. Of course, if no exchange takes place, the ligand in the bulk solution is completely insensitive to the presence of the paramagnetic metal ion.

In the intermediate situation the measured parameters are complicated functions of the full effects experienced in the paramagnetic site ($\Delta\omega_M$, R_{1M}, R_{2M}) and the exchange rate τ_M^{-1}. The τ_M^{-1} dependence of such functions has been summarized in Figs. 4.2 and 4.3. There are, however, particular regions in which the measured parameters are simpler functions of τ_M^{-1}. These regions correspond to cases in which Eqs. (4.8), (4.10) and (4.11) can be approximated by

Case I: $$\Delta\omega_p = \frac{f_M}{\tau_M^2 \Delta\omega_M} \qquad (\Delta\omega_M \gg R_{2M}, \tau_M^{-1}) \tag{4.14}$$

Case I occurs when the difference in shift between the paramagnetic and the diamagnetic site, in angular frequency units, is large compared with both transverse relaxation and the exchange rate. In this case the diamagnetic line is only slightly shifted, and the shift is inversely proportional to the paramagnetic shift.

Case II: $$\Delta\omega_{\mathrm{p}} = \frac{f_M \Delta\omega_M}{\tau_M^2 R_{2M}} \qquad (R_{2M} \gg \tau_M^{-1}, \Delta\omega_M) \tag{4.15}$$

Case II occurs when the transverse relaxation rate is faster than both exchange rate and paramagnetic shift. Here the observed shift is directly proportional to the paramagnetic shift.

Case III: $$R_{1\mathrm{p}} = f_M \tau_M^{-1} \qquad (R_{1M} \gg \tau_M^{-1}) \tag{4.16}$$

Case IV: $$R_{2\mathrm{p}} = f_M \tau_M^{-1} \qquad (R_{2M} \gg \tau_M^{-1}) \tag{4.17}$$

Cases III and IV represent the slow exchange limits for longitudinal (III) and transverse (IV) relaxation enhancements. The observed enhancements are proportional to the exchange rate, independently of the values of R_{1M} and R_{2M}.

Case V: $$R_{2\mathrm{p}} = f_M (\Delta\omega_M)^2 \tau_M \qquad (\tau_M^{-1} \gg (\Delta\omega_M)^2 \tau_M \gg R_{2M}) \tag{4.18}$$

Case V is an intermediate exchange case that can be encountered for transverse relaxation. Here, the dependence on τ_M is opposite to that in Case IV. Case V occurs in the presence of relatively large paramagnetic shift, as long as $(\Delta\omega_M)^2$ is small compared with τ_M^{-2} but large compared with $R_{2M}\tau_M^{-1}$.

In Cases III and IV, $R_{1\mathrm{p}}$ and $R_{2\mathrm{p}}$ are a direct measure of the exchange rate τ_M^{-1}.

Eqs. (4.11)–(4.18) also hold for $R_{1\rho \mathrm{p}}$ by simply substituting $R_{1\rho M}$ for R_{2M}.

4.3.3 Temperature and exchange

The measurement of the exchange time τ_M may provide useful kinetic information on the system. Kinetic parameters for the dissociation process may be obtained by performing relaxation measurements as a function of temperature. If it is assumed that the dissociation of the ligand from the paramagnetic site is a first order kinetic process, the dissociation rate constant τ_M^{-1} is given by the Eyring relationship

$$\boxed{\tau_M^{-1} = \frac{kT}{h} \exp\left(\frac{-\Delta G^{\neq}}{RT}\right)} \tag{4.19}$$

where $\Delta G^{\neq}$ is the free energy of activation for the dissociation process.

In the normal range of temperatures used in NMR experiments, the major source of variation of τ_M^{-1} with temperature is contained in the exponential part. In other words, a plot of $\log \tau_M$ against $1/T$ (Arrhenius plot) will give a fairly straight line. Given the linear relationship between τ_M^{-1} and the relaxation rates,

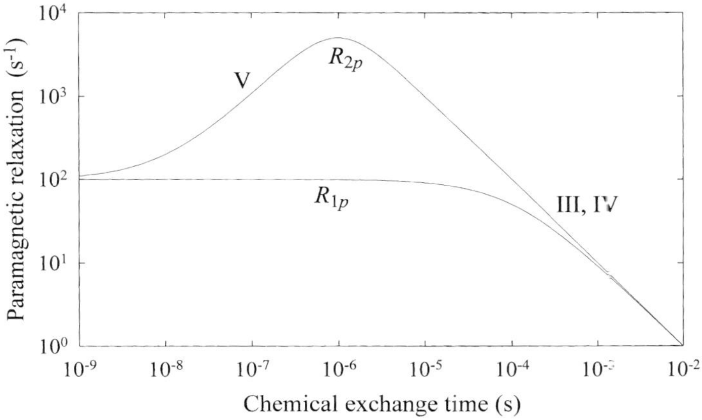

Fig. 4.4. Linear dependence of R_{1p} and R_{2p} (log scale) on τ_M (log scale) in the exchange regions corresponding to Cases III–V. Conditions: $R_{iM} = 10^4\ s^{-1}$, $f_M = 10^{-2}$, and $\Delta\omega_M = 10^6$ rad s^{-1}.

straight lines will also be obtained by plotting log R_{1p} or log R_{2p} (Cases III and IV) against $1/T(\propto \log \tau_M)$ (Fig. 4.4). Note that a straight line of opposite slope is obtained in Case V. The appropriate equations can then be found by simply taking the log of expressions III, IV and V, with τ_M given by Eq. (4.19) (and remembering that $\Delta G = \Delta H - T\Delta S$). From Eqs. (4.16) and (4.17) we thus obtain

$$\log R_{1p} = \log R_{2p}$$

$$= \log\left(f_M \frac{kT}{h}\right) - \frac{\Delta H^{\neq}}{RT} + \frac{\Delta S^{\neq}}{R} \cong \text{const} - \frac{\Delta H^{\neq}}{RT} \tag{4.20}$$

while from Eq. (4.18) we obtain:

$$\log R_{2p} = \log\left(f_M \Delta\omega_M^2 \frac{h}{kT}\right) + \frac{\Delta H^{\neq}}{RT} - \frac{\Delta S^{\neq}}{R} \cong \text{const}' - \frac{\Delta H^{\neq}}{RT} \tag{4.21}$$

where $\Delta H^{\neq}$ and $\Delta S^{\neq}$ are the enthalpy and entropy of activation for the dissociation process.

Temperature dependent experiments are often performed just to check whether R_{1p} in Eq. (4.10) is in the fast exchange or in the exchange limited region. A strong increase of R_{1p} with temperature is a clear indication of τ_M being the limiting factor.

4.3.4 *Saturation transfer*

As already mentioned (Sections 1.7.4 and 3.13), when a nuclear spin system experiences fluctuating magnetic fields, relaxation occurs. If these fields are produced by other nuclear spins, there is a reciprocal influence that leads to a complicated picture. However, if these fluctuating magnetic fields are produced by particles which belong to the lattice, i.e. their energy is negligibly affected by nuclear relaxation (for example, unpaired electrons), the description of the effects is much simpler (Chapter 3) and the return to equilibrium after a perturbation is exponential. Under these circumstances we are going to analyze the effect of chemical exchange.

We have already defined the equilibrium magnetization of a spin I in a given magnetic field B_0 as $M_z(\infty)$, where the (∞) refers to the fact that the sample must have been exposed to the field for time sufficiently long for equilibrium magnetization to be virtually achieved. After any perturbation from equilibrium of the nuclear spin system such that, at time zero after the perturbation, $M_z(0) \neq M_z(\infty)$, the system will tend to return to equilibrium with a simple rate law of the type

$$\frac{\mathrm{d}M_z(t)}{\mathrm{d}t} = -R_1 \left[M_z(t) - M_z(\infty)\right] \tag{4.22}$$

which integrates to an exponential magnetization recovery

$$M_z(t) - M_z(\infty) = \left[M_z(0) - M_z(\infty)\right] \exp(-R_1 t) \tag{4.23}$$

R_1 being the rate constant for the longitudinal relaxation process. In the case of an inversion recovery experiment, $M_z(0) = -M_z(\infty)$, and Eq. (4.23) reduces to

$$M_z(t) = M_z(\infty) - 2M_z(\infty) \exp(-R_1 t) \tag{4.24}$$

which is Eq. (3.40).

The rate equations for a spin system in chemical exchange between two sites

$$I^{\mathrm{A}} \underset{k_{-1}}{\overset{k_1}{\rightleftharpoons}} I^{\mathrm{B}}$$

(where $k_1 = \tau_{\mathrm{A}}^{-1}$, $k_{-1} = \tau_{\mathrm{B}}^{-1}$), are given by[2]

$$\frac{\mathrm{d}M_z^{\mathrm{A}}(t)}{\mathrm{d}t} = -R_1^{\mathrm{A}} \left[M_z^{\mathrm{A}}(t) - M_z^{\mathrm{A}}(\infty)\right] - k_1 \left[M_z^{\mathrm{A}}(t) - M_z^{\mathrm{A}}(\infty)\right] + k_{-1} \left[M_z^{\mathrm{B}}(t) - M_z^{\mathrm{B}}(\infty)\right] \tag{4.25}$$

$$\frac{\mathrm{d}M_z^{\mathrm{B}}(t)}{\mathrm{d}t} = -R_1^{\mathrm{B}} \left[M_z^{\mathrm{B}}(t) - M_z^{\mathrm{B}}(\infty)\right] - k_{-1} \left[M_z^{\mathrm{B}}(t) - M_z^{\mathrm{B}}(\infty)\right] + k_1 \left[M_z^{\mathrm{A}}(t) - M_z^{\mathrm{A}}(\infty)\right] \tag{4.26}$$

[2] The transfer of spin population between sites A and B can be written as magnetization transfer through Eqs. (1.26), (1.27) and (1.33).

As anticipated in Section 4.1, solution of these two coupled differential equations gives biexponential behavior, therefore preventing the definition of R_1 for each site (although the two rate constants for the two components are known (Eq. 4.2). However, Eqs. (4.25) and (4.26) explicitly contain the rate constants for the exchange process, k_1 and k_{-1}, suggesting that they can be used to obtain information on the exchange dynamics through some kind of perturbation on the equilibrium populations of the spins [7]. Suppose that the signal I^{A} is saturated by applying a weak r.f. on the A resonance for a reasonably long period of time. Then, the steady state intensity of signal B can be obtained from Eq. (4.26) by setting $\mathrm{d}M_z^{\mathrm{B}}(t)/\mathrm{d}t$ and $M_z^{\mathrm{A}}(t)$ to zero:

$$0 = (R_1^{\mathrm{B}} + k_{-1})\left[M_z^{\mathrm{B}}(t) - M_z^{\mathrm{B}}(\infty)\right] - k_1 M_z^{\mathrm{A}}(\infty) \tag{4.27}$$

from which we obtain an expression for the fractional change in intensity of signal B upon saturation of A:

$$\frac{M_z^{\mathrm{B}}(t) - M_z^{\mathrm{B}}(\infty)}{M_z^{\mathrm{B}}(\infty)} = \frac{M_z^{\mathrm{A}}(\infty)}{M_z^{\mathrm{B}}(\infty)} \frac{-k_1}{R_1^{\mathrm{B}} + k_{-1}} \tag{4.28}$$

Therefore, by knowing the longitudinal relaxation rate constant of the nucleus in the B site R_1^{B} in the absence of exchange and by measuring the fractional change in intensity of signal B (called saturation transfer), the rate constants can be obtained. We recall that the two rate constants are related by the fractional populations of the two sites (Section 4.2), in turn proportional to the equilibrium intensities of the two signals

$$k_1 M_z^{\mathrm{A}}(\infty) = k_{-1} M_z^{\mathrm{B}}(\infty) \tag{4.29}$$

thus, Eq. (4.28) can be further simplified as

$$\boxed{\frac{M_z^{\mathrm{B}}(t) - M_z^{\mathrm{B}}(\infty)}{M_z^{\mathrm{B}}(\infty)} = \frac{-k_{-1}}{R_1^{\mathrm{B}} + k_{-1}}} \tag{4.30}$$

These equations hold for any population distribution of the two sites. Eq. (4.30) is the equation to be used when a saturation transfer experiment is going to be planned.

From the functional form of Eq. (4.30) it is easy to predict the behavior of saturation transfer as a function of the exchange rate. When the rate constant for the B $\rightarrow$ A transformation (k_{-1}) is much smaller than the longitudinal relaxation rate of the nucleus in the B site (R_1^{B}), the saturation transfer tends to zero. When the rate constant is much higher, the saturation transfer tends to -1, i.e. there is a total transfer of magnetization to the B site when the A site is kept saturated. Note that these conditions are referred to as fast exchange, even if the exchange is still slow with respect to the chemical shift separation. Fast exchange conditions on the relaxation time scale are often reached before the signals are actually coalesced, so that there is still a reasonable range of exchange rates for which saturation transfer can be observed.

Throughout this section it has been assumed that relaxation in the A and B sites is intrinsically exponential. Warnings about this assumption have been made elsewhere (Sections 3.13 and 7.2.2).

The methods to detect and analyze saturation transfer in 2D experiments are described in Section 8.2.

4.4 EQUILIBRIUM CONSTANTS

4.4.1 NMR of metal ligands

Consider a metal complex CM, where C is a multidentate ligand that leaves an empty coordination position on the metal, in the presence of a monodentate ligand L. CM could also be a metalloenzyme interacting with a substrate or an inhibitor L. The paramagnetic effects observed on a nucleus of L can then be used to obtain information on its dissociation constant:

$$K = \frac{[CM][L]}{[CML]} = \frac{(C_{CM} - [CML])(C_L - [CML])}{[CML]} \tag{4.31}$$

where C_{CM} and C_L are the total concentrations of all the metal-containing and ligand-containing species respectively, and $[CML]$ is the equilibrium concentration of the adduct.

Under fast exchange conditions, the molar fraction of bound ligand f_M can be expressed in terms of Eqs. (4.13) and (4.31) as

$$\begin{aligned} f_M &= \frac{[CML]}{C_L} = \frac{\Delta\omega_{\mathrm{p}}}{\Delta\omega_M} = \frac{R_{1\mathrm{p}}}{R_{1M}} = \frac{R_{2\mathrm{p}}}{R_{2M}} \\ &= \frac{K + C_{CM} + C_L - \left[(K + C_{CM} + C_L)^2 - 4C_{CM}C_L\right]^{1/2}}{2C_L} \end{aligned} \tag{4.32}$$

Therefore, measurements of either $\Delta\omega_{\mathrm{p}}$, $R_{1\mathrm{p}}$, or $R_{2\mathrm{p}}$ at various concentrations of C_L and/or C_{CM} allow a two-parameter fitting of the data through Eq. (4.32) in terms of K and $\Delta\omega_M$, R_{1M}, or R_{2M}.

If the experimental conditions are such that $[CML]$ is always much smaller than C_L (i.e. large excess of L with respect to CM), then Eq. (4.31) becomes

$$K = \frac{(C_{CM} - [CML])\, C_L}{[CML]} \tag{4.33}$$

and Eq. (4.32) then becomes

$$f_M = \frac{[CML]}{C_L} = \frac{C_{CM}}{K + C_L}. \tag{4.34}$$

Expressing f_M in the form of Eq. (4.13) gives

$$C_L = C_{CM}\frac{\Delta\omega_M}{\Delta\omega_p} - K = C_{CM}\frac{R_{1M}}{R_{1p}} - K = C_{CM}\frac{R_{2M}}{R_{2p}} - K \tag{4.35}$$

For constant C_{CM}, a plot of $1/\Delta\omega_p$, $1/R_{1p}$ ($= T_{1p}$), or $1/R_{2p}$ ($= T_{2p}$) against C_L, gives a straight line; $\Delta\omega_M$, R_{1M}, or R_{2M} can then be obtained from the slope, and K can be found from the intercept on the y axis.

Eq. (4.35), like Eq. (4.32), is valid when the ligand is in fast exchange; it is also valid under the exchange-limited conditions, which are described in Eqs. (4.8), (4.10) and (4.11), as long as τ_M is not dependent on C_L. In this case, K can still be obtained, in addition to the limit values of R_{1p}, R_{2p}, or $\Delta\omega_p$ at $f_M = 1$.

4.4.2 NMR of water protons (the enhancement factor)

If one or more ligands L in large excess interact with a metal ion in a metal complex CM in the presence of free metal ions M in solution, then the exchange of ligand L among three sites should be considered. A typical case is when L is a coordinating solvent molecule, e.g. water. The molar fraction of water nuclei is given by

$$f_M = \frac{p[M]}{C_{H_2O}} \qquad f_{CM} = \frac{q[CM]}{C_{H_2O}} \qquad f_{\text{bulk}} = \frac{[H_2O]}{C_{H_2O}} \approx 1 \tag{4.36}$$

where p and q are the numbers of water molecules interacting with free and bound metal ions respectively. The water proton R_{1p} is given by the contribution of the metal, $(R_{1p})_M$, and of the complex $(R_{1p})_{CM}$, both of which can be expressed through Eq. (4.10):

$$\begin{aligned} R_{1p} &= f_M R_{1M}\frac{\tau_M^{-1}}{R_{1M}+\tau_M^{-1}} + f_{CM}R_{1CM}\frac{\tau_{CM}^{-1}}{R_{1CM}+\tau_{CM}^{-1}} \\ &= \frac{p[M]}{C_{H_2O}}R_{1M}\frac{\tau_M^{-1}}{R_{1M}+\tau_M^{-1}} + \frac{q[CM]}{C_{H_2O}}R_{1CM}\frac{\tau_{CM}^{-1}}{R_{1CM}+\tau_{CM}^{-1}}. \end{aligned} \tag{4.37}$$

The two limiting cases of the metal completely free ($K \to \infty$) or completely bound ($K \to 0$), where K is the dissociation constant of the CM complex, would give

$$\begin{aligned} R_{1p}(\infty) &= \frac{pC_M}{C_{H_2O}}R_{1M}\frac{\tau_M^{-1}}{R_{1M}+\tau_M^{-1}} \\ R_{1p}(0) &= \frac{qC_M}{C_{H_2O}}R_{1CM}\frac{\tau_{CM}^{-1}}{R_{1CM}+\tau_{CM}^{-1}} \end{aligned} \tag{4.38}$$

where C_M is the total metal concentration. The first case can be simulated by simply not adding any ligand C to the solution. It is then customary to define an enhancement factor as [8]

$$\varepsilon = \frac{R_{1p}}{R_{1p}(\infty)} \tag{4.39}$$

where R_{1p} is measured in a solution containing both the metal ion and the ligand C, and $R_{1p}(\infty)$ is obtained from a solution containing the same concentration of metal ion but in the absence of the ligand C. From Eqs. (4.37) and (4.38)

$$\begin{aligned}\varepsilon &= \frac{[M]}{C_M} + \frac{q[CM]}{pC_M}\frac{R_{1CM}\tau_{CM}^{-1}}{R_{1M}\tau_M^{-1}}\frac{R_{1M}+\tau_M^{-1}}{R_{1CM}+\tau_{CM}^{-1}} \\ &= \frac{[M]}{C_M} + \frac{q[CM]}{pC_M}\frac{(T_{1CM}+\tau_{CM})^{-1}}{(T_{1M}+\tau_M)^{-1}} = \frac{[M]}{C_M} + \frac{[CM]}{C_M}\varepsilon_0 \\ &= f_f + f_b\varepsilon_0 = 1 + f_b(\varepsilon_0 - 1)\end{aligned} \tag{4.40}$$

where f_f and f_b are the molar fractions of free and bound metal ion and

$$\varepsilon_0 = \frac{q}{p}\frac{(T_{1CM}+\tau_{CM})^{-1}}{(T_{1M}+\tau_M)^{-1}} = \frac{R_{1p}(0)}{R_{1p}(\infty)}.$$

When the CM complex is fully formed, $[M] = 0$ and $[CM] = C_M$; therefore, from Eq. (4.40), $\varepsilon = \varepsilon_0$. *$\varepsilon_0$ is thus defined as the enhancement factor measured in a solution where all the metal is complexed.* Since usually $q < p$, ε_0 should be smaller than unity if the intrinsic nuclear relaxation times are the same in the metal complex and in the aquaion. However, as often $T_{1CM} < T_{1M}$ owing to a longer correlation time τ_c in the complex (Chapter 3), ε_0 can be larger than unity. This is particularly true when C is a macromolecule (e.g. a protein) and M is a metal ion with long electronic relaxation times.

As it appears in Eq. (4.40), ε_0 is defined only in terms of the molar fraction of bound metal ion; that is, independently of the actual concentrations. By using an equation for f_b analogous to Eq. (4.32), the enhancement factor can be expressed as:

$$\varepsilon = 1 + \frac{K + C_M + C_C - \left[(K + C_M + C_C)^2 - 4C_MC_C\right]^{1/2}}{2C_M}(\varepsilon_0 - 1). \tag{4.41}$$

Both K and ε_0 can be obtained through a two-parameter fitting of the ε data obtained at various C_C and/or C_M concentrations.

Eq. (4.41) is valid irrespective of the rate at which water exchanges from the two paramagnetic sites. In fact, the enhancement factor is defined in terms of the quantities $T_{1M} + \tau_M$ and $T_{1CM} + \tau_{CM}$, which are likely to be constant and, in particular, independent of the concentration of the various species in solution. Furthermore, when the exchanging ligand is the solvent, as in the above examples, its concentration is virtually constant under any circumstances.

4.5 BEYOND THE CONCEPT OF BINDING SITE

4.5.1 τ_M as correlation time

Let us consider a ligand nucleus (e.g. a proton in a water molecule) and its interaction with a paramagnetic center (e.g. a metal ion) to which the ligand is bound. In the absence of exchange, the hyperfine shifts and relaxation rates are given by the equations developed in Chapters 2 and 3. In the case of fast exchange of ligands (e.g. bulk water), we have seen in Section 4.3.2. that the shift and relaxation rates are the weighted averages between the free and bound ligands. This means that *exchange does not alter the shift and relaxing capabilities of the metal site*. Whereas this is always true as far as the shift is concerned, it may not be true for relaxation rates. We have seen that nuclear relaxation in the bound species is proportional to the average squared interaction energy (either dipolar or contact) and to a function of the correlation time that modulates the interaction. We have also seen that the correlation time for the dipolar interaction is given by the reciprocal of the sum of the electronic relaxation rate, the rotational correlation rate, and the chemical exchange rate, whereas in the case of contact relaxation it is given by the reciprocal of the sum of the electronic relaxation rate and the chemical exchange rate (Eqs. (3.6) and (3.7)). We have learned that relaxation arises from the modulation of the interaction energy by whatever random process. If chemical exchange is present, clearly the interaction energy can be modulated by exchange, because the interaction (both dipolar and contact) is lost upon ligand detachment. In the limit situation where τ_M becomes shorter than τ_s (for contact) or shorter than both τ_s and τ_r (for dipolar relaxation) then it becomes the correlation time (Eqs. (3.6) and (3.7)). Of course, the R_{1M} and R_{2M} profiles still maintain the same shape as shown in Figs. 3.10 and 3.12.

4.5.2 Outer sphere relaxation

So far we have assumed that the molecule bearing the nucleus under investigation spends a finite (although small) time in a well-defined binding site, and the rest of the time in the bulk solution at a distance from the metal which may be considered infinite. In this process, we have considered negligible the time spent in approaching and leaving the binding site (i.e. spent at a finite metal–nucleus distance). As τ_M becomes shorter and shorter, it will eventually reach the point where it is comparable with, or shorter than, the diffusional correlation time of the molecule. Under these conditions, the metal–nucleus interaction during the approach and departure of the ligand becomes a substantial fraction of the total interaction. A limit situation can be reached where a binding site no longer exists and the metal–nucleus interaction is only exerted by random encounters between the molecules, regulated only by diffusion processes. The correlation time for the latter, termed τ_D, then becomes an important parameter. In the only presence of diffusion-controlled interactions, contact shifts no longer exist, and dipolar shifts average to zero if the approach to the ion can occur in every direction. Therefore, hyperfine shifts are zero (although an effect on the shifts can still be detected, see Section 4.5.3), whereas relaxation enhancements do not drop to zero. This situation goes under the name of 'outer sphere' relaxation. At variance with the correlation functions encountered so far, the correlation function for molecular diffusion is not exponential [8,9]. Furthermore, its form depends on the assumed diffusional model.

The diffusional correlation time τ_D depends on the size of both the metal and the ligand-containing moieties, according to their diffusion coefficients, D_M and D_L, and on the minimal distance that can be achieved between the ligand and the metal ion, called *distance of closest approach*, d [8,9].

$$\boxed{\tau_D = \frac{d^2}{D_M + D_L}} \tag{4.42}$$

In turn, the diffusion coefficients are defined by assuming that the molecules behave as rigid spheres in a medium of viscosity η:

$$D_M = \frac{kT}{6\pi a_M \eta} \qquad D_L = \frac{kT}{6\pi a_L \eta} \tag{4.43}$$

where a_M and a_L are the radii of the metal-containing and ligand-containing molecules. Note that, when one of the two molecules has a much larger size than the other, its diffusion coefficient is much smaller than the other, and does not contribute appreciably to the denominator of Eq. (4.42). In other words, the diffusional correlation time is only dependent on the size of the small molecule.

In outer sphere relaxation two limiting situations may thus occur, depending on whether the electronic relaxation time τ_s is shorter or longer than the diffusional correlation time τ_D. Since the metal ion and the interacting nucleus are not held together in the same molecular framework, the rotational correlation time τ_r is ineffective in modulating the electron–nucleus interaction and needs not be considered further. τ_D is typically in the range 10^{-9} to 10^{-11} s and seldom reaches 10^{-12} s (for instance, water molecules in water have $\tau_D \approx 3 \times 10^{-12}$ s). Thus, τ_s can still be the dominant correlation time when the metal ion undergoes fast electron relaxation. In such a case, one can figure out that, for each metal–ligand distance between d and infinity, the metal–ligand system can be considered as frozen on the time interval over which the electron undergoes many transitions between the spin levels. Therefore, R_{1p} and R_{2p} can be evaluated by simply integrating the Solomon equations (Eqs. (3.19) and (3.20)) over the distance range between d and infinity [8–10]:

$$R_{1p} = \frac{2}{15}\left(\frac{\mu_0}{4\pi}\right)^2 1000N_A[M]\,\frac{4}{3}\pi\,\frac{\gamma_I^2 g_e^2 \mu_B^2 S(S+1)}{d^3}\left(\frac{7T_{2e}}{1+\omega_S^2 T_{2e}^2} + \frac{3T_{1e}}{1+\omega_I^2 T_{1e}^2}\right) \tag{4.44}$$

$$R_{2p} = \frac{1}{15}\left(\frac{\mu_0}{4\pi}\right)^2 1000 N_A[M]\frac{4}{3}\pi \frac{\gamma_I^2 g_e^2 \mu_B^2 S(S+1)}{d^3} \times \left(4T_{1e} + \frac{13T_{2e}}{1+\omega_S^2 T_{2e}^2} + \frac{3T_{1e}}{1+\omega_I^2 T_{1e}^2}\right) \quad (4.45)$$

where N_A is the Avogadro constant and $[M]$ is the concentration of the molecule bearing the paramagnetic center. Note that only the paramagnetic enhancements R_{1p} and R_{2p} can be evaluated, since R_{1M} and R_{2M} cannot be defined. Also, at variance with normal chemical exchange situations, R_{1p} and R_{2p} do not depend on ligand concentration but only on the concentration of the metal-containing species. Indeed, when no binding site exists, each nucleus may interact with more than one metal at a time, independently of the concentration of nuclei; R_{1p} and R_{2p} are therefore only proportional to the number of metal ions per unit volume, which is expressed by $N_A[M] \times 1000$ in SI units.

In the diffusion-controlled regime different equations should be derived, taking into account that the interaction energy is now modulated by fluctuations in r between d and infinity. In this case the kind of integration to be performed depends on the model assumed for the diffusional behavior of the system. According to one of the most commonly used models for diffusion [11,12], the following equations have been derived when τ_D is the dominant correlation time:

$$\boxed{R_{1p} = \frac{32}{405}\pi\left(\frac{\mu_0}{4\pi}\right)^2 \frac{1000 N_A[M]\gamma_I^2 g_e^2 \mu_B^2 S(S+1)}{d(D_M + D_L)} \times \left\{7J(\omega_S) + 3J(\omega_I)\right\}} \quad (4.46)$$

$$\boxed{R_{2p} = \frac{16}{405}\pi\left(\frac{\mu_0}{4\pi}\right)^2 \frac{1000 N_A[M]\gamma_I^2 g_e^2 \mu_B^2 S(S+1)}{d(D_M + D_L)} \times \left\{4J(0) + 13J(\omega_S) + 3J(\omega_I)\right\}} \quad (4.47)$$

where the spectral density functions are given by

$$J(\omega) = \frac{1 + 5z/8 + z^2/8}{1 + z + z^2/2 + z^3/6 + 4z^4/81 + z^5/81 + z^6/648} \quad (4.48)$$

with

$$z = (2\omega\tau_D)^{1/2} \quad (4.49)$$

It should be noted that, because the interaction energy is averaged in a different way with respect to the case of shortest τ_s discussed above, the equations look very different. In particular, the $J(\omega)$ do not have the usual Lorentzian form (Fig. 4.5).

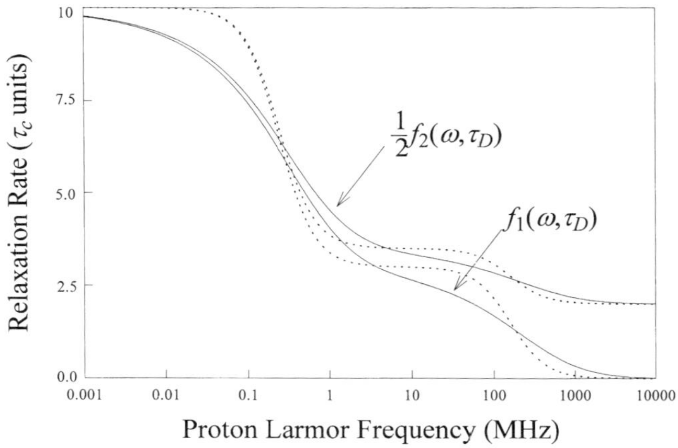

Fig. 4.5. Plot of the spectral density functions of Eqs. (4.46) and (4.47) as a function of magnetic field (expressed as proton Larmor frequency; log scale). $\tau_D = 2 \times 10^{-9}$ s. The Solomon profiles obtained for $\tau_c = 2 \times 10^{-9}$ s are also reported (dotted lines) for comparison purposes.

Equations are also available for the case of τ_D and τ_s having comparable values [13]:

$$z = \sqrt{2\left(\omega\tau_D + \frac{\tau_D}{\tau_s}\right)}. \tag{4.50}$$

Calculations were also performed for electron spin Hamiltonian dominated by zero field splitting [14,15].

4.5.3 Bulk susceptibility shift

In the diffusion limit, the hyperfine shift, defined as the shift difference between a paramagnetic and a diamagnetic environment both measured with respect to the same internal standard, is zero. However, the absolute shifts of all nuclei in the paramagnetic solution are all offset by the same extent with respect to an analogous diamagnetic solution because of the bulk paramagnetism of the sample. As the effect is the same for all nuclei in the sample, when chemical shifts are reported relative to an internal reference signal the effect is therefore canceled. However, it is possible to measure this effect, which is proportional to the magnetic susceptibility of the paramagnetic solute and to its concentration, by referring the shifts to an *external* reference signal. The experimental procedure is referred to as the Evans method [16], and constitutes one of the best ways of measuring solute magnetic susceptibilities in solution around room temperature.

The contribution to the absolute chemical shift from the magnetic susceptibility of a sample is given by the following equation [17]:

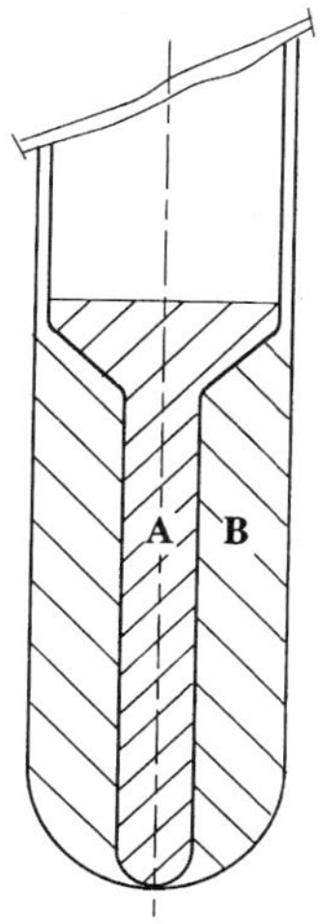

Fig. 4.6. Coaxial NMR tubes for the measurement of magnetic susceptibility of a paramagnetic solute. Solution A contains the paramagnetic solute and an inert probe substance. Solution B contains the probe substance in the same solvent. The measurement is then repeated substituting the paramagnetic solute in A with a diamagnetic analog of the same concentration.

$$\delta_\chi = m\left(\tfrac{1}{3} - \alpha\right)\left[\chi^{\text{para}} + \chi^{\text{dia}} - \chi^{\text{solv}}\left(1 + \frac{\rho_{\text{solv}} - \rho_{\text{soltn}}}{m}\right)\right] \tag{4.51}$$

where m is the mass of solute per unit volume, α is a demagnetization factor, χ^{para} and χ^{dia} are the paramagnetic and diamagnetic contributions to the mass susceptibility of the solute, χ^{solv} is the mass susceptibility of the solvent, and ρ_{solv} and ρ_{soltn} are the densities of the solvent and the solution respectively. χ^{dia} is negative and usually smaller than χ^{para} in absolute value, unless when dealing with large macromolecules. The demagnetization factor α depends on the geometry of the sample and on its orientation relative to the external magnetic field. For a spherical sample $\alpha = 1/3$ and the susceptibility effect on the chemical shift vanishes. For cylindrical samples, $\alpha = 1/2$ if the magnetic field is perpendicular to the cylinder's axis (as in most electromagnets) and $\alpha = 0$ if the magnetic field is parallel to the cylinder axis (as is usual in superconducting magnets). The quantity $1/3 - \alpha$ is therefore equal to $-1/6$ in the former case and $1/3$ in the latter; that is, opposite in sign and double in magnitude. Therefore, a high field cryomagnet would be preferable both because of α and higher resolution.

The experimental setup consists of two coaxial tubes (Fig. 4.6) [18], one of which, e.g. the inner one, containing a solution of an inert probe substance and the paramagnetic solute, and the other containing a solution of only the inert probe substance in the same solvent. The shifts of the probe substance differ in the two solutions, and two different signals are observed. Their chemical shift separation is measured, and the experiment repeated with the same solution in the outer tube and a diamagnetic analog in the inner tube. The chemical shift separation of the two signals is again measured. The difference between the two values, if the concentrations of the paramagnetic and diamagnetic solutes are the same, is directly related to χ^{para}

$$\Delta\delta = m\left(\tfrac{1}{3} - \alpha\right)\chi^{\text{para}} \tag{4.52}$$

Note that the terms containing the solvent susceptibility in Eq. (4.51) cancel because the two solutions have the same density. Eq. (4.52) can also be rewritten as

$$\Delta\delta = 1000M\left(\tfrac{1}{3} - \alpha\right)\chi_M^{\text{para}} \tag{4.53}$$

where $1000M$ is the concentration (mol m^{-3}) of the paramagnetic solute and χ_M^{para} is the paramagnetic contribution to the molar susceptibility, which is directly related to the effective magnetic moment of the paramagnetic center (Chapter 1):

$$\mu_{\text{eff}}^2 = \frac{3kT}{N_A\mu_0}\chi_M^{\text{para}} \tag{4.54}$$

Typical values of χ_M^{para} range from 2×10^{-8} to 3×10^{-7} m^3 mol^{-1}, so that for a millimolar solution in a superconducting magnet ($1000M = 1, \alpha = 0$), $\Delta\delta$ goes from 0.007 to 0.1 ppm, that is from 6 to 90 Hz on a 900 MHz instrument.

Possible sources of error may arise if the probe is not completely inert. In fact, even a very weak interaction of the probe with the paramagnetic center may cause specific effects on the chemical shift, which may substantially alter the measurement, given the relatively small effects being measured. The best approach is to repeat the measurements using different probes. Another source of error, for very large macromolecules, lies in the sizable χ^{dia} term, whose absolute value may be of the same order of magnitude or even larger than χ^{para}, making the subtraction more critical.

4.6 REFERENCES

[1] H.S. Gutowsky, D.M. McCall, C.P. Slichter (1953) J. Chem. Phys. 21, 279.
[2] H.S. Gutowsky, C.H. Holm (1956) J. Chem. Phys. 25, 1228.
[3] A.C. McLaughlin, J.S. Leigh, Jr. (1973) J. Magn. Reson. 9, 296.
[4] J.S. Leigh, Jr. (1971) J. Magn. Reson. 4, 308.
[5] T.J. Swift, R.E. Connick (1962) J. Chem. Phys. 37, 307.
[6] T.J. Swift (1973) In: La Mar G.N., Horrocks W.D., Jr., Holm R.H. (eds), NMR of Paramagnetic Molecules. Academic Press, New York: pp. 53.
[7] E.R. Johnston (1995) Conc. Magn. Reson. 7, 219.
[8] R.A. Dwek (1973) Nuclear Magnetic Resonance in Biochemistry: Applications to Enzyme Systems. Oxford University Press, London.
[9] A. Abragam (1961) The Principles of Nuclear Magnetism. Oxford University Press, Oxford.

[10] Z. Luz, S. Meiboom (1964) J. Chem. Phys. 40, 2686.
[11] L.P. Hwang, J.H. Freed (1975) J. Chem. Phys. 63, 4017.
[12] C.F. Polnaszek, R.G. Bryant (1978) J. Chem. Phys. 68, 4034.
[13] J.H. Freed (1978) J. Chem. Phys. 68, 4034.
[14] T. Bayburt, R.R. Sharp (1990) J. Chem. Phys. 92, 5892.
[15] J. Kowalewski, personal communication.
[16] D.F. Evans (1959) J. Chem. Soc. 2003.
[17] W.C. Dickinson (1951) Phys. Rev. 81, 717.
[18] W.D. Phillips, M. Poe (1972) Methods Enzymol. 24, 304.

Chapter 5

Transition Metal Ions: Shift and Relaxation

The aim of this chapter is that of providing the reader with guidelines for the interpretation of the spectra of compounds with some common metal ions. Here, only mononuclear complexes are considered. When dealing with the NMR spectra of a paramagnetic compound in solution, one has first of all to figure out the electron relaxation times (and electron relaxation mechanisms) and the electron–nucleus correlation time, then has to guess nuclear relaxation, in order to set the experiments, and finally has to gain structural and dynamic information from the nuclear relaxation properties and from the hyperfine shifts. The experience of the authors and a few other pieces of knowledge will be presented in the following sections without any goal of being comprehensive or reviewing the field.

5.1 IRON

Iron compounds are known with positive oxidation numbers ranging from +2 to +6, the most stable being by far the +2 and +3 states. In complexes, iron(III) commonly gives rise to high ($S = 5/2$) or low ($S = 1/2$) spin states, depending on geometry and ligand field strength. Iron(II) tetrahedral or pseudotetrahedral complexes are always high spin ($S = 2$), octahedral or pseudooctahedral complexes may have either high spin ($S = 2$) or low spin ($S = 0$) ground states. These are the cases discussed hereafter.

5.1.1 Iron(III), high spin

High spin iron(III) complexes occur with some or all weak donor atoms. High spin iron(III) has one unpaired electron in each of the five d orbitals and every orbital can contribute to the overall spin density. The ground state is a sextuplet with an orbitally non degenerate ground state. The orbitals and their occupancy in various symmetries are reported in Fig. 5.1. There are no excited levels with the same spin multiplicity, since moving one electron in an excited d orbital requires spin pairing, and thus electron relaxation is not efficient.

Electron relaxation times have been reported in the range 10^{-11} s to 10^{-9} s. Spin–orbit coupling with an excited quartet causes the splitting of the $S = 5/2$ state [1] (Fig. 5.2). The larger the splitting, the faster the electron relaxation rate and the

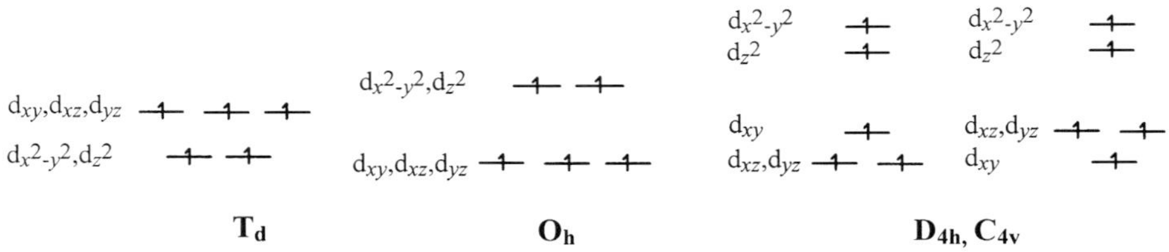

Fig. 5.1. Common d-orbital splitting patterns in high spin iron(III) complexes of tetrahedral (T_d), octahedral (O_h) and tetragonal (D_{4h} or C_{4v}) symmetries.

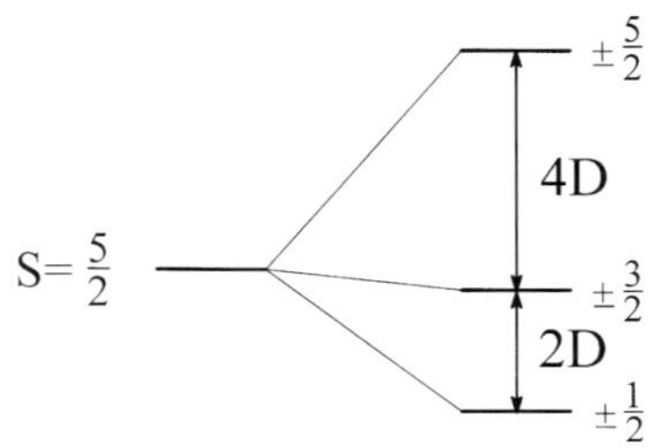

Fig. 5.2. ZFS of the spin sextet levels of high spin iron(III).

sharper the proton NMR lines [2]. When the zero field splitting, D, is small, as it may occur in quasi-symmetrical complexes, the effective electron relaxation time τ_s can be as large as 10^{-9} s [3]. D may be of the order of 10 cm^{-1} in porphyrin complexes with one axial ligand or with two weak ligands, and then τ_s decreases to about 10^{-11} s. In the latter case, hyperfine coupled proton NMR signals can be detected.

5.1.1.1 Water proton relaxation

In order to investigate electron and nuclear properties, it is convenient to study the interactions of a paramagnetic metal ion with water protons, by measuring longitudinal water proton relaxation rates as a function of magnetic field. Details on the technique used to perform relaxation measurements as a function of magnetic field (Field-Cycling relaxometry) are reported in Section 9.6. This type of measurements, called Nuclear Magnetic Relaxation Dispersions (NMRD), permit the determination of the field dependence of R_{1M} (as depicted in Sections 3.4–3.6) by exploiting the chemical exchange of the bound water with the bulk water molecules according to Eq. (4.10). The ^{1}H NMRD profiles of water solutions of $Fe(H_2O)_6^{3+}$ in 1 M perchloric acid at 278, 288, 298, 308 K are shown in Fig. 5.3 [4]. Only one dispersion is displayed at about 7 MHz. It corresponds to a correlation time $\tau_c \approx 3 \times 10^{-11}$ s. The small increase of R_{1p} above 20 MHz at the lowest temperatures makes evident that: (1) τ_s is field dependent; and (2) τ_s is influent in the determination of τ_c. Furthermore, since R_{1p} increases at low

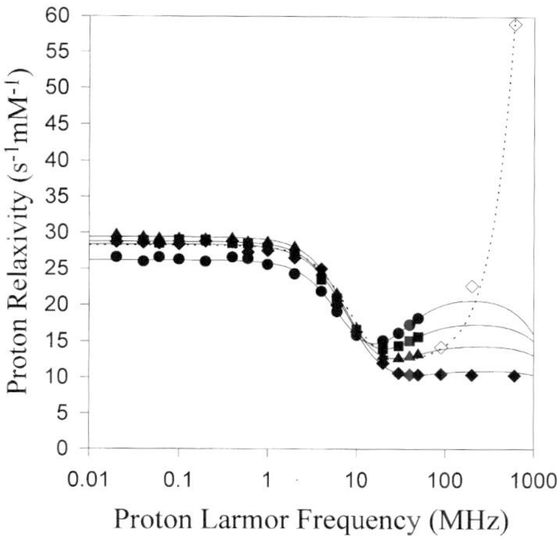

Fig. 5.3. Water proton longitudinal relaxivity as a function of proton Larmor frequency (^{1}H NMRD profiles) for solutions of $Fe(OH_2)_6^{3+}$ at (•) 278 K, (■) 288 K, (▲) 298 K, (◆) 308 K. High field transverse relaxivity data at 308 K (◇) are also shown. The lines represent the best fit curves using the Solomon–Bloembergen–Morgan equations (Eqs. (3.11), (3.12), (3.16), (3.17), (3.26) and (3.27)) [4].

field with increasing the temperature, τ_M may contribute to the nuclear relaxation, according to Eq. (4.10). From independent measurements it was possible to set $\tau_M = 3.8 \times 10^{-7}$ s at 298 K [5]. Both dipolar and contact relaxation are present, although only one dispersion is observed presumably because at low magnetic fields τ_s is similar to or smaller than τ_r, i.e. around 5×10^{-11}–10^{-10} s. The electron lattice relaxation time, τ_v is estimated around 5×10^{-12} s at room temperature, as commonly found for small complexes in water solution (see Table 5.6 in Section 5.4). It is responsible for the high field increase of R_{1p} and R_{2p} (see Section 3.7.2). The evidence of contact relaxation is confirmed by the large increase of the R_{2p} values at high magnetic field, due to the field dependence of τ_s (Fig. 5.3). The constant of contact interaction, A/h, is found to be equal to 0.43 MHz. The parameters obtained by applying the fitting procedure to the profiles at each temperature are reported in Table 5.1. The obtained value of $r = 2.62$ Å is slightly shorter than the distance evaluated with X-ray structure analysis. This could be due to the fact that outer-sphere relaxation has been neglected (it should contribute for about 1 s^{-1} mM^{-1} in the proton relaxivity [6]) as well as any contribution of the second coordination sphere. It may also be noted that τ_r for the water solution is somewhat longer than the value of about 3×10^{-11} s calculated by using the Stokes–Einstein hydrodynamic model (Eq. (3.8)). This may suggest the presence of some second sphere waters. τ_v is of the order expected for the mean lifetime between collisions of molecules and is consistent with those from ESR studies [7,8], as well as Δ_t is [7,8]. On the contrary, we note that the value of A/h obtained from chemical shift measurements is much larger (1.2–1.3

TABLE 5.1
Best fit parameters of the NMRD profiles of hexaaquairon(III) [4]

	τ_r (ps)	τ_v (ps)	τ_M (μs)	τ_{s0} (ps)
278 K	110	6.2	1.7	78
288 K	70	6.1	0.78	79
298 K	53	5.3	0.38	90
308 K	39	3.5	0.20	130

$A/h = 0.43$ MHz, $r_{\text{Fe–H}} = 262$ pm, $\Delta_t = 0.095$ cm^{-1}

MHz [4,5]) than the present one. The ^{1}H NMRD profiles of Fe(III) aqua ions decrease markedly in overall amplitude above pH 3 as a result of the formation and precipitation of a variety of hydroxides.

By increasing the viscosity through glycerol–water mixtures (Fig. 5.4) it is shown that the relative influence of τ_r in τ_c with respect to τ_s becomes lower and lower and the hump in the high field region is thus more and more evident. Because the frequency at which $R_{1\text{p}}$ begins to increase moves gradually to lower field with increasing the viscosity, also τ_v must increase with viscosity. By assuming that r, A and Δ_t (see Eqs. (3.11), (3.16) and (3.26)) are not affected by the presence of glycerol, the fitting of the profile in Fig. 5.4, correspondent to a concentration of 60% of glycerol in solution, provides the values for $\tau_v = 1.4 \times 10^{-11}$ s (instead of 5.3×10^{-12} s, obtained for water solution) and $\tau_r = 2 \times 10^{-10}$ s (instead of 5.3×10^{-11} s), while the number of exchanging protons goes from 12 to 8. This indicates that glycerol replaces the water molecules in the coordination sphere.

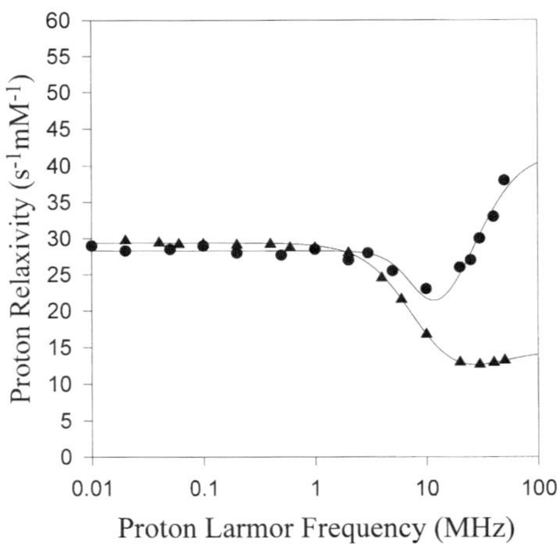

Fig. 5.4. Water ^{1}H NMRD profiles of $Fe(OH_2)_6^{3+}$ at 298 K with (▲) pure water and (•) 60% glycerol. The lines represent the best fit curves using the Solomon–Bloembergen–Morgan equations (Eqs. (3.11), (3.12), (3.16), (3.17), (3.26) and (3.27)) [4].

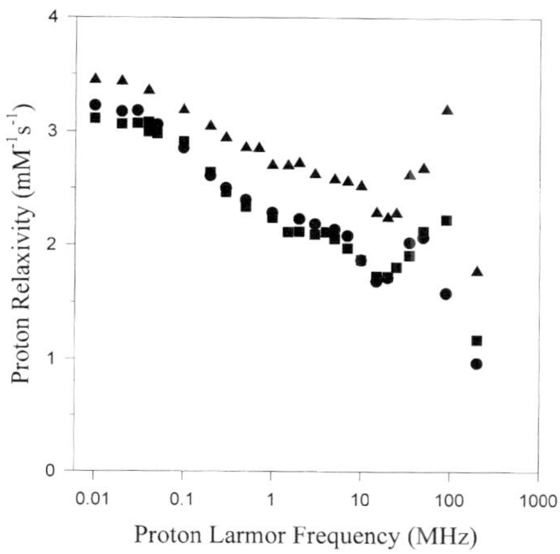

Fig. 5.5. Water ^{1}H NMRD profiles of diferric transferrin solution at (●) 278 K, (■) 293 K, (▲) 308 K [3].

The ^{1}H NMRD profile of the diferric transferrin solution (Fig. 5.5) [3] is also instructive for the case of a macromolecule containing a Fe(III) atom. The profile shows four inflections: the first is ascribed to the ω_S dispersion, the second one to the transition from the dominant ZFS limit to the dominant Zeeman limit (see Section 3.7.1), the following increase is due to the field dependent electron relaxation time (see Section 3.7.2) and finally the ω_I dispersion appears. The best fit analysis provides the presence of a rhombic ZFS with $D = 0.2$, $E/D = 1/3$, in accordance with EPR spectra [9]. The analysis suggests that two sets of electron relaxation times must be considered, in the range $0.3\text{–}1 \times 10^{-9}$ s. In fact, Eqs. (3.11) and (3.12) are inadequate to describe the field dependence of the electron relaxation over the whole range of frequencies due to the presence of static ZFS [10].

Although the $S = 5/2$ system with ZFS is difficult to understand completely in terms of electron relaxation as several different electron transitions are operative, we can conclude that the effective electron relaxation time (as defined in Eqs. (3.11) and (3.12)) is of the order of 10^{-10} s at low fields and that it increases with increasing the field. Under these circumstances there are no hopes to investigate iron(III) compounds with small ZFS by high resolution NMR. On the contrary, Fe(III) complexes have also been investigated as possible contrast agents for MRI [11].

When water is directly coordinated to iron(III) in a macromolecule, slow exchange effects often quench the relaxivity (Fig. 5.6) [12,13]. However, the fluoride derivative of MetMb displays a clear example of a fast exchanging water molecule interacting with the fluoride i.e. by H-bond (Fig. 5.6) [12,13]. In fact, while for methemoglobin the value of the relaxation rate is related to one water molecules coordinated to the paramagnetic center in slow exchange regime, besides the outer-sphere contribution, in fluoro-methemoglobin the water

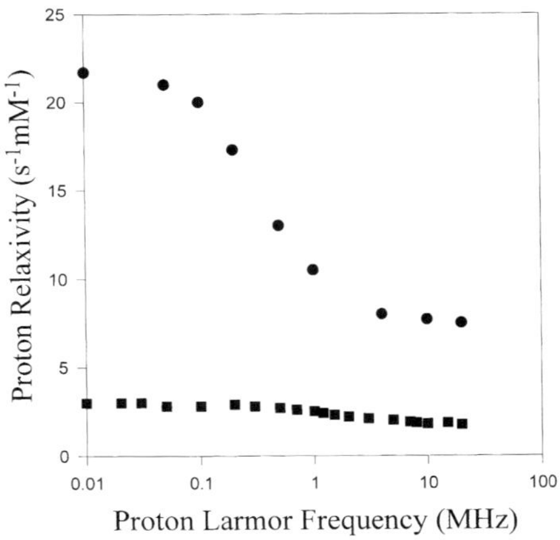

Fig. 5.6. Water ^{1}H NMRD profiles for a solution of methemoglobin (■) and fluoro-methemoglobin (•) at 279 K. In the latter case, fast exchange is responsible for water proton relaxation enhancements which are quenched by slow exchange in the former case [12].

molecule is replaced by fluoride, which is H-bonded to a water molecule. This fast exchanging water molecule is responsible for the overall shape of the NMRD profile, according to the Solomon equation.

5.1.1.2 High resolution NMR

From the above information it appears that high spin iron(III) is not an ideal metal ion to be studied by high resolution NMR because of the dramatic effects on linewidths. In small complexes, τ_r may be the correlation time especially if it is considered that τ_s may increase with the external magnetic field [14].

NMR studies of porphyrin-containing iron(III) complexes are very many owing to their importance in biological systems. The porphyrin systems are tetradentate dianionic ligands and essentially planar, as reported in Fig. 5.7, and allows easy access of monodentate ligands to both the axial coordination positions [15].

The large tetragonal component due to strong equatorial ligands provides relatively large ZFS and, as mentioned, relatively short electron relaxation times. In five-coordinated complexes with a halide in the apical position, the order of the D value is $I^- \geq Br^- > Cl^- > NCS^- > F^-$ [16]. Typical values range from a few wavenumbers to ca. 15 cm^{-1}. Spin–orbit coupling may also mix the $S = 5/2$ ground state with the excited $S = 3/2$ state [17]. This is known in the literature as quantum mechanical spin admixing, which is reported to be relevant either in five-coordinated complexes or six-coordinated with very weak apical ligands [18] (see Section 5.1.4). In porphyrins the electron correlation time contributes to the overall correlation time.

The occupancy of the d orbitals in high spin porphyrin compounds is of the type shown in Fig. 5.1 under the heading D_{4h}. Such occupancy accounts for both

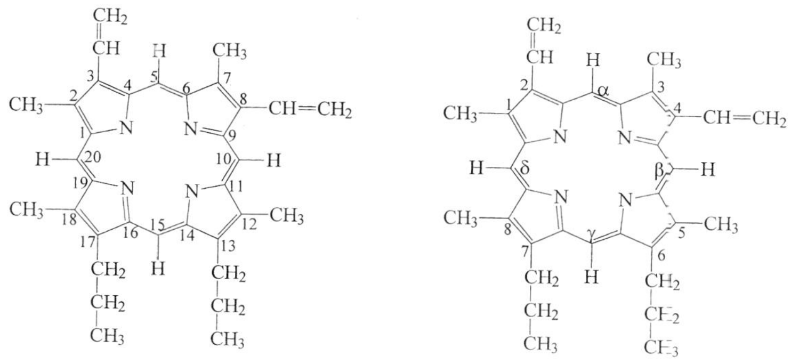

Fig. 5.7. Labeling of the positions in the porphyrin ring. The porphyrin shown is protoporphyrin IX. A is the IUPAC recommendation, while B is still commonly used. The two numbering systems will be used in this book interchangeably.

σ and π spin density distribution on the porphyrin ring. In fact, whereas the $d_{x^2-y^2}$ orbital has the correct symmetry to form σ bonds, the d_{xz} and d_{yz} have the correct symmetry to give rise to π bonds. Therefore, two mechanisms contribute to transfer unpaired electron spin on the resonating nucleus H: delocalization through σ bonds and through the pπ heme orbitals. The $d_{x^2-y^2}$ orbital has the correct symmetry to overlap with the σ bonds of the porphyrin and directly transmits spin density to the resonating protons. The unpaired electrons in the d_{xz} and d_{yz} orbitals delocalize along the pyrrole rings of the heme, through π bonds, and reach the peripheral heme carbon atoms. Thus, the 1s orbital of protons of CH_3 groups slightly overlaps with the pπ of the pyrrole carbons. This overlap depends on the dihedral angle between the pyrrole plane and the plane formed by the pyrrole carbon-side group carbon-attached proton (see Section 2.4), but in the case of methyl protons it is averaged by their rotation around the C–C axis. Finally, through the d_{z^2} orbital, the unpaired electron delocalizes on the coordinated axial ligand.

The σ spin density transfer mechanism accounts for the large downfield shifts (Fig. 5.8A) of signals 2, 3, 7, 8, 12, 13, 17, 18 (numbering as in Fig. 5.7A), or *pyrrole* signals. Consistently, CH_3 and CH_2 substituents in these positions are also downfield shifted. This is nicely shown in the proton NMR spectrum of FeTPPCl [19] (where TPP = 5,10,15,20-tetraphenylporphyrin) and of Fe(protoporphyrin IX dimethylester)Cl [20], as shown in Fig. 5.8. The protons of the phenyl groups at the 5, 10, 15, 20 (or *meso*) positions of FeTPPCl experience some alternating hyperfine shifts, which indicate that some π spin density is present on the phenyl rings as a result of a π spin density at the *meso* position. ^{13}C NMR data confirm the presence of π spin density of positive sign on the *meso* carbons [21]. *Meso* protons are generally downfield in six-coordinated complexes (about 50 ppm) (Fig. 5.9) [20] and upfield in five-coordinated complexes (Fig. 5.12) [22]. Such

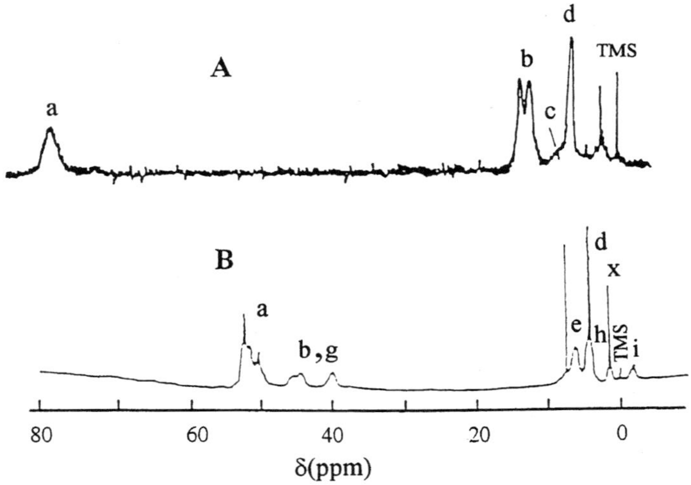

Fig. 5.8. ^{1}H NMR spectra of high spin Fe(III) porphyrins. (A) Fe TPP-Cl (no substituent at the pyrrole positions). Signal a refers to pyrrole protons; signals b, c and d refer to the *meta*, *ortho* and *para* phenyl protons respectively [19]. (B) Fe (protoporphyrin IX)-Cl (see Fig. 5.7A for the ligand). Signals a belong to the methyl groups, signals b and g to the 13,17 α-CH_2 and the 3,8 α-CH; signals d to the COOH; signals e to the 13,17 β-CH_2; signals h and i to the 3,8 β-CH *cis* and β-CH *trans* respectively [20].

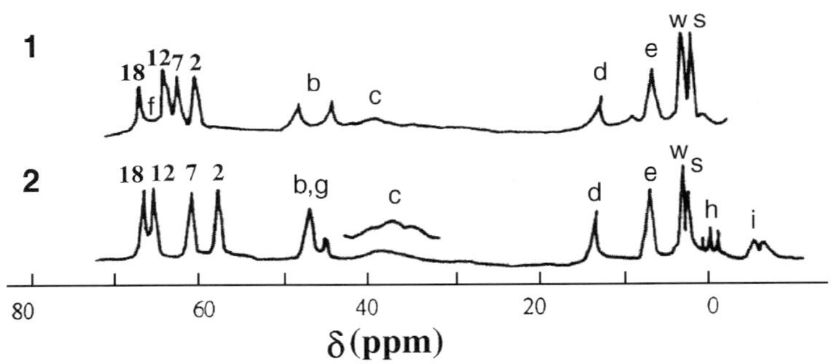

Fig. 5.9. ^{1}H NMR spectra of a six-coordinated iron(III) porphyrin, 1 = 3,8-H; 2 = 3,8-vinyl porphyrin IX (Fig. 5.7A). Numbers 2, 7, 12, 18 refer to methyl positions on the heme; b refers to 13- and 17-propionate α-CH_2 signals; c to the 5, 10, 15 and 20 protons; d to the COOH protons; e to the 13,13 propionate β-CH_2 signals; f to the 3,8 H; g to the 3,8 α-CH; h to the 3,8 β-CH *cis*; i to the 3,8 β-CH trans [20].

difference is ascribed to the coplanarity of the metal with the ligand in the former case which is lost in the latter [20]. A downfield pyrrole proton and methyl shifts, with the methyl shifts smaller, as found for the five-coordinate compounds, is indicative of a dominant σ delocalization. Contact shift arising from delocalized π spin density, on the other hand, yield upfield pyrrole proton and downfield methyl shifts. If both σ and π spin transfers take place, the added effect of the π mechanism on a dominant σ mechanism would be to decrease the downfield pyrrole-H and increase the downfield pyrrole-CH_3 shifts. Experimental data support the conclusion that π spin transfer is more important in six- than in five-coordinate complexes [20,23].

In macromolecules [24,25], τ_s definitely dominates and the linewidths depend dramatically on its value. In turn, τ_s decreases with increasing the magnitude of ZFS [26]. When ZFS is large, as for instance in heme proteins, reasonably sharp lines can be obtained. However the resolution is limited by the fact that, if τ_s becomes shorter than about 10^{-10} s, Curie relaxation, which is always present, becomes the main source of linebroadening. Eqs. (3.30) and (3.27) shows that Curie relaxation decreases with decreasing the rotational correlation time, and thus with the size of the protein, whereas the contact contribution to nuclear relaxation, present for iron coordinated amino acids, does not depend on it. On the other side, contact relaxation increases with increasing electron relaxation time, whereas Curie relaxation is independent on it. These features, sketched in Fig. 5.10, indicate that for relatively small values of τ_r and large values of τ_s the contact term is responsible for signal linewidth, and for relatively large values of τ_r and small values of τ_s Curie relaxation is responsible for signal linewidth. Dipolar relaxation may be important only in a small range between the above described regions. Examples, described below, are shown in Figs. 5.12 and 5.13.

The ^{1}H NMR spectra of ferricytochromes c' are typical of high spin iron(III). The iron atom is pentacoordinated, with four ligands provided by the nitrogen atoms of a porphyrin (see Fig. 5.7) and the fifth ligand being a histidine residue exposed to the solvent. Being a cytochrome of c type, the heme moiety is covalently bound to two cysteinyl residues by means of thioether links (Fig. 5.11).

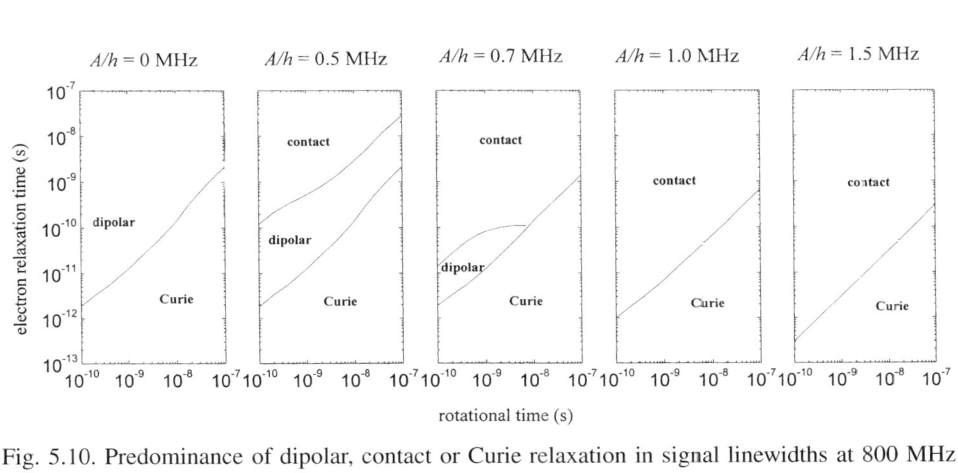

Fig. 5.10. Predominance of dipolar, contact or Curie relaxation in signal linewidths at 800 MHz for different rotational and electron relaxation times, and for different constants for the contact interaction. Calculations have been performed for protons at 5 Å from a $S = 5/2$ ion.

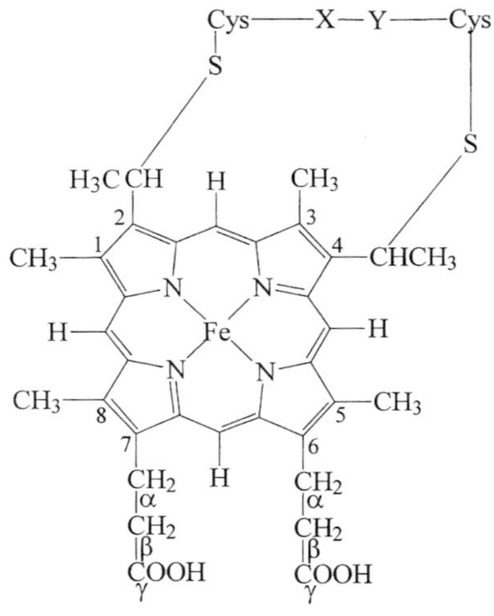

Fig. 5.11. Schematic drawing of the heme moiety in cytochromes *c′* (labeling as in Fig 5.7B).

The spectrum at pH 5.0 consists of four strongly downfield shifted signals each of intensity three (which are due to the four heme methyl groups) and eight downfield signals each of one proton intensity. These signals are due to the two α-CH protons establishing the heme thioether bridge, to the four α-CH_2 protons of the propionate heme side chains, and to the two β-CH_2 of the proximal histidine bound to the iron atom (Fig. 5.12) [27]. A sizable line broadening is observed, induced by Curie relaxation, since the protein is a dimer of 28 kDa. Broad upfield signals observed in the low pH species are attributable to *meso* protons, as expected for five-coordinated high spin Fe(III). Cytochromes *c′* display pH-modulated transitions that can be monitored by means of NMR [27,28]. The spreading of well resolved peaks in the NMR spectrum in fact allows us to follow the changes in the hyperfine shifts by changing pH. The pair corresponding to the α-CH_2 propionate of carbon 7 experiences a dramatic pH dependence at low pH, in accordance with the fact that it belongs to a propionate and should be influenced by the presence of Glu-10, which is another ionizable group in the vicinity of the heme [27]. At high pH the histidine becomes a histidinato ion, thus providing a different shift pattern.

The ^1H NMR spectrum of met-myoglobin [29] is another example of high spin iron(III) (Fig. 5.13). The assignment of the signals experiencing large isotropic shifts is obtained from a combination of isotope labeling and NOEs (see Chapter 7 and Section 9.3) [22]: A to D are assigned to the ring methyl groups in the order 8, 5, 3, and 1; W, Y and Z to three of the four *meso* protons, and all other signals to the other protons of the porphyrin ring. The shifts of the *meso* protons are indicative of six coordination.

The shifts of the nuclei of the heme axial ligand, which often is an imidazole

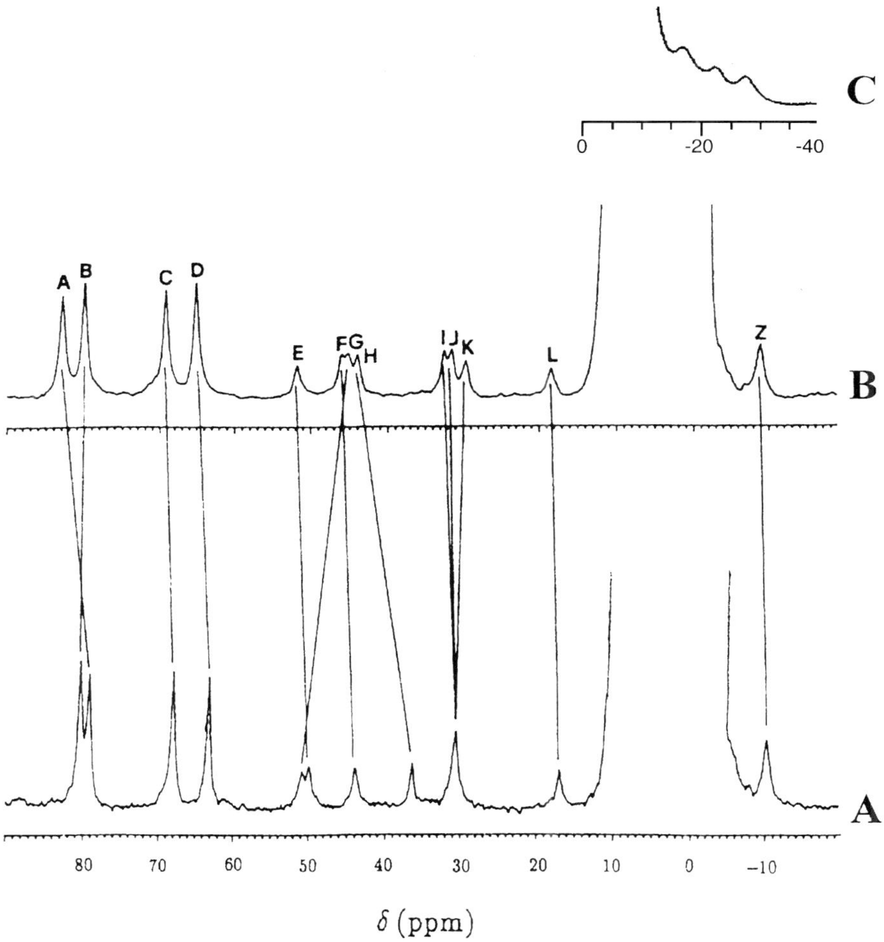

Fig. 5.12. 600 MHz ^1H NMR spectra of a five-coordinated iron(III) porphyrin, ferricytochrome *c′* from *R. gelatinosus* at 300 K in H_2O at (A) pH 7.4 and (B) pH 4.0 (adapted from [27]). Signals have been assigned: A, B, C, D, 1-, 8-, 5- 3-CH_3; E, 6-CHα′; F, 4-CHα; G, 7-CHα′; H, 7-CHα; I, 6-CHα; J, His β; K, His β′; L, 2-CHα. The upfield signals belonging to *meso* protons of cytochrome *c′* from *R. palustris*, at pH 5, are shown in (C) [23] (labeling as in Fig. 5.7B).

ring, are far downfield, indicating the predominance of σ spin delocalization. This is consistent with the presence of an unpaired electron in the d_{z^2} orbital.

In oxidized rubredoxin there is a pseudotetrahedral iron(III) coordinated to four cysteine sulfurs. The cysteine protons are broad beyond detection. The βC^2H_2 signals of ^2H labeled cysteines are located far downfield (300–900 ppm) (Figs. 5.14 and 5.15) [30], which correspond to hyperfine coupling constants of about 1–3 MHz for protons. Two of the four $^2H_\alpha$ protons appear downfield (180 and 150 ppm), while the other two appear upfield (−10 ppm, overlapped). Either

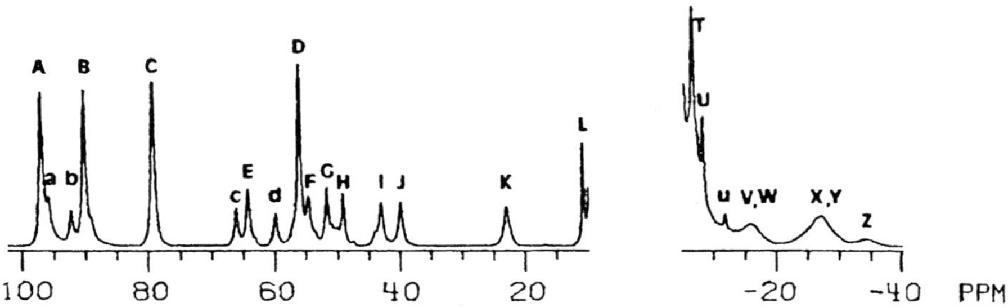

Fig. 5.13. 360 MHz ^{1}H NMR spectrum of *Aplysia* met-myoglobin in ^2H_2O at 298 K and pH 5.9 (adapted from [22]). Chemical shifts are in ppm from DSS. Signal assignments: A, B, C, D, 8-, 5-, 3-, 1-CH_3; E, 7-CHα; F, 6-CHα; G, 4-CHα; H, 6-CHα'; I, 7-CHα'; J, 2-CHα; U, 4-CHβ; W, Y, Z, *meso* H; X, Val-E11 γ-CH_3. a–d methyl peaks correspond to the reversed heme orientation (labeling as in Fig. 5.7B).

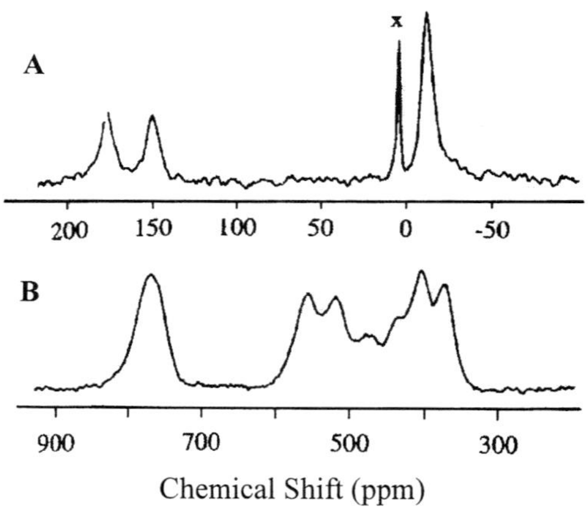

Fig. 5.14. ^{2}H NMR spectra of oxidized [$^2H^\alpha$]Cys-labeled rubredoxin (A) and oxidized [$^2H^{\beta2,\beta3}$]Cys-labeled rubredoxin (B) at 308 K, pH 6.0 (adapted from [30]). The peak labeled x arises from residual $^1H^2HO$.

dipolar or contact contributions of opposite sign as those operative on the βC^2H_2 protons may be responsible for the upfield shifts [30].

5.1.2 *Iron(III), low spin*

Low spin iron(III) occurs with strong ligands and often with hexacoordination. A typical complex is $Fe(CN)_6^{3-}$. The short electron relaxation times have allowed the ^{13}C and ^{15}N investigation of the complexes [31]. Its NMRD in water indicates a $\tau_s \leq 10^{-13}$ s [32]. The electronic configuration of a generic low symmetry low spin iron(III) complex is shown in Fig. 5.16. In distorted octahedral coordination,

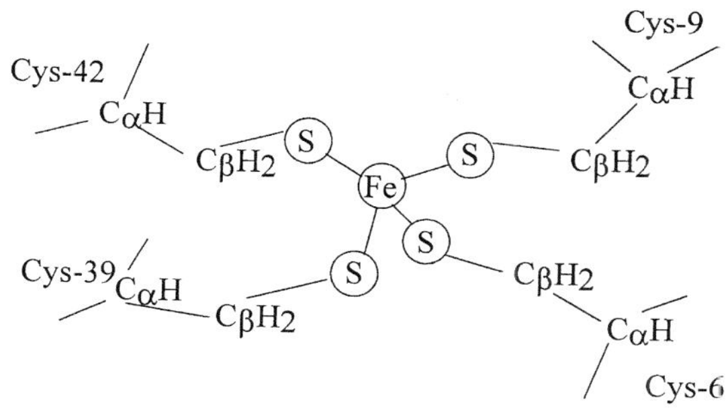

Fig. 5.15. The iron core of rubredoxin from *Clostridium pasteurianum*.

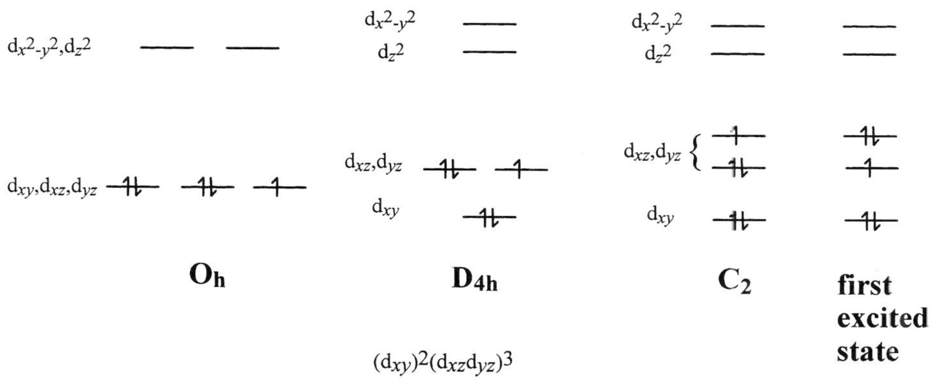

Fig. 5.16. d-orbital splitting in various low spin iron(III) complexes. In distorted symmetries the ground state is $(d_{xy})^2(d_{xz}d_{yz})^3$.

the unpaired electron is in the lowest orbital after filling two other orbitals with two electrons each. The first and second excited states are obtained by promoting one or two electrons so that the unpaired electron resides in the second lowest or lowest orbital. The vicinity of the levels is responsible for fast electron relaxation. Consistent with it, the g values of the ground state significantly deviate from 2 (typically, from 0.9 to 3.5) [1,33]. In heme complexes the effective τ_s is about 10^{-12} s when the apical ligands are His/His, His/Cys or His/CN. In the cases of His/OH^- and Cys/H_2O τ_s is somewhat larger and the g anisotropy decreases.

In D_{4h} (tetragonal) symmetry there is only one unpaired electron in two degenerate orbitals of correct symmetry to give rise to π bonds with a π orbital of the porphyrin moiety [34]. The ^{1}H NMR spectrum of Fe(protoporphyrin IX)–imidazole–cyanide is reported in Fig. 5.17 [35]. The free rotation of the imidazole ring about the metal–nitrogen bond, which is fast on the NMR timescale, simulates a tetragonal symmetry as far as the chemical shifts are concerned [36]. The four methyls are all downfield, though to a quite smaller value than in the case of

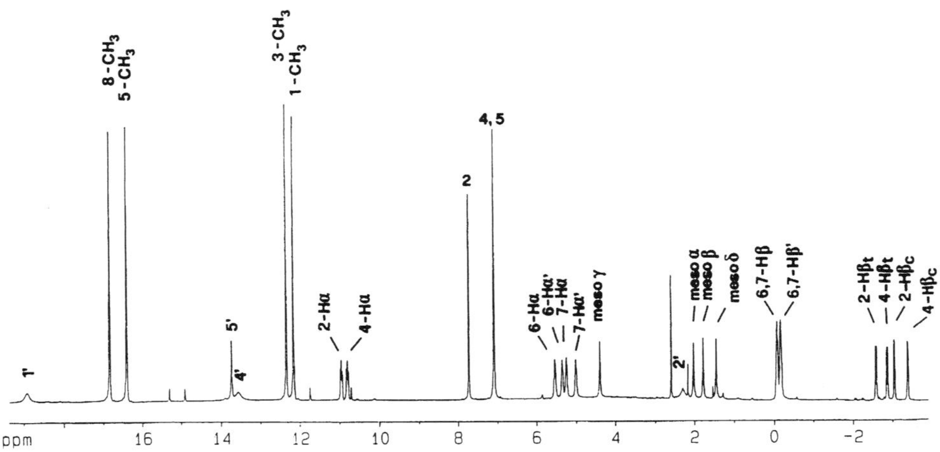

Fig. 5.17. ^{1}H NMR spectrum of Fe(protoporphyrin IX)–imidazole–cyanide (adapted from [35, 60]) (labeling as in Fig. 5.7B).

high spin iron(III) complexes. Now the pseudocontact shifts are larger than in the case of high spin iron(III). An estimate for bis-imidazole systems is reported in Table 5.2 [37,38].

In proteins, the apical ligand is fixed and removes the degeneracy of the d_{xz} and d_{yz} orbitals, thus introducing large anisotropy in the peripheral substituents of the heme ring. As a consequence, in heme proteins, different pyrrole substituents may experience different unpaired electron delocalization depending on the orientation of the axial ligands. The splitting of the d_{xz} and d_{yz} orbitals can be as large as several times kT. When it is of the order of kT, the temperature dependence of the shifts for the heme protons is complicated because of the Boltzmann population of the excited level [39,40].

The NMR spectrum of the heme group of oxidized horse heart cytochrome *c* is shown in Fig. 5.18 [41,42]. Cytochrome *c* is a heme protein where the iron atom is hexacoordinated, with a histidine and a methionine as axial ligands. As common for low spin Fe(III) systems, it presents relatively sharp signals and a

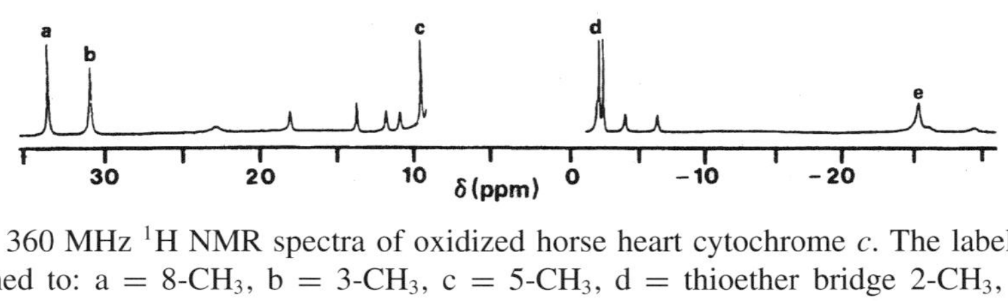

Fig. 5.18. 360 MHz ^{1}H NMR spectra of oxidized horse heart cytochrome *c*. The labeled signals are assigned to: a = 8-CH_3, b = 3-CH_3, c = 5-CH_3, d = thioether bridge 2-CH_3, e = axial methionine S-CH_3; the resonances at 7.4 ppm (1-CH_3) and 3.1 ppm (thioether bridge 4-CH_3) are not shown. Chemical shifts are in ppm from DSS (adapted from [42]) (labeling as in Fig. 5.7B).

TABLE 5.2
Separation of contact and pseudocontact shifts in low-spin bis-imidazole iron(III) porphyrins [37,38] (labeling as in Fig. 5.7B)

Position	Hyperfine shift (ppm)	Pseudocontact shift (ppm)	Contact shift (ppm)
Heme			
meso o-H [a]	−3.09	−3.09	0
meso m-H [a]	−1.49	−1.44	~0
meso p-H [a]	−1.37	−1.27	~0
meso p-CH_3 [b]	−0.94	−0.94	0
pyrrole H [a,b]	−25.4	−5.8	−19.5
meso α-CH_2 [c]	0.6	−4.5	5.1
meso H [d]	−7.0	−9.3	2.3
pyrrole α-CH_2 [d]	2.0	−3.2	5.2
Axial imidazole			
1-H [a]	~2	−9.6	11.6
1-CH_3 [e]	17.2	10.3	6.9
2-H [a]	−9.5	−28.0	18.5
2-CH_3 [e]	~12	~12	~0
4-H [a]	9.7	−8.2	17.9
5-H [a]	4.0	−7.6	11.6
5-CH_3 [e]	15.5	9.0	6.5

[a] For Fe α,β,γ,δ-tetraphenylporphyrin-$(Im)_2^+$.
[b] For Fe α,β,γ,δ-tetratoluylporphyrin-$(Im)_2^+$.
[c] For Fe α,β,γ,δ-tetrapropylporphyrin-$(Im)_2^+$.
[d] For Fe 1,2,3,4,5,6,7,8-octaethylporphyrin-$(Im)_2^+$.
[e] For Fe α,β,γ,δ-tetraphenylporphyrin-$(CH_3Im)_2^+$.

narrow range of isotropic shifts. On the basis of their intensity, the four signals at about 34, 31, 10 and 7 ppm downfield are assigned to the ring methyl groups. A noticeable feature of this spectrum is the larger spreading of the ring methyl resonances with respect to isolated low spin Fe(III) porphyrins. The dipolar shifts are relatively small for all but the *meso* positions, and therefore the spreading of the ring methyl shifts is due to differences in the contact contributions, which may be ascribed to a protein-induced asymmetry of the unpaired electron spin density distribution in the heme group. Hyperfine shifts are provided by both contact and pseudocontact contributions. The latter are absolutely dominant for protons many bonds away from the metal ion, and permit the determination of magnetic susceptibility anisotropies together with the principal direction of the χ tensor and a refinement of the solution structure. Once the correct magnetic susceptibility tensor is available, the contact and pseudocontact contributions to the hyperfine shifts of the heme moiety and of the iron ligands can be separated (Table 5.3). From the knowledge of the g and $\Delta\chi$ values, the contact shift of the heme methyl protons can also be estimated through a ligand field analysis together with the

TABLE 5.3
Separation of contact and pseudocontact contributions to the hyperfine shift for horse heart cytochrome *c* at 293 K [45] (labeling as in Fig. 5.7B)

Atom name	Residue name	Chemical shift		Hyperfine shift (ppm)	Calculated pseudo-contact shift (ppm)	Contact shift (ppm)
		Oxidized species (ppm)	Reduced species (ppm)			
Hα	Met	2.77	3.09	−0.32	−0.03 ± 0.30	−0.29
ε-CH_3	Met	−24.7	−3.30	−21.4	4.42 ± 1.4	−25.8
Hα	His	9.10	3.50	5.30	4.28 ± 0.04	1.04
Hδ2	His	24.6	0.14	24.5	25.7 ± 0.45	−1.20
Hε1	His	−25.7	0.51	−26.2	12.9 ± 0.64	−39.0
8-CH_3	heme	35.7	2.16	33.5	−1.86 ± 0.04	35.4
meso-δH	heme	2.11	9.06	−6.95	−9.06 ± 0.15	2.11
1-CH_3	heme	6.81	3.46	3.35	−4.38 ± 0.07	7.73
2-Hα	heme	−1.33	5.20	−6.53	−4.73 ± 0.05	−1.8
2-CH_3	heme	−2.63	1.46	−4.09	−3.09 ± 0.02	−1.0
meso-βH	heme	1.40	9.30	−7.90	−6.31 ± 0.09	−1.6
3-CH_3	heme	32.8	3.84	29.0	−1.15 ± 0.01	30.1
4-Hα	heme	2.09	6.30	−4.21	−2.70 ± 0.09	−1.51
4-CH_3	heme	3.05	2.57	0.48	−1.37 ± 0.02	1.85
meso-βH	heme	−0.92	9.61	−10.5	−10.63 ± 0.13	0.13
5-CH_3	heme	9.72	3.58	6.14	−4.86 ± 0.03	11.0
meso-γH	heme	7.50	9.64	−2.14	−6.25 ± 0.09	4.11

Kurland and McGarvey approach [43] and an angular dependence of the contact coupling constant (as provided by the following Eq. (5.2)), by taking into account the contributions of the various electronic levels [44].

Homonuclear correlation spectroscopy (COSY) experiments (see Chapter 9) substantiate the theoretical predictions, based on molecular orbital calculation, of the pattern of spin delocalization in the $3e_\pi$ orbitals of low-spin Fe(III) complexes of unsymmetrically substituted tetraphenylporphyrins [46]. Furthermore, the correlations observed show that this π electron spin density distribution is differently modified by the electronic properties of a mono-*ortho*-substituted derivative, depending on the distribution of the electronic effect over both sets of pyrrole rings or only over the immediately adjacent pyrrole rings [46]. No NOESY cross peaks are detectable, consistently with expectations of small NOEs for relatively small molecules and effective paramagnetic relaxation [47].

In the case of His/CN systems, structurally pertinent to myoglobin cyanide, peroxidase cyanide, cytochrome *c* cyanide and the CN^- derivative of a cytochrome *c* mutant, where the axial ligand methionine is substituted with alanine (Ala80–cyt *c*–CN), the simple assignment of the four methyl protons, which is an almost trivial task nowadays, provides direct structural information on the axial ligands, which can be used for structural analysis in solution. The chemical shifts for each

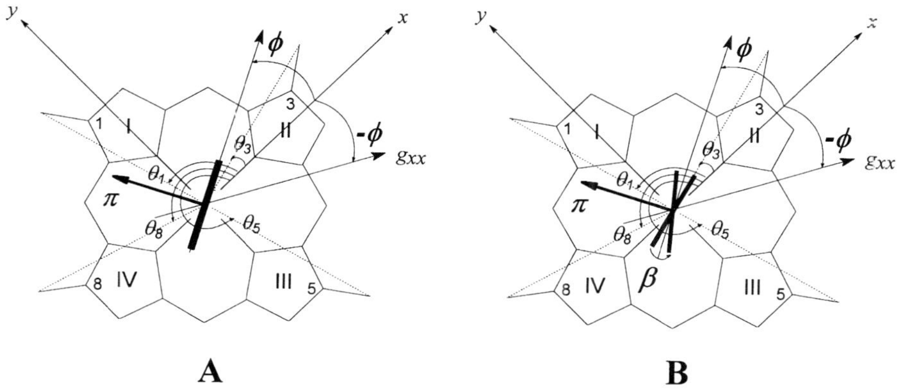

Fig. 5.19. Schematic representation of the heme moiety. The x axis is taken along the metal–pyrrole II direction. The θ_i angles for the four methyl groups are defined as the angles between the metal–methyl ith vector and the x axis. The φ angle defines (A) the direction of the histidine ring plane, in histidine–cyanide systems; and (B) the direction of the bisector of the dihedral angle β formed by the two axial histidines, in bis-histidine systems.

methyl proton at 298 K appear to be dependent on the angle φ between the metal–pyrrole II direction and the direction of the histidine π interaction according to the heuristic equation [48]:

$$\delta_i = a \sin^2(\theta_i - \varphi) + b \cos^2(\theta_i + \varphi) + c + k_i$$
$$a \cong 18.4 \qquad b \cong -0.8 \qquad c \cong 6.1 \tag{5.1}$$

where θ_i is the angle between the metal–ith-methyl direction and the metal–pyrrole II axis (see Fig. 5.19A) and k_i is a correction term to account for an average higher shift value of the protons of methyls 5 and 8 and an average lower shift value of the protons of methyls 1 and 3 ($k_i = \pm 0.7$ and ± 2.7 ppm for c- and b-type heme, respectively). Such equation finds its ground on the dependence on φ of contact and pseudocontact shifts for the methyl protons. The spin density on the pyrrole carbons determines the value of the contact shift of the methyl protons, which is maximal when the histidine plane forms an angle of 90° with the Fe-carbon methyl line and zero when it points at the methyl group (i.e., the methyl is in the histidine plane). This may be expressed as:

$$\delta_i^{\text{con}} \propto \sin^2(\theta_i - \varphi) \tag{5.2}$$

From Eq. (2.20), the pseudocontact shift is proportional to a function of the type

$$\delta^{\text{pc}} \propto k \cos 2\Omega - 1$$

where k is the ratio between the rhombic and the axial susceptibility anisotropy, k

being negative if $g_x < g_y$ as is always the case in the present systems. In these low spin heme systems, the g_x axis is rotated clockwise from a metal–pyrrole direction by an angle which is the same in magnitude but opposite in direction to the angle φ [49]. Therefore, $\Omega = \theta_i + \varphi$, and the sum of contact and pseudocontact shift is consistent with the heuristic Eq. (5.1).

As an example, from the chemical shifts of methyl protons of the cyanide derivative of Met80Ala cytochrome *c*, an angle φ of $57 \pm 9°$ is found by using Eq. (5.1) for the orientation of the histidine plane with respect to the pyrrole II direction, to be compared to the value obtained from the NMR solution structure, equal to 49° (see Fig. 5.20A).

Analogously, in bis-histidine ferriheme proteins, cytochromes b_5 (see Fig. 5.20B) [50,51], cytochromes c_3 [52] and cytochromes c_7 [53], the chemical shifts for each methyl proton at 298 K appear to be dependent on the φ and β structural angles according to the heuristic equation [48]:

$$\begin{aligned} &\delta_i = \cos\beta\left[a\sin^2(\theta_i - \varphi) + b\cos^2(\theta_i + \varphi) + c\right] + d\sin\beta + k_i \\ &a \cong 38.8 \qquad b \cong -10.5 \qquad c \cong -1.1 \qquad d \cong 9.4 \end{aligned} \tag{5.3}$$

where β is the acute angle between the two histidine planes and φ is the angle between the bisector of the angle β and the metal–pyrrole II axis (see Fig. 5.19B).

The $(d_{xz}d_{yz})^4d_{xy}^1$ state in porphyrins is often the excited state (see Fig. 5.16) at about 1500 cm^{-1}. Weak axial ligands decrease its energy relative to the d_{xz}, d_{yz} orbitals. When there is distortion of the porphyrin ring due to bulky axial ligands (with weak σ donating and strong π accepting character), the $(d_{xz}d_{yz})^4d_{xy}^1$ state may become the ground state (Fig. 5.21). In this case, the anisotropy of g is $g_\perp > 2 > g_\parallel$ and the spin delocalization which occurs through π orbitals involving d_{xz} and d_{yz} is small [54–57]. As a consequence, little hyperfine shifts are observed on the β-pyrrole positions. Conversely, large π spin delocalization is observed on the *meso* positions, probably due to spin delocalization from d_{xy} to the porphyrin ring by means of in plane π-bonding to the four porphyrin nitrogen atoms [54].

5.1.3 *Iron(II)*

High spin iron(II) complexes are obtained with weak or medium strength ligands. The electron relaxation times are rather short ($\tau_s \approx 10^{-12}$ s) as the electron configuration probably is as shown in Fig. 5.22 and the excited levels are close in energy for the same reasons as in the case of low spin iron(III). Consistently, the NMRD profile of $Fe(OH_2)_6^{2+}$, obtained from Mohr salt ($(NH_4)_2Fe(SO_4)_2 \cdot 6H_2O$), reported in Fig. 5.23, does not exhibit any dispersion below 50 MHz of proton Larmor frequency.

For $S = 2$ systems, a predominance of σ delocalization is expected as well as moderate π delocalization at the *meso* positions [58]. A spectrum is reported in Fig. 5.24 [58], where the pyrrole protons are downfield.

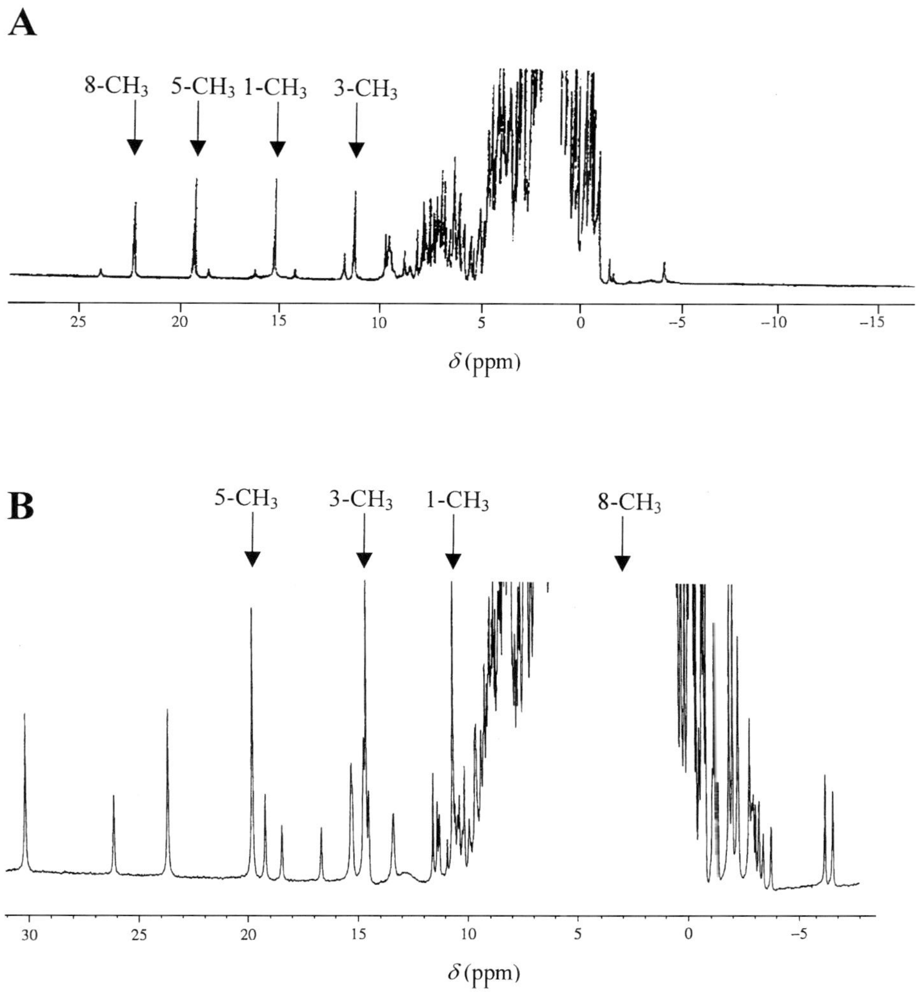

Fig. 5.20. Hyperfine shifts of methyl protons in (A) Met80Ala cytochrome *c*–CN$^-$, and (B) cytochrome b_5. The former is a histidine–cyanide ferriheme protein, since the axial ligand methionine is substituted with alanine, the latter is a bis-histidine ferriheme protein (labeling as in Fig. 5.7B).

In proteins, the position of the axial histidine ring determines the inequivalence of protons in the porphyrin plane. Probably as a result of this, a variety of patterns is observed with either one or two of the four methyl substituents of protoporphyrin IX being downfield [2,27,59–62]. The orientation of the χ_x and χ_y axes in the porphyrin plane depends on the orientation of the axial π interaction, with a behavior analogous to the low spin Fe(III) case, but with the x and y axes interchanged. In other words, when the π interaction is along a metal–pyrrole nitrogen direction, it defines the χ_y and not the χ_x direction. When the π interaction rotates

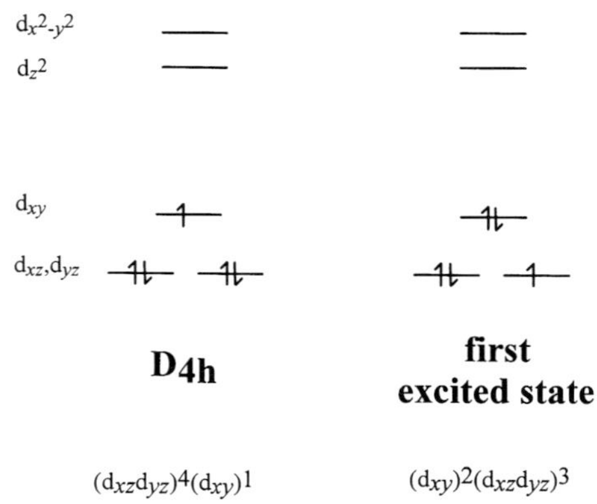

Fig. 5.21. d-orbital splitting in the iron(III) porphyrins with $(d_{xz}d_{yz})^4(d_{xy})^1$ ground state.

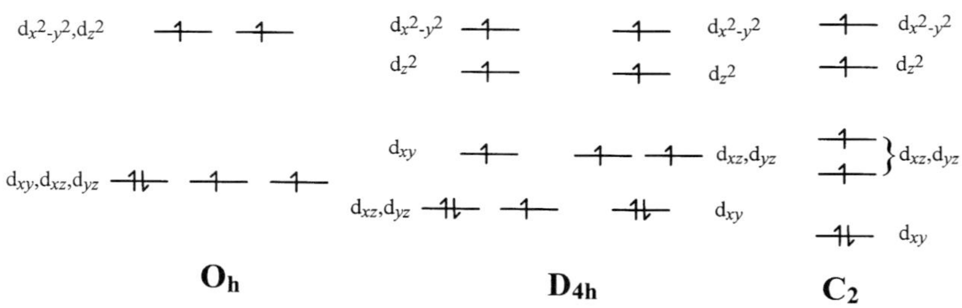

Fig. 5.22. d-orbital splitting in various high spin iron(II) complexes.

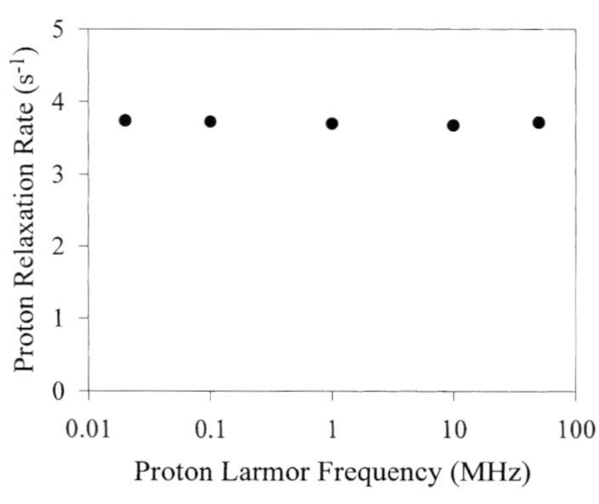

Fig. 5.23. Water ^1H NRMD profiles for 10 mM solution of $(NH_4)_2Fe(SO_4)_2 \cdot 6H_2O$.

clockwise away from the metal pyrrole direction by an angle α, it is χ_y that rotates counterclockwise by α. Therefore, in analogy with the low spin Fe(III) case, the pseudocontact shift can be calculated as a function of the orientation of the axial

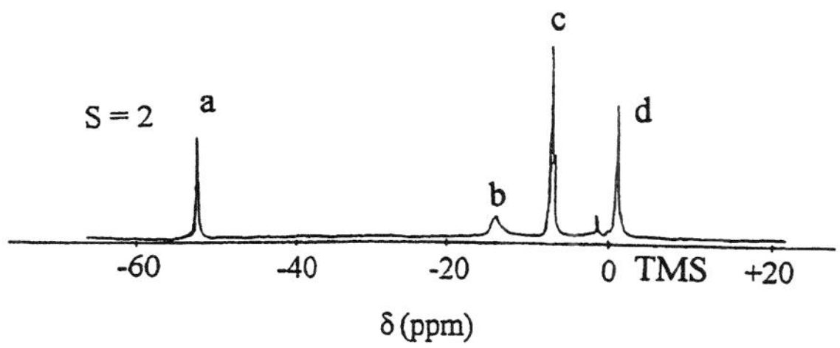

Fig. 5.24. ^{1}H NMR spectrum of $S = 2$ Fe 5,10,15,20-tetraphenylporphyrin (2-CH_3Im) (adapted from [196]). a = 1,2,3,4,5,6,7,8-H; b = 2-CH_3Im 4,5-H; c = phenyl *ortho*-, *meta*- and *para*-H; d = 2-CH_3ImCH_3 (labeling as in Fig. 5.7B).

histidine for all methyl protons, by using Eq. (2.20). The anisotropy of the contact shifts has been predicted to display the same dependence on the orientation of the axial π interaction, being maximal along the direction of the π interaction (i.e. perpendicular to the histidine plane) and zero at 90° from the direction of the π interaction. The angle dependence is thus the same with respect to the low spin Fe(III) case. Therefore, hyperfine shifts result to have a different angle dependence of that observed in low spin Fe(III) systems, as the dependence is the same for the contact contribution and shifted by 90° for the pseudocontact contribution. The overall effect is shown in Fig. 5.25. As an example, in reduced cytochrome c' the shift pattern of the methyls is three signals upfield and one downfield. This is due to the relative weight of contact (which causes downfield shifts) and pseudocontact (which causes upfield shifts if the axial ligand field strength is smaller than the equatorial and therefore $\chi_{\parallel} > \chi_{\perp}$) contribution. These experimental data are in good agreement with predictions, the histidine orientation being along the β–δ *meso* axis, and thus forming relatively small angles with methyl groups 1, 5 and 8, and a large angle with methyl group 3 [63]. In sperm whale deoxymyoglobin, methyl 1 and 5 experience upfield pseudocontact shifts whereas methyls 3 and 8 experience downfield pseudocontact shift [64]. This pattern is in agreement with a position of the histidine along the pyrrole II–pyrrole IV axis [65].

By measuring the linewidth at different magnetic fields (from 90 to 360 MHz) for deoxymyoglobin and for deoxyhemoglobin, the Curie contribution to the overall line broadening was separated, and the rotational correlation time was found to be 5×10^{-9} s for deoxymyoglobin and 2.5×10^{-8} s for deoxyhemoglobin [66]. Unlike the linewidth, longitudinal relaxation times are only negligibly affected by Curie relaxation mechanism, and therefore an independent estimate of the electron relaxation time can be made, provided r is known and the Solomon equation (Eq. (3.19)) holds. Such a value for deoxyhemoglobin was found to be 7×10^{-13} s [67].

Fig. 5.26 shows the ^{1}H-NMR spectrum of iron(II) bleomycin [68]. The high resolution and the relatively narrow line widths observed in the spectrum are as expected for high-spin Fe(II) complexes. Paramagnetically shifted resonances out

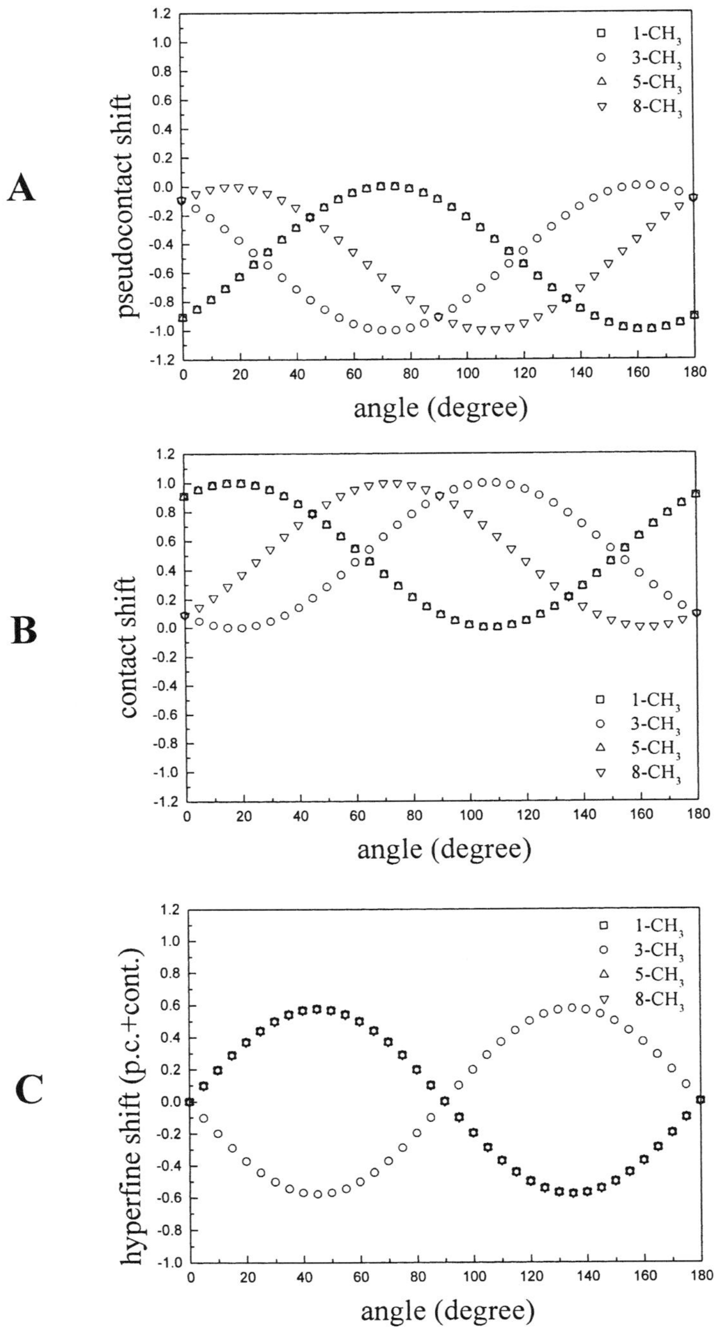

Fig. 5.25. (A) Dependence of the heme methyl pseudocontact shifts as a function of the angle α between the pyrrole II–pyrrole IV molecular axis and the projection of the His plane on the heme. (B) Dependence of the contact shift as a function of the same α angle. (C) Dependence of the sum of the pseudocontact plus contact shifts on the α angle. The sum in the example is obtained by giving the same weight to both contributions [63].

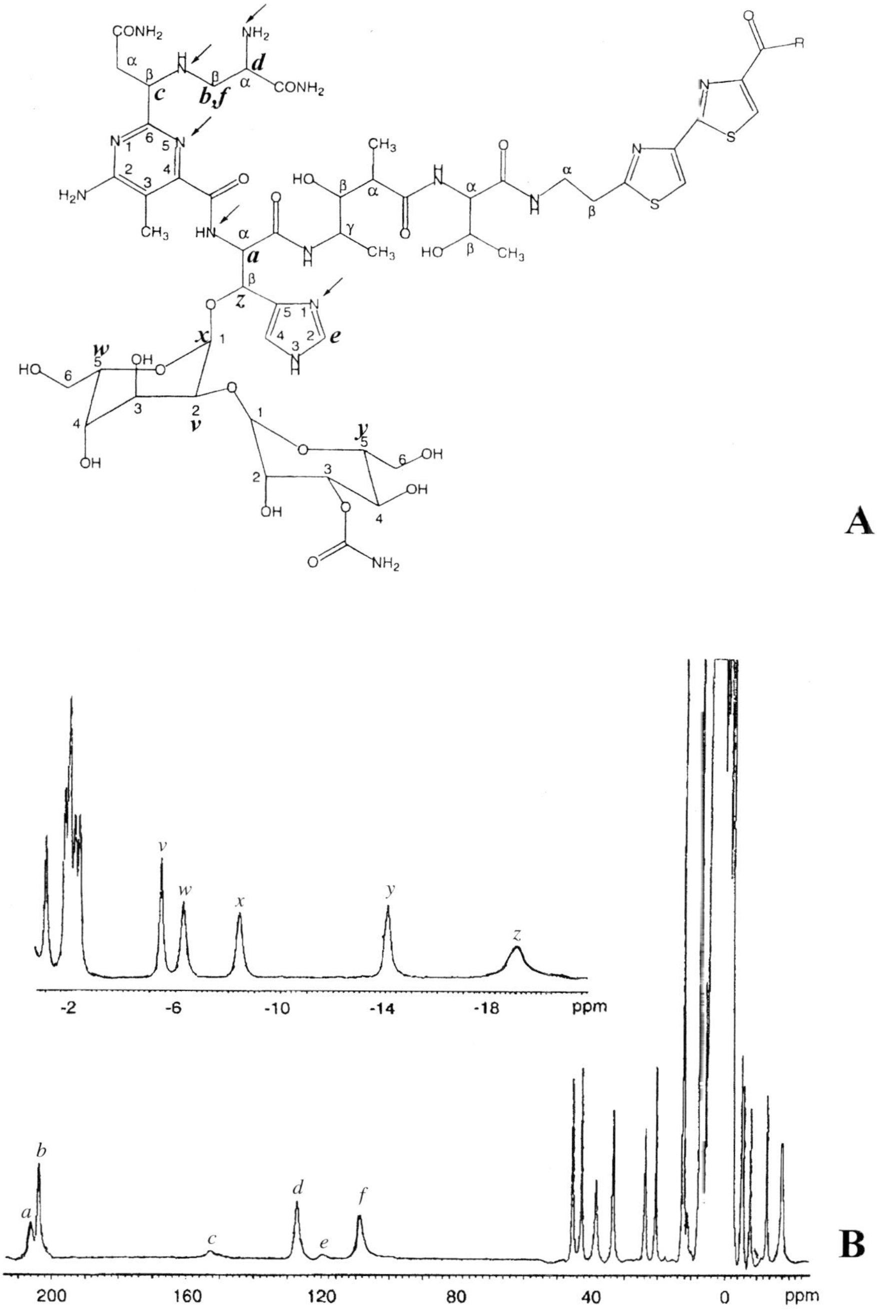

Fig. 5.26. (A) Schematic structure of bleomycin. The arrows indicate the ligands to the metal center. (B) 300 MHz ^{1}H NMR spectrum of Fe(II)–bleomycin at 298 K (adapted from [68]).

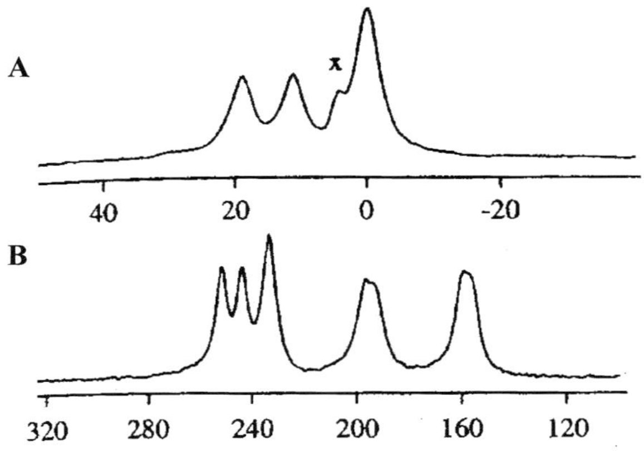

Fig. 5.27. ^{2}H NMR spectra of reduced [^{2}H$^\alpha$]Cys-labeled rubredoxin (A) and reduced [^{2}H$^{\beta2,\beta3}$]Cys-labeled rubredoxin (B) at 308 K, pH 6.0 (adapted from [30]).

to 230 ppm have been observed, and two-dimensional NMR experiments (see Chapter 9), together with the measurement of the T_1 values, allowed the authors to estimate metal–proton distances and to identify the ligands to the metal center (arrows in Fig. 5.26).

In reduced rubredoxin, the eight H$^\beta$ of the four coordinated cysteines are at 280–150 ppm, whereas the four H$^\alpha$ (Fig. 5.27) are little shifted (19 ppm, 16 ppm and 0 ppm).

5.1.4 Spin-admixed Fe(III)-P and high spin–low spin equilibria

High spin Fe(III)–PX compounds, where X is a very weak anion, have spin admixed ground states of $S = {}^5\!/_2$ and $S = {}^3\!/_2$ species. The pure $S = {}^5\!/_2$ and $S = {}^3\!/_2$ states are closely spaced [69], so spin–orbit coupling provides a mechanism for mixing them. Rather than create the commonly observed thermal equilibrium between the two spin states (so called spin crossover), the selection rules of quantum mechanics and spin–orbit coupling allow the two states to mix and to create a new, discrete, admixed ground state. Such admixed $S = {}^3\!/_2, {}^5\!/_2$ states give rise to magnetic properties that lie along a continuum between the extremes of the pure $S = {}^5\!/_2$ and $S = {}^3\!/_2$ states [70]. The difference between the two spin states lies in the fact that the $d_{x^2-y^2}$ orbital is empty for $S = {}^3\!/_2$ species. Since unpaired spin in the $d_{x^2-y^2}$ orbital is associated with predominant σ spin delocalization and downfield pyrrole proton isotropic shifts, whereas unpaired spin in d_{xz} and d_{yz} orbitals results in upfield pyrrole proton isotropic shifts through π spin delocalization, proton NMR resonances are quite sensitive to the $S = {}^5\!/_2$ and $S = {}^3\!/_2$ contributions in a spin-admixed complex. Therefore, in tetraphenylporphyrinate (TPP) complexes,

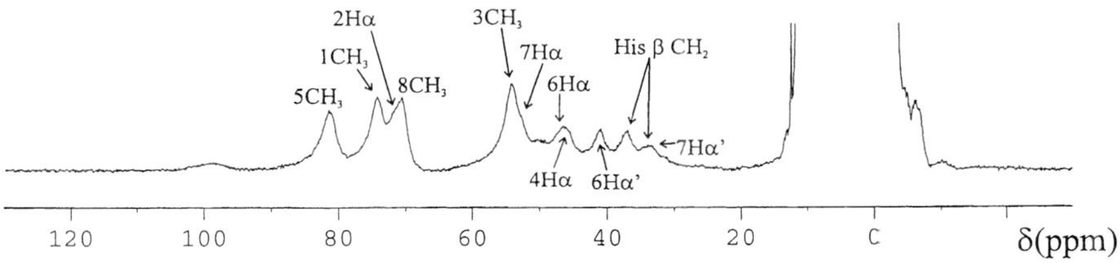

Fig. 5.28. 800 MHz ^{1}H NMR spectrum of horseradish peroxidase in 50 mM phosphate buffer. Shifts are in ppm from DSS (labeling as in Fig. 5.7B).

^{1}H NMR δ_{pyrrole} values shift dramatically upfield with increasing $S = 3/2$ character. High spin species such as FeCl(TPP) have large downfield shift (+80 ppm) of the eight pyrrole protons on the periphery of the porphyrin macrocycle, whereas species approaching pure intermediate spin have upfield shifts that can be as large as −62 ppm [70]. In addition, temperature changes affect the mixing of the two states with dramatic effects on the pyrrole proton shifts [18].

The spectrum of horseradish peroxidase (HRP) is reported in Fig. 5.28, together with the assignment, mainly obtained through the detection of proton–proton dipolar connectivities [71]. Methyl groups are shifted in the range 70–80 ppm downfield. The NH resonance of the histidine ring of the axial ligand occurs at around 100 ppm downfield, while the Hε1 and Hε2 resonances of the same histidine are too broad to be detected. The general pattern of the isotropic shifts is thus indicative of high spin iron(III); however, when the protein is substituted with deuterohemin, in which the two vinyl groups are substituted by pyrrole protons, the signals of the latter are not observed in the usual region around 60–80 ppm downfield, but most probably lie within the diamagnetic region. This has been taken to be consistent with the proposed admixture of $S = 5/2$ and $S = 3/2$ ground states [72]. This also accounts for the fact that the magnetic moment (5.2 μ_B) is less than the spin-only value (5.92 μ_B) for $S = 5/2$, indicating a contribution from a $S < 5/2$ state [71].

Chemical equilibria between species with different spin states are common. When there is chemical equilibrium between $S = 5/2$ and $S = 1/2$, if the equilibrium is fast on the NMR time scale, i.e. the exchange rate between the two states is faster than the difference in resonance frequency of the two states, a weighted average shift is observed (see Section 4.2). When the equilibrium is slow on the NMR time scale, saturation transfer can be observed if the exchange rate between the two states is slower than the difference in resonance frequency of the two states, but faster than the longitudinal relaxation rates of the nucleus in the two environments (Sections 4.3.4 and 8.2). As an example, saturation transfer experiments have been performed to assign heme proton signals in ferric low spin cyanide horseradish peroxidase [73,74]. In ferric high spin horseradish peroxidase large scalar interaction leads to the resolution of most heme signals.

The interconversion between horseradish peroxidase and cyanide horseradish peroxidase in a sample containing both the forms can lead to magnetization transfer between resonances in the two states, due to CN^- exchange. The exchange rate is 2.5 s^{-1} at 328 K, which allows saturation transfer experiments to be performed, and allowed to assign the heme signals of horseradish peroxidase through the assignment of the signals of cyanide horseradish peroxidase, upon saturation of the former. Cyanide horseradish peroxidase, on the other hand, possesses considerable magnetic anisotropy and hence exhibits more favorable relaxation properties and it is easily assigned.

Some iron(II) compounds undergo a spin transition over a small temperature interval from a diamagnetic system at low temperature to a paramagnetic system at higher temperature (spin-crossover transition). Due to the short τ_s in the paramagnetic state of Fe(II), it is possible to detect the proton NMR in the paramagnetic state as well as in the diamagnetic state. The width of the paramagnetic spectrum increases with decreasing temperature following the Curie law, until the spin-crossover transition, where a strong decrease in linewidth occurs and the system becomes diamagnetic [75–77].

5.2 COBALT

Cobalt(II) is a d^7 ion which can be high spin or low spin. Generally, it is low spin in planar or some square pyramidal compounds. The effective electron relaxation times in the low spin state are long enough (10^{-9}–10^{-10} s) so that EPR spectra can be recorded at room temperature [78] and the proton NMR lines are broad. This is due to the high energy of the first excited state (Fig. 5.29). In

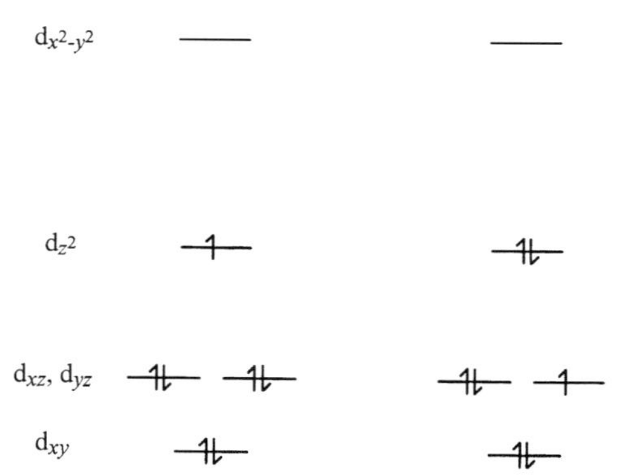

Fig. 5.29. Electron configurations for low spin cobalt(II). The first excited state is often high in energy, and thus the electron relaxation times are at least one order of magnitude longer than in high spin cobalt(II).

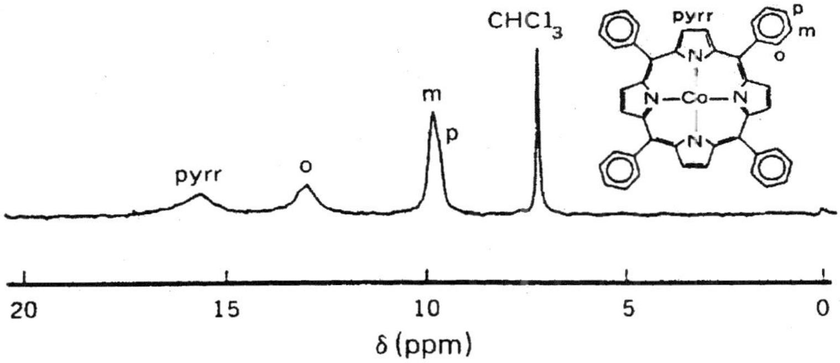

Fig. 5.30. 90 MHz ^{1}H NMR spectrum of tetraphenylporphynato cobalt(II) in $CDCl_3$ at 298 K. Chemical shifts are in ppm from TMS. The signal assignment is also shown (adapted from [80]).

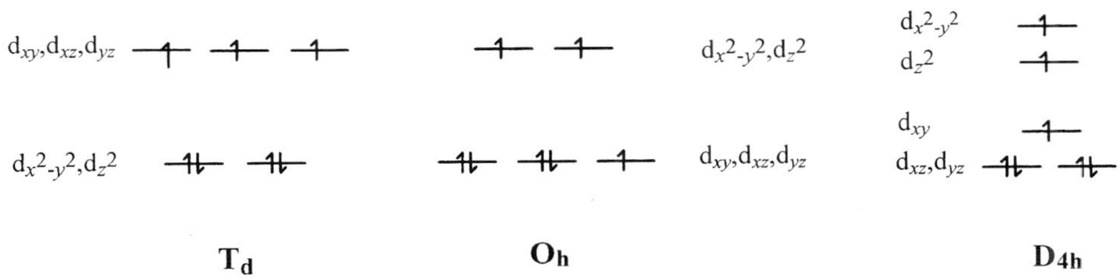

Fig. 5.31. Electron configurations for high spin cobalt(II).

small compounds the electron–nucleus correlation time is given by the rotational correlation time which can be as small as 10^{-11}–10^{-10} s and some proton NMR spectra on systems like cobalt(II) porphyrins have been extensively studied by NMR spectroscopy. The ^{1}H NMR spectrum of cobalt(II) tetraphenylporphyrin, which is reported in Fig. 5.30 [79,80], is consistent with a mainly dipolar nature of the hyperfine shifts. This finding is expected in view of the slight tendency of the unpaired electron in the d_{z^2} orbital to delocalize into the ligand *xy* plane. ^{13}C NMR spectra [79,80], however, show that there is also some spin density on the carbon nuclei of the porphyrin ring. Nitrogen-containing ligands coordinate in the axial position and give rise to five-coordinated adducts [79,80]: the quality of the spectra for such compounds decreases because the electronic relaxation times become longer. Indeed, the first excited state contains two electrons in the d_{z^2} orbital, which is destabilized by the fifth ligand (Fig. 5.29). Therefore, the first excited state becomes higher in energy upon base binding to cobalt(II) porphyrins and the electronic relaxation times become longer.

High spin cobalt(II), a $S = 3/2$ ion, is more interesting from the NMR point of view as the electron relaxation times are shorter that in the low spin case. The electron configurations are reported in Fig. 5.31. High spin cobalt(II) is commonly encountered in four- (tetrahedral), five- (either square pyramidal or

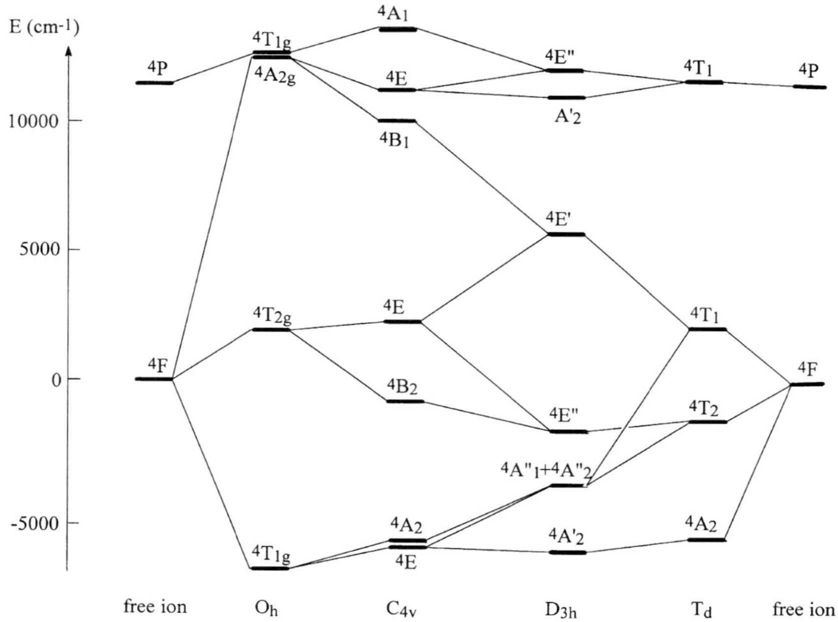

Fig. 5.32. Splitting of the ^{4}F and ^{4}P free ion terms of high spin cobalt(II) in octahedral (O_h), square pyramidal (C_{4v}), trigonal bipyramidal (D_{3h}), and tetrahedral (T_d) geometries [90,191].

trigonal bipyramidal), and six-coordinated (octahedral) complexes. The electronic ground state in octahedral geometry is triply degenerate ($^4T_{1g}$), i.e. is made by three orbitals at the same energy, in square pyramidal geometry is doubly degenerate (^4E), while in trigonal bipyramidal and tetrahedral geometries is non-degenerate ($^4A_2$) (Fig. 5.32). When the ion is in low symmetry situations, for instance when it is in a protein environment, it can be still referred to one of the above high-symmetry situations. The low symmetry removes orbital degeneracy, so that in pseudooctahedral and pseudosquarepyramidal geometries two or one excited states, respectively, are always close to the ground state. Under these conditions, efficient Orbach relaxation mechanisms (see Section 3.3) are operative and electron relaxation is very fast. On the other hand, in pseudotetrahedral (and sometimes in trigonal bipyramidal) complexes the separation in energy between the ground and the excited states is relatively large, and electron relaxation times are one order of magnitude longer.

The water proton NMRD of $Co(OH_2)_6^{2+}$ is reported in Fig. 5.33. The pseudo-octahedral cobalt(II) complex provides almost field-independent water proton R_1 values in the 0.01–60 MHz region [81]. By assuming the validity of the Solomon equation (Eq. (3.16)), both the $\omega_S\tau_c = 1$ and $\omega_I\tau_c = 1$ dispersions can be placed at fields higher than 60 MHz, and therefore an upper limit for τ_c equal to 10^{-12} s can be set. Since the rotational correlation time, τ_r, is likely to be very similar

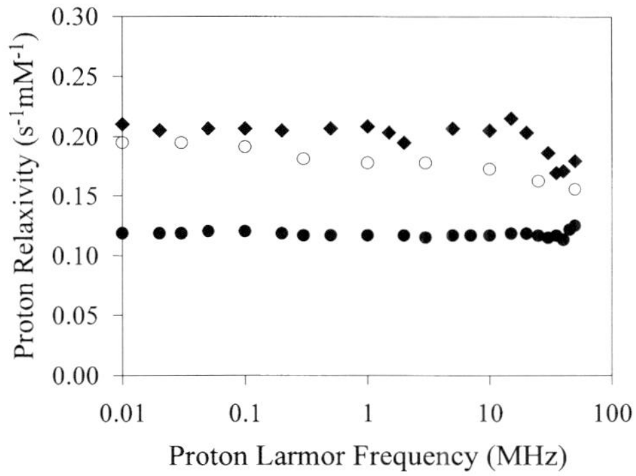

Fig. 5.33. Solvent ^{1}H NMRD profiles for water solution of $Co(ClO_4) \cdot 6H_2O$ at 298 K (○) as compared to those of ethyleneglycol solutions at 264 K (♦) and 298 K (•) [81].

to that of $Fe(OH_2)_6^{2+}$, $Cu(OH_2)_6^{2+}$ and $Mn(OH_2)_6^{2+}$ — about 3×10^{-11} s — a shorter correlation time must dominate the interaction; thus it must be the effective electronic relaxation time, τ_s (see Table 5.6). Such a low τ_s value is consistent with the low water proton R_1 values. If the number of interacting water molecules is the same as for the other aqua ions, τ_s at 298 K can be estimated to be about 3×10^{-12} s from the Solomon equation, or up to 6×10^{-12} s from the equation including the effects of probable static ZFS [81]. When measurements are performed in highly viscous ethyleneglycol the observed rates are similar to those obtained in water. This indicates that τ_s is also similar and that nuclear relaxation is rotation-independent [81]. The resulting picture is fully consistent with very efficient electron relaxation due to Orbach mechanisms (see Table 5.6 in Section 5.4).

With such short τ_s values, in 5 or 6 coordinated cobalt(II) proteins, the metal ion provides negligible paramagnetic effects with respect to the diamagnetic contribution. Therefore, NMRD measurements are uninformative. When the protein contains tetracoordinated cobalt(II), then NMRD becomes again relevant. Water [1]H R_1 measurements of the high pH species of cobalt(II)-substituted carbonic anhydrase (MW 30,000), which is tetrahedral with three coordinated histidines,

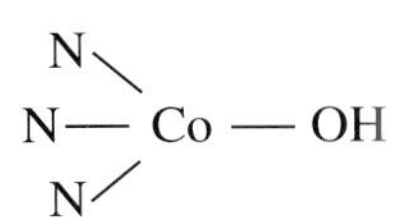

reveal the coordinated hydroxide ion [82]. The NMRD profile (Fig. 5.34) shows a $\omega_S \tau_c = 1$ dispersion centered around 10 MHz, which qualitatively sets the τ_c value around 10^{-11} s. As the rotational correlation time of the molecule is much longer, this value is a measure of the effective electronic relaxation time. Compared to the hexaaqua complex, the nuclear-relaxing capability of this pseudotetrahedral cobalt(II) protein complex is thus relatively large.

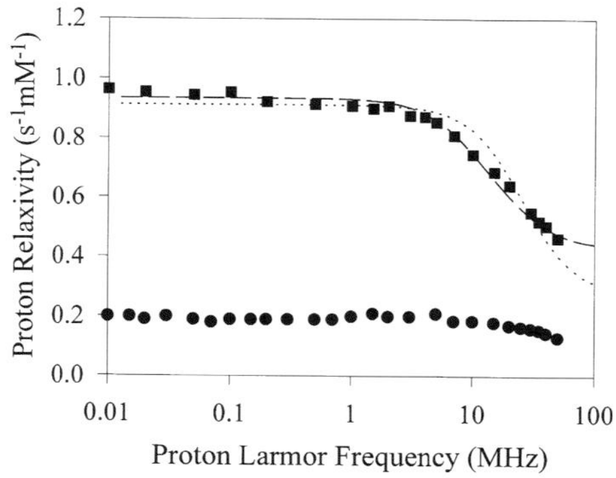

Fig. 5.34. Water ^{1}H NMRD profiles for cobalt(II) human carbonic anhydrase I at pH 9.9 and 298 K (■) [82,83] and for solutions of the nitrate adduct of cobalt(II) bovine carbonic anhydrase II at pH 6.0 and 298 K (•) [90]. The dashed line shows the best fit profile of the former data calculated by including the effect of ZFS, whereas the dotted line shows the best fit profile calculated without the effect of ZFS.

A more quantitative analysis of the data would require consideration of the possible effects of zero field splitting, which is known to be sizeable in cobalt(II) complexes. Such effects have been taken into account to explain the smoother shape of the dispersion curve with respect to what is predicted by the Solomon equation, and to obtain more accurate values of τ_s (see Section 3.7.1) [83].

If a ligand is added in the fifth coordination site of the above system and water is maintained in the coordination sphere [84], the water ^{1}H NMRD profile decreases because the effective electron relaxation time decreases of at least an order of magnitude [81,85]. This is shown by the profile of the nitrate derivative (Fig. 5.34).

Summarizing, data indicate that the six-coordinated $Co(OH_2)_6^{2+}$ chromophore has a τ_s of the order of 10^{-12} s, the tetrahedral CoN_3O chromophore has τ_s values of the order of 10^{-11} s, and five-coordinated chromophores have a τ_s of about 3 to 5 $\times$ 10^{-12} s. These values of τ_s follow the order of availability of low-lying excited states: the closer the excited states, the shorter τ_s. Therefore, six- and five-coordinated high spin cobalt(II) chromophores are expected to display relatively narrow proton NMR signals, whereas tetrahedral complexes provide broader lines [86–91].

In the protein systems, cobalt has been used as a probe for other metal ions like the paramagnetic copper(II) and the diamagnetic zinc(II). For instance, cobalt(II) has been successfully used as spectroscopic probe replacing the copper ion in many blue copper proteins (see Section 5.3) [92–98]. The ^{1}H NMR spectrum of Co(II)-azurin shows several well-resolved signals sizably downfield and upfield shifted with relatively short T_1 values and belonging to residues directly coordinated to the cobalt ion (Fig. 5.35). Numerous other shifted signals are observed closer to the diamagnetic region of the spectrum, due to residues near

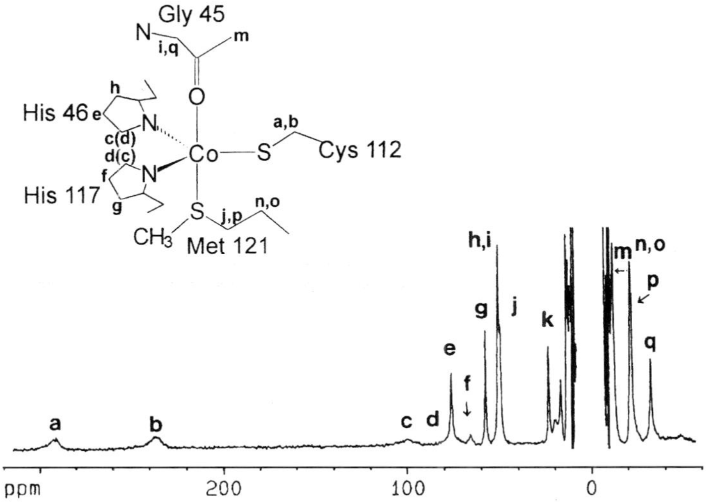

Fig. 5.35. 1D ^{1}H NMR spectrum (200 MHz, 298 K) of Co(II)–azurin in H_2O (adapted from [96]). A schematic drawing of the metal site in *Pseudomonas aeruginosa* native azurin is shown in the upper left corner.

Co(II) but not bound to it. The two β-CH_2 protons of the coordinated cysteine are found far downfield, as generally found for metal-coordinated cysteines, i.e. in the 200–300 ppm region [92,96,99,100]. As always observed for histidines coordinated to high spin Co(II) [92,93], the ring signals are always downfield, between 30 and 80 ppm [89], the sharper being the N−H and the proton in *meta*-like position with respect to the coordinating nitrogen.

The hyperfine shifts of groups bound to the donor atom are largely dominated by the contact interaction, even if pseudocontact shift contributions are sizable and any quantitative use of the shifts should rely on the separated contributions. Longitudinal nuclear relaxation times can be used, and have been used in the case of cobalt substitute stellacyanin, to determine metal–proton distances [101]. The contribution of Curie relaxation, estimated from the field dependence of the linewidths, can be used both for assignment and to determine structural constrains [101].

In the case of cobalt substituted Zn-fingers [102], the differences between the chemical shifts for corresponding resonances in the Co(II) and Zn(II) complexes allow the determination of the orientation and anisotropy of the magnetic susceptibility tensor [103]. Similar studies are available for pseudotetrahedral Co(II) in the zinc site of superoxide dismutase [104] and five coordinated carbonic anhydrase derivatives [105].

Solution structures of RNA and DNA are more difficult to determine, compared to those of proteins, due to the lack of medium and long range NOEs. Long range constraints can be provided by the pseudocontact shifts obtained in the presence of a Co(II) ion (see Section 2.9). The metal ion can be bound to a drug, which then binds the fragment DNA. The resonances range from 50 to −60 ppm [106]. Pseudocontact shifts have been used to obtain global structural information [107].

5.3 COPPER

Copper(II) has a $3d^9$ electronic configuration. In principle, pure octahedral and tetrahedral symmetries can never be observed because Jahn–Teller distortions (see Section 3.3.1) remove the orbital degeneracy of the ground state. The separation of the electronic energy levels depends on the coordination number and stereochemistry, as well as on the nature of the ligands. However, the ground state orbital is always well isolated from the excited states, and therefore the electronic relaxation mechanisms are relatively inefficient. Copper(II) complexes have thus relatively sharp EPR signals, and it is often possible to record these spectra at room temperature.

The water proton NMRD profile of Cu(II) aqua ion at 298 K [108] (Fig. 5.36) is in excellent accordance with what expected from the dipole–dipole relaxation theory, as described by the Solomon equation (Eq. (3.16)). The best fitting procedure applied to a configuration of 12 water protons bound to the metal ion provides a distance between water protons and the paramagnetic center equal to 2.7 Å, and a correlation time equal to 2.6×10^{-11} s, which defines the position of the ω_s dispersion. The correlation time is determined by rotation as expected from the Stokes–Einstein equation (Eq. (3.8)). The electron relaxation time is in fact expected to be one order of magnitude longer (see Table 5.6). This also ensures

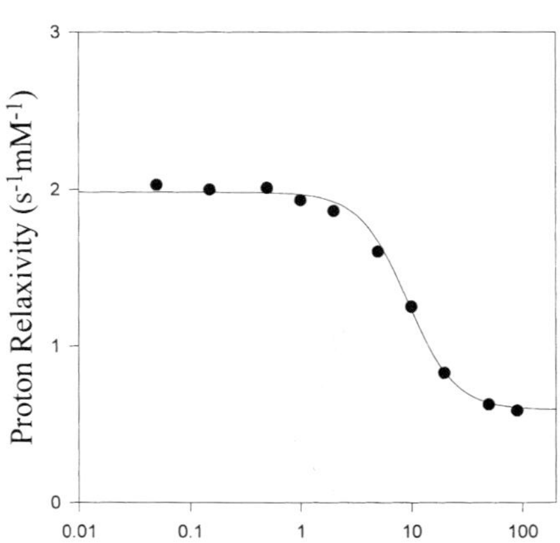

Fig. 5.36. Water ^{1}H NMRD profiles for an aqueous solution of $Cu(OH_2)_6^{2+}$ at 298 K. The solid line represents the best-fit curve obtained using the Solomon equation (Eq. (3.16)), with a Cu–H distance of 2.7 Å and $\tau_c = 2.6 \times 10^{-11}$ s.

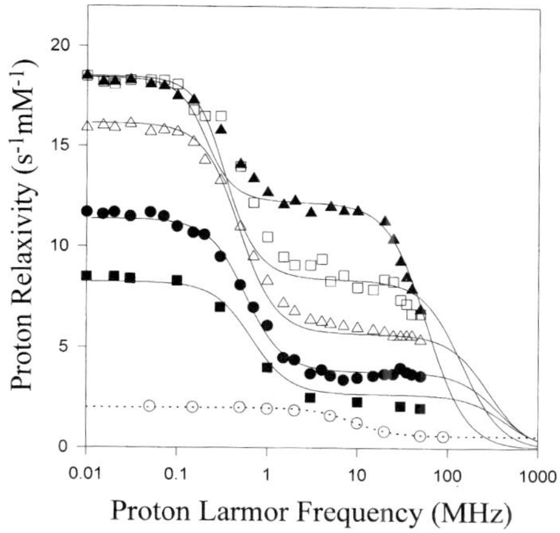

Fig. 5.37. Solvent ^{1}H NMRD profiles for ethyleneglycol solutions of $Cu(ClO_4) \cdot 6H_2O$ at 264 (▲), 278 (□), 288 (△), 298 (•) and 312 (■) K as compared to those of water solution at 298 K (○) [81]. The solid lines are best fit curves obtained using an isotropic A value.

the absence of any contact relaxation contribution as otherwise the corresponding dispersion would have been present in the NMRD profile.

When the viscosity of the solution increases by using ethyleneglycol or glycerol water mixtures as solvent, the rotational correlation time increases. This determines: (1) higher relaxivity values at low frequencies; (2) a shift toward lower frequencies of the ω_s dispersion; (3) the appearance of a second dispersion (ascribed to the ω_I dispersion) at high fields. Temperature dependence studies show that the observed rates are not controlled by exchange, but arise from variation of the rotational correlation time.

Fig. 5.37 shows the ^{1}H NMRD profiles of Cu(II) in ethyleneglycol at different temperatures [81]. The profile maintains the shape predicted by the Solomon equation (Eq. (3.16)) until τ_r is increased so much that the correlation time is determined by the electron relaxation. At this point a sizable change in the profile is induced by the presence of the hyperfine coupling between the metal nucleus ($I = 3/2$) and the unpaired electron ($S = 1/2$), in the range of frequencies where the hyperfine coupling Hamiltonian is larger than the Zeeman Hamiltonian and thus $A \geq g_e \mu_B B_0$, i.e. $A \geq \hbar \tau_c^{-1}$, where A is the electron–copper nucleus hyperfine coupling constant (see Section 3.7.1). The profiles have been fitted with a isotropic hyperfine constant of 0.0026 cm^{-1}. The presence of such term can be easily recognized by observing that the ratio in relaxation rate before and after the ω_s dispersion is different from the expected 10/3 value.

The shape of the ^{1}H NMRD profiles can be very different according to the different values of $A_\parallel$, $A_\perp$ and the position of the proton with respect to the molecular frame defined by the hyperfine $\boldsymbol{A}$ tensor. As an example, the ^{1}H NMRD profiles of the copper(II) protein superoxide dismutase are shown in Fig. 5.38 for

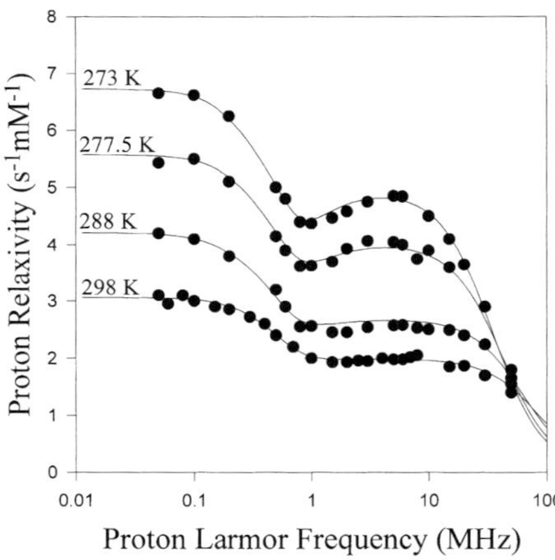

Fig. 5.38. Water ^{1}H NMRD profiles for superoxide dismutase solutions at various temperatures [197]. The solid lines are best-fit curves obtained with the inclusion of the effect of hyperfine coupling with the metal nucleus [109].

different temperatures [109]. Copper(II) in this protein sits in a distorted tetragonal coordination environment (type 2 copper protein). The fits indicate the presence of an axial water molecule, coordinated to the copper ion with a Cu–H distance of 3.2 Å, τ_s values ranging from 4.6×10^{-9} s at 273 K to 1.8×10^{-9} s at 298 K, and $A_{\parallel} = 0.0137$ cm^{-1} and $A_{\perp}$ negligible, as obtained from EPR measurements [110]. Best fit values for several copper proteins are given in Table 5.4. The distance r is reported for two water protons coordinated to the metal ion. If no exchangeable proton is present in the first coordination sphere, any paramagnetic effect would be due to a second sphere and/or outer-sphere relaxation.

The electron relaxation time is always found to be field independent, and has values in the 10^{-8}–10^{-9} s range, i.e. one order of magnitude longer than the aqua ion (Table 5.6). Electron relaxation can occur through modulation of g and A anisotropy, which is τ_r dependent, or through the interaction with collective motions in the case of proteins.

When copper is strongly bound to two sulfur atoms of cysteines and a nitrogen of a histidine in an essentially trigonal ligand environment with one or two weakly bound axial ligands (type 1 copper proteins, Fig. 5.39), the $A_{\parallel}$ values are about half of those of type 2 copper proteins and the τ_s values are reduced to about 5×10^{-10} s [111] (see azurin in Table 5.4). Other examples are ceruloplasmin and copper(II) substituted liver alcohol dehydrogenase [112–114]. For this type of copper center the relatively small energy gap between the ground and the excited states in fact causes the electronic relaxation times to decrease about one order of magnitude.

As far as NMR is concerned, the hyperfine coupled proton lines of copper(II) macromolecules are broad beyond detection, due to the long electronic relax-

TABLE 5.4
Best fit values of NMRD parameters for several copper proteins

Complex	T (K)	$r_{M\text{–}H}$ [a] (Å)	$\tau_c \approx \tau_s$ (ns)	$A_\parallel$ (cm^{-1})	$A_\perp$ (cm^{-1})	Reference
CuBCA II	298	2.8	1.9	0.0131	0.0020	[168,169]
CuBCA II + HCO_3^-	298	3.4	2.1	0.0131	0.0020	[168,169]
CuBCA II + N_3^-	298	2.7	2.6	0.0124	0.0040	[168,169]
CuBCA II + $C_2O_4^{2-}$	298	3.6	3.1	0.0150		[170]
Cu_2Zn_2SOD	298	3.2	1.8	0.0137	0.0040	[109]
	288	3.2	2.5	0.0137	0.0040	[109]
	278	3.2	3.8	0.0137	0.0040	[109]
	273	3.2	4.6	0.0137	0.0040	[109]
Cu_2Zn_2SOD + NCO^-	298	3.9	3.2	0.0158		[171]
Cu_2Zn_2SOD + NCS^-	298	3.6	2.4	0.0148		[171]
Cu_2Zn_2SOD + N_3^-	298	4.9	3.9	0.0157		[171]
Cu_2Zn_2SOD + CN^-	298	5.1	7.7	0.0188		[171]
Cu_2Cu_2SOD	298	3.4	4.2	0.0143	0.0020	[172]
Cu_2E_2SOD	298	3.2	3.6	0.0148	0.0020	[172]
Cu_2TRN	281	3.9	7.6	0.0167	0.0020	[140]
	298	3.7	5.7	0.0167	0.0020	[140]
	311	3.7	5.4	0.0167	0.0020	[140]
Cu_2AP	298	3.0	3.0	0.0164		[173]
Cu_2Mg_4AP + 2P_i	298	3.5	3.0	0.0129		[173]
Benzylamide oxidase	298	2.7	7.0	0.0165		[174]
CuPDO	298	3.4	5.4	0.0153	0.0027	[175]
CuPDO+phthalate	298	2.5	13	0.0168		[175]
Azurin (type 1)	293	5	0.8	0.0062		[114]
Azurin His46Glu (type 2)	293	3.2	15	0.0160		[114]

[a] Assuming two equivalent protons.
Abbreviations: BCA II: bovine carbonic anhydrase II; SOD: superoxide dismutase; TRN: transferrin; AP: alkaline phosphatase; P_i: inorganic phosphate; PDO: phthalate dioxygenase; E = 'empty'.

ation times. Only partial signal assignments of oxidized amicyanin [115] and azurin [116], two blue copper proteins, were possible with standard techniques. The complete spectra of oxidized blue copper proteins plastocyanin, azurin and stellacyanin (see Fig. 5.40) were assigned through saturation transfer with the reduced diamagnetic species (see Table 5.5). To detect the β-CH_2 signals of the cysteine strongly bound to copper(II) in the trigonal plane, an audacious technique has been applied which is based on irradiation of regions where such signals are expected but not detected, since too broad, and the corresponding saturation transfer on the reduced species is observed. In this way, the protons of the copper(II)-coordinated cysteine have been located. The Cys84 β-CH_2 protons of plastocyanin were located at 650 and 489 ppm, with linewidths of 519 and 329 kHz [117]. Analogously, the hyperfine shifts and linewidths of the latter signals

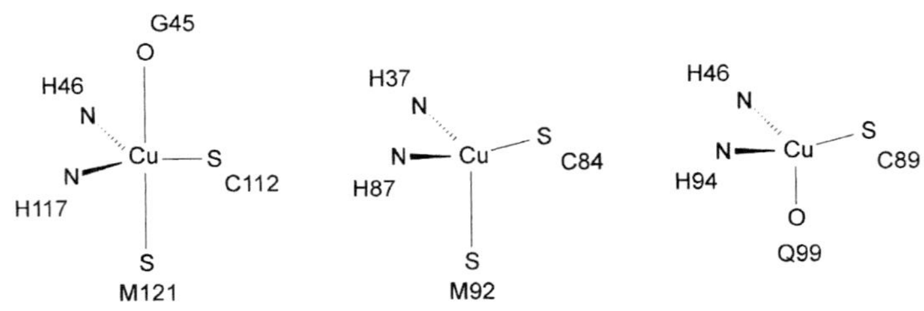

Azurin Plastocyanin Stellacyanin

Fig. 5.39. Schematic drawing of the active site of *P. aeruginosa* azurin, spinach plastocyanin and cucumber stellacyanin (H = histidine, C = cysteine, M = methionine, G = glycine, Q = glutamine) (adapted from [198]).

TABLE 5.5

Assignments of the signals corresponding to copper-ligands in Cu(II) and Cu(I) azurin and stellacyanin recorded at 800 MHz

Assignment	δ_{OX} [a] (ppm)	δ_{RED} (ppm)	$\Delta\nu_{OX}$ (Hz)	A/h (MHz)
Azurin				
Hβ1/2 Cys-112	850	3.48	1.2×10^6	28/27
Hβ2/1 Cys-112	800	2.91	1.2×10^6	27/28
Hδ2 His-117	54.0	6.91	6000	1.61
Hδ2 His-46	49.1	5.92	5500	1.49
Hε1 His-117/46	46.7	6.78		1.45/1.02
Hε1 His-46/117	34.1	6.87		1.06/1.48
Hε2 His-117	27	11.69		
Hε2 His-46	26.9	11.46	1400	0.56
Hα Asn-47	19.9	4.69	1200	0.52
Hα Cys-112	−7.0	5.79		−0.38
NH Asn-47	−30	10.68		−1.3
Stellacyanin				
Hβ1/2 Cys-89	450	2.61	2.8×10^5	16/13
Hβ2/1 Cys-89	375	2.43	2.1×10^5	13/16
Hδ2 His-94/46	55.0	7.01	3800	1.77/1.51
Hδ2 His-46/94	48.0	7.10	3000	1.53/1.78
Hε1 His-94/46	41.2	7.60	3750	1.40/1.01
Hε1 His-46/94	29.8	7.42	2700	0.70/1.09
Hε2 His-46	26	10.10		0.62
Hα Asn-47	16.9	4.46	420	0.47
Hα Cys-89	−7.5	5.10		−0.41
NH Asn-47	−15	10.50		−0.8

[a] The estimated errors are ±0.2 ppm for signals detected directly and ±10% for signals detected indirectly through saturation transfer.

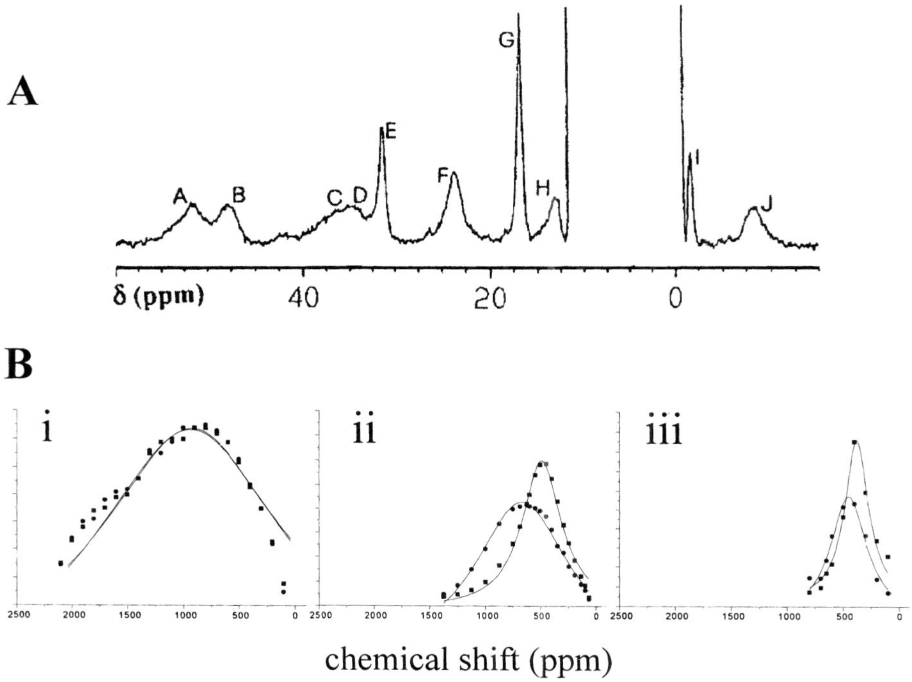

Fig. 5.40. (A) ^{1}H NMR spectra at 298 K of oxidized spinach plastocyanin at 800 MHz (adapted from [117]). (B) Far downfield region of the ^{1}H NMR spectra of oxidized (i) *P. aeruginosa* azurin, (ii) spinach plastocyanin and (iii) cucumber stellacyanin containing signals not observable in direct detection (adapted from [198]). The positions and line widths of the signals were obtained using saturation transfer experiments by plotting the intensity of the respective exchange connectivities with the reduced species as a function of the decoupler irradiation frequency.

for oxidized azurin and stellacyanin were obtained, and it was observed that they differ dramatically from one protein to another: average hyperfine shifts of about 850, 600 and 400 ppm, and average linewidths of 1.2, 0.45 and 0.25 MHz are observed for azurin, plastocyanin and stellacyanin, in that order (see Fig. 5.40B). The contact hyperfine coupling constants for protons belonging to copper(II)-bound protein residues were calculated from the contact shifts, after correcting for the small pseudocontact contributions. The dependence of the contact shift on the Cu–S–C–Hβ dihedral angle is of the $\sin^2\theta$ type, as expected for a spin density on β protons depending from an overlap between the sulfur p orbital and the 1s orbital of hydrogen [118]. The variations among the different proteins examined are interpreted as a measure of the out-of-plane displacement of the copper ion, related to the strength of the axial ligand(s), which increases on passing from azurin to plastocyanin to stellacyanin (Fig. 5.39). Passing from 600 to 800 MHz unambiguously indicates that the Curie contribution is negligible, as expected for small proteins containing $S = 1/2$ metal ions. The 800 MHz ^{1}H NMR spectrum of plastocyanin clearly shows eight downfield and two upfield hyperfine shifted signals, each of them accounting for one proton (see Fig. 5.40A), with very short nuclear relaxation times. They were assigned either through saturation transfer or 1D NOE (see Chapter 7).

The linewidths of the hyperfine shifted signals are determined by dipolar and contact contributions. The former depends on the reciprocal sixth power of the metal–nucleus distance, while the latter depends on the square of the hyperfine coupling constant, which in turn is proportional to the contact shift. Indeed, it was observed [117] that the linewidths of some signals (e.g. the βCH_2 protons of the bound cysteine) are dominated by contact relaxation while some others (e.g. the bound His Hε1 signals) are dominated by dipolar relaxation. In any case, both dipolar and contact contributions are proportional to the electron longitudinal relaxation time τ_s. For signals with similar hyperfine shifts arising from protons at similar distance from the metal ion in the three proteins, the linewidths should hold the same ratio as the τ_s values. According to this reasoning, and referring to the histidine Hδ2 protons, it was qualitatively stated that τ_s is the shortest for plastocyanin. It then should be slightly longer in stellacyanin, and about two times longer in azurin (for which a value of 0.8 ns was estimated [114] (see Table 5.4)).

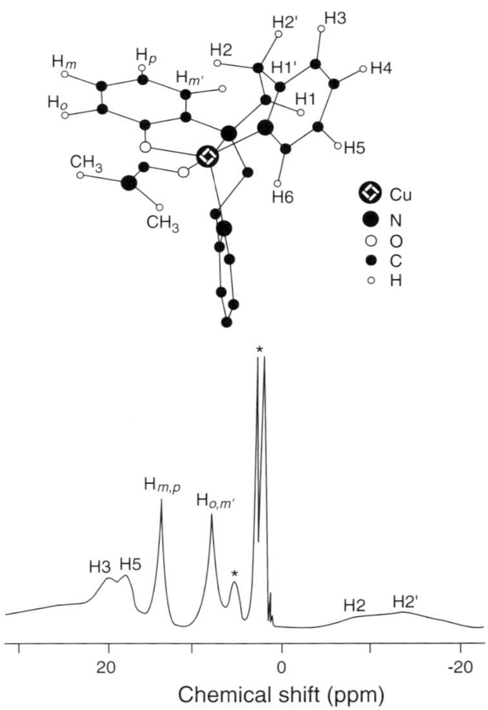

Fig. 5.41. ^{1}H NMR spectrum at 400 MHz of a Cu(II) complex (inset) illustrating the modest spectral resolution (adapted from [119]). The asterisk represents peaks due to DMF

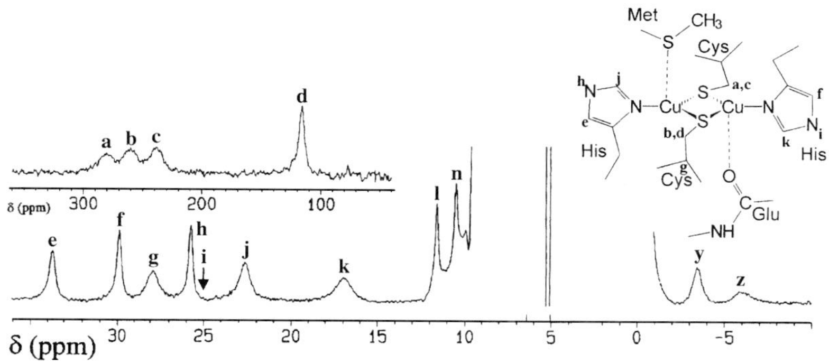

Fig. 5.42. 600 MHz ^1H NMR spectrum of water solutions of the *Thermus* Cu_A domain at pH 8 and 278 K. Signal i is observable at lower pH (adapted from [120]).

A Cu(II) complex with square-pyramidal geometry, and its ^1H NMR spectrum are shown in Fig. 5.41. The relatively broad signals, in the range from 20.4 to -13 ppm, as typical for mononuclear copper(II) systems, are related to a correlation time of 4×10^{-11} s, which is the rotational time, as the electron relaxation time is about two orders of magnitude longer [119]. The reader is referred to Section 6.3.2 for a comparison with a dimeric copper species with similar ligands.

The Cu_A site, common in biology (inset in Fig. 5.42), is dinuclear with two copper atoms bridged by the thiolate sulfurs of two cysteine ligands. One unpaired electron is delocalized over two metals, which are thus $Cu^{1.5+}$. The NMR spectra show narrow lines from the copper ligands (Fig. 5.42) [120,121], corresponding to an electron relaxation time of 10^{-11} s, as in Cu^{2+}–Cu^{2+} dimers (see Section 6.3.2). However, in Cu_A there is no magnetic coupling between the two centers, as they contain only one unpaired electron just as an isolated Cu^{2+} ion. Electron relaxation of Cu_A may be fast because the orbital overlap between the two copper centers provides new relaxation mechanisms not available to a monomer (as Orbach or Raman relaxation).

5.4 MANGANESE

5.4.1 Manganese(II)

Manganese(II) has a $3d^5$ electronic configuration, giving rise to an orbitally non-degenerate ^6S ground term and thus to high spin $S = 5/2$ compounds, just like in the high spin Fe(III) case already discussed (Fig. 5.1). The six-fold spin degeneracy of the ground state in O_h symmetry is only removed through spin–orbit coupling with excited levels. This perturbation splits the spin degeneracy into three levels in a symmetry lower than cubic (Fig. 5.2). Typical ZFS values for manganese(II) are in the range 0 to 1 cm^{-1} [78], i.e. they are very small.

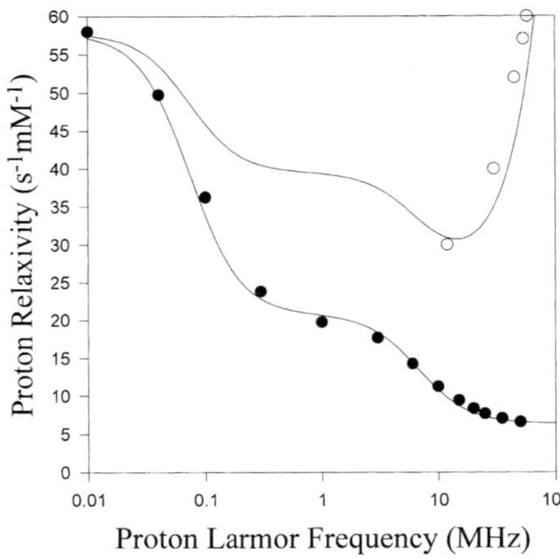

Fig. 5.43. Water ¹H NMRD longitudinal (•) and transverse (○) profiles of $Mn(OH_2)_6^{2+}$ solutions at 298 K.

The relaxation mechanisms for such an ion are bound to the zero field splitting modulation (Table 5.6), which may arise from rotation of the complex or, more probably, from distortions of the coordination sphere as a result of collisions with solvent molecules (Eqs. (3.11) and (3.12)).

The ¹H NMRD profiles of $Mn(OH_2)_6^{2+}$ in water solution show two dispersions (Fig. 5.43). The first (at ca. 0.05 MHz, at 298 K) is attributed to the contact relaxation and the second (at ca. 7 MHz, at 298 K) to the dipolar relaxation. From the best fit procedure, the electron relaxation time, given by $\tau_{s0} = 3.5 \times 10^{-9}$ and $\tau_v = 5.3 \times 10^{-12}$ s, is consistent with the position of the first dispersion, the rotational correlation time $\tau_r = 3.2 \times 10^{-11}$ s is consistent with the position of the second dispersion and is in accordance with the value expected for hexaaquametal(II) complexes, the water proton–metal center distance is 2.7 Å and the constant of contact interaction is 0.65 MHz (see Table 5.6). The impressive increase of R_2 at high fields is due to the field dependence of the electron relaxation time and to the presence of a non-dispersive τ_s term in the equation for contact relaxation (see Section 3.7.2). If it were not for the finite residence time, τ_M, of the water molecules in the coordination sphere, the increase in R_2 could continue as long as the electron relaxation time increases.

Fig. 5.44 reports the NMRD profiles of Mn(II) at different concentrations of d^8 glycerol and the best fit profiles (assuming $\Delta_t = 0.03$ cm^{-1}, $A/h = 0.64$ MHz and $r = 2.78$ Å) [122]. Notably, both τ_r and τ_v increase linearly with viscosity, pointing out that modulation of the quadratic ZFS by collision with solvent is the dominant source of relaxation up to very high viscosity. As a consequence τ_{s0} decreases with increasing viscosity. The range for τ_r spreads from 3×10^{-11} s for water solution to 3×10^{-10} s for a solution containing 65% w/w of glycerol; for τ_v, from 6×10^{-12} to 5×10^{-11} s, at 288 K. The fitting was obtained by

TABLE 5.6
Summary of the ^{1}H NMRD parameters for some aqua ion complexes at 298 K [a]

Metal ion	I	S	A/h (MHz)	τ_{s0} (ps)	τ_v (ps)	τ_c	Main electron relaxation mechanism	Reference
Cu(II)	3/2	1/2	0–0.2	300	–	τ_r	Raman, Orbach, spin-rotation	[108,176–181]
VO(IV)	7/2	1/2	2.1	500	6	τ_r	A-anisotropy	[139]
Ti(III)		1/2	4.5	40	–	τ_r-τ_s	Orbach	[142]
Mn(II)	5/2	5/2	0.6–1.0	3500	5.3	τ_r	ZFS modulation	[122,182,183]
Fe(III)		5/2	0.4–1.3	90	5.3	τ_r-τ_s	ZFS modulation	[4,182,184,185]
Fe(II)		2		≈ 1	–	τ_s	Orbach	
Cr(III)		3/2	2.0–2.3	400	2.3	τ_r	ZFS modulation	[180,182]
Co(II)	7/2	3/2	0.4	3–6	–	τ_s	Orbach	[81,180,183,186]
Ni(II)		1	0.2	3–10	2.2	τ_s	ZFS modulation	[130,180,187,188]
Gd(III)		7/2	0	120	16	τ_r	ZFS modulation	[81,145,186]
Ln(III)			0	0.1–1	–	τ_s	Orbach	[189–194]

[a] See also Sections 3.2, 3.3 and 3.5.

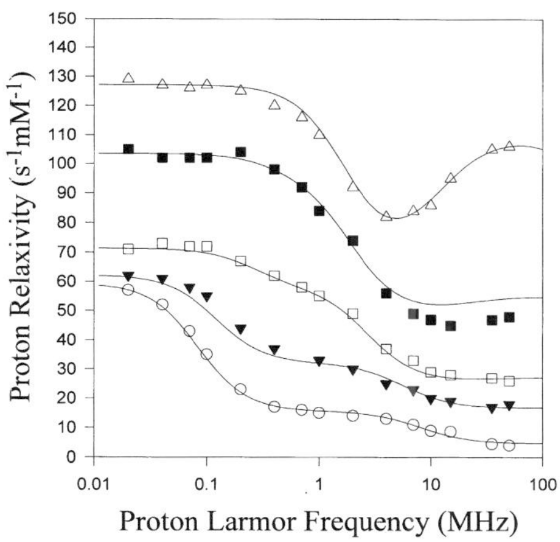

Fig. 5.44. Water ^{1}H NMRD profiles for $Mn(OH_2)_6^{2+}$ solutions at 308 K in pure water (○) and with increasing amounts of d^8-glycerol: 35% (▼), 55% (□), 65% (■), 75% (△) [122].

keeping the number of bound waters free to change with the concentration of d^8 glycerol, because of the possible substitution of water by glycerol, and by taking into account the outer-sphere contribution, theoretically estimated around 5–10% of the total relaxivity [122].

Fig. 5.45 shows the NMRD profiles obtained with Mn(II) bound to the protein concanavalin A. Both contact and dipolar relaxation are now dispersed at the same frequency, the rotational correlation time being longer than the electron relaxation time. After the beginning of the ω_s dispersion, the profile is dominated by the

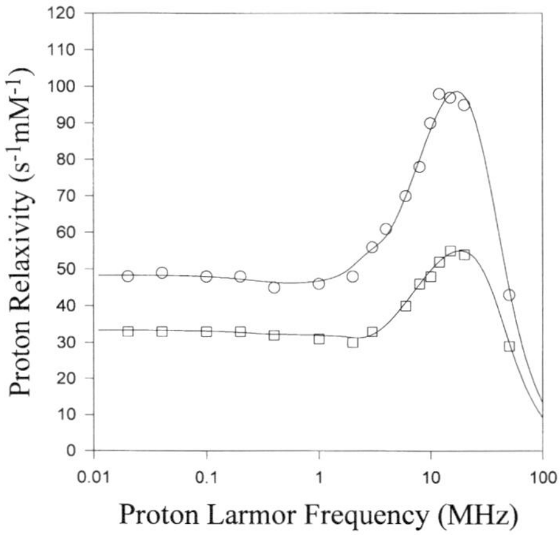

Fig. 5.45. Water ^{1}H NMRD profiles for solutions of Mn^{2+} concanavalin A at 298 (○) and 278 (□) K [199]. The solid curves are calculated with $D = 0.04$ cm^{-1}.

increase of the electron relaxation time, with τ_v of about 6×10^{-11} s. The final dispersion is then due to the ω_I dispersion. Inclusion of ZFS is also required, as found from EPR measurements [123].

In all Mn(II) proteins and in most complexes the contact interaction is found negligible. In fact, the ^{1}H NMRD profile of MnEDTA, for instance, indicates the presence of the dipolar contribution only, and one water bound to the complex. The relaxation rate of manganese(II) complexes with DTPA (see Fig. 5.56) is instead provided by outer-sphere relaxation only, since no water molecules are bound to the complex (see Section 4.5.2).

Proton NMR spectra for Mn(II) systems are expected to have very broad signals, broader than for high spin Fe(III) systems, since, the electronic configuration being the same, ZFS is smaller and thus the electron relaxation time longer.

5.4.2 Manganese(III)

Manganese(III) is a d^{4} ion and generally gives rise to high spin $S = 2$ compounds. ZFS is in general $\leq$1 cm^{-1}, except for porphyrin derivatives, where it is larger [124]. The electron relaxation time in hexaaqua complexes is $\leq 10^{-11}$ s [125], and about 10^{-10} s in Mn(III)–porphyrin [26]. In the latter complexes the electron relaxation time is relatively longer since the two more excited 3d orbitals (which contain one electron only) are not very close in energy (Fig. 5.46A).

The NMRD profiles of manganese(III) porphyrins (Fig. 5.46B) are often characterized by an increase in the relaxivity above about 2 MHz, so that relaxivity at 10–100 MHz is larger than for Mn(II), Fe(III), Cu(II) complexes. Such a hump is probably caused by a magnetic field dependent electron relaxation time.

^{1}H NMR spectra of manganese(III)porphyrin derivatives are available [26,36,

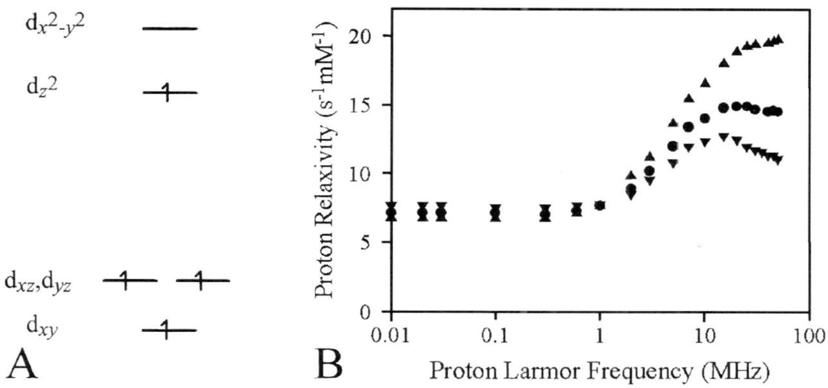

Fig. 5.46. Electronic configurations for high spin manganese(III) (A) and ^{1}H NMRD profiles of a water solution of the tetraphenylsulfanyl porphyrin ($TPPS_4$) manganese(III) complex at 278 (▲), 293 (•) and 308 (▼) K [200] (B).

126] in five- and six-coordinated adducts. The ^{1}H NMR spectrum in $CDCl_3$ of the five-coordinated tetra-*p*-tolyl porphynato manganese(III) (Table 5.7) chloride shows alternation of sign of the isotropic shifts of the *meta* and *para* protons and methyl substituents of the aryl moiety. These shifts, although very small, rule out the possibility of sizeable pseudocontact contributions. The large upfield shifts for the pyrrole protons, which compare with downfield shifts for α-CH_2 substituents, indicate π spin density on pyrrole rings. Predominance of σ spin density is proposed for the *meso* protons [36]. The π spin density distribution is consistent with porphyrin-to-metal π charge transfer as the delocalization mechanism.

5.5 CHROMIUM

Chromium(III) is generally six coordinated. The NMRD profiles for hexaaqua chromium(III) are shown in Fig. 5.47. The position of the first dispersion, in the higher temperature profile, indicates a correlation time equal to 4×10^{-10} s. Since it is too long to be the rotational correlation time, it has to be ascribed to the electron relaxation time (see Table 5.6). Therefore, such dispersion must be due to contact relaxation. The constant of contact interaction, A/h, was found equal to 2 MHz. The high field dispersion is related to dipolar relaxation, modulated by the rotational correlation time $\tau_r = 3 \times 10^{-11}$ s. According to Stokes–Einstein law, this time increases with decreasing temperature, and correspondingly the position of the dispersion moves toward lower fields. The exchange of the coordinated water molecules is very slow (10^5 to 10^6 s), whereas proton exchange is relatively fast ($\tau_M \approx 5 \times 10^{-6}$ s). Therefore, R_1 is not affected by τ_M only above 300 K, while for lower temperatures the exchange rate causes the disappearing of the first dispersion.

TABLE 5.7
Isotropic shift data (ppm) for manganese(III) porphyrins in $CDCl_3$ solutions [36,126,195]

Complex	Temperature (K)	Pyrrole-H	Phenyl protons				Propyl protons			Ethyl protons		Methyl protons	*meso*-H
			o-H [a]	*m*-H	*p*-H	CH_3	α-CH_2	β-CH_2	γ-CH_3	α-CH_2	β-CH_3		
TPPMnCl	308	−30.3		+0.4	−0.4								
o-CH_3 TPPMnCl	308	−29.0		+0.5	−0.7	−0.4							
m-CH_3 TPPMnCl	308	−30.0		+0.5	−0.6	−0.22							
p-CH_3 TPPMnCl	308	−30.2		+0.6		+0.29							
T-*n*-PrPmnCl	308	−29.3					+4.7	−0.5	≈0				
OEPMnCl	308									+18.2	+0.7		+41.4 [b]
EPMnCl	294									+18.5	+0.7	+31.7	−20.6
MPDMEMnCl	294									+17.0	+0.9	+31.9, +36.0	−20.5, −23.5

TPP, tetraphenylporphyrin; T-*n*-PrP, tetra-*n*-propylporphyrin; OEP, octaethylporphyrin; EP, etioporphyrin; MPDME, mesoporphyrin IX dimethylester.
[a] Unresolved signal.
[b] Assignment controversial [36].

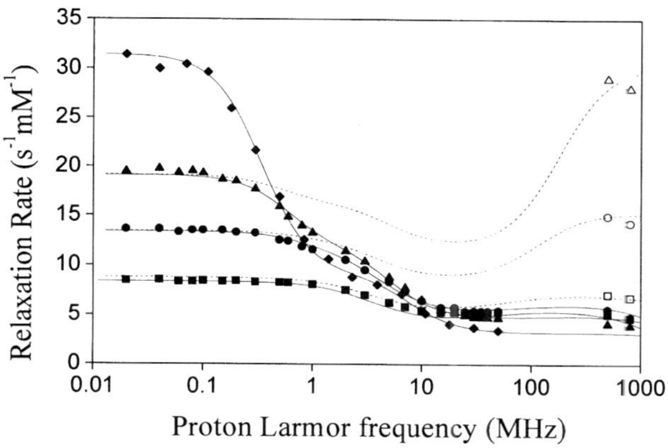

Fig. 5.47. Water ^{1}H NMRD profiles for $Cr(OH_2)_6^{2+}$ solutions at pH 0 and 278 (■), 298 (•), 313 (▲) and 333 (♦) K. Solid symbols indicate R_1 measurements and open symbols R_2 measurements. The solid lines represent the best fit profiles of R_1; dashed lines indicate the best fit profiles of R_2 [201].

The relative long values of τ_s arises from the orbitally non-degenerate ground state of the chromium(III) ion, which makes Orbach relaxation inefficient. Electron relaxation is attributed to modulation of the ZFS of the $S = 3/2$ ground state (Table 5.6).

In chromium(III) complexes, proton hyperfine shifted signals are expected to be broad beyond detection. Actually, the binding of Cr(III) is followed in proteins by the disappearance of signals [127–129]. ^{2}H NMR spectroscopy can be used in the study of small chromium(III) complexes to reduce the line broadening.

5.6 NICKEL

Nickel(II) is a 3d^8 ion and has two unpaired electrons ($S = 1$) when it is six coordinated or four coordinated pseudotetrahedrally. In the latter configuration, sharp proton NMR signals are obtained. When it is planar four coordinated nickel(II) is always low spin diamagnetic ($S = 0$). Five-coordinated nickel(II) complexes can either be high ($S = 1$) or low spin ($S = 0$) depending on the nature of the donor atoms.

The proton relaxivity for the hexaaqua Ni(II) complex is independent on the magnetic field up to 140 MHz, as shown in Fig. 5.48. The increase noted at higher fields is due to a field dependence of the electron relaxation time, caused by fluctuations of the quadratic zero field splitting (Table 5.6). From this inflection, a value of τ_{s0} around 3×10^{-12} s is calculated in the Solomon limit, or around 10^{-11} s if ZFS is taken into account. ZFS is estimated to be around 3 cm^{-1} [130]. No ω_s dispersion appears, as theoretically expected for $S = 1$ complexes with large ZFS [131–134]. The ω_I dispersion occurs outside the accessible frequency range. No

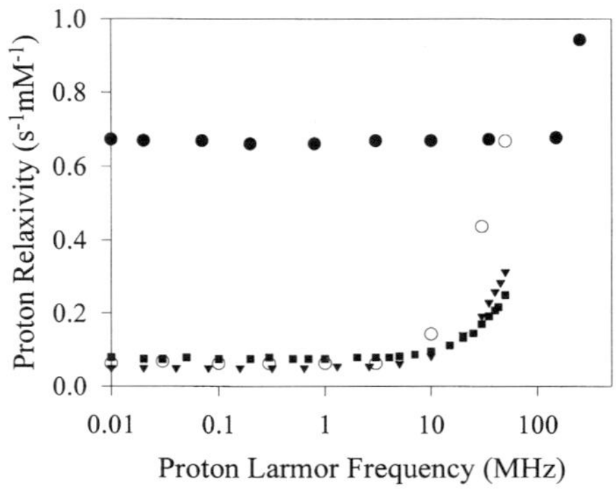

Fig. 5.48. Solvent ^{1}H NMRD profiles for $Ni(OH_2)_6^{2+}$ solutions in water at 298 K (•), and in ethyleneglycol at 264 (▼) and 298 (■) K [81], and for nickel(II) bovine carbonic anhydrase II at pH 6.0 and 298 K (○) [132].

contribution from contact relaxation is expected. Outer-sphere relaxation has been estimated to contribute about 10% of the total [130].

If the number of interacting protons is similar in water and in ethyleneglycol solutions, as it is for other aqua ions in viscous solvents, the large difference in relaxivity indicates that τ_{s0} must be one order of magnitude shorter for Ni(II) in ethyleneglycol. This means that larger distortions of the coordination sphere of the ion upon collisions with solvent molecules must occur. Therefore, rotation and viscosity affect the mechanisms of electron relaxation.

Water ^{1}H R_1 values have been measured for nickel(II)-substituted bovine carbonic anhydrase II [132,135] (Fig. 5.48), for which the following coordination polyhedron has been proposed [136,137]:

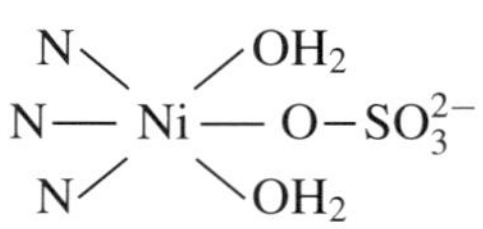

Although the low field profile is flat, a sharp rise occurs at high field. The absence of the $\omega_s\tau_s = 1$ dispersion is accounted for by the presence of a static ZFS, large with respect to the Zeeman splitting. As in the aqua complex, the increase in relaxivity is due to the field dependence of the electron relaxation. As for other metal ions, the correlation time, τ_v, for the modulation of the ZFS in proteins is longer than that in the aquaion.

A change in coordination number from six to four perturbs the electronic relaxation times of nickel(II) complexes, in a sense opposite to that observed in cobalt(II) complexes. Nickel(II) complexes, when six coordinated, have an orbitally nondegenerate ground state in octahedral symmetry, with the first excited state at several thousands wave numbers; in pseudotetrahedral symmetry, on the contrary, there are always closely spaced levels originating from a threefold

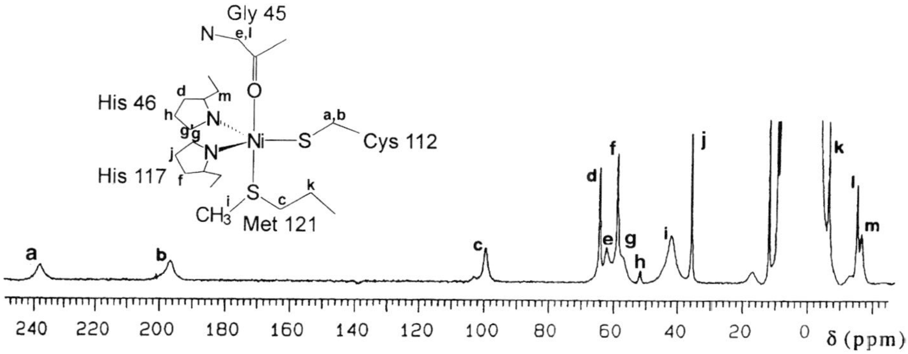

Fig. 5.49. 300 MHz ^{1}H NMR spectrum of nickel(II)-substituted azurin at pH 7.0 and 303 K, and a schematic drawing of the metal coordination polyhedron (adapted from [97]).

orbitally degenerate ground state. Such levels cause a shortening of the electronic relaxation times through Orbach relaxation.

An excellent example of the sharp NMR signals observed for pseudotetrahedral nickel(II) complexes is the ^{1}H NMR spectrum of nickel(II)-substituted azurin [97,138]. Azurin is a small protein (MW 14000) containing a single type 1 copper (see Section 5.3) forced into distorted trigonal environment with weak oxygen and sulfur axial ligands, as shown in Fig. 5.39. The spectrum of the nickel(II) derivative shows several well resolved signals [97]. The two most downfield signals arise from the βCH_2 protons of the equatorial cysteine, signal c from one of the γCH_2 protons of the axial methionine, signal e from one of the αCH protons of the axial glycine, while the signals from the ring protons of the two equatorial histidines are all in the 30–70 ppm region (Fig. 5.49).

5.7 OTHER METAL IONS

5.7.1 Vanadium

As vanadium(III) has a d^2 electron configuration with a triply degenerate ground state in octahedral symmetry, pseudooctahedral vanadium(III), in analogy to pseudotetrahedral nickel(II), is expected to display short electronic relaxation times.

Vanadium (IV) is a d^1 ion. The electron relaxation times are long and high resolution NMR is hardly performed. The NMRD profiles of $VO(H_2O)_5^{2+}$ at different temperatures are shown in Fig. 5.50 [139]. The first dispersion in the profiles is ascribed to the contact relaxation and corresponds to an electron relaxation time of about 5×10^{-10} s (Table 5.6), the second to the dipolar relaxation and corresponds to the rotational correlation time of about 5×10^{-11} s. The value of the correlation time connected to the first dispersion cannot be

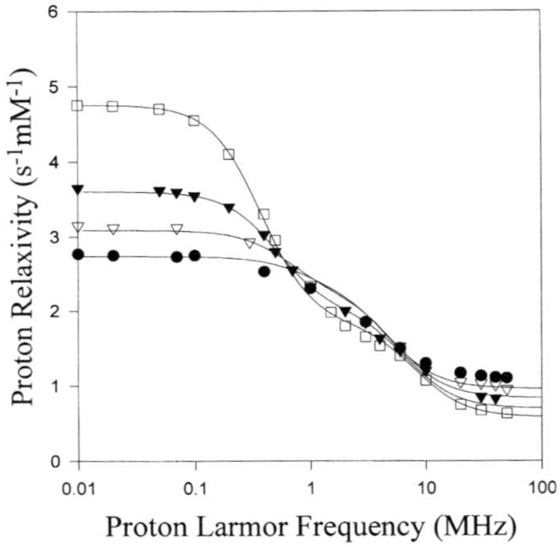

Fig. 5.50. Water ^{1}H NMRD profiles for solutions of $VO(H_2O)_5^{2+}$ at (●) 278 K, (▽) 288 K, (▼) 298 K and (□) 308 K. The solid lines represent the best-fit curves using the Solomon–Bloembergen–Morgan equations (Eqs. (3.11), (3.12), (3.16) and (3.26)) [139].

ascribed to the presence of exchange, because the hydrogen exchange rate of the equatorial water molecules is much longer than 10^{-10} s. Therefore, it was concluded that this value corresponds to τ_s even if it is shorter than expected on the basis of the EPR spectra.

The protons of the four water molecules in the equatorial plane are found at a distance of 2.6 Å from the paramagnetic center, those of the fifth axial water at 2.9 Å. The constant of contact interaction for the equatorial water molecules is found, by fitting the NMRD data, equal to 2.1 MHz.

By increasing viscosity, the longer value of τ_r moves the second dispersion toward lower frequency, at the same time increasing the relative contribution of the dipolar relaxation with respect to the contact relaxation [26]. The nonlinear increase of relaxation rate with τ_r at low fields provides evidence that τ_M is affecting the correlation time. When τ_r becomes longer than the electron relaxation time, the latter becomes the correlation time for nuclear relaxation and the field dependence of τ_s is revealed by the hump in the high field region.

Values of τ_s consistent with expectations are found for VO-protein compounds. In bis-oxovanadium(IV) transferrin, water protons are sensitive to the paramagnetic metal through second sphere water, the oxogroup occupying the only solvent-accessible coordination position. Furthermore, its EPR spectra are very sharp and characterized by a well-resolved hyperfine structure. The NMRD profile displays two dispersions (Fig. 5.51). The one at high frequency (ca. 10 MHz) is attributed to the ω_I dispersion, providing a value for τ_c of 2×10^{-8} s. Since the rotational time of the protein is of the order of $2–3 \times 10^{-7}$ s, τ_c is mainly determined by τ_s, with possible contributions by τ_M. The low field region is sizably affected by the presence of the hyperfine coupling between unpaired electrons and

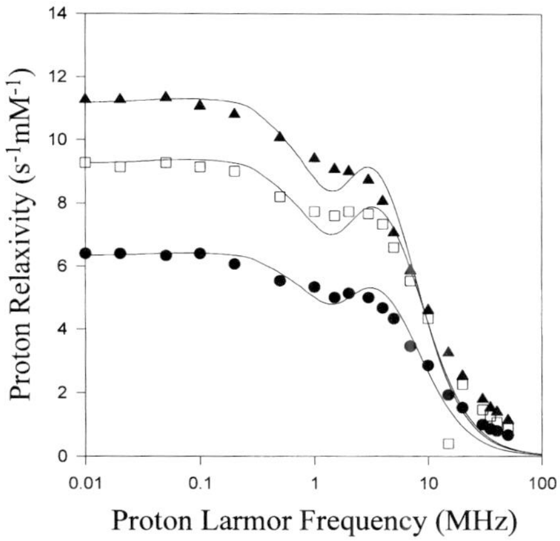

Fig. 5.51. Water ^{1}H NMRD profiles for VO^{2+}-transferrin solutions at pH 8, for three temperatures: (▲) 281 K, (□) 298 K and (•) 311 K [140].

the $I = 7/2$ metal nucleus. The constants of this interaction are $A_{\parallel} = 170 \times 10^{-4}$ and $A_{\perp} = 60 \times 10^{-4}$ cm^{-1} [140].

5.7.2 *Titanium*

Titanium(III) is a d^1 metal ion. The first excited state for titanium hexaaqua ion has been estimated to be around 2000 cm^{-1} higher than the ground state [141], while in the case of Cu(II) and VO(IV) — which have longer τ_s values — the first excited states are several thousands of cm^{-1}. The ^{1}H NMRD profiles of water solution of $Ti(H_2O)_6^{2+}$ at 293 K is shown in Fig. 5.52. Their analysis provides similar values of τ_s and τ_r, around 3×10^{-11} s, and a constant of contact interaction of 4.5 MHz [142], so that the contact and dipolar dispersions are overlapped. This is confirmed by the ratio between the relaxivity values before and after the dispersion, which is different from 10/3. Twelve water protons are found to be at 2.62 Å from the metal ion. The electron relaxation time is found to be field independent up to 600 MHz, and to be independent of τ_r, suggesting an Orbach relaxation mechanism (Table 5.6). The increase at high fields in the R_2 values provides values for τ_M equal to 4.2×10^{-7} s and 1.2×10^{-7} s at 293 and 308 K, respectively (see Eq. (4.11)). In fact such increase can only be due to the chemical exchange contribution, as R_1 does not reveal a field dependence of τ_s. Measurements on a solution containing 65% glycerol-d$_8$ indicate an electron relaxation time 6 times longer than in pure water. Such increase has been considered consistent with the increase in viscosity, being capable of affecting the efficiency of the Orbach mechanism [141,142].

In general, NMR spectra of titanium(III) complexes are expected to provide quite broad lines for hyperfine coupled protons.

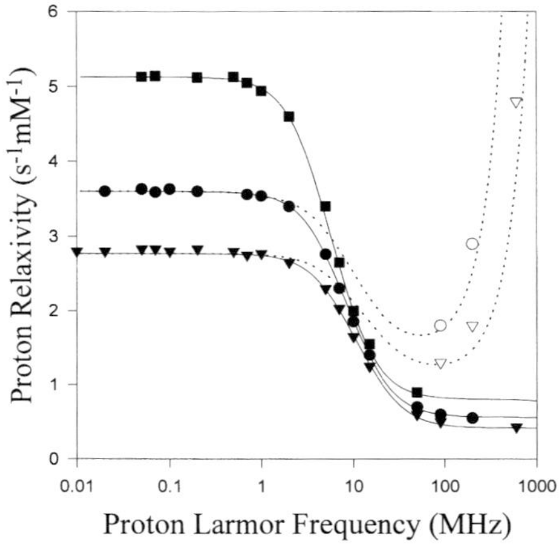

Fig. 5.52. Water ^{1}H NMRD longitudinal profiles for $Ti(H_2O)_6^{3+}$ solutions at 278 K (■), 293 K (●) and 308 K (▼) and transverse profiles at 293 K (○) and 308 K (▽) [142].

5.7.3 Gadolinium

Gadolinium(III) is an f^7 ion with the seven electrons being distributed one electron per f orbital. In analogy to Mn(II), which has one electron per d orbital and displays the longest electronic relaxation times among the 3d metal ions, gadolinium(III) also has the longest electronic relaxation times of all the lanthanides. Indeed, it is the only ion in the lanthanide series that can be investigated through EPR, which provides values for $\tau_s \geq 2 \times 10^{-10}$ s, at room temperature [143,144]. Electron relaxation is due to ZFS modulation (Table 5.6). Despite the difference in unpaired electrons, the high spin–orbit coupling of gadolinium(III) results in a nuclear relaxing capability close to that of manganese(II).

Fig. 5.53 shows the NMRD profile of Gd(III) water solution [81]. The observed dispersion corresponds to a τ_c equal to 5×10^{-11} s, which can be ascribed to the rotational correlation time, but the shape of the dispersion at high fields can be reproduced only by introducing the contribution of a field dependent τ_s. The rotational correlation time and the electron relaxation time have thus to be of the same order of magnitude above 10 MHz. This means that a field dependence of the electron relaxation time must be considered. The fitting provides $\tau_{s0} = 1.2 \times 10^{-10}$ s, $\tau_v = 1.6 \times 10^{-11}$ s and 8–9 coordinated water molecules with metal proton distances of 3.1–3.2 Å.

The proton NMRD profile of an ethyleneglycol solution containing $GdCl_3$ is reported in Fig. 5.53 at two different temperatures. The correlation time for electron relaxation, τ_v, is longer than in water. This could indicate that collisions of solvent molecules with the ion are slowed down in viscous solvents. τ_{s0}, which is related to the magnitude of the instantaneous ZFS induced by collisions, instead, does not change much. Therefore, the decrease of thermal motion as well as the

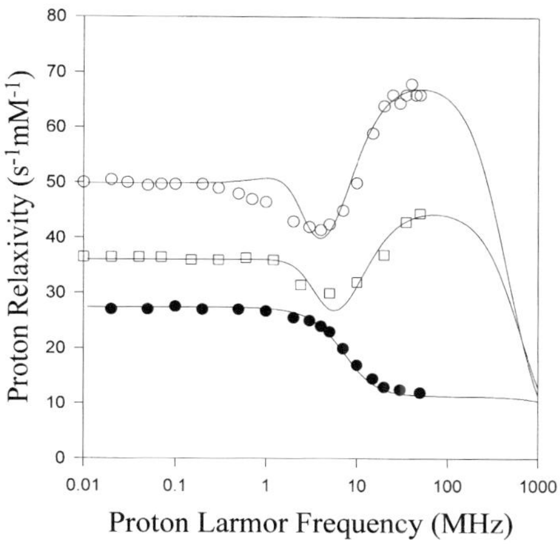

Fig. 5.53. Solvent ^{1}H NMRD profiles for water solutions of $GdCl_3$ at 298 K (•) and for ethyleneglycol solutions of $GdCl_3$ at 298 K (○) and 312 K (□) [81].

increase in viscosity of the solution are responsible of the increase of both τ_v and τ_r. In macromolecules, as also found for manganese(II) systems, τ_v seems not to depend on the rotational correlation time, i.e. on the molecular weight. The presence of a peak in the high field region of the NMRD profile of solution containing Gd(III) complexed with the protein concanavalin A confirms the field dependence of τ_S.

Gadolinium(III) ion is extensively investigated from a relaxometric point of view because its complexes are the best contrast agents for Magnetic Resonance Imaging. In order to avoid toxic effects for the human body, very stable complexes, as Gd-DTPA, Gd-DOTA (see Fig. 5.56) and other derivatives, have been designed. Both inner- and outer-sphere contributions are present (Fig. 5.54). Fig. 5.55 shows the NMRD profile of the complex of Gd-EDTA and of the complex covalently linked to the protein BSA [145]. It is apparent that in the latter the correlation time for nuclear relaxation is governed by the electron relaxation time, the hump in the high field region being due to its field dependence, and the relaxivity is much higher than in the free complex. Therefore, the strategy for the development of contrast agents with high relaxivity consists in selectively increasing the nuclear correlation time when the molecule is in a particular environment, taking advantage of the hump in relaxivity caused by the field dependent electron relaxation, and in increasing the rotational correlation time due to the formation of adducts with proteins, macromolecules or tissues. The complexes are usually functionalized with groups that increase their ability to bind non-covalently to macromolecules, for instance with the human serum albumin present in the plasma, in order to increase their rotational time, but at the same time to still permit dissociation and excretion of the complex. As seen in Chapter 4, relaxivity is also strongly influenced by the exchange time of coordinated water protons with

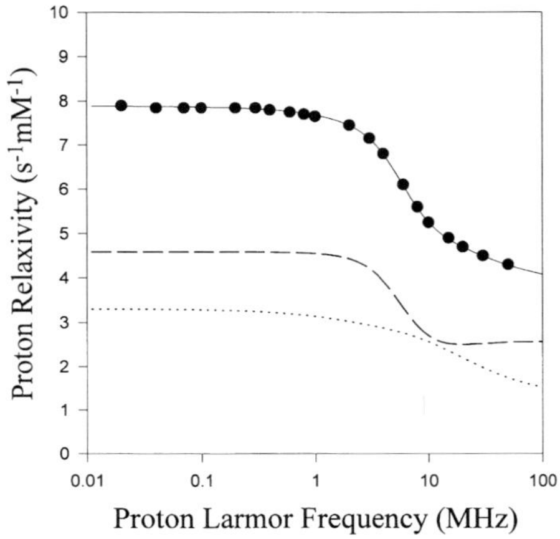

Fig. 5.54. Water ^{1}H NMRD profiles for Gd–DTPA solutions at 298 K. The dashed line indicates the inner-sphere contribution and the dotted line the outer-sphere contribution [145].

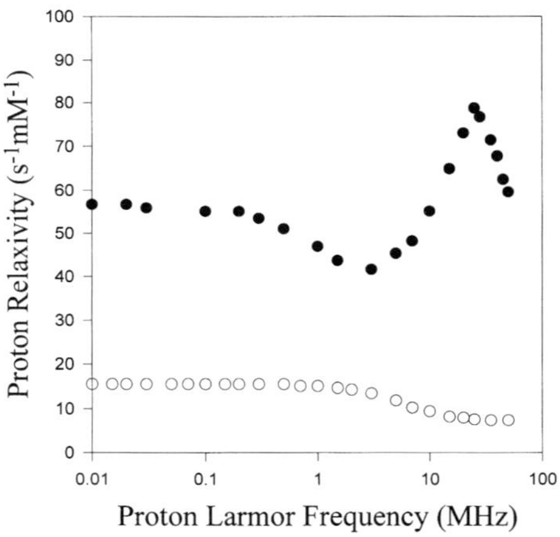

Fig. 5.55. Water ^{1}H NMRD profiles for Gd–EDTA (○) and for its covalent adduct with bovine serum albumin (●) [145].

protons of bulk water. It is calculated that the optimal value of τ_M in order to have the maximum value of relaxivity is about 10^{-8} s [146]. The rate of water exchange between coordinated water and the bulk can usually be estimated by ^{17}O NMR by measuring the transverse relaxation rate of water in the presence of the Gd(III) chelate as a function of temperature, reminding that, since the oxygen is directly coordinated to the Gd(III) ion, R_{2M} is dominated by the scalar term [147,148]. Also the charge of the complex can influence the exchange time and thus relaxivity, as τ_M decreases with increasing the negative charge of the complex [148,149]. A strategy to synthesize a selective contrast agent is to functionalize

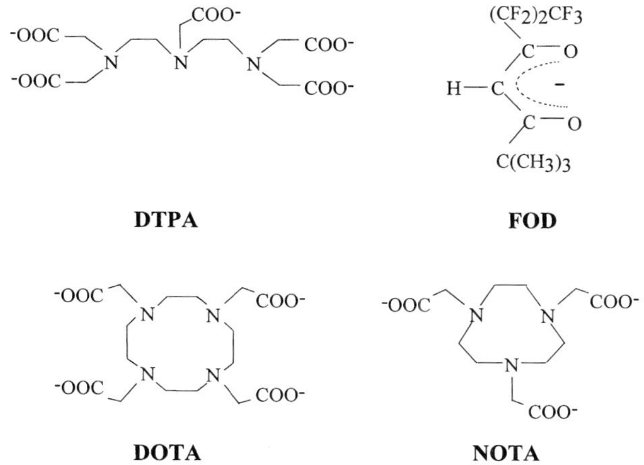

Fig. 5.56. Some lanthanide-chelating ligands used as pseudocontact shift reagents.

it in such a way that it is able to specifically react with a biological target. For example, a gadolinium complex has been synthesized holding a galactopyranose group through a glycosidyc bond. This group occupies the free coordination position on the metal ion, preventing the binding of water. The complex is targeted at the enzyme β-galactosidase. When the contrast agent encounters the enzyme *in vivo*, the glycosidyc bond is cleaved, thus exposing gadolinium to the solvent and enhancing relaxivity, which then signals the presence of this enzymatic activity [150].

5.7.4 *Other lanthanides*

The f^n configurations of lanthanides(III) give rise to several free ion terms that upon spin–orbit coupling provide, except for f^7 configurations (Gd^{3+}, Eu^{2+}), several closely spaced energy levels. The multiplicity of the ground levels varies from 6 to 17 (Table 2.10), and is further split by crystal field effects. According to the electronic relaxation mechanisms discussed in Section 3.3, the availability of low-lying excited levels accounts for short electronic relaxation times, again with the exception of metal ions with an f^7 configuration. For these fast-relaxing systems, the correlation times are determined by the electronic relaxation times. Values for the latter have been estimated from nuclear relaxation measurements (see Sections 3.4 and 3.6).

The NMRD profiles show small and field independent relaxivity values for proton frequencies lower than 50 MHz (Fig. 5.57), thus indicating that τ_c is indeed provided by τ_s. The fitting made with Eqs. (3.32)–(3.37) of different sets of nuclear relaxation measurements provides the values for τ_s reported in Fig. 5.58. These very short values confirm an efficient Orbach relaxation

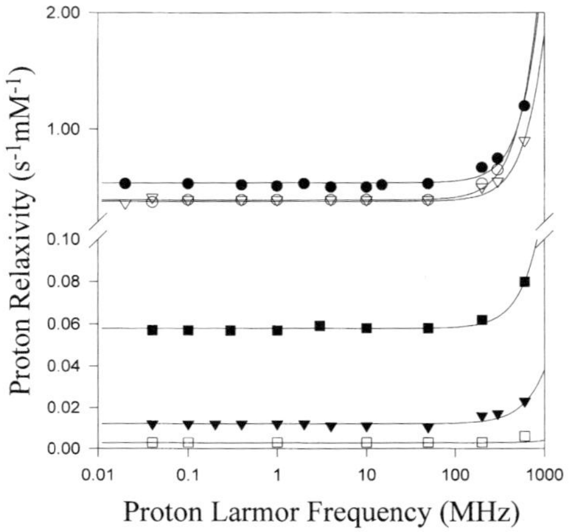

Fig. 5.57. Water ^1H NMRD profiles of some lanthanide aqua ions at 298 K: (□) Sm, (▼) Pr, (■) Yb, (▽) Er, (○) Ho, (●) Dy [189].

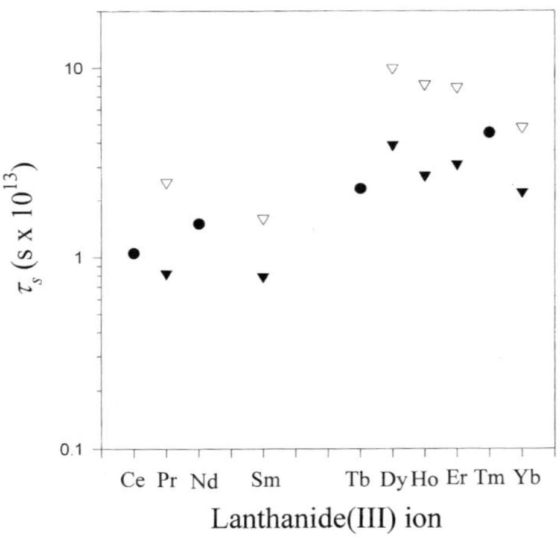

Fig. 5.58. τ_s values for lanthanide(III) aqua ions obtained from the fit of the NMRD profiles to Eqs. (3.32) and (3.36) (▼) [189] or estimated from single field measurements (●) [192]. τ_s values estimated by including the effect of ZFS (▽) [189] (using the J and g_J values reported in Table 2.10) are also reported.

mechanism (Table 5.6). In the case of lanthanides, Curie relaxation effects had been observed also in small complexes [41,151]. Lanthanides constitute the only case in which the field dependence of nuclear longitudinal relaxation (Fig. 5.57) is due to Curie relaxation for complexes as small as the aqua ion. Indeed, Curie relaxation on small complexes can only be expected when $\tau_s \ll \tau_r$, as it occurs for lanthanides. From ^{17}O NMR measurements, τ_M was found in the range 10^{-6}–10^{-8} s; it can be thus neglected. This is confirmed by the temperature dependence of the ^{1}H NMRD profiles. Paramagnetic shift measurements of ^{13}C, ^{17}O and ^{1}H nuclei in some lanthanide chelating agents point out that the contact interaction between the metal ion and the hydrogen nuclei can be neglected [152]. The coordination number of water molecules changes from nine to eight on passing from Ce(III) to Yb(III) [153].

Short electronic relaxation times result in very little line broadening, besides Curie relaxation (Section 3.6), of NMR signals. The hyperfine shifts are mainly pseudocontact in origin and the contact contribution, which arises from covalent bonds, is small owing to the strong radial contraction of the 4f orbitals. The hyperfine coupling constant presumably has a lower value than that of the 3d transition metal ions, for the same type of nuclei in the same geometry. Although some ions have been found to give rise to larger contact contributions than other ions of the series, such contributions are generally neglected for nuclei not directly coordinated to the metal. Separation of the hyperfine shifts can in principle be successfully achieved by utilizing different lanthanides and assuming that the pseudocontact shifts have axial symmetry (see Section 2.8.4). The combined effects of small line broadening and dominant pseudocontact shifts, which depend on the geometrical coordinates described in Eqs. (2.36) and (2.37), justify the use of these ions and their complexes as shift reagents since the early days of NMR in organic chemistry. As these systems have fast electronic relaxation times, the availability of high magnetic fields enables exploitation of Curie relaxation (Section 3.6) for an independent estimate of the distance of the resonating nucleus from line broadening measurements.

The successful use of lanthanides as pseudocontact shift reagents is also due to their tendency to interact with molecules containing oxygen and nitrogen as donor atoms. For example, lanthanide complexes with ligands of the type shown in Fig. 5.56 often interact with organic molecules, possibly of biological interest. Pseudocontact shifts (and relaxation measurements) may then allow mapping of the nuclei in the molecule of interest according to the guidelines given in Sections 2.9 and 3.12. Progress in this direction is being made. It has been shown that solution structures of lanthanide complexes can indeed be achieved by using pseudocontact shifts and relaxation data together with a structural model as a starting point [154]. By allowing the internal coordinates to change within a relatively narrow range around the average values obtained from crystallographic data, a best fit can be achieved of the experimental pseudocontact shifts. The resulting set of coordinates is then checked against the known constraints imposed by the chemical bonds and bond angles in the complex [154].

Lanthanides have been used as substitutes for calcium in calcium binding proteins since the early days in NMR [155,156]. Remarkable information was obtained on systems such as Yb(III)-substituted parvalbumin (which contains a typical calcium binding site called EF-hand [157,158]) from 1D spectroscopy alone [159–164], later complemented by 2D spectroscopy [165]. More recently, pseudocontact shifts and longitudinal proton relaxation times have been used to

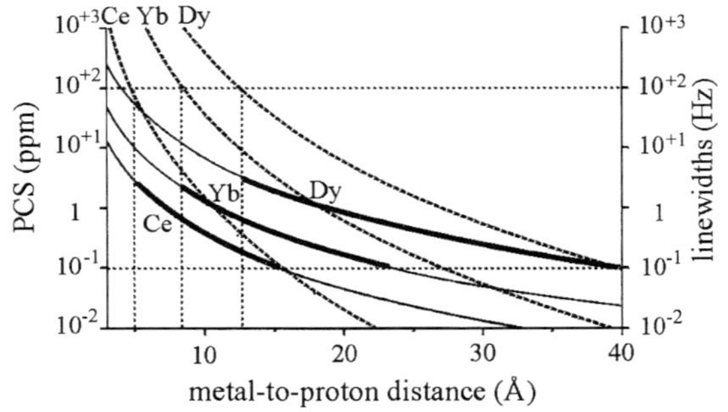

Fig. 5.59. Useful ranges of pseudocontact shifts (solid lines) for Ce(III)-, Yb(III)-, and Dy(III)-containing proteins. Dashed lines indicate observed line widths for the three metals. The lower distance limit is estimated from the observability of HSQC peaks (see Section 8.7) in Calbindin, and the upper distance limit from the observed pseudocontact shifts being smaller than 0.1 ppm in absolute value. The lower distance limits can be further lowered by including shifts measured from 1D ^{1}H NMR spectra (adapted from [167]).

locate metal ions within a protein structure. In fact, pseudocontact shifts may provide information on the immediate neighborhood of the paramagnetic center, where only the shortest range proton–proton dipolar couplings are still observable. The solution structure of the N-terminal fragment of calmodulin was thus obtained where the two calcium ions were substituted by Ce(III), which has one of the smallest magnetic moments among lanthanides (about 2.5 μ_B) and a relatively large magnetic anisotropy, yielding a pseudocontact shift/Curie line broadening ratio more favorable than other lanthanides for macromolecules at high field [166]. At 600 MHz, in fact, Curie relaxation becomes important and the ion determining the larger dipolar contribution and the lower Curie contribution must be selected. The resolution improvement in the regions close to the paramagnetic centers determined by the inclusion of the pseudocontact shifts constraints was sizable.

Analogously, the pseudocontact shifts were used to refine the solution structure of the Ce(III), Dy(III) and Yb(III) derivatives of monolanthanide-substituted Calbindin D_{9k}. Since the three lanthanides span a wide range of magnetic anisotropies, the refinement was effective in shells from the metal of ~5–15 Å for Ce(III), ~9–25 Å for Yb(III), ~13–40 Å for Dy(III), as useful pseudocontact shifts were observed in these shells (Fig. 5.59) [167]. Therefore, by using different lanthanides it was possible to enlighten shells at variable distances from the metal itself.

5.8 REFERENCES

[1] A. Abragam, B. Bleaney (1970) Electron Paramagnetic Resonance of Transition Metal Ions. Clarendon Press, Oxford.

[2] G.N. La Mar, F.A. Walker (1973) J. Am. Chem. Soc. 95, 1782.

[3] I. Bertini, O. Galas, C. Luchinat, L. Messori, G. Parigi (1995) J. Phys. Chem. 99, 14217.

[4] I. Bertini, F. Capozzi, C. Luchinat, Z. Xia (1993) J. Phys. Chem. 97, 1134.
[5] Z. Luz, R.G. Shulman (1965) J. Chem. Phys. 43, 3750.
[6] S.H. Koenig, R.D. Brown III (1987) In: Sigel H. (ed) Metal ions in biological systems, Vol. 21. Marcel Dekker, New York.
[7] H. Levanon, S. Charbinsky, Z. Luz (1970) J. Chem. Phys. 53, 3056.
[8] H. Levanon, G. Stein, Z. Luz (1970) J. Chem. Phys. 53, 876.
[9] L.P. Hwang, J.H. Freed (1975) J. Chem. Phys. 63, 4017.
[10] I. Bertini, J. Kowalewski, C. Luchinat, T. Nilsson, G. Parigi (1999) J. Chem. Phys. 111, 5795.
[11] S.H. Koenig, R.D. Brown III, J.F. Gibson, R.J. Ward, T.J. Peters (1986) Magn. Reson. Med. 3, 755.
[12] S.H. Koenig, R.D. Brown III, T.R. Lindstrom (1981) Biophys. J. 34, 397.
[13] T.E. Zewert, H.B. Gray, I. Bertini (1994) J. Am. Chem. Soc. 116, 1169.
[14] R.H. Heistand, R.B. Lauffer, E. Fikrig, L. Que, Jr. (1982) J. Am. Chem. Soc. 104, 2789.
[15] H. Tsurumaki, Y. Watanabe, I. Morishima (1993) J. Am. Chem. Soc. 115, 11784.
[16] G.N. La Mar, G.R. Eaton, R.H. Holm, F.A. Walker (1973) J. Am. Chem. Soc. 95, 63.
[17] M.M. Maltempo, T.H. Moss, M.A. Cusanovich (1974) Biochim. Biophys. Acta 351, 237.
[18] A.D. Boersma, H.M. Goff (1982) Inorg. Chem. 21, 581.
[19] F.A. Walker, G.N. La Mar (1973) Ann. New York Acad. Sci. 206, 328.
[20] D.L. Budd, G.N. La Mar, K.C. Langry, K.M. Smith, R. Nayyir-Mazhir (1979) J. Am. Chem. Soc. 101, 6091.
[21] J. Mispelter, M. Momenteau, J.-M. Lhoste (1981) J. Chem. Soc., Dalton Trans. 1729.
[22] U. Pande, G.N. La Mar, J.T.J. Lecomte, F. Ascoli, M. Brunori, K.M. Smith, R.K. Pandey, D.W. Parish, V. Thanabal (1986) Biochemistry 25, 5638.
[23] G.N. La Mar, J.T. Jackson, L.B. Dugad, M.A. Cusanovich, R.G. Bartsch (1990) J. Biol. Chem. 265, 16173.
[24] R.B. Lauffer, L. Que, Jr. (1982) J. Am. Chem. Soc. 104, 7324.
[25] L. Que, Jr., R.B. Lauffer, J.B. Lynch, B.P. Murch, J.W. Pyrz (1987) J. Am. Chem. Soc. 109, 5381.
[26] G.N. La Mar, F.A. Walker (1973) J. Am. Chem. Soc. 95, 6950.
[27] I. Bertini, G. Gori, C. Luchinat, A.J. Vila (1993) Biochemistry 32, 776.
[28] G.N. La Mar, J.T. Jackson, R.G. Bartsch (1981) J. Am. Chem. Soc. 103, 4405.
[29] G.N. La Mar, D.L. Budd, K.M. Smith, K.C. Langry (1980) J. Am. Chem. Soc. 102, 1822.
[30] B. Xia, W.M. Westler, H. Cheng, J. Meyer, J.-M. Moulis, J.L. Markley (1995) J. Am. Chem. Soc. 117, 5347.
[31] R.M. Golding, R.O. Pascual, J. Vrbancich (1976) J. Mol. Phys. 31, 731.
[32] I. Bertini, C. Luchinat, G. Parigi, unpublished results.
[33] B.R. McGarvey (1998) Coord. Chem. Rev. 170, 75.
[34] A. Mayer, S. Ogawa, R.G. Shulman, T. Yamane, J.A.S. Cavaleiro, A.M.d. Rocha Gonsalves, G.W. Kenner, K.M. Smith (1974) J. Mol. Biol. 86, 749.
[35] I. Bertini, F. Capozzi, C. Luchinat, P. Turano (1991) J. Magn. Reson. 95, 244.
[36] G.N. La Mar, F.A. Walker (1979) In: Dolphin D. (ed) The Porphyrins. Academic Press, New York: p. 61.
[37] W.D. Horrocks, Jr., E.S. Greenberg (1973) Biochim. Biophys. Acta 322, 38.
[38] J. Trehwella, P.E. Wright (1980) Biochim. Biophys. Acta 625, 202.
[39] D.L. Turner, R.J.P. Williams (1993) Eur. J. Biochem. 211, 555.
[40] N.V. Shokhirev, F.A. Walker (1995) J. Phys. Chem. 99, 17795.
[41] P.D. Burns, G.N. La Mar (1982) J. Magn. Reson. 46, 61.
[42] K. Wüthrich, R.M. Keller (1978) Biochim. Biophys. Acta 533, 195.
[43] R.J. Kurland, B.R. McGarvey (1970) J. Magn. Reson. 2, 286.
[44] I. Bertini, C. Luchinat, G. Parigi (2000) Eur. J. Inorg. Chem. 2473.

[45] L. Banci, I. Bertini, H.B. Gray, C. Luchinat, T. Reddig, A. Rosato, P. Turano (1997) Biochemistry 36, 9867.
[46] Q. Lin, U. Simonis, A.R. Tipton, C.J. Norvell, F.A. Walker (1992) Inorg. Chem. 31, 4216.
[47] U. Simonis, J.L. Dallas, F.A. Walker (1992) Inorg. Chem. 31, 5349.
[48] I. Bertini, C. Luchinat, G. Parigi, F.A. Walker (1999) JBIC 4, 515.
[49] N.V. Shokhirev, F.A. Walker (1998) J. Am. Chem. Soc. 120, 981.
[50] L. Banci, I. Bertini, F. Ferroni, A. Rosato (1997) Eur. J. Biochem. 249, 270.
[51] F. Arnesano, L. Banci, I. Bertini, I.C. Felli (1998) Biochemistry 37, 173; F. Arnesano, L. Banci, I. Bertini, I.C. Felli, D. Koulougliotis (1999) Eur. J. Biochem. 260, 347.
[52] R.O. Louro, I.J. Correia, L. Brennan, I.B. Coutinho, A.V. Xavier, D.L. Turner (1998) J. Am. Chem. Soc. 120, 13240.
[53] L. Banci, I. Bertini, M. Bruschi, P. Sompornpisut, P. Turano (1996) Proc. Natl. Acad. Sci. USA 93, 14396.
[54] M.K. Safo, F.A. Walker, A.M. Raitsimiring, W.P. Walters, D.P. Dolata, P.G. Debrunner, W.R. Scheidt (1994) J. Am. Chem. Soc. 116, 7760.
[55] M. Nakamura, T. Ikeue, H. Fujii, T. Yoshimura (1997) J. Am. Chem. Soc. 119, 6284.
[56] F.A. Walker, H. Nasri, I. Turowska-Tyrk, K. Mohanrao, C.T. Watson, N.V. Shokhirev, P.G. Debrunner, W.R. Scheidt (1996) J. Am. Chem. Soc. 118, 12109.
[57] S. Wolowiec, L. Latos-Grazynski, D. Toronto, J.-C. Marchon (1998) Inorg. Chem. 37, 724.
[58] G.N. La Mar, D.L. Budd, H.M. Goff (1977) Biochem. Biophys. Res. Commun. 77, 104.
[59] L.P. Yu, G.N. La Mar, K. Rajarathnam (1990) J. Am. Chem. Soc. 112, 9527.
[60] S.D. Emerson, G.N. La Mar (1990) Biochemistry 29, 1556.
[61] J.D. Satterlee, G.N. La Mar (1976) J. Am. Chem. Soc. 98, 2804.
[62] L. Banci, I. Bertini, S. Marconi, R. Pierattelli (1993) Eur. J. Biochem. 215, 431.
[63] I. Bertini, A. Dikiy, C. Luchinat, R. Macinai, M.S. Viezzoli (1998) Inorg. Chem. 37, 4814.
[64] C.M. Bougault, Y. Dou, M. Ikeda-Saito, K.C. Langry, K.M. Smith, G.N. La Mar (1998) J. Am. Chem. Soc. 120, 2113.
[65] F. Yang, G.N. Phillips Jr. (1996) J. Mol. Biol. 256, 762.
[66] M.E. Johnson, L.W.M. Fung, C. Ho (1977) J. Am. Chem. Soc. 99, 1245.
[67] J. Hochmann, H. Kellerhals (1980) J. Magn. Reson. 38, 23.
[68] T.E. Lehmann, L.-J. Ming, M.E. Rosen, L. Que, Jr. (1997) Biochemistry 36, 2807.
[69] P. Bertrand, F.X. Theodule, J.-P. Gayda, J. Mispelter, M. Momenteau (1983) Chem. Phys. Lett. 102, 442.
[70] C.A. Reed, F. Guiset (1996) J. Am. Chem. Soc. 118, 3281.
[71] J.S. de Ropp, P. Mandal, S.L. Brauer, G.N. La Mar (1997) J. Am. Chem. Soc. 119, 4732.
[72] G.N. La Mar, J.S. de Ropp, K.M. Smith, K.C. Langry (1980) J. Biol. Chem. 255, 6646.
[73] V. Thanabal, J.S. de Ropp, G.N. La Mar (1987) J. Am. Chem. Soc. 109, 265.
[74] L. Banci, I. Bertini, P. Turano, J.C. Ferrer, A.G. Mauk (1991) Inorg. Chem. 30, 4510.
[75] P.S. Rao, P. Ganguli, B.R. McGarvey (1981) Inorg. Chem. 20, 3682.
[76] A. Ozarowski, Y. Shunzhong, B.R. McGarvey, A. Mislankar, J.E. Drake (1991) Inorg. Chem. 30, 3167.
[77] E. Breuning, M. Ruben, J.-M. Lehn, F. Renz, Y. Garcia, V. Ksenofontov, P. Gütlich, E. Wegelius, K. Rissanen (2000) Angew. Chem. Int. Ed 39, 2504.
[78] A. Bencini, D. Gatteschi (1982) In: Melson GA, Figgis BN (eds) Transition Metal Chemistry. Marcel Dekker, New York, p. 1.
[79] A. Shirazi, H.M. Goff (1982) Inorg. Chem. 21, 3420.
[80] G.P. Fulton, G.N. La Mar (1976) J. Am. Chem. Soc. 98, 2119.
[81] L. Banci, I. Bertini, C. Luchinat (1985) Inorg. Chim. Acta 100, 173.
[82] S.H. Koenig, R.D. Brown III, I. Bertini, C. Luchinat (1983) Biophys. J. 41, 179.
[83] I. Bertini, C. Luchinat, M. Mancini, G. Spina (1984) J. Magn. Reson. 59, 213.

[84] I. Bertini, G. Canti, C. Luchinat, A. Scozzafava (1978) J. Am. Chem. Soc. 100, 4873.
[85] I. Bertini, G. Canti, C. Luchinat (1981) Inorg. Chim. Acta 56, 99.
[86] C. Benelli, I. Bertini, D. Gatteschi (1972) J. Chem. Soc. Dalton Trans. 661.
[87] I. Bertini, G. Canti, C. Luchinat, L. Messori (1982) Inorg. Chem. 21, 3426.
[88] I. Bertini, D. Gatteschi, L.J. Wilson (1970) Inorg. Chim. Acta 629
[89] I. Bertini, G. Canti, C. Luchinat, F. Mani (1981) J. Am. Chem. Soc. 103, 7784.
[90] I. Bertini, C. Luchinat (1985) Adv. Inorg. Biochem. 6, 71.
[91] I. Bertini, C. Luchinat (1983) In: Sigel H. (ed) Metal ions in biological systems. Marcel Dekker, New York: p. 101.
[92] I. Bertini, M. Gerber, G. Lanini, C. Luchinat, W. Maret, S. Rawer, M. Zeppezauer (1984) J. Am. Chem. Soc. 106, 1826.
[93] I. Bertini, C. Luchinat, L. Messori, A. Scozzafava (1984) Eur. J. Biochem. 141, 375.
[94] D.L. Tennent, D.R. McMillin (1979) J. Am. Chem. Soc. 101, 2307.
[95] C.O. Fernández, A.I. Sannazzaro, A.J. Vila (1997) Biochemistry 36, 10566.
[96] M. Piccioli, C. Luchinat, T.J. Mizoguchi, B.E. Ramirez, H.B. Gray, J.H. Richards (1995) Inorg. Chem. 34, 737.
[97] J. Salgado, H.R. Jimenez, J.M. Moratal Mascarell, S.J. Kroes, G.C.M. Warmerdam, G.W. Canters (1996) Biochemistry 35, 1810.
[98] J.M. Moratal, J. Salgado, A. Donaire, H.R. Jimenez, J. Castells (1993) Inorg. Chem. 32, 3587.
[99] I. Bertini, C. Luchinat, L. Messori, M. Vasak (1989) J. Am. Chem. Soc. 111, 7300.
[100] I. Moura, M. Teixeira, J. LeGall, J.J.G. Moura (1991) J. Inorg. Biochem. 44, 127.
[101] S. Dahlin, B. Reinhammar, J. Ångstrom (1989) Biochemistry 28. 7224.
[102] B.A. Krizek, B.T. Amann, V.J. Kilfoil, D.L. Merkle, J.M. Berg (1991) J. Am. Chem. Soc. 113, 4518.
[103] L.V. Harper, B.T. Amann, V.K. Vinson, J.M. Berg (1993) J. Am. Chem. Soc. 115, 2577.
[104] I. Bertini, C. Luchinat, M. Piccioli, M. Vicens Oliver, M.S. Viezzoli (1991) Eur. Biophys. J. 20, 269.
[105] L. Banci, L.B. Dugad, G.N. La Mar, K.A. Keating, C. Luchinat, R. Pierattelli (1992) Biophys. J. 63, 530.
[106] M. Gochin (1998) J. Biomol. NMR 12, 243.
[107] M. Gochin (1997) J. Am. Chem. Soc. 119, 3377.
[108] R. Hausser, F. Noack (1964) Z. Phys. 182, 93.
[109] I. Bertini, F. Briganti, C. Luchinat, M. Mancini, G. Spina (1985) J. Magn. Reson. 63, 41.
[110] R.A. Liebermann, R.H. Sands, J.A. Fee (1982) J. Biol. Chem. 257, 336.
[111] R.H. Holm, P. Kennepohl, E.I. Solomon (1996) Chem. Rev. 96, 2239.
[112] S.H. Koenig, R.D. Brown (1973) Ann. New York Acad. Sci. 222, 752.
[113] I. Andersson, W. Maret, M. Zeppezauer, R.D. Brown III, S.H. Koenig (1981) Biochemistry 20, 3424.
[114] S.J. Kroes, J. Salgado, G. Parigi, C. Luchinat, G.W. Canters (1996) JBIC 1, 551.
[115] A.P. Kalverda, J. Salgado, C. Dennison, G.W. Canters (1996) Biochemistry 35, 3092.
[116] J. Salgado, S.J. Kroes, A. Berg, J.M. Moratal, G.W. Canters (1998) J. Biol. Chem. 273, 177.
[117] I. Bertini, S. Ciurli, A. Dikiy, R. Gasanov, C. Luchinat, G. Martini, N. Safarov (1999) J. Am. Chem. Soc. 121, 2037.
[118] K.W. Penfield, A.A. Gewirth, E.I. Solomon (1985) J. Am. Chem. Soc. 107, 4519.
[119] N.N. Murthy, K.D. Karlin, I. Bertini, C. Luchinat (1997) J. Am. Chem. Soc. 119, 2156.
[120] I. Bertini, K.L. Bren, A. Clemente, J.A. Fee, H.B. Gray, C. Luchinat, B.G. Malmström, J.H. Richards, D. Sanders, C.E. Slutter (1996) J. Am. Chem. Soc. 46, 11658.
[121] C. Dennison, A. Berg, G.W. Canters (1997) Biochemistry 36, 3262.
[122] I. Bertini, F. Briganti, C. Luchinat, Z. Xia (1993) J. Magn. Reson. 101, 198.

[123] E. Meirovitch, Z. Luz, A.J. Kalb (1974) J. Am. Chem. Soc. 96, 7538.
[124] G.C. Brackett, P.L. Richards, W.S. Caughey (1971) J. Chem. Phys. 54, 4383.
[125] Z. Luz, D. Fiat (1967) J. Chem. Phys. 46, 469.
[126] T.R. Janson, L.J. Bouchet, J.J. Katz (1973) Inorg. Chem. 12, 940.
[127] S.-C. Im, G. Liu, C. Luchinat, A.G. Sykes, I. Bertini (1998) Eur. J. Biochem. 258, 465.
[128] J.R. Bocarsly, J.K. Barton (1992) Inorg. Chem. 31, 2827.
[129] J.R. Bocarsly, J.K. Barton (1989) Inorg. Chem. 28, 4189.
[130] J. Kowalewski, T. Larsson, P.-O. Westlund (1987) J. Magn. Reson. 74, 56.
[131] V.U. Lindner (1965) Ann. Phys. 16, 319.
[132] I. Bertini, C. Luchinat, M. Mancini, G. Spina (1985) In: Gatteschi D., Kahn O., Willett R.D. (eds) Magneto-structural correlations in exchange-coupled systems. Reidel, Dordrecht: p. 421.
[133] J. Kowalewski, L. Nordenskiöld, N. Benetis, P.-O. Westlund (1985) Progr. NMR Spectrosc. 17, 141.
[134] N. Benetis, J. Kowalewski, L. Nordenskiöld, Wennerström, P.-O. Westlund (1983) Mol. Phys. 48, 2.
[135] I. Bertini, E. Borghi, C. Luchinat, R. Monnanni (1982) Inorg. Chim. Acta 67, 99.
[136] I. Bertini, E. Borghi, C. Luchinat (1978) Bioinorg. Chem. 9, 495.
[137] K. Hakansson, A. Wehnert, A. Liljas (1994) Acta Cryst. D50, 93.
[138] J.A. Blaszak, E.L. Ulrich, J.L. Markley, D.R. McMillin (1982) Biochemistry 21, 6253.
[139] I. Bertini, C. Luchinat, Z. Xia (1992) J. Magn. Reson. 99, 235.
[140] I. Bertini, F. Briganti, S.H. Koenig, C. Luchinat (1985) Biochemistry 24, 6287.
[141] R.C. Wilson, R.J. Myers (1976) J. Chem. Phys. 64, 2208.
[142] I. Bertini, C. Luchinat, Z. Xia (1992) Inorg. Chem. 31, 3152.
[143] E.M. Stephens, C.M. Grisham (1979) Biochemistry 18, 4876.
[144] M. Epstein, J. Reuben, A. Levitzki (1977) Biochemistry 16, 2449.
[145] S.H. Koenig, R.D. Brown III (1990) Progr. NMR Spectrosc. 22, 487.
[146] S. Aime, M. Botta, M. Fasano, E. Terreno (1998) Chem. Soc. Rev. 27, 19.
[147] K. Micskei, L. Helm, E. Brücher, A.E. Merbach (1993) Inorg. Chem. 32, 3844.
[148] G. Gonzalez, D.H. Powell, V. Tissieres, A.E. Merbach (1994) J. Phys. Chem. 98, 53.
[149] A.E. Merbach (1987) Pure Appl. Chem. 59, 161.
[150] R.A. Moats, S.E. Frazer, T.J. Meade (1997) Angew. Chem. Int. (Ed. Engl.) 36, 726.
[151] S. Aime, L. Barbero, M. Botta (1991) Magn. Res. Imaging 9, 843.
[152] J.A. Peters, J. Huskens, D.J. Raber (1996) Progr. NMR Spectrosc. 28, 283.
[153] L. Helm, A.E. Merbach (1991) Eur. J. Solid State Inorg. Chem. 28, 245.
[154] S. Aime, M. Botta, G. Ermondi (1992) Inorg. Chem. 31, 4291.
[155] J. Sowadski, G. Cornick, R.H. Kretsinger (1978) J. Mol. Biol. 124, 123.
[156] W.D. Horrocks, Jr., D.R. Sudnick (1979) J. Am. Chem. Soc. 101, 334.
[157] R.H. Kretsinger, C.E. Nockolds (1973) J. Biol. Chem. 248, 3313.
[158] P.C. Moews, R.H. Kretsinger (1975) J. Mol. Biol. 91, 201.
[159] L. Lee, B.D. Sykes (1980) In: Darnall D.W., Wilkins R.G. (eds) Advances in Inorganic Biochemistry, Vol. 2. Elsevier, New York: p. 183.
[160] L. Lee, B.D. Sykes (1980) Biochemistry 19, 3208.
[161] L. Lee, B.D. Sykes (1981) Biochemistry 20, 1156.
[162] L. Lee, B.D. Sykes (1983) Biochemistry 22, 4366.
[163] L. Lee, B.D. Sykes (1980) J. Magn. Reson. 41, 512.
[164] L. Lee, B.D. Sykes (1980) Biophys J. 32, 193.
[165] F. Capozzi, M.A. Cremonini, C. Luchinat, M. Sola (1993) Magn. Reson. Chem. 31, S118–S127.
[166] D. Bentrop, I. Bertini, M.A. Cremonini, S. Forsén, C. Luchinat, A. Malmendal (1997) Biochemistry 36, 11605.

[167] M. Allegrozzi, I. Bertini, M.B.L. Janik, Y.-M. Lee, G. Liu, C. Luchinat (2000) J. Am. Chem. Soc. 122, 4154.
[168] I. Bertini, G. Canti, C. Luchinat, E. Borghi (1983) J. Inorg. Biochem. 18, 221.
[169] I. Bertini, C. Luchinat (1986) In: Karlin K.D., Zubieta J. (eds) Biological and Inorganic Copper Chemistry. Adenine Press, New York: p. 23.
[170] I. Bertini, G. Canti, C. Luchinat (1981) Inorg. Chim. Acta 56, 1.
[171] L. Banci, I. Bertini, C. Luchinat, R. Monnanni, A. Scozzafava (1988) Inorg. Chem. 27, 107.
[172] I. Bertini, L. Banci, R.D. Brown III, S.H. Koenig, C. Luchinat (1988) Inorg. Chem. 27, 951.
[173] I. Bertini, C. Luchinat, A. Scozzafava, A. Maldotti, O. Traverso (1983) Inorg. Chim. Acta 78, 19.
[174] R. Barker, N. Boden, G. Cayley, S.C. Charlton, R. Henson, M.C. Holmes, I.D. Kelly, P.F. Knowles (1979) Biochem. J. 177, 289.
[175] I. Bertini, C. Luchinat, G. Mincione, G. Parigi, G.T. Gassner, D.P. Ballou (1996) JBIC 1, 468. 5181.
[176] C. Luchinat, Z. Xia (1992) Coord. Chem. Rev. 120, 281.
[177] S. Fugiwara, H. Hayashu (1965) J. Chem. Phys. 43, 23.
[178] D.H. Powell, L. Helm, A.E. Merbach (1991) J. Chem. Phys. 95(12), 9258.
[179] I. Bertini, C. Luchinat, R.D. Brown III, S.H. Koenig (1989) J. Am. Chem. Soc. 111, 3532.
[180] Z. Luz, R.G. Shulman (1965) J. Chem. Phys. 43, 3750.
[181] W.B. Lewis, M. Alei, L.O. Morgan (1966) J. Chem. Phys. 44, 2409.
[182] N. Bloembergen, L.O. Morgan (1961) J. Chem. Phys. 34, 842.
[183] Y. Ducommun, K.E. Newman, A. Merbach (1980) Inorg. Chem. 19, 3696.
[184] R.T. Boere, R. Kidel (1982) Annu. Rep. NMR Spectrosc. 13, 319.
[185] S.D. Kennedy, R.G. Bryant (1985) Magn. Reson. Med. 2, 14.
[186] T.J. Swift, R.E. Connick (1962) J. Chem. Phys. 37, 307.
[187] H.L. Friedman, M. Holz, H.G. Hertz (1979) J. Chem. Phys. 70, 3369.
[188] Y. Ducommun, W.L. Earl, A.E. Merbach (1979) Inorg. Chem. 18, 2754.
[189] I. Bertini, F. Capozzi, C. Luchinat, G. Nicastro, Z. Xia (1993) J. Phys. Chem. 101, 198.
[190] L. Banci, I. Bertini, C. Luchinat (1991) Nuclear and electron relaxation. The magnetic nucleus-unpaired electron coupling in solution. VCH, Weinheim.
[191] I. Bertini, C. Luchinat (1986) NMR of paramagnetic molecules in biological systems. Benjamin/Cummings, Menlo Park, CA.
[192] B.M. Alsaadi, F.J.C. Rossotti, R.J.P. Williams (1980) J. Chem. Soc., Dalton Trans. 2147.
[193] R. Orbach (1961) Proc. R. Soc. London, Ser. A264, 458.
[194] P.D. Burs, G.N. La Mar (1982) J. Magn. Reson. 46, 61.
[195] G.N. La Mar, F.A. Walker (1975) J. Am. Chem. Soc. 97, 5103.
[196] H. Goff, G.N. La Mar, C.A. Reed (1977) J. Am. Chem. Soc. 99, 3641.
[197] B.P. Gaber, R.D. Brown III, S.H. Koenig, J.A. Fee (1972) Biochim. Biophys. Acta 271, 1.
[198] I. Bertini, C.O. Fernández, B.G. Karlsson, J. Leckner, C. Luchinat, B.G. Malmström, A.M. Nersissian, R. Pierattelli, E. Shipp, J.S. Valentine, A.J. Vila (2000) J. Am. Chem. Soc. 122, 3701.
[199] L. Banci, I. Bertini, F. Briganti, C. Luchinat (1986) J. Magn. Reson. 66, 58.
[200] S.H. Koenig, R.D. Brown III, M. Spiller (1987) Magn. Reson. Med. 4, 252.
[201] I. Bertini, M. Fragai, C. Luchinat, G. Parigi (2001) Inorg. Chem., in press.

Chapter 6

Magnetic Coupled Systems

Several polymetallic systems experience magnetic coupling, either ferromagnetic or antiferromagnetic. Such magnetic coupling may affect the hyperfine shift and electron relaxation. Beneficial effects are always expected and observed on nuclear relaxation. A series of examples are discussed.

6.1 THE INDUCED MAGNETIC MOMENT PER METAL ION IN POLYMETALLIC SYSTEMS, THE HYPERFINE CONTACT SHIFT, AND THE NUCLEAR RELAXATION RATES

The compounds containing more than one metal ion with unpaired electrons deserve special attention because NMR is a powerful technique to relate the properties of the whole systems with those of each metal ion. Magnetic coupling occurs when the electronic spin magnetic moments of each metal ion interact with one another. The interaction is again quantized; for example, given S_1 and S_2 as two interacting spins, new spin levels are established, with total S' varying in unitary steps from $S_1 + S_2$ to $|S_1 - S_2|$. A major effect of magnetic spin coupling is the occurrence of new energy levels. Their energy separations vary from almost zero to the order of 10^3 cm^{-1}. At any laboratory temperature, kT may be of the order of the above separation and so the problem of the temperature-dependent population of the levels is introduced. The population P_i of any spin level of energy E_i is given by the Boltzmann law (Eq. (1.25)), which we recall here for convenience

$$P_i = \frac{\exp(-E_i/kT)}{\sum_i \exp(-E_i/kT)}.$$

When kT is large with respect to the energy gap, the population of each level is just one over the number of levels or functions. When kT is small with respect to the energy separation, then only the lowest level is occupied. The new energy levels S' are linear combinations of the $|S, M_S>$ functions of each metal ion. The functions and energies can be calculated by the simple Heisenberg Hamiltonian, that for a dimer is

$$H = J\mathbf{S_1} \cdot \mathbf{S_2} \tag{6.1}$$

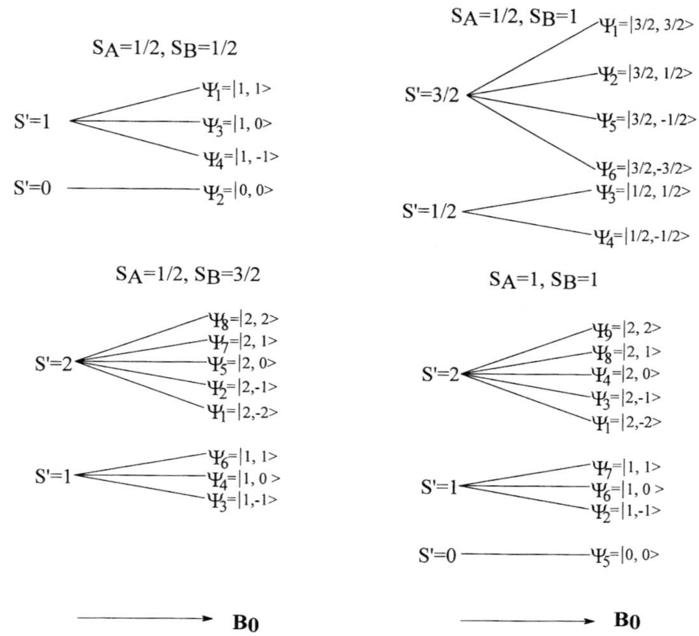

Fig. 6.1. Electronic spin energy levels for dimetallic systems ½–½, ½–1, ½–3/2, and 1–1.

where J is a constant and $\boldsymbol{S_1}$ is the spin operator which operates only on the functions of metal 1 [1]. The eigenvalues of Eq. (6.1) and the energy separation between S' and the neighbor $S' - 1$ levels are, respectively [1],

$$E = \frac{J}{2} S'(S' + 1) \quad \text{and} \quad \Delta E = JS'. \tag{6.2}$$

When J is positive, the ground state has $S' = |S_1 - S_2|$ and the highest level has $S' = S_1 + S_2$ (see Fig. 6.1). We say that the coupling is *antiferromagnetic*. The contrary holds when J is negative and the coupling is *ferromagnetic*.

When more metal ions are coupled, the following Hamiltonian can be used:

$$H = \sum_{j \neq k} J_{jk} \boldsymbol{S_j} \cdot \boldsymbol{S_k} \tag{6.3}$$

where all j–k pairwise interactions are taken into consideration. For each ith

[1] The eigenvalues of Eq. (6.1) can be derived by squaring the equality $\boldsymbol{S'} = \boldsymbol{S_1} + \boldsymbol{S_2}$: $\boldsymbol{S'}^2 = \boldsymbol{S_1}^2 + \boldsymbol{S_2}^2 + 2\boldsymbol{S_1} \cdot \boldsymbol{S_2}$ (since $\boldsymbol{S_1}$ and $\boldsymbol{S_2}$ commute) and, by rearranging, $\boldsymbol{S_1} \cdot \boldsymbol{S_2} = \frac{1}{2}[\boldsymbol{S'}^2 - \boldsymbol{S_1}^2 - \boldsymbol{S_2}^2]$. The eigenvalues of the Hamiltonian $J\boldsymbol{S_1} \cdot \boldsymbol{S_2}$ are thus given by: $E = J/2[S'(S' + 1) - S_1(S_1 + 1) - S_2(S_2 + 1)]$ (see Section 1.2). Because $S_1(S_1 + 1)$ and $S_2(S_2 + 1)$ are the same for all S' levels, they introduce only an energy offset and can be dropped. The splitting of the levels is thus given by Eq. (6.2).

S', the contribution to the total induced magnetic moment is (in analogy to Eq. (1.33)):

$$\langle \mu_i \rangle = -\mu_B g_e \langle S'_z \rangle_i P_i \tag{6.4}$$

where P_i is the population, the total induced magnetic moment being

$$\langle \mu \rangle = -\mu_B g_e \sum_i \langle S'_z \rangle_i P_i. \tag{6.5}$$

The contact hyperfine shift experienced by a nucleus is proportional to the induced magnetic moment of each S' level times the hyperfine coupling constant

$$\delta^{\text{con}} = -\sum_i \frac{A_i}{\hbar \gamma_I B_0} \langle S'_z \rangle_i P_i. \tag{6.6}$$

This equation would be of little help if further considerations were not made. In fact, we do not know the several A_i values. However, the $|S', M_{S'}\rangle$ wavefunctions are linear combinations of the $|S, M_S\rangle$ wavefunctions of each metal ion; therefore we want to show that it is possible to express A_i through the hyperfine constant A_M of the single ions when they are not magnetically coupled. Let us assume that, without loss of generality, a nucleus senses only one metal ion of the cluster, M, i.e. it senses only the induced magnetic moment of that ion. If we are able to project out the $|S, M_S\rangle$ part for that metal from each $|S', M_{S'}\rangle$ wavefunction, then we can use the hyperfine constant A_M that the nucleus would experience if the metal ion were not involved in the magnetic coupling, i.e. if $J = 0$. Indeed the following relationship holds for each S' level:

$$A_i \langle S'_z \rangle_i = A_M \langle S_{Mz} \rangle_i \tag{6.7}$$

where $\langle S_{Mz} \rangle_i$ is the expectation value of the contribution of metal M to the whole wavefunction S'_i and $\langle S'_z \rangle_i$ is the usual expectation value of S'_z for each i level [2]. This relationship is based on the assumption that the establishment of magnetic interactions within a cluster does not change the spin density distribution on the ligands nor on the metal ion. This is apparently the case in the investigated systems. If $|J| \ll kT$, all the levels are equally populated and

$$\sum_i A_i \langle S'_z \rangle_i P_i = A_M \sum_i \langle S_{Mz} \rangle_i P_i = A_M \langle S_{Mz} \rangle \tag{6.8}$$

where $\langle S_{Mz} \rangle$ is the expectation value of S_z for metal M in the absence of magnetic interactions. It follows that the energy of interaction between a nucleus and the induced electronic magnetic moment sensed by it ($\propto A_M \langle S_{Mz} \rangle$) does not change upon establishment of magnetic coupling. The energy of interaction does change if only some of the S_i levels are occupied.

TABLE 6.1

Coefficients relating the hyperfine coupling of an S_1 or S_2 spin system in a monomer to that in a coupled system, for each S_i spin level [3]

S_1	S_2: 1/2			1			3/2			2			5/2			3			7/2		
	S	C_1	C_2	S	C_1	C_2	S	C_1	C_2	S	C_1	C_2	S	C_1	C_2	S	C_1	C_2	S	C_1	C_2
1/2	1	1/2	1/2	3/2	1/3	2/3	2	1/4	3/4	5/2	1/5	4/5	3	1/6	5/6	7/2	1/7	6/7	4	1/8	7/8
	0	–	–	1/2	−1/3	4/3	1	−1/4	5/4	3/2	−1/5	6/5	2	−1/6	7/6	5/2	−1/7	8/7	3	−1/8	9/8
1				2	1/2	1/2	5/2	2/5	3/5	3	1/3	2/3	7/2	2/7	5/7	4	1/4	3/4	9/2	2/9	7/9
				1	1/2	1/2	3/2	4/15	11/15	2	1/6	5/6	5/2	4/35	31/35	3	1/12	11/12	7/3	4/63	59/63
				0	–	–	1/2	−2/3	5/3	1	−1/2	3/2	3/2	−2/5	7/5	2	−1/3	4/3	5/2	−2/7	9/7
3/2							3	1/2	1/2	7/2	3/7	4/7	4	3/8	5/8	9/2	1/3	2/3	5	3/10	7/10
							2	1/2	1/2	5/2	13/35	22/35	3	7/24	17/24	7/2	5/21	16/21	4	1/5	4/5
							1	1/2	1/2	3/2	1/5	4/5	2	1/12	11/12	5/2	1/35	34/35	3	0	1
							0	–	–	1/2	−1	2	1	−3/4	7/4	3/2	−3/5	8/5	2	−1/2	3/2
2										4	1/2	1/2	9/2	4/9	5/9	5	2/5	3/5	11/2	4/11	7/11
										3	1/2	1/2	7/2	26/63	37/63	4	7/20	13/20	9/2	10/33	23/33
										2	1/2	1/2	5/2	12/35	23/35	3	1/4	3/4	7/2	4/21	17/21
										1	1/2	1/2	3/2	2/15	13/15	2	0	1	5/2	−2/35	37/35
										0	–	–	1/2	−4/3	7/3	1	−1	2	3/2	−4/5	9/5
5/2													5	1/2	1/2	11/2	5/11	6/11	6	5/12	7/12
													4	1/2	1/2	9/2	43/99	56/99	5	23/60	37/60
													3	1/2	1/2	7/2	25/63	38/63	4	13/40	27/40
													2	1/2	1/2	5/2	11/35	24/35	3	5/24	19/24
													1	1/2	1/2	3/2	1/15	14/15	2	−1/12	13/12
													0	–	–	1/2	−5/3	8/3	1	−5/4	9/4
3																6	1/2	1/2	13/2	6/13	7/13
																5	1/2	1/2	11/2	64/143	79/143
																4	1/2	1/2	9/2	14/33	19/33
																3	1/2	1/2	7/2	8/21	13/21
																2	1/2	1/2	5/2	2/7	5/7
																1	1/2	1/2	3/2	0	1
																0	–	–	1/2	−2	3
7/2																			7	1/2	1/2
																			6	1/2	1/2
																			5	1/2	1/2
																			4	1/2	1/2
																			3	1/2	1/2
																			2	1/2	1/2
																			1	1/2	1/2
																			0	–	–

A useful formula to express the contact shift, in the case of a nucleus sensing only one metal ion, is

$$\delta^{\mathrm{con}} = A_M \frac{g_e \mu_B}{\hbar \gamma_I 3kT} \frac{\sum_i C_i S_i'(S_i'+1)(2S_i'+1)\exp(-E_i/kT)}{\sum_i (2S_i'+1)\exp(-E_i/kT)} \tag{6.9}$$

where

$$C_i = \langle S_{Mz} \rangle_i / \langle S_z' \rangle_i . \tag{6.10}$$

Eq. (6.9) reduces to Eq. (2.5) in the high temperature limit, as shown in Appendix VIII.

If the nucleus senses another metal ion with a different A_M value, then the total shift will be given by the sum of two terms of the type of Eq. (6.9), each containing its own A_M and C_i values. A_{M1} and A_{M2} can be estimated from the monomeric analogs. The C_{i1} and C_{i2} values can be calculated once and for all for each system with the help of Hamiltonian Eq. (6.3). Their values are reported in Table 6.1 for dimers involving S_1 and S_2 values from ½ to 7/2 [3]. These coefficients can be calculated from the following equations:

$$\begin{aligned} C_{i1} &= [S_i'(S_i'+1) + S_1(S_1+1) - S_2(S_2+1)]/[2S_i'(S_i'+1)] \\ C_{i2} &= [S_i'(S_i'+1) + S_2(S_2+1) - S_1(S_1+1)]/[2S_i'(S_i'+1)]. \end{aligned} \tag{6.11}$$

Note that in the ground state of an antiferromagnetic dimer the metal 2 with smaller S_2 always has a negative coefficient, whereas metal 1 has a positive coefficient. This observation is of paramount importance to detect the establishment of antiferromagnetic coupling.

There are no treatments for the pseudocontact term, since the orbital part has never been considered when dealing with magnetic coupled systems. However, Eqs. (6.7) and (6.10) hold for the hyperfine splitting in EPR spectra of both solids and solutions [1]. Therefore the same reasoning is likely to apply to the pseudocontact shift as well. The major complication arises from the point-dipolar nature of the pseudocontact shift treatment, which contrasts with the idea of a polymetallic center.

By analogy with the shift, in dimetallic systems the relaxation rate enhancement of a nucleus sensing one metal ion is given by:

$$R_{1M} \propto E^2 S'(S'+1) f(\omega, \tau_c) \sum_i C_i^2 P_i \tag{6.12}$$

where $E^2 S'(S'+1)$ is a generic notation for the average square of the interaction energy between S' and I, whose modulation with time constant τ_c causes relaxation [2], and P_i is the Boltzmann partition function.

Eq. (6.12) for the case of dipolar relaxation takes the form:

$$R_{1M} = \frac{2}{15}\left(\frac{\mu_0}{4\pi}\right)^2 \frac{\gamma_I^2 g_e^2 \mu_B^2}{r^6} \times \frac{\sum_i \left[C_i^2 S_i'(S_i'+1)(2S_i'+1)\exp(-E_i/kT)\left(\frac{7\tau_{c2}}{1+\omega_S^2\tau_{c2}^2}+\frac{3\tau_{c1}}{1+\omega_I^2\tau_{c1}^2}\right)\right]}{\sum_i \left[(2S_i'+1)\exp(-E_i/kT)\right]} \tag{6.13a}$$

where τ_{ci}^{-1} is defined as usual as the sum of rotational, exchange and electronic relaxation (τ_{c1}^{-1} = longitudinal, τ_{c2}^{-1} = transverse) rates.

Analogously, the equation for contact relaxation takes the form:

$$R_{1M} = \frac{2}{3}\frac{A_M^2}{\hbar^2} \frac{\sum_i \left[C_i^2 S_i'(S_i'+1)(2S_i'+1)\exp(-E_i/kT)\left(\frac{\tau_{c2}}{1+\omega_S^2\tau_{c2}^2}\right)\right]}{\sum_i \left[(2S_i'+1)\exp(-E_i/kT)\right]} \tag{6.13b}$$

where τ_{ci}^{-1} is now defined as the sum of exchange and electronic relaxation rates.

6.2 ELECTRON RELAXATION AND MAGNETIC COUPLING

A consequence of magnetic coupling can be a change in the electronic relaxation times of the involved metal ions. Magnetic coupling gives rise to new energy levels which can provide new relaxation pathways. New transitions involving spin levels can occur through coupling with the lattice, i.e. solvent collisions, solvent fluctuations, molecular tumbling, phonons when appropriate. This is indeed what happens in polymetallic systems. In this case there is more than one level with the same S' value which allow electronic transitions.

[2] $E^2S(S+1)$ is a generic notation that holds for dipolar, contact or Curie relaxation. In the equation for Curie relaxation there is a $S^2(S+1)^2$ term because a further $S(S+1)$ term is contained in the E^2 term.

6.2.1 Homodimers

In homodimers, all the levels have different S' values and in principle no further relaxation pathway is occurring. Experiments have been performed aiming at comparing the τ_s value in copper(II) dimetallic systems with respect to monometallic systems [4]. A protein frame has been used in which either only one copper(II) ion is present or two magnetically coupled copper(II) ions are present. In the latter case J is 52 cm^{-1} and all the S' levels are almost equally populated at room temperature. The protein is superoxide dismutase, whose metal ligands are shown in Fig. 6.2. We may have either copper–zinc or copper–copper systems. Nuclear longitudinal relaxation was measured for the water protons since one water molecule interacts with one copper ion. It is shown that in the two cases the copper(II) ion bound to water maintains similar τ_s values. The nuclear relaxation in the dimer is, however, one-half that in the monomer on account of Eq. (6.13a). In the case of two $S = \frac{1}{2}$ ions ($C_i = \frac{1}{2}$) and $|J| \ll kT$, $P = 3/4$, because only the $S' = 1$ level which is threefold degenerate contributes to paramagnetic relaxation, the $S' = 0$ level being magnetically silent. Since R_{1M} for the monomer depends on $S(S+1)$ with $S = \frac{1}{2}$, and for the dimer it depends on $S'(S'+1)$ with $S' = 1$ (Eq. (6.13a)), the ratio between the R_{1M} values of the dimer for a nucleus

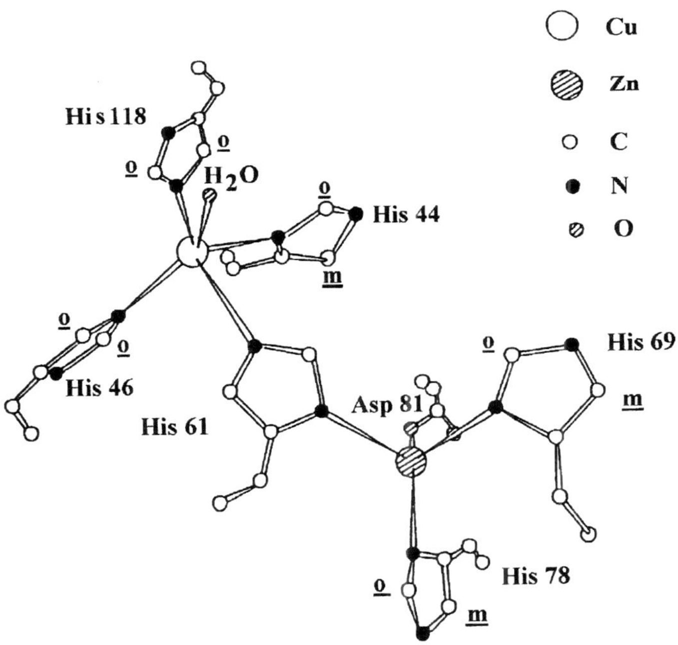

Fig. 6.2. Schematic structure of the active site of copper–zinc superoxide dismutase [72]. The zinc ion can be substituted, among others, by copper(II), cobalt(II) and nickel(II) ions.

sensing only one metal ion and of the monomer is

$$\frac{C^2 S'(S'+1)P}{S(S+1)} = \frac{1/4 \times 2 \times 3/4}{3/4} = \frac{1}{2}.$$

The constancy of τ_s upon establishing a magnetic coupling in homodimers can be also demonstrated theoretically [5]. It is possible, however, that in the presence of large ZFS of the $S' = 1$ level, new mechanisms of electron relaxation are effective, and τ_s shortens up. That is why sometimes the NMR linewidths of copper(II) dimers are sharper than in monomers by more than a factor 2 [6].

Even for dimetallic centers with each metal ion having $S > ½$, Eq. (6.12) provides a value of R_{1M} which is one-half of that provided by the analogous equation for the monomer. If we repeat the calculation for the hyperfine shifts (Eqs. (2.6) and (6.9)), at variance with nuclear relaxation, we see that the shifts for the coupled and uncoupled systems are the same as long as $|J| \ll kT$. Note that ferromagnetic coupling would have provided the same results because our reasoning was not based on the order of the S' levels.

When $|J|$ is of the order of kT, then the excited S' levels are not fully populated. In the case of antiferromagnetic coupling the magnetic susceptibility decreases and both shifts and nuclear relaxation values decrease. This is a significant advantage when systems with long electron relaxation times are investigated. Typically, copper(II) complexes are not easily investigated by high resolution NMR because τ_s is of the order of 10^{-9} s and τ_r for small complexes is about 10^{-10} s. Therefore, the linewidth at 500 MHz for a proton dipole–dipole coupled at 4 Å is 700 Hz in the absence of contact relaxation. In the case of a copper dimer with a J of 100 cm^{-1}, the shift of a proton sensing only one copper ion according to Eq. (6.9) is 86% of the monomeric case, and the linewidth (which behaves analogously to R_{1M} in Eq. (6.13a)) is 43% of the monomer at room temperature.

Ferromagnetic coupling generally has small $|J|$ values. For the sake of completeness, we consider the case in which there is a ferromagnetic coupling of -100 cm^{-1}. By repeating the calculations, the shifts are 111% of the monomer and the linewidths 55% of the monomer.

6.2.2 *Heterodimers*

When we deal with two different paramagnetic ions, it should be kept in mind that in the absence of magnetic coupling one will have a longer electronic relaxation time and the other will have a shorter one, according to their electronic structure. Such differences are in general very large. The slow relaxing metal ion will be referred to as M_1 and the fast relaxing one as M_2. It is reasonable to believe that, when the absolute value of J (expressed as $|J|/\hbar$) is smaller than τ_{s1}^{-1} (i.e. the electronic relaxation rate of the slow relaxing M_1 ion) no effect on the electronic relaxation of the pair will take place. When $|J|/\hbar > \tau_{s1}^{-1}$ but smaller than τ_{s2}^{-1}, we can approach the system through a first order perturbation treatment. The coupling between S_1 and S_2 is very similar to that between I and S already discussed (Sections 3.4 and 3.5). In both cases we have a scalar coupling which is the contact coupling in the case of the electron–nucleus coupling and the exchange or superexchange coupling in the case of magnetic interactions. Analogous to the Bloembergen Eq. (3.26) for the contact nuclear relaxation we can write the following relationship:

$$\Delta\tau_{s1}^{-1} = \frac{2}{3}S_2(S_2+1)\left(\frac{J}{\hbar}\right)^2 \frac{\tau_{s2}}{1+(\omega_{S1}-\omega_{S2})^2\tau_{s2}^2} \tag{6.14}$$

where $\Delta\tau_{s1}^{-1}$ is the enhancement in the electronic relaxation rate of the slow relaxing ion, ω_{S1} and ω_{S2} are the Larmor frequencies of the M_1 and M_2 ions, and τ_{s2} is the correlation time for the interaction which is given by the electronic relaxation rate of the fast relaxing ion.

When the coupling between S_1 and S_2 is dipolar in origin the equation should again be analogous to the dipolar coupling in the case of nuclear relaxation (Eq. (3.16)):

$$\Delta\tau_{s1}^{-1} = \frac{2}{15}\left(\frac{\mu_0}{4\pi}\right)^2\left(\frac{1}{\hbar}\right)^2 \frac{g_e^4\mu_B^4 S_2(S_2+1)}{\langle r^3\rangle^2}$$
$$\times\left(\frac{\tau_{s2}}{1+(\omega_{S1}-\omega_{S2})^2\tau_{s2}^2} + \frac{3\tau_{s2}}{1+\omega_{S1}^2\tau_{s2}^2} + \frac{6\tau_{s2}}{1+(\omega_{S1}+\omega_{S2})^2\tau_{s2}^2}\right) \tag{6.15}$$

where $\langle r^3\rangle$ is the average cube of the interelectronic distance.

It is apparent that the dipolar coupling is a short range coupling and rapidly vanishes with the distance. This does not hold for the scalar coupling. It is also apparent that small $|J|$ values can affect τ_{s1} under favorable conditions. In other words, small $|J|$ values can be revealed through the linewidth of the EPR signal or through NMR relaxation measurements.

Eqs. (6.14) and (6.15) hold as long as $|J|/\hbar < \tau_{s2}^{-1}$. This is a general requirement in relaxation theory called Redfield limit [7]. The coupling energy between the relaxing spin and the lattice (which can be another spin) has to be smaller than the product of $\hbar$ times the reciprocal correlation time. Equivalent to this statement is the following: the relaxation rate of a spin cannot be higher than the reciprocal of the correlation time. In other words, τ_{s1}^{-1} approaches τ_{s2}^{-1} when $|J|$ increases. Outside the Redfield limit, i.e. outside the perturbative limit, each S' level should be regarded as a single spin. The transitions between $M_{S'}$ levels should be evaluated in a manner similar to that discussed for the $S > 1$ multiplets in isolated ions (Section 3.3). If reference is made to the τ_{s1} and τ_{s2} of the single ions in the coupled system they cannot be easily predicted except for their becoming similar and close to the shorter one (see Section 6.4 for further discussion of this aspect).

The absolute value of J, which we have used all over the treatment within the Redfield limit, implies that the theory is the same for ferro- and antiferromagnetic coupling. When $|J| \ll kT$, the sign of J is irrelevant for the nuclear shifts, the electronic relaxation times and nuclear relaxation times. When J is of the order of kT, presumably only electronic relaxation times within the various S' level can be defined, and they are independent of the sign of J, but the nuclear shifts and relaxation rates are different because the various S'_i are differently populated according to the sign of J. Eqs. (6.9) and (6.13a) should be used to predict the particular behavior.

The above treatment is probably valid only in ideal cases. In practical cases spin–orbit coupling causes ZFS, and the S' levels are split at zero magnetic field just like the S levels are. This consideration requires that each S' level will have its own effective electronic relaxation time. Therefore, the above quantitative considerations should be regarded as useful guidelines.

6.2.3 Polymetallic systems

In the case of polyionic systems the picture becomes still more complicated. There are no examples of prediction of τ_s values. However, we have empirically learned that τ_s of the entire system is shorter than in the absence of magnetic coupling. High resolution NMR experiments are, in general, successful with magnetically coupled polyionic systems.

6.3 NMR OF DIMETALLIC SYSTEMS

It is convenient to divide this section into systems containing equal or different metal ions. Each of the two cases will be treated according to whether $|J|$ is larger or smaller than kT.

6.3.1 Systems containing equal metal ions, $|J| < kT$

We take now as an illustrative example the dimers with PMK as ligand. Such a ligand is capable of giving bimetallic complexes of formula $[(PMK)_3M_2]^{4+}$ and with the structure shown in Fig. 6.3 [8]. The ligand provides a pathway for weak magnetic coupling. When $|J| \ll kT$, the shift experienced by a nucleus is simply the sum of the contributions of the two metal ions, i.e. of two terms given by Eq. (6.9) and containing A_{M1} and C_{i1}, and A_{M2} and C_{i2}, respectively. This is indeed the case, as proven by using mixed complexes with the diamagnetic zinc ion [9–11]. The correlation time for the nucleus electron coupling is given by the rotational correlation time in the case of copper(II) complexes ($\tau_r \approx 2 \times 10^{-10}$ s) and by the electronic relaxation time ($\tau_s \approx 6 \times 10^{-13}$ s) in the case of cobalt(II) complexes [12,13]. As explained in the previous section, a decrease is expected

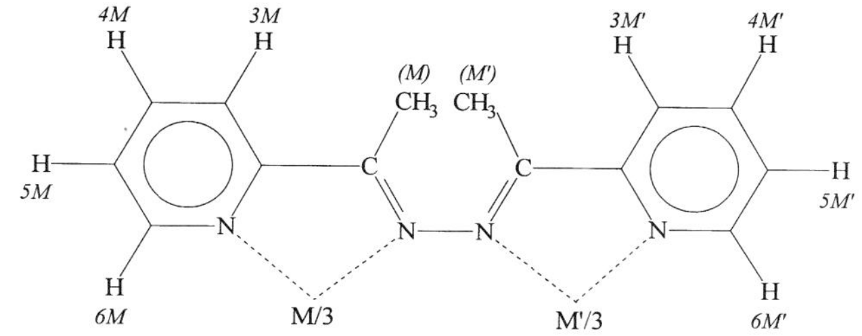

Fig. 6.3. Schematic structure of $[(PMK)_3M_2]^{4+}$ bimetallic complexes [8].

in the linewidth, which would be of a factor of two if the proton senses only one paramagnetic ion. Indeed, some decrease in the linewidth is observed. In Table 6.2 the NMR parameters for the CoCo system are reported, together with those of the CoZn, NiZn, CuZn, CoNi, and CoCu systems (see Section 6.3.3) [13].

6.3.2 Systems containing equal metal ions, $|J| \approx kT$

The first systems of this kind investigated in the literature are dicopper(II) dimers [14,15], where J is positive and of the order of hundreds of wavenumbers. The correlation time is τ_r; the nuclear relaxation rates, as well as the shifts, are decreased according to the Boltzmann partition coefficients of Eqs. (6.13a) and (6.9), respectively. A factor larger than two in nuclear relaxation rate between coupled and uncoupled systems should be operative if only one metal ion is sensed by the resonating nucleus. The decrease in magnetic susceptibility due to the population of the lower part of the energy ladder makes these systems more suitable for NMR investigation than the uncoupled ones. The same holds for μ-oxo iron(III) dimers [16].

In Fig. 6.4 the spectra of a copper(II) dimer are shown [15]. The temperature dependence of the shifts is consistent, according to Eq. (6.9), with a J value of about 500 cm^{-1}. In the inset, the ^2H NMR spectrum of a selectively deuterated derivative is reported [15], showing the smaller linewidth of the ^2H signal. In an increasing number of copper dimers the NMR lines have been found to be far too sharp to be explained [15,17–22], thus allowing high resolution NMR investigations, being two orders of magnitude sharper than the mononuclear analogs (see Fig. 6.5). Since the hyperfine shifts are very large, as expected for the mononuclear analog, and display an essentially normal Curie temperature dependence, the line sharpening must be due to a decrease in the electron relaxation time (from 10^{-9} to 10^{-11} s), and cannot be due to a reduction of paramagnetism caused by a depopulation of the excited states. Hyperfine shifts are in fact very large, spanning from +230 to −14 ppm. This confirm the absence of strong antiferromagnetic coupling (which would cause the paramagnetism to be strongly reduced due to depopulation of the excited $S = 1$ state). What causes

TABLE 6.2
Proton hyperfine shifts (ppm) and T_{1M}^{-1} values (s^{-1}, estimated errors in parentheses) for $MM'(PMK)_3^{4+}$ complexes [13]

		3-H	4-H	5-H	6-H	$-CH_3$
$CoZn(PMK)_3^{4+}$						
Shifts	(Co)	66.81	2.74	42.31	141.47	44.99
	(Zn)	−9.88	2.41	−10	−0.69 [a]	8.08
T_{1M}^{-1}	(Co)	36.1 (0.9)	11.5 (0.3)	29.2 (0.7)	670 (220)	71.9 (1.6)
	(Zn)	6.9 (0.3)	1.41 (0.02)	1.78 (0.06)	3.97 (0.06) [a]	37.3 (0.6)
$NiZn(PMK)_3^{4+}$						
Shifts	(Ni)	62.98	8.54	40.29		39.23
	(Zn)	−9.15	2.76	−8.07	2.11	−10.51
T_{1M}^{-1}	(Ni)	2000 (500)	590 (5)	1630 (110)		5900 (350)
	(Zn)	370 (15)	120 (10)	140 (10)	310 (15)	2600
$CuZn(PMK)_3^{4+}$						
Shifts	(Cu)	39.68	5.68	23.43		17.26
	(Zn)	−4.19	1.31	−3.67	0.87	−3.55
T_{1M}^{-1}	(Cu)		235 (3)	526 (60)		1200 (40)
	(Zn)	118 (10)	43.3 (1)	64 (6)	121 (2)	709 (10)
$Co_2(PMK)_3^{4+}$						
Shifts		56.46	5.08	31.99	141.21	51.88
T_{1M}^{-1}		48.5 (0.5)	17.0 (0.1)	35.8 (0.3)	588 (35)	114 (2)
$CoNi(PMK)_3^{4+}$						
Shifts	(Co)	56.54	4.74	34.47	149.55	33.75
	(Ni)	50.75	11.05	28.91	148.8 [b]	44.65
T_{1M}^{-1}	(Co)	52.1 (1.4)	18.6 (0.2)	38.0 (0.6)	645 (17)	180 (0.4)
	(Ni)	112 (3)	35.2 (0.5)	88.5 (1.6)	909 (83) [b]	225 (7)
$CoCu(PMK)_3^{4+}$						
Shifts	(Co)	62.43	3.78	38.61	143.84	41.35
	(Cu)	27.60	8.06	12.21	87.38	24.65
T_{1M}^{-1}	(Co)	40.8 (1.8)	13.5 (0.4)	30.3 (1.2)	529 (28)	94.3 (3.6)
	(Cu)	37.7 (1.8)	8.33 (0.14)	24.2 (0.4)	307 (10)	75.8 (2.3)

[a] Direct overlap with 5-H (Zn_2).
[b] Broad and overlapping with the other 6-H signal.

this decrease is still under investigation. It was suggested that new relaxation pathways can be created upon dimer formation, as a result of modulation of the J value by solvent collision, modulation of the ZFS of the $S' = 1$ state or electron delocalization [22]. Also the so-called Cu_A centers exhibit a similar behavior (see Section 5.3).

When the dimetallic center is embedded in a protein, the correlation time is given by the electronic relaxation time. A studied case is provided by the

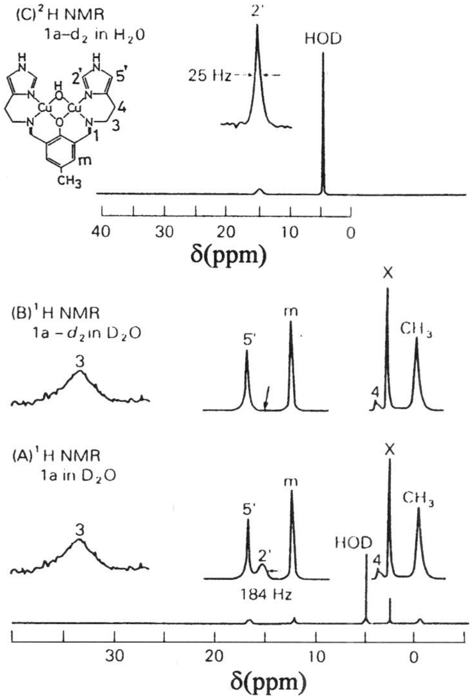

Fig. 6.4. ^{1}H NMR spectra of a copper(II) dimer (A) and of its selectively deuterated derivative (B) (the arrow pointing to the missing signal). The ^{2}H NMR spectrum of the latter is shown in (C), together with the structure of the compound [15].

$[Fe_2S_2]^{2+}$ unit represented in Fig. 6.6 and found in oxidized ferredoxins. The large correlation time makes the system difficult to investigate by NMR [23,24] despite the sizable decrease in paramagnetism due to a J of about 300 cm^{-1}. The sharpening of the lines of oxidized ferredoxins with respect to the monomeric iron model provided by the oxidized rubredoxin thus arises from the decreased Boltzmann population of the paramagnetic excited states.

6.3.3 Systems containing different metal ions, $|J| < kT$

With the PMK ligand a CoCu derivative has been obtained [7] (Fig. 6.7). From the temperature dependence of the shifts (and magnetic susceptibility measurements in solution), the value of J appears to be positive and much smaller than kT [13]. As expected, the hyperfine shifts are the sum of those of the CuZn and ZnCo systems for each proton (Table 6.2). The NMR lines of the copper domain are now quite sharp, even sharper than those of the cobalt domain (Fig. 6.7). Qualitatively, the data can be accounted for if τ_s of copper is sizably reduced and approaches that of cobalt (and thus the Redfield limit is reached).

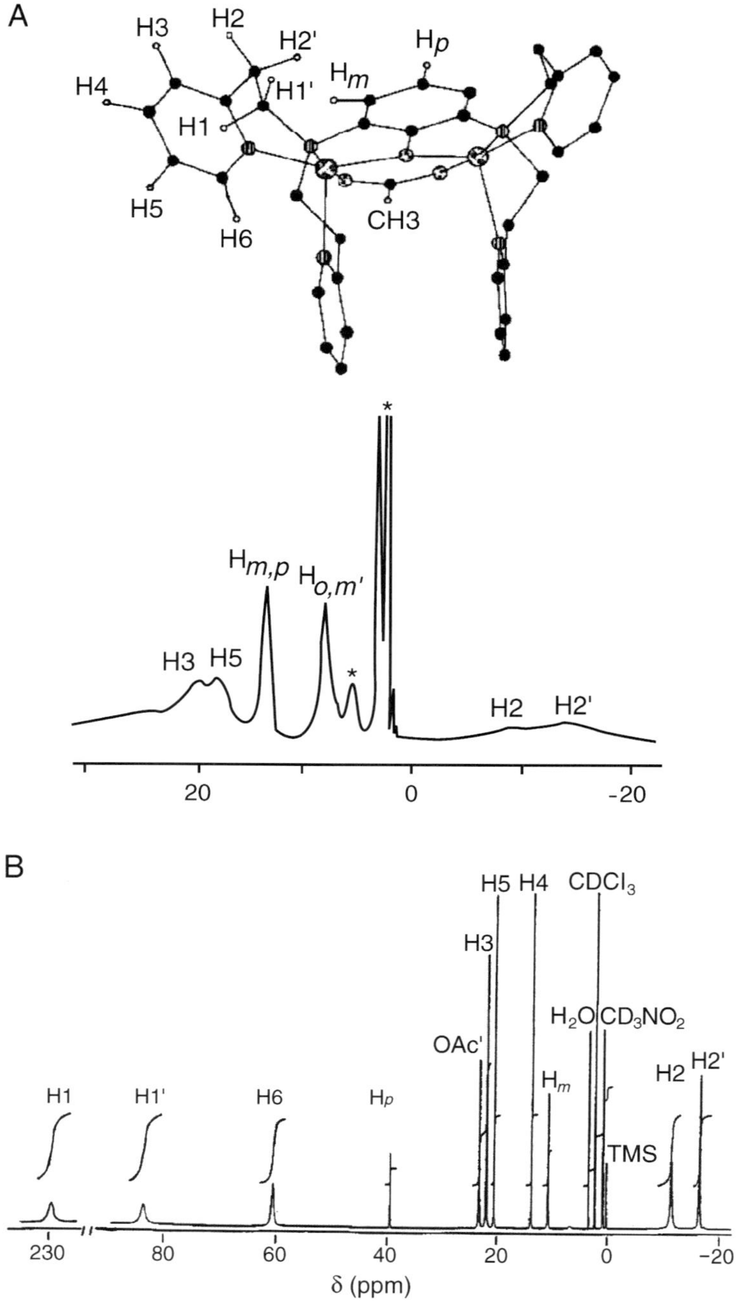

Fig. 6.5. Crystal structures and ^{1}H NMR spectra of a copper monomer (A) and the corresponding copper dimer (B). The signal assignment is also shown [22].

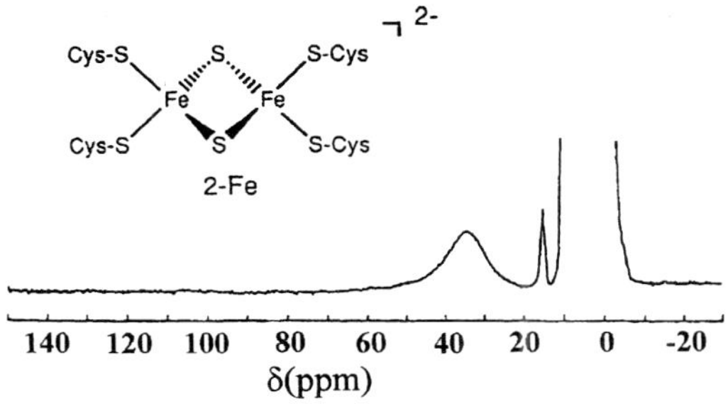

Fig. 6.6. Schematic structure of the $[Fe_2S_2](SR)_4^{2-}$ unit as a found in oxidized ferredoxins and ^{1}H NMR spectrum showing the β-CH_2 proton signals at about 40 ppm.

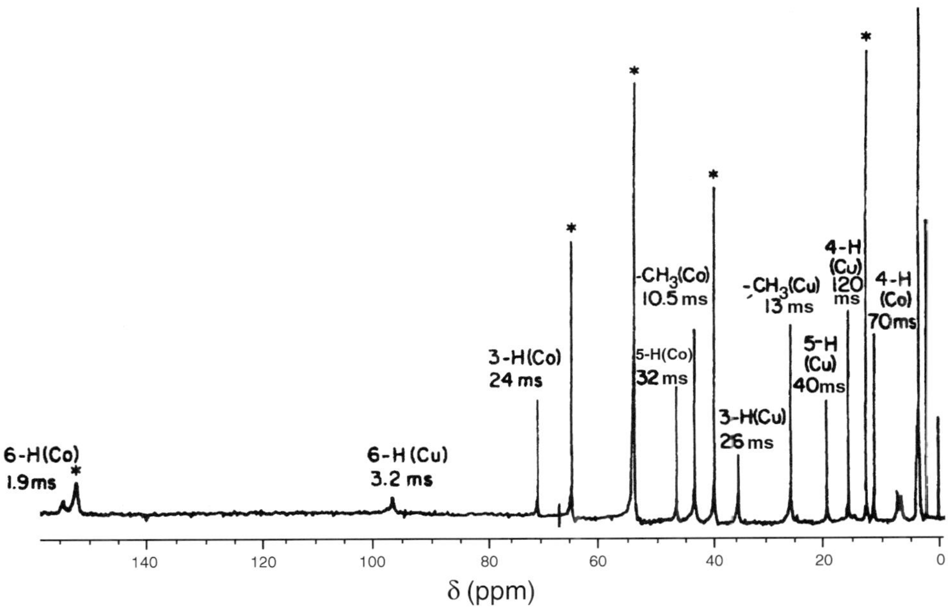

Fig. 6.7. ^{1}H NMR spectra of the CoCu derivative of PMK (Fig. 6.3) [13].

The magnetic coupling is antiferromagnetic with an $S' = 1$ ground state and an $S' = 2$ excited state. The C_i coefficients are those of Table 6.1, i.e. $-1/4$ and $1/4$ for copper and $5/4$ and $3/4$ for cobalt. Such coefficients provide a ratio in nuclear R_{1M} of ca. 12 for the protons of the cobalt domain with respect to those of the copper domain, if τ_s were the same (Eq. (6.13a)). The experimental ratio is about two. The discrepancy can be understood by considering: (i) that the electronic relaxation times for the two metal ions are not identical; (ii) that copper(II) tends to give rise to ligand-centered effects more than cobalt, and therefore to experience higher nuclear relaxation rates than expected on the basis of a pure

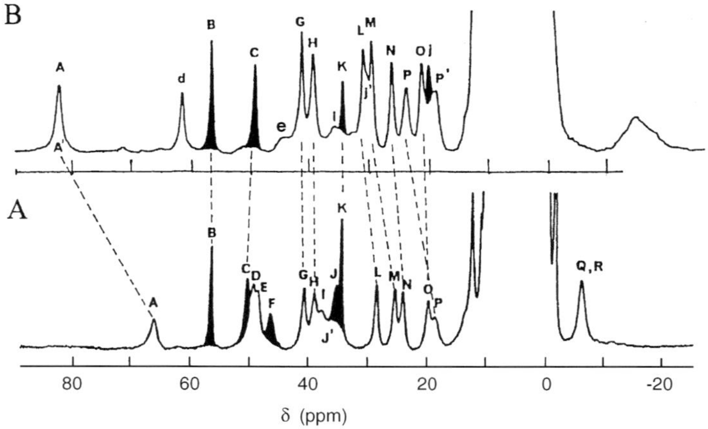

Fig. 6.8. ^{1}H NMR spectra of Cu_2Co_2 (A) [28] and Cu_2Ni_2 (B) [29] superoxide dismutase. The dashed lines relate signals belonging to the various histidine ligands of the copper domain. The black-shaded signals disappear when the spectra are recorded in D_2O.

dipolar mechanism; (iii) that the presence of ZFS complicates the prediction strategies.

The above system is similar to that built into the protein superoxide dismutase (Fig. 6.2). The zinc can be replaced by several metal ions. Cobalt(II) and nickel(II) are pertinent here. The cobalt–copper system has $J \approx 17$ cm^{-1} with $S' = 1$ ground state [25]. The electronic relaxation times for the two ions are expected to be either equal or similar. The ^{1}H NMR spectrum is shown in Fig. 6.8A. The assignment has been performed through several steps.

(a) The histidine NH protons have been recognized because their signals disappear when the spectrum is recorded in D_2O (Fig. 6.8A) [26].

(b) The protons of the copper domain have smaller R_1 than those of the cobalt domain if the metal–proton distances are similar [26].

(c) The linewidths have large Curie relaxation contributions which can be factorized out by recording the spectra at different fields. Such contributions and the R_1 values should provide a qualitative order of distances of the protons from the metal ions. However, large spin delocalization effects may provide larger linewidths to protons farther from the metal ion and this has led to a misassignment [27].

(d) NOE measurements have allowed the assignment of all histidine protons and of the β-CH_2 of the Asp group [28].

(e) NOESY spectra have largely confirmed the previous assignment and have permitted the assignment of the β-CH_2 protons and of some α-CH protons of the histidines [28].

The resulting assignment, which is one of the most successful in paramagnetic metalloproteins, is reported in Table 6.3. The nuclear relaxation rates, which are

TABLE 6.3

Assignment of ^{1}H NMR signals of metal coordinated residues in Cu_2Co_2SOD (298 K, pH 5.5, buffer 50 mM acetate) [a]

Signal	Shift (ppm)	Assignment	Signal (ppm)	Shift	Assignment
A	67.4	His 63 $H\delta_2$	a′	12.30	His 48 $H\beta_2$
B	57.0	His 120 $HN\delta_1$	b′	11.21	His 120 $H\beta_2$
C	50.3	His 46 $NH\varepsilon_2$	c′	8.44	His 120 NH
D	49.6	His 71 $H\delta_2$	d′	6.40	His 48 $H\beta_1$
E	49.0	His 80 $H\delta_2$	e′	4.43	His 120 $H\alpha$
F	46.7	His 80 $HN\varepsilon_2$	f′	3.73	His 48 $H\alpha$
G	41.0	His 46 $H\delta_2$	g′	3.13	His 120 $H\beta_1$
H	39.5	His 120 $H\varepsilon_1$	h′	1.23	Ala 140 βCH_3
I	38.7	Asp 83 $H\beta_1$	i′	0.56	Val 118 $\gamma_1 CH_3$
J′	37.0	Asp 83 $H\beta_2$	j′	−0.28	Arg 143 $H\gamma_1$
J	35.6	His 71 $HN\varepsilon_2$	k′	−1.51	Val 118 $\gamma_1 CH_3$
K	34.7	His 48 $HN\delta_1$			
L	28.4	His 48 $H\delta_2$			
M	25.7	His 46 $H\varepsilon_1$			
N	24.3	His 120 $H\delta_2$			
O	19.8	His 48 $H\varepsilon_1$			
P	18.7	His 46 $H\beta_1$			
Q	−6.2	His 71 $H\beta_2$			
R	−6.2	His 46 $H\beta_2$			

[a] The tentative assignment of some signals involving residues not directly bound to the metal ions, and falling in the diamagnetic region of Fig. 6.8(A) is also reported [28].

in a ratio similar to those of the CuCoPMK system, have been interpreted on the basis of large ligand-centered effects on the copper ligands [28].

The value of J is not known in the case of the copper–nickel derivative. However, the same line of discussion should hold for this derivative, whose spectrum is shown in Fig. 6.8B [29]. It is noteworthy that the R_1 values of the protons of the copper domain are smaller in this case than in the cobalt case. This is probably due to a shorter electronic relaxation time of tetrahedral nickel(II) than tetrahedral cobalt(II) (see Section 3.3). In this respect it may be interesting to note that in the case of the cobalt–cobalt derivative we have a tetrahedral and a square pyramidal cobalt(II) ion [30,31]. The former has longer electronic relaxation times. Upon establishment of magnetic coupling, the electronic relaxation times of the tetrahedral cobalt(II) ion decrease and tend to reach the values of the square pyramidal cobalt(II) ion. This is evident by comparing the spectra of the protein containing only the tetrahedral cobalt(II) ion with those containing both ions (Fig. 6.9).

Many systems of this kind are available in the literature. Only a small part has been studied by NMR, and sometimes not all the structural and dynamic

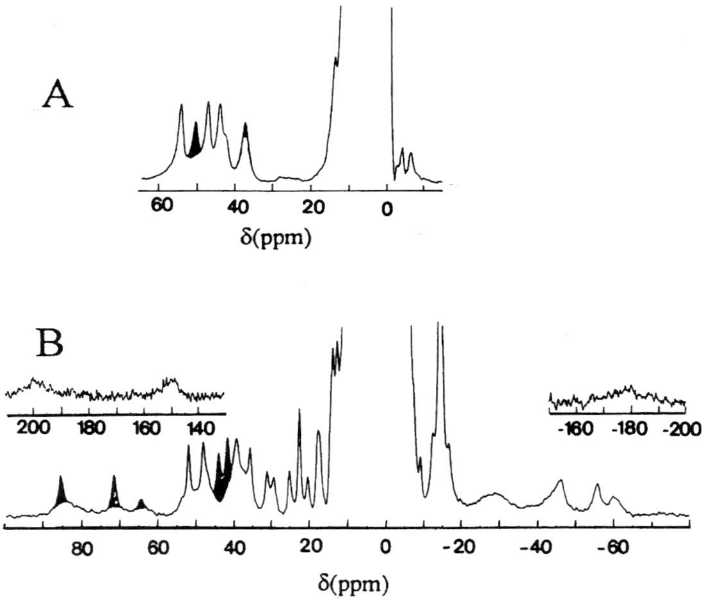

Fig. 6.9. ^{1}H NMR spectra of E_2Co_2 (A) (E = empty) and Co_2Co_2 superoxide dismutase (B). The black-shaded signals disappear when the spectra are recorded in D_2O [30,31].

information has been extracted from the spectra with the support of the available theory.

The results are expected to be the same if the coupling is of ferromagnetic type.

6.3.4 *Systems containing different metal ions, $|J| \approx kT$*

If $|J|$ were much larger than kT, only the ground state would be populated. This limit condition is very useful to illustrate the peculiarities of asymmetric dimeric systems. In these systems, one ion will have a spin larger than the other. In the ground state the larger spin will always be aligned along the external magnetic field. Now we can distinguish between ferro- and antiferromagnetic coupling. In the former case the smaller spin is aligned along with the larger spin; in the latter case the smaller spin is forced to be aligned opposite to the external magnetic field. The second case is the most common. The larger spin provides a ‘normal’ additional magnetic field to the resonating nucleus in a way similar to the case of absence of magnetic coupling. The smaller spin is forced to provide an additional field of opposite sign. From the theory we do expect a negative sign for the coefficient related to the smaller spin (see Eqs. (6.10) and (6.11)). Therefore, if the nuclei coupled to the larger spin experience downfield shift, the nuclei coupled to the smaller spin will experience upfield shifts (Fig. 6.10). In the limit of $|J| \ll kT$ the shifts of the two kinds of nuclei are the same as if there were no

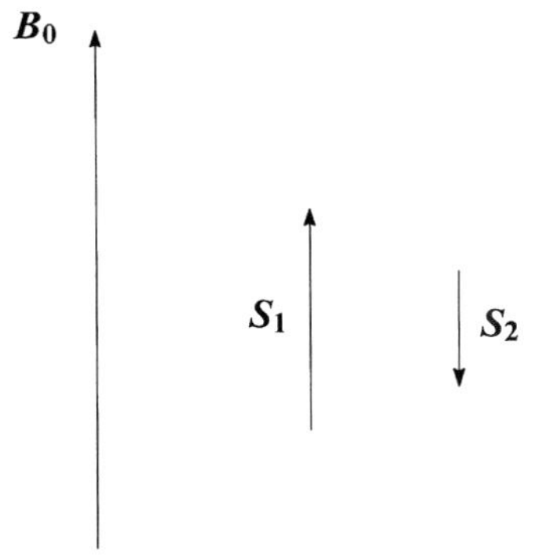

Fig. 6.10. Pictorial scheme of the magnetic moments corresponding to S_1 and S_2 induced by a magnetic field B_0 in an antiferromagnetically coupled ground state, S_1 being larger than S_2, and $g_e \mu_B B_0 \ll J$.

magnetic coupling at all. In the intermediate cases the shifts of the nuclei coupled to the smaller spin can be anywhere from downfield to upfield. However, if they are downfield, they will increase with increasing temperature, as the system will tend towards $|J| \ll kT$. The shifts of the nuclei coupled to the larger spin will decrease with increasing temperature similarly to a normal paramagnet (see Eqs. (2.5) and (2.6)). The expected temperature dependence is illustrated in Fig. 6.11, as calculated from Eq. (6.9).

From the point of view of relaxation we expect similar values of τ_s for the two spins, unless ZFS of the magnetically coupled S' levels introduces unpredictable effects.

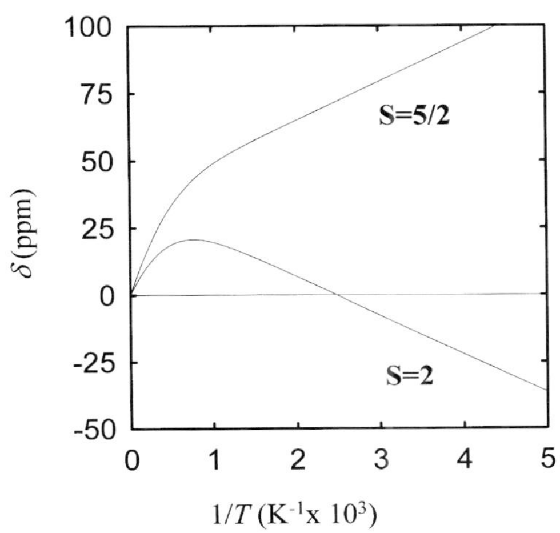

Fig. 6.11. Predicted temperature dependence of the shifts of nuclei sensing either the $S = 5/2$ or the $S = 2$ spin of an antiferromagnetically coupled pair with $J = 300\ cm^{-1}$.

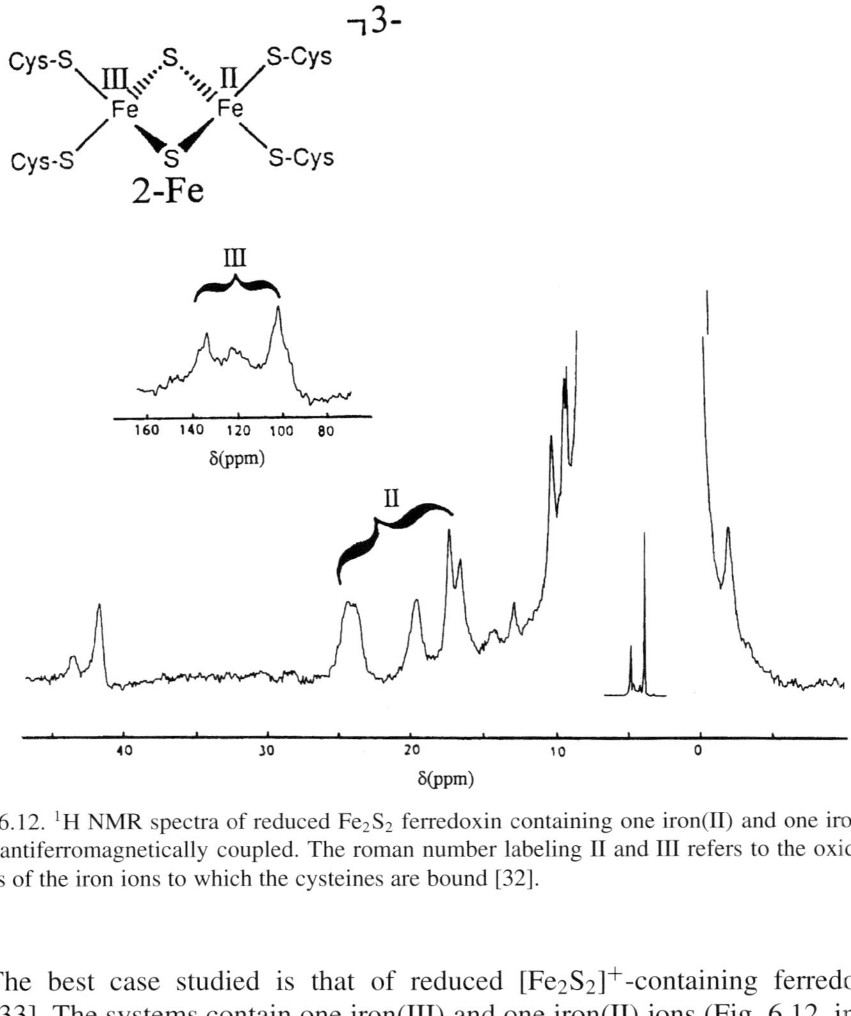

Fig. 6.12. ^{1}H NMR spectra of reduced Fe_2S_2 ferredoxin containing one iron(II) and one iron(III) ions antiferromagnetically coupled. The roman number labeling II and III refers to the oxidation states of the iron ions to which the cysteines are bound [32].

The best case studied is that of reduced $[Fe_2S_2]^+$-containing ferredoxins [32,33]. The systems contain one iron(III) and one iron(II) ions (Fig. 6.12, inset). Both of them are tetrahedral and high spin. The iron(III) has $S = {}^5\!/\!_2$ and the iron(II) $S = 2$. The ^{1}H NMR spectrum of one protein is reported in Fig. 6.12 [32]. The four signals far downfield belong to the β-CH_2 protons of the two cysteines bound to iron(III). Such signals have a temperature dependence of Curie type, i.e. the hyperfine shifts increase with decreasing temperature. The sharp signal around 40 ppm is assigned to an α-CH proton of one cysteine of the iron(III) domain. The four sharp signals between 30 and 15 ppm belong to the β-CH_2 protons of the cysteines of the iron(II) domain. The latter four signals have a temperature dependence of antiCurie type, i.e. the hyperfine shifts increase with increasing temperature.

NOE studies have shown the pairwise nature of the β-CH_2 protons of the cysteines of the iron(II) domain [34]; the β-CH_2 proton signals of the iron(III) domain are too broad to observe any dipolar connectivity. Inter-3-CH_2 NOEs have also been observed between two signals of two β-CH_2 of the iron(II) domain, thus providing a key information to assign the cysteines bound to iron(II) to Cys-41 and Cys-46 and, by exclusion, the cysteines bound to iron(III) to Cys-49 and Cys-79. In fact, by analyzing the X-ray structure it appears that Cys-41 and Cys-46 are close enough to one another to give inter-residue NOE. The detection of the iron(III) and iron(II) domains is a piece of information that only NMR could have provided.

6.4 BEYOND THE REDFIELD LIMIT: $|J|/\hbar > \tau_s^{-1}$

A fast relaxing metal ion in magnetic exchange coupled dimers shortens the electron relaxation time of the slower relaxing metal ion, often to the value of the fast relaxing ion. Electron–electron interactions are thus so strong that the Redfield limit is reached. To describe this situation, linewidths and nuclear relaxation rates can be calculated as a function of the electron relaxation rates of the isolated spins in the hypothesis that the coupling frequency between the spins is large in absolute value compared to the electronic relaxation rates but small compared to the modulation of the electron lattice interaction:

$$|J|\hbar \geq \tau_{s1}^{-1}, \qquad |J|\hbar \geq \tau_{s2}^{-1}$$
$$|J|\hbar \leq \tau_{v1}^{-1}, \qquad |J|\hbar \leq \tau_{v2}^{-1}.$$

It is further assumed that each electron spin relaxes with its own mechanisms independently of the occurrence of electron spin coupling. The calculations result in different electronic relaxation times for each level and for each transition in the coupled system (see Table 6.4) [5]. The electron relaxation rates for the pair are the sum of the rates of the two spins, weighted by coefficients that differ from one transition to another. When one rate is much smaller than the other, the rate of the pair only depends on the rate of the fast relaxing system, being proportional to it through a coefficient. When the fast relaxing system is the one with the larger spin (as for CuCoSOD), the overall rates are larger than for the isolated fast relaxing ion, while they are smaller when the fast relaxing system is the one with the smaller spin.

The obtained electron relaxation times for each level and each transition can then be introduced in equations analogous to Eq. (6.13a), thus providing as many spectral density functions as the number of the levels and transitions in the system. However, it turns out that the spectral densities containing longitudinal electron relaxation times (lifetimes), as well as those containing transverse electron relaxation times (transition linewidths), can be lumped together, for $J \ll kT$ or $J \gg kT$, providing effective average τ_{s1} and τ_{s2} values (see Table 6.5), that can be simply introduced in Eq. (6.13a) as such [5]. For instance, for CuCoSOD, the

TABLE 6.4

Inverse level lifetimes and total transition linewidths in some magnetic exchange coupled systems [a]

$S_1 = 1/2$, $S_2 = 1/2$

	1	2	3	4
1	$1/2R_A + 1/2R_B$	$3/4R_A + 3/4R_B$	$3/4R_A + 3/4R_B$	$R_A + R_B$
2	$3/4R_A + 3/4R_B$	$3/4R_A + 3/4R_B$	$3/4R_A + 3/4R_B$	$3/4R_A + 3/4R_B$
3	$3/4R_A + 3/4R_B$	$3/4R_A + 3/4R_B$	$3/4R_A + 3/4R_B$	$3/4R_A + 3/4R_B$
4	$R_A + R_B$	$3/4R_A + 3/4R_B$	$3/4R_A + 3/4R_B$	$1/2R_A + 1/2R_B$

$S_1 = 1/2$, $S_2 = 1$

	1	2	3	4
1	$1/2R_A + R_B$	$2/3R_A + 5/3R_B$	$5/6R_A + 4/3R_B$	$2/3R_A + 8/3R_B$
2	$2/3R_A + 5/3R_B$	$13/18R_A + 17/9R_B$	$7/9R_A + 16/9R_B$	$13/18R_A + 20/9R_B$
3	$5/6R_A + 4/3R_B$	$7/9R_A + 16/9R_B$	$13/18R_A + 14/9R_B$	$7/9R_A + 7/3R_B$
4	$2/3R_A + 8/3R_B$	$13/18R_A + 14/9R_B$	$7/9R_A + 7/3R_B$	$13/18R_A + 14/9R_B$
5	$5/6R_A + 7/3R_B$	$7/9R_A + 19/9R_B$	$13/18R_A + 20/9R_B$	$7/9R_A + 16/9R_B$
6	$R_A + 3R_B$	$5/6R_A + 7/3R_B$	$2/3R_A + 8/3R_B$	$5/6R_A + 4/3R_B$

$S_1 = 1/2$, $S_2 = 3/2$

	1	2	3	4
1	$1/2R_A + 3/2R_B$	$5/8R_A + 21/8R_B$	$7/8R_A + 57/32R_B$	$3/4R_A + 15/4R_B$
2	$5/8R_A + 21/8R_B$	$11/16R_A + 51/16R_B$	$13/16R_A + 87/32R_B$	$9/8R_A + 15/4R_B$
3	$7/8R_A + 57/32R_B$	$13/16R_A + 87/32R_B$	$11/16R_A + 2R_B$	$3/4R_A + 117/32R_B$
4	$3/4R_A + 15/4R_B$	$9/8R_A + 15/4R_B$	$3/4R_A + 117/32R_B$	$3/4R_A + 15/4R_B$
5	$3/4R_A + 15/4R_B$	$9/8R_A + 15/4R_B$	$3/4R_A + 117/32R_B$	$3/4R_A + 15/4R_B$
6	$5/8R_A + 177/32R_B$	$11/16R_A + 147/32R_B$	$13/16R_A + 41/16R_B$	$3/4R_A + 117/32R_B$
7	$7/8R_A + 39/8R_B$	$13/16R_A + 69/16R_B$	$11/16R_A + 147/32R_B$	$9/8R_A + 15/4R_B$
8	$R_A + 6R_B$	$7/8R_A + 39/8R_B$	$5/8R_A + 177/32R_B$	$3/4R_A + 15/4R_B$

$S_1 = 1$, $S_2 = 1$

	1	2	3	4
1	$R_A + R_B$	$3/2R_A + 3/2R_B$	$3/2R_A + 3/2R_B$	$15/8R_A + 15/8R_B$
2	$3/2R_A + 3/2R_B$	$7/4R_A + 7/4R_B$	$7/4R_A + 7/4R_B$	$15/8R_A + 15/8R_B$
3	$3/2R_A + 3/2R_B$	$7/4R_A + 7/4R_B$	$7/4R_A + 7/4R_B$	$15/8R_A + 15/8R_B$
4	$15/8R_A + 15/8R_B$	$15/8R_A + 15/8R_B$	$15/8R_A + 15/8R_B$	$7/4R_A + 7/4R_B$
5	$2R_A + 2R_B$	$2R_A + 2R_B$	$2R_A + 2R_B$	$15/8R_A + 15/8R_B$
6	$15/8R_A + 15/8R_B$	$15/8R_A + 15/8R_B$	$15/8R_A + 15/8R_B$	$7/4R_A + 7/4R_B$
7	$5/2R_A + 5/2R_B$	$9/4R_A + 9/4R_B$	$9/4R_A + 9/4R_B$	$15/8R_A + 15/8R_B$
8	$5/2R_A + 5/2R_B$	$9/4R_A + 9/4R_B$	$9/4R_A + 9/4R_B$	$15/8R_A + 15/8R_B$
9	$3R_A + 3R_B$	$5/2R_A + 5/2R_B$	$5/2R_A + 5/2R_B$	$15/8R_A + 15/8R_B$

[a] The numbers indicate the labeling of the ψ functions as reported in Fig. 6.1.

average τ_{s1}^{-1} is $9/16\ R_{Cu} + 179/96\ R_{Co}$, where R_{Cu} and R_{Co} are the electron relaxation rate of the uncoupled ions Cu(II) and Co(II), respectively, i.e. about twice R_{Co}, as R_{Cu} and R_{Co} for the isolated ions are 5×10^8 and 10^{11} s^{-1}, respectively. The experimental τ_{s1}^{-1} value ranges between 1 and 2×10^{11} s^{-1}, in good agreement with predictions.

When the above treatment is applied to homodimers, it shows that, if the electron relaxation rates are the same, they cannot strongly influence each other

5	6
$5/6R_A + 7/3R_B$	$R_A + 3R_B$
$7/9R_A + 19/9R_B$	$5/6R_A + 7/3R_B$
$13/18R_A + 20/9R_B$	$2/3R_A + 8/3R_B$
$7/9R_A + 16/9R_B$	$5/6R_A + 4/3R_B$
$13/18R_A + 17/9R_B$	$2/3R_A + 5/3R_B$
$2/3R_A + 5/3R_B$	$1/2R_A + R_B$

5	6	7	8
$3/4R_A + 15/4R_B$	$5/8R_A + 177/32R_B$	$7/8R_A + 39/8R_B$	$R_A + 6R_B$
$9/8R_A + 15/4R_B$	$11/16R_A + 147/32R_B$	$13/16R_A + 69/16R_B$	$7/8R_A + 39/8R_B$
$3/4R_A + 117/32R_B$	$13/16R_A + 41/16R_B$	$11/16R_A + 147/32R_B$	$5/8R_A + 177/32R_B$
$3/4R_A + 15/4R_B$	$3/4R_A + 117/32R_B$	$9/8R_A + 15/4R_B$	$3/4R_A + 15/4R_B$
$3/4R_A + 15/4R_B$	$3/4R_A + 117/32R_B$	$9/8R_A + 15/4R_B$	$3/4R_A + 15/4R_B$
$3/4R_A + 117/32R_B$	$11/16R_A + 2R_B$	$13/16R_A + 87/32R_B$	$7/8R_A + 57/32R_B$
$9/8R_A + 15/4R_B$	$13/16R_A + 87/32R_B$	$11/16R_A + 51/16R_B$	$5/8R_A + 21/8R_B$
$3/4R_A + 15/4R_B$	$7/8R_A + 57/32R_B$	$5/8R_A + 21/8R_B$	$1/2R_A + 3/2R_B$

5	6	7	8	9
$2R_A + 2R_B$	$15/8R_A + 15/8R_B$	$5/2R_A + 5/2R_B$	$5/2R_A + 5/2R_B$	$3R_A + 3R_B$
$2R_A + 2R_B$	$15/8R_A + 15/8R_B$	$9/4R_A + 9/4R_B$	$9/4R_A + 9/4R_B$	$5/2R_A + 5/2R_B$
$2R_A + 2R_B$	$15/8R_A + 15/8R_B$	$9/4R_A + 9/4R_B$	$9/4R_A + 9/4R_B$	$5/2R_A + 5/2R_B$
$15/8R_A + 15/8R_B$	$7/4R_A + 7/4R_B$	$15/8R_A + 15/8R_B$	$15/8R_A + 15/8R_B$	$15/8R_A + 15/8R_B$
$2R_A + 2R_B$	$15/8R_A + 15/8R_B$	$2R_A + 2R_B$	$2R_A + 2R_B$	$2R_A + 2R_B$
$15/8R_A + 15/8R_B$	$7/4R_A + 7/4R_B$	$15/8R_A + 15/8R_B$	$15/8R_A + 15/8R_B$	$15/8R_A + 15/8R_B$
$2R_A + 2R_B$	$15/8R_A + 15/8R_B$	$7/4R_A + 7/4R_B$	$7/4R_A + 7/4R_B$	$3/2R_A + 3/2R_B$
$2R_A + 2R_B$	$15/8R_A + 15/8R_B$	$7/4R_A + 7/4R_B$	$7/4R_A + 7/4R_B$	$3/2R_A + 3/2R_B$
$2R_A + 2R_B$	$15/8R_A + 15/8R_B$	$3/2R_A + 3/2R_B$	$3/2R_A + 3/2R_B$	$R_A + R_B$

(see Table 6.5) in the absence of other electron relaxation mechanisms. As $S_1 = S_2$, all C_i are equal to $1/2$ and therefore a decrease of a factor 2 is expected in nuclear relaxation for $J \ll kT$. On the other hand, for $J \gg kT$ and antiferromagnetic coupling, nuclear relaxation will decrease substantially, as the ground state is diamagnetic ($S' = 0$).

TABLE 6.5

Effective electron relaxation rates for J-coupled systems experienced by nuclei sensing the S_1 (N-S_1) or the S_2 (N-S_2) spins in the HT (high temperature), AF (antiferromagnetic, $|J| \gg kT$ and $J > 0$) and F (ferromagnetic, $|J| \gg kT$ and $J < 0$) limits

	N-S_1			N-S_2		
	HT	AF	F	HT	AF	F
$S_1 = 1/2$, $S_2 = 1/2$						
τ_{s1}^{-1}	$1/2R_A + 1/2R_B$	diamagnetic state	$1/2R_A + 1/2R_B$	$1/2R_A + 1/2R_B$	diamagnetic state	$1/2R_A + 1/2R_B$
τ_{s2}^{-1}	$3/4R_A + 3/4R_B$		$3/4R_A + 3/4R_B$	$3/4R_A + 3/4R_B$		$3/4R_A + 3/4R_B$
$S_1 = 1/2$, $S_2 = 1$						
τ_{s1}^{-1}	$107/198R_A + 112/99R_B$	$13/18R_A + 14/9R_B$	$47/90R_A + 49/45R_B$	$73/126R_A + 77/63R_B$	$13/18R_A + 14/9R_B$	$47/90R_A + 49/45R_B$
τ_{s2}^{-1}	$71/99R_A + 187/99R_B$	$7/9R_A + 7/3R_B$	$32/45R_A + 83/45R_B$	$46/63R_A + 125/63R_B$	$7/9R_A + 7/3R_B$	$32/45R_A + 83/45R_B$
$S_1 = 1/2$, $S_2 = 3/2$						
τ_{s1}^{-1}	$9/16R_A + 179/96R_B$	$11/16R_A + 2R_B$	$43/80R_A + 147/80R_B$	$331/560R_A + 2123/1120R_B$	$11/16R_A + 2R_B$	$43/80R_A + 147/80R_B$
τ_{s2}^{-1}	$43/48R_A + 215/64R_B$	$3/4R_A + 117/32R_B$	$37/40R_A + 33/10R_B$	$69/80R_A + 7677/2240R_B$	$3/4R_A + 117/32R_B$	$37/40R_A + 33/10R_B$
$S_1 = 1$, $S_2 = 1$						
τ_{s1}^{-1}	$5/4R_A + 5/4R_B$	diamagnetic state	$23/20R_A + 23/20R_B$	$5/4R_A + 5/4R_B$	diamagnetic state	$23/20R_A + 23/20R_B$
τ_{s2}^{-1}	$7/4R_A + 7/4R_B$		$69/40R_A + 69/40R_B$	$7/4R_A + 7/4R_B$		$69/40R_A + 69/40R_B$

6.5 POLYMETALLIC SYSTEMS

In polymetallic systems, the larger the number of coupled ions, the larger the spreading of the S' levels. As a consequence, even relatively small $|J|$ values give rise to large separations of the S' levels and therefore to depopulation of the highest levels. The general theory is still the same. Analytical solutions, as in dimeric systems, are seldom possible. Often, numerical solutions are possible.

The general Hamiltonian is given by Eq. (6.3), where the sum is over all the j–k pairs. The treatment is in fact based on bimetallic coupling. The S' levels are obtained and the hyperfine coupling for each S' level is given by

$$A_{ij} = A_j \frac{\langle S_{jz} \rangle_i}{\langle S'_z \rangle_i} = A_j C_{ij} \tag{6.16}$$

where A_{ij} is the hyperfine coupling between the ith S' level and a nucleus sensing the jth metal. A_j is the hyperfine coupling between the nucleus and the spin on the jth metal (S_j) in the absence of magnetic coupling. $\langle S_{jz} \rangle_i$ is the contribution of metal j (S_j) to the expectation value of S'_z of the ith level. This equation is quite analogous to Eq. (6.7).

Under certain symmetry conditions, Eq. (6.3) can be rearranged in order to allow for analytical solutions [3]. In trimetallic systems, when $J_{12} \neq J_{13} = J_{23}$, Eq. (6.3) becomes [1]

$$H = J(\mathbf{S_1} \cdot \mathbf{S_2} + \mathbf{S_1} \cdot \mathbf{S_3} + \mathbf{S_2} \cdot \mathbf{S_3}) + \Delta J_{12} \mathbf{S_1} \cdot \mathbf{S_2} \tag{6.17}$$

where ΔJ_{12} indicates the difference between J_{12} and the other two J values. The energies are

$$E(S'_{12}, S') = \tfrac{1}{2}[JS'(S'+1) + \Delta J_{12} S'_{12}(S'_{12}+1)] \tag{6.18}$$

where S'_{12} is the subspin value coming out from the coupling of spins 1 and 2.

---

[3] In case all J values are equal, the eigenvalues of the Hamiltonian $H = J(\mathbf{S_1}\cdot\mathbf{S_2}+\mathbf{S_1}\cdot\mathbf{S_3}+\mathbf{S_2}\cdot\mathbf{S_3})$ can be obtained by trivial extension of the dimer case. As $\mathbf{S}' = \mathbf{S_1} + \mathbf{S_2} + \mathbf{S_3}$, we can square the total $\mathbf{S}'$ value to obtain $\mathbf{S}'^2 = \mathbf{S_1}^2+\mathbf{S_2}^2+\mathbf{S_3}^2+2\mathbf{S_1}\cdot\mathbf{S_2}+2\mathbf{S_1}\cdot\mathbf{S_3}+2\mathbf{S_2}\cdot\mathbf{S_3}$, and write $H = (J/2)(\mathbf{S}'^2 - \mathbf{S_1}^2 - \mathbf{S_2}^2 - \mathbf{S_3}^2)$, whose eigenvalues are $E = (J/2)[S'(S'+1) - S_1(S_1+1) - S_2(S_2+1) - S_3(S_3+1)]$. Since $S_1(S_1+1)$, $S_2(S_2+1)$ and $S_3(S_3+1)$ are the same for all levels, they introduce only an energy offset and can be dropped. The splitting of the levels is just given by $E = (J/2)S'(S'+1)$. Analytical solutions can still be obtained when one J value (e.g. J_{12}) differs from the other two. The first term of Hamiltonian (6.17), is just as in the above case. Then a subspin S'_{12} can be defined such that $\mathbf{S}'_{12} = \mathbf{S_1} + \mathbf{S_2}$ and, from its square, $\mathbf{S_1}\cdot\mathbf{S_2} = \frac{1}{2}(\mathbf{S_{12}}'^2 - \mathbf{S_1}^2 - \mathbf{S_2}^2)$. Then, from the previous definitions, $H = (J/2)(\mathbf{S}'^2 - \mathbf{S_1}^2 - \mathbf{S_2}^2 - \mathbf{S_3}^2) + (\Delta J_{12}/2)(\mathbf{S_{12}}^2 - \mathbf{S_1}^2 - \mathbf{S_2}^2)$. Dropping again the constant terms, the eigenvalues are: $E = (1/2)[JS'(S'+1) + \Delta J_{12} S'_{12}(S'_{12}+1)]$ (Eq. (6.18)).

The coefficients C_{ij} in Eq. (6.16) are then given by [35]

$$\begin{aligned}
C_{i1} &= [S'(S'+1) + S'_{12}(S'_{12}+1) - S_3(S_3+1)]/2S'(S'+1) \\
&\quad \times [S'_{12}(S'_{12}+1) + S_1(S_1+1) - S_2(S_2+1)]/2S'_{12}(S'_{12}+1) \\
C_{i2} &= [S'(S'+1) + S'_{12}(S'_{12}+1) - S_3(S_3+1)]/2S'(S'+1) \\
&\quad \times [S'_{12}(S'_{12}+1) + S_2(S_2+1) - S_1(S_1+1)]/2S'_{12}(S'_{12}+1) \\
C_{i3} &= [S'(S'+1) + S_3(S_3+1) - S'_{12}(S'_{12}+1)]/2S'(S'+1).
\end{aligned} \tag{6.19}$$

For tetraionic systems, analytical solutions are possible in the two following cases [36–38]:

(a) $J_{12} \neq J_{34} \neq J_{13} = J_{14} = J_{23} = J_{24}$

$$\begin{aligned}
H &= J(\mathbf{S_1}\cdot\mathbf{S_2} + \mathbf{S_1}\cdot\mathbf{S_3} + \mathbf{S_1}\cdot\mathbf{S_4} + \mathbf{S_2}\cdot\mathbf{S_3} + \mathbf{S_2}\cdot\mathbf{S_4} + \mathbf{S_3}\cdot\mathbf{S_4}) \\
&\quad + \Delta J_{12}(\mathbf{S_1}\cdot\mathbf{S_2}) + \Delta J_{34}(\mathbf{S_3}\cdot\mathbf{S_4})
\end{aligned} \tag{6.20}$$

$$\begin{aligned}
E(S'_{12}, S'_{34}, S') = \tfrac{1}{2}\big[&JS'(S'+1) + \Delta J_{12}S'_{12}(S'_{12}+1) \\
&+ \Delta J_{34}S'_{34}(S'_{34}+1)\big]
\end{aligned} \tag{6.21}$$

$$\begin{aligned}
C_{i1} &= [S'(S'+1) + S'_{12}(S'_{12}+1) - S'_{34}(S'_{34}+1)]/2S'(S'+1) \\
&\quad \times [S'_{12}(S'_{12}+1) + S_1(S_1+1) - S_2(S_2+1)]/2S'_{12}(S'_{12}+1) \\
C_{i2} &= [S'(S'+1) + S'_{12}(S'_{12}+1) - S'_{34}(S'_{34}+1)]/2S'(S'+1) \\
&\quad \times [S'_{12}(S'_{12}+1) + S_2(S_2+1) - S_1(S_1+1)]/2S'_{12}(S'_{12}+1) \\
C_{i3} &= [S'(S'+1) + S'_{34}(S'_{34}+1) - S'_{12}(S'_{12}+1)]/2S'(S'+1) \\
&\quad \times [S'_{34}(S'_{34}+1) + S_3(S_3+1) - S_4(S_4+1)]/2S'_{34}(S'_{34}+1) \\
C_{i4} &= [S'(S'+1) + S'_{34}(S'_{34}+1) - S'_{12}(S'_{12}+1)]/2S'(S'+1) \\
&\quad \times [S'_{34}(S'_{34}+1) + S_4(S_4+1) - S_3(S_3+1)]/2S'_{34}(S'_{34}+1).
\end{aligned} \tag{6.22}$$

Here S'_{12}, S'_{34} and S' are good quantum numbers for the relative subspins.

(b) $J_{12} \neq J_{13} = J_{23} \neq J_{14} = J_{24} = J_{34}$

$$\begin{aligned}
H &= J(\mathbf{S_1}\cdot\mathbf{S_2} + \mathbf{S_1}\cdot\mathbf{S_3} + \mathbf{S_1}\cdot\mathbf{S_4} + \mathbf{S_2}\cdot\mathbf{S_3} + \mathbf{S_2}\cdot\mathbf{S_4} + \mathbf{S_3}\cdot\mathbf{S_4}) \\
&\quad + \Delta J_{12}(\mathbf{S_1}\cdot\mathbf{S_2}) + \Delta J_{123}(\mathbf{S'_{12}}\cdot\mathbf{S_3})
\end{aligned} \tag{6.23}$$

$$E(S'_{12}, S'_{123}, S') = \tfrac{1}{2}\big[JS'(S'+1) + \Delta J_{12}S'_{12}(S'_{12}+1) + \Delta J_{123}S'_{123}(S'_{123}+1)\big] \tag{6.24}$$

$$\begin{aligned}
C_{i1} &= [S'(S'+1) + S'_{123}(S'_{123}+1) - S_4(S_4+1)]/2S'(S'+1) \\
&\quad\times [S'_{123}(S'_{123}+1) + S'_{12}(S'_{12}+1) - S_3(S_3+1)]/2S'_{123}(S'_{123}+1) \\
&\quad\times [S'_{12}(S'_{12}+1) + S_1(S_1+1) - S_2(S_2+1)]/2S'_{12}(S'_{12}+1) \\
C_{i2} &= [S'(S'+1) + S'_{123}(S'_{123}+1) - S_4(S_4+1)]/2S'(S'+1) \\
&\quad\times [S'_{123}(S'_{123}+1) + S'_{12}(S'_{12}+1) - S_3(S_3+1)]/2S'_{123}(S'_{123}+1) \\
&\quad\times [S'_{12}(S'_{12}+1) + S_2(S_2+1) - S_1(S_1+1)]/2S'_{12}(S'_{12}+1) \\
C_{i3} &= 1 - C_{i1} - C_{i2} - C_{i4} \\
C_{i4} &= [S'(S'+1) + S_4(S_4+1) - S'_{123}(S'_{123}+1)]/2S'(S'+1).
\end{aligned} \tag{6.25}$$

Here ΔJ_{12} indicates the difference between J_{12} and J, ΔJ_{123} the difference between J_{13} $(= J_{23})$ and J, and S'_{123} the subspin resulting from the coupling of S'_{12} and S_3.

All of these cases have been analyzed in NMR studies [4,39].

When the same metals differ in the oxidation states, i.e. Fe^{3+} and Fe^{2+}, the treatment may be even more complicated [36–42]. We will deal with these systems later.

6.5.1 *The $[(RS)_3Fe_3S_4]^{2-}$ case*

In this case there are three $S = 5/2$ Fe^{3+} ions antiferromagnetically coupled [35]. One J value (e.g. J_{12}) is larger than the other two (Fig. 6.13). It follows that S'_{12} must be smaller than S_3; the total spin is $S' = 1/2$, and S'_{12} from Mössbauer results to be equal to 2 [35], i.e. S_1 and S_2 are not completely antiferromagnetically coupled.

The spectrum of a protein containing such a cluster is reported in Fig. 6.14 [43]. There are four signals downfield which are assigned to one β-CH_2 pair of the cysteine bound to iron 3 and to one β-CH_2 proton for each of the two cysteines bound to irons 1 and 2. The geminal protons of the latter are found in the diamagnetic region. The β-CH_2 pair of the cysteine coordinated to iron 3 ($S = 5/2$) experiences a Curie temperature dependence, whereas the other four β-CH_2 protons are antiCurie, in agreement with their belonging to the cysteines coordinated to the $S'_{12} = 2$ pair which is smaller than $S_3 = 5/2$. Other proteins containing Fe_3S_4 clusters have similar NMR properties [44].

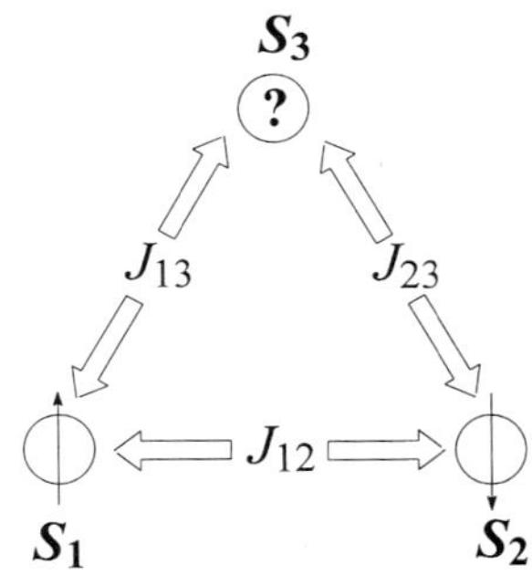

Fig. 6.13. Magnetic coupling scheme in a Fe_3 system. If $J_{12} > J_{13}$, J_{23}, S_1 and S_2 are antiferromagnetically coupled, and S_3 cannot be antiferromagnetically coupled to both S_1 and S_2.

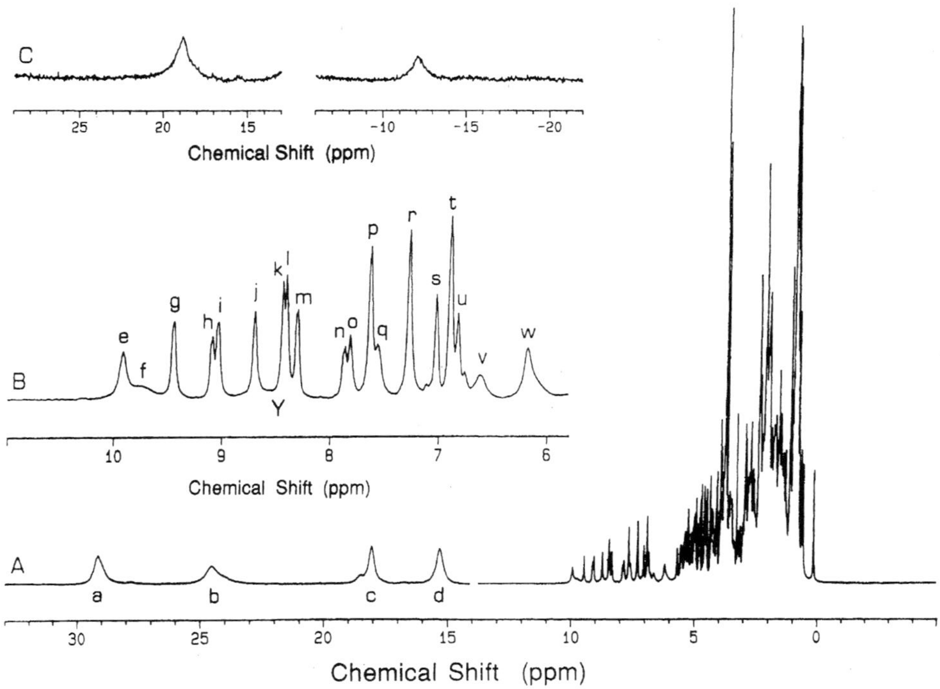

Fig. 6.14. ^{1}H NMR spectrum of oxidized *D. gigas* ferredoxin II [43]. The two most downfield-shifted signals belong to a cysteine β-CH_2 pair experiencing Curie-type temperature dependence.

6.5.2 The $[(RS)_{11}Co_4]^{3-}$ case

The protein metallothionein (MT) has a MW of 6000 and is rich in thiolate groups. It binds up to seven zinc(II), cadmium(II) and cobalt(II) ions. The seven ions are organized in two clusters. The so-called M_4S_{11} cluster is shown in

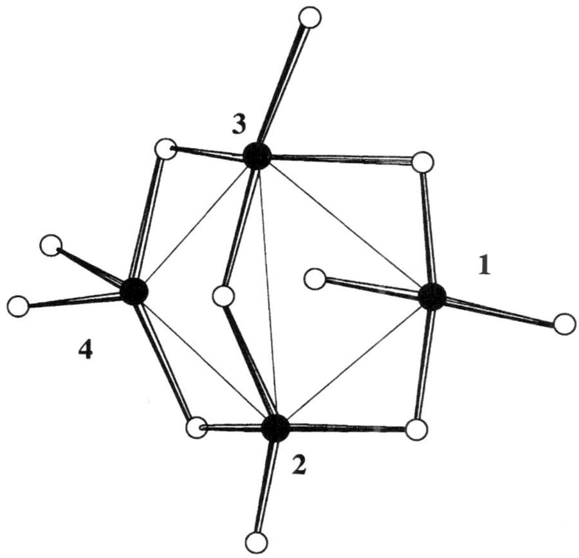

Fig. 6.15. Schematic structure of the M_4S_{11} cluster in metallothioneins (M = Zn, Cd, Co) [45–48]. There is no bridging ligand between metal 1 and metal 4.

Fig. 6.15. As suggested from X-ray and NMR measurements, metals 1 and 4 are coordinated by two terminal and two bridging cysteines, whereas metals 2 and 3 are coordinated by one terminal and three bridging cysteines [45–48]. As a consequence, $J_{14} = 0$. If the ions are antiferromagnetically coupled, metals 2 and 3 are overall more antiferromagnetically coupled than metals 1 and 4. This is going to explain some features of the NMR spectra [49,50]. Indeed, it has been proposed that the spectrum of Co_7MT shown in Fig. 6.16 is due only to the Co_4S_{11} cluster. This conclusion is reached on the basis of titration of the apoprotein with cobalt(II) and comparison with the Cd_4Co_3MT species. The spectrum is fascinating, in that signals are spread over 400 ppm. The pairwise nature of nine out of 11 cysteine β-CH_2s has been demonstrated through NOE and NOESY [51] (see Sections 7.1 and 8.3). The connected signals are shown in Fig. 6.16.

The temperature dependence of the shifts shows that most of the signals follow a Curie-like behavior, whereas six signals (k, m, p, q, r, s) exhibit a more or less pronounced antiCurie behavior. The ground state of the system is $S' = 0$, since four equal ions are antiferromagnetically coupled. With a J of about 50 cm^{-1} several excited levels are occupied in the investigated range of temperature. If all the J values were equal we would expect an antiCurie behavior of all the signals. If one J value (namely J_{14}) is zero, we expect, from calculations performed using either Hamiltonian (6.20) or (6.23), the four terminal cysteines bound to metals 1 and 4 to exhibit normal Curie behavior; the two terminal cysteines bound to

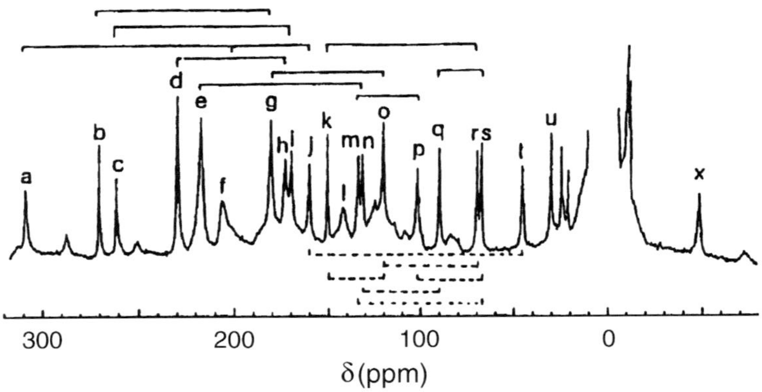

Fig. 6.16. ^{1}H NMR spectrum of Co_7MT showing the signals of the Co_4S_{11} cluster. The pairwise assignment of the β-CH_2 protons, obtained through NOE and NOESY is also shown [49–51].

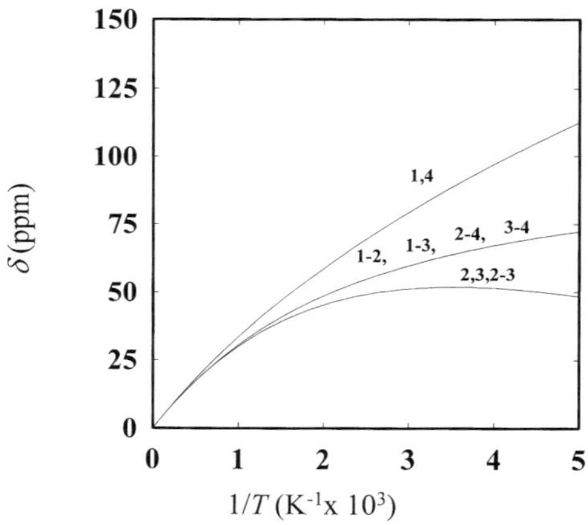

Fig. 6.17. Calculated hyperfine shifts for cysteine β-CH_2 protons of the Co_4S_{11} cluster from a magnetic coupling model with $J_{14} = 0$ and all other J values equal to 50 cm^{-1}. Numbers refer to the metals to which the cysteines are coordinated (bridging cysteines being identified by two numbers).

metals 2 and 3, and the cysteine bridging metals 2 and 3 to exhibit antiCurie behavior; and the four cysteines bridging metals 1 and 2, 1 and 3, 2 and 4, and 3 and 4 to experience contrasting effects, and thus to have a less pronounced temperature dependence (Fig. 6.17). As already mentioned, there are indeed six signals exhibiting antiCurie behavior [49,50].

6.5.3 The $[(RS)_4Fe_4S_4]^{2-}$ case

The $[(RS)_4Fe_4S_4]^{2-}$ case is encountered as inorganic compounds with other thiolate ligands and in proteins. The four iron atoms and the four sulfur atoms

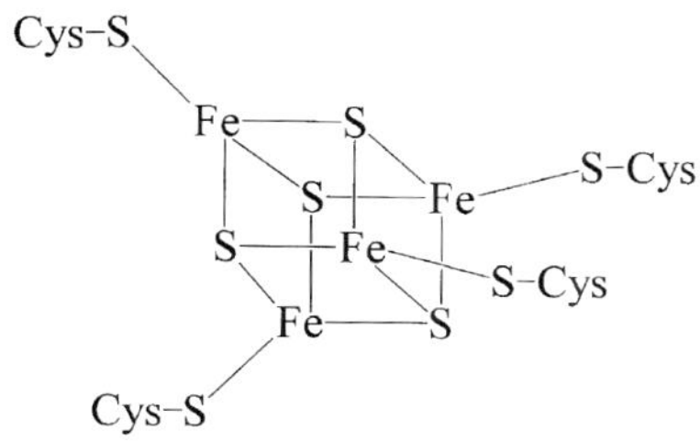

Fig. 6.18. Cubane-type Fe_4S_4 cluster. The total charge can be +3, +2, or +1 depending on the oxidation states of the iron ions. The cluster shown in the figure is coordinated to four protein cysteines, giving total charge of −1, −2, or −3, respectively.

occupy the vertices of a cube (see Fig. 6.18). The tetracoordination of iron is reached through coordination to a further sulfur atom of a cysteine provided by the protein or of an organic thiolate in model complexes. Formally the cluster is constituted by two iron(III) and two iron(II) ions; actually Mössbauer spectroscopy showed that all iron ions are equivalent [52] and that the oxidation number is 2.5. The ground state is $S' = 0$, but magnetically active states are populated at room temperature. The experimental temperature dependence of the shifts is of antiCurie type. NMR has shown that the organic thiolate ligand exchanges fast on the time scale of the difference in shift between the various positions. Only one signal for the CH_2 protons of ethyl thiolate is observed [53,54]. In the case of proteins the rigidity of the ligand makes the two protons distinguishable. As discussed in Section 2.4, the shifts depend on the dihedral angles between the Fe–S–C and the S–C–H planes. NOE, NOESY and COSY spectra are available which show the connectivities between the β-CH_2 protons, and between the latter and the α-CH, as well as with various amino acidic residues [55,56]. The solution structures of several proteins containing these systems are available.

6.5.4 *The $[(RS)_4Fe_4S_4]^-$ case*

This cluster formally contains three iron(III) and one iron(II). It is present in a class of proteins called high potential iron–sulfur proteins (HiPIP). It has also been prepared through oxidation of $[(RS)_4Fe_4S_4]^{2-}$ model compounds [57]. Both in the model compound at low temperatures and in proteins there is electron delocalization on one mixed valence pair [58–62]. Therefore, the polymetallic center is constituted by two iron ions at the oxidation state +2.5 and two iron ions at the oxidation state +3. Hamiltonian (6.20), or a more complicated one [40, 41,43], can be used to describe the electronic structure. Indeed, a delocalization operator is sometimes needed in the Hamiltonian [40,41,43]. Consistently with magnetic Mössbauer data the S'_{34} subspin involving the mixed valence pair is 9/2, whereas the S'_{12} subspin involving the iron(III) ions is 4. Mössbauer and EPR data do not exclude 7/2 and 3, respectively, for the two pairs [57]; in any case, the

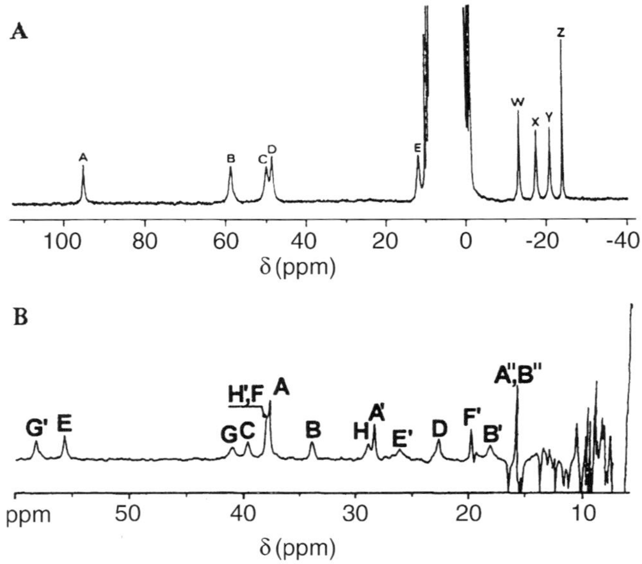

Fig. 6.19. ^{1}H NMR spectra of (A) oxidized HiPIP II from *E. halophila* [63] and (B) reduced ferredoxin from *C. acidi urici* [56].

subspin value of the mixed valence pair is larger than that of the iron(III) pair. This information is vital for the interpretation of the NMR spectra. The β-CH_2 of the cysteines of the mixed valence domain go downfield because they sense the larger spin, whereas the β-CH_2 protons of the cysteines of the iron(III) domain are upfield. Note the similarity with the reduced Fe_2S_2 case (see Section 6.3.4). The total spin of the ground state is $S' = 1/2$, and levels with larger S' values are available at room temperature. The T_1 values of the hyperfine shifted signals are accordingly quite short (4–25 ms). The spectrum of the oxidized HiPIP II from *E. halophila* is shown in Fig. 6.19A. The β-CH_2 protons are pairwise assigned through NOEs. The picture is absolutely consistent with what is discussed above [63]. The temperature dependence is consistent with expectations (Fig. 6.20). It should be noted now that electron delocalization may require a further operator in Hamiltonian (6.20). Its introduction may be relevant for the interpretation of fine aspects of oxidized HiPIPs, but we deliberately avoid this further treatment [39].

The next step in the knowledge of the protein structure is that of identifying the cysteines bound to the two kinds of iron ions. This goal is reached with the procedures illustrated in the next chapters.

6.5.5 *The $[(RS)_4Fe_4S_4]^{3-}$ case*

This center formally contains three iron(II) and one iron(III) ions. The Mössbauer data indicate that in most cases we are in the presence of two irons at the

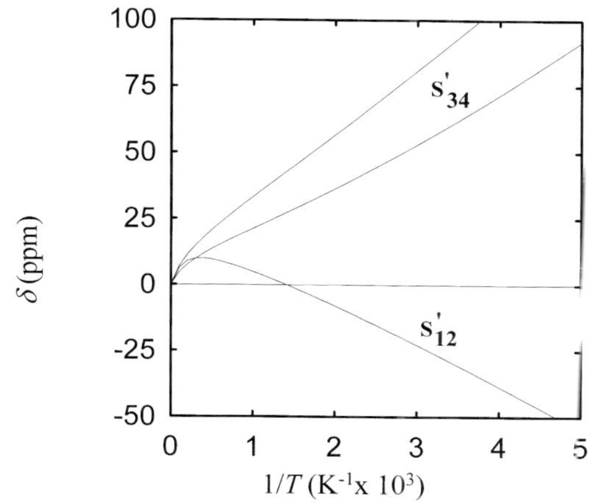

Fig. 6.20. Calculated temperature dependence of the hyperfine shifts of the cysteine β-CH_2 protons of HiPIP II from *E. halophila* [63]. S'_{12} refers to the protons sensing the ferric pair and S'_{34} to those sensing the mixed valence pair.

oxidation state +2.5 and two irons at the oxidation state +2 [64–66]. The system is present in ferredoxins and has been synthesized as model compounds. The ground state in proteins seems to be $S' = \frac{1}{2}$. The ^{1}H NMR spectrum of the two reduced clusters present in *C. acidi urici* ferredoxin is shown in Fig. 6.19B. Note the differences in chemical shifts with respect to Fig. 6.19A; such differences are attributed to different degrees of electron delocalization. In reduced Fe_4S_4 ferredoxins the shifts are all downfield, half of them with Curie and half with antiCurie temperature dependence, just like in the case of the reduced Fe_2S_2 ferredoxins discussed in Section 6.3.4. The successful strategy for the investigation of these systems [55,56] is similar to that of the oxidized HiPIPs. In model compounds both $\frac{1}{2}$ and $\frac{3}{2}$ ground states have been found [67]. When S is substituted with Se a $\frac{7}{2}$ ground state has been claimed [68]. Surely there will be fun in the investigation of these systems by NMR.

6.5.6 The $[(RS)_3LFe_3NiS_4]^{3-}$ case

In this case one iron(II) has been substituted by a nickel(II) ion. As a model complex this is a nice system to investigate [69]. In the model complex a triphenylphosphine is bound to nickel (if L is a neutral ligand the negative charge of the title compound should be decreased by one unit). Interestingly, the shift pattern of the triphenylphosphine moiety is reversed with respect to a nickel(II) complex without magnetic coupling [69]. This can be seen as the result of a spin $S' = \frac{5}{2}$, resulting from the Fe_3S_4 moiety, antiferromagnetically coupled to $S = 1$ of nickel(II). Alternatively, the $S' = \frac{9}{2}$ of the iron(III)–iron(II) mixed valence pair is antiferromagnetically coupled to the iron(II)–nickel(II) pair [70].

This system has been chosen just to pinpoint the variety of systems and of spin coupling schemes available in heteropolymetallic systems.

6.6 SUPERPARAMAGNETISM

Superparamagnets are small ferromagnetic crystals (magnetite, for instance), of the diameter of 10–30 nm. The electron spins are all coupled, so that one single large spin can be considered per particle. Total S' values as high as 10^4 can be reached. As the coupling is ferromagnetic, the ground state is the one with largest S' (Section 6.1). The level immediately above $(S' - 1)$ is separated by and energy $S'|J|$ (Eq. (6.2)), which is much larger than kT. Therefore, only the ground state S' is populated. In addition, S' is split by the so-called 'anisotropy energy', which is analogous to a negative ZFS. Again, this energy is larger than kT, so that only the most negative M_S state is populated. The resulting magnetic moment aligns along one of the so-called directions of easy magnetization and, in the absence of magnetic field, jumps from one easy direction to another, with a correlation time of the order of few nanoseconds (Néel relaxation process). Fig. 6.21 shows the typical ^{1}H NMRD profile acquired in water solution containing superparamagnet particles [71]. A dispersion is observed when the Zeeman energy equals the anisotropy energy (analogously to what happens for ZFS). A typical feature of superparamagnetism is that the average magnetic moment is proportional to the magnetic field until saturation is reached: this causes increase of the relaxation rate with increasing the field, until saturation is reached, which value is proportional to M_S^2 (Curie relaxation). The following dispersion is then related to the diffusion time of water molecules diffusing nearby the particles (outer-sphere relaxation, see Section 4.5.2), and is related to the particle size.

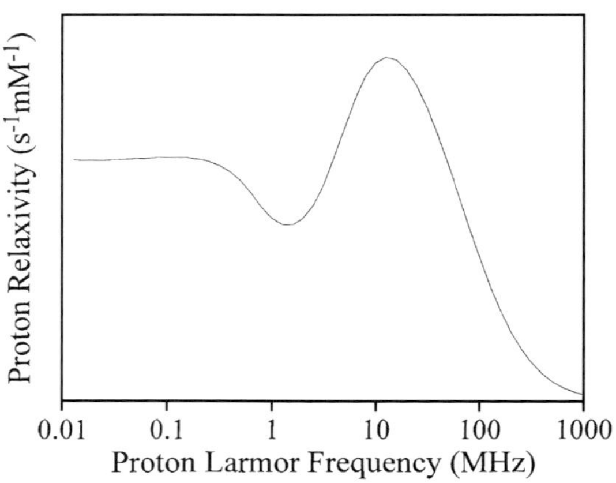

Fig. 6.21. Water ^{1}H NMRD profile of a superparamagnet.

6.7 REFERENCES

[1] A. Bencini, D. Gatteschi (1990) Electron Paramagnetic Resonance of Exchange-Coupled Systems. Springer-Verlag, Berlin.
[2] W.R. Dunham, G. Palmer, R.H. Sands, A.J. Bearden (1971) Biochim. Biophys. Acta 253, 373; L. Banci, I. Bertini, C. Luchinat (1990) Struct. Bonding 72, 113.
[3] C. Luchinat, S. Ciurli (1993) Biol. Magn. Reson. 12, 357.
[4] I. Bertini, L. Banci, R.D. Brown III, S.H. Koenig, C. Luchinat (1988) Inorg. Chem. 27, 951.
[5] I. Bertini, O. Galas, C. Luchinat, G. Parigi, G. Spina (1998) J. Magn. Reson. 130, 33.
[6] V. Clementi, C. Luchinat, (1998) Acc. Chem. Res. 31, 351.
[7] A.G. Redfield (1957) Res. Dev. 1, 19.
[8] W.J. Stratton (1970) Inorg. Chem. 9, 517.
[9] P.D.W. Boyd, M. Gerlock, G.M. Sheldrick (1974) J. Chem. Soc., Dalton Trans. 1097.
[10] D. Gatteschi, A. Dei, E. Piergentili (1979) Inorg. Chem. 18, 89.
[11] C. Benelli, A. Dei, D. Gatteschi (1982) Inorg. Chem. 108, 3298.
[12] C. Owens, R.S. Drago, I. Bertini, C. Luchinat, L. Banci (1986) J. Am. Chem. Soc. 108, 3298.
[13] I. Bertini, C. Luchinat, C. Owens, R.S. Drago (1987) J. Am. Chem. Soc. 109, 5208.
[14] W. Byers, R.J.P. Williams (1973) J. Chem. Soc. 11, 134.
[15] M. Maekawa, S. Kitagawa, M. Munakata, H. Masuda (1989) Inorg. Chem. 28, 1904.
[16] G.N. La Mar, G.R. Eaton, R.H. Holm, F.A. Walker (1973) J. Am. Chem. Soc. 95, 63.
[17] R.C. Holz, J.M. Brink (1994) Inorg. Chem. 33, 4609.
[18] P.K. Mandal, P.T. Manoharan (1995) Inorg. Chem. 34, 270.
[19] L. Lubben, R. Hage, A. Meetsma, K. Byma, B.L. Feringa (1995) Inorg. Chem. 34, 2217.
[20] J.H. Satcher, A.L. Balch (1995) Inorg. Chem. 34, 3371.
[21] J.M. Brink, R.A. Rose, R.C. Holz (1996) Inorg. Chem. 35, 2878.
[22] N.N. Murthy, K.D. Karlin, I. Bertini, C. Luchinat (1997) J. Am. Chem. Soc. 119, 2156.
[23] M. Poe, W.D. Phillips, J.D. Glickson, C.C. McDonald, A. San Pietro (1971) Proc. Natl. Acad. Sci. USA 68, 68.
[24] I. Saalmen, G. Palmer (1972) Arch. Biochem. Biophys. 150, 767.
[25] I. Morgenstern-Badarau, D. Cocco, A. Desideri, G. Rotilio, J. Jordanov, N. Dupre' (1986) J. Am. Chem. Soc. 108, 300.
[26] I. Bertini, G. Lanini, C. Luchinat, L. Messori, R. Monnanni, A. Scozzafava (1985) J. Am. Chem. Soc. 107, 4391.
[27] L. Banci, I. Bertini, C. Luchinat, A. Scozzafava (1987) J. Am. Chem. Soc. 109, 2328.
[28] I. Bertini, C. Luchinat, M. Piccioli (1994) Progr. NMR Spectrosc. 26, 91.
[29] I. Bertini, C. Luchinat, L.-J. Ming, M. Piccioli, M. Sola, J.S. Valentine (1992) Inorg. Chem. 31, 4433.
[30] L. Banci, I. Bertini, C. Luchinat, R. Monnanni, A. Scozzafava, B. Salvato (1986) Gazz. Chim. Ital. 116, 51.
[31] L. Banci, I. Bertini, C. Luchinat, R. Monnanni, A. Scozzafava (1987) Inorg. Chem. 26, 153.
[32] I. Bertini, G. Lanini, C. Luchinat (1984) Inorg. Chem. 23, 2729.
[33] L. Skjeldal, J.L. Markley, V.M. Coghlan, L.E. Vickery (1991) Biochemistry 30, 9078.
[34] L.B. Dugad, G.N. La Mar, L. Banci, I. Bertini (1990) Biochemistry 29, 2263.
[35] T.A. Kent, B.H. Huynh, E. Munk (1980) Proc. Natl. Acad. Sci. USA 77, 6574.
[36] L. Noodleman (1988) Inorg. Chem. 27, 3677.
[37] L. Noodleman (1991) Inorg. Chem. 30, 256.
[38] L. Noodleman (1991) Inorg. Chem. 30, 246.
[39] I. Bertini, S. Ciurli, C. Luchinat (1995) Struct. Bonding 83, 1.

[40] G. Blondin, J.-J. Girerd (1990) Chem. Rev. 90, 1359.
[41] E.L. Bominaar, S.A. Borshch, J.-J. Girerd (1994) J. Am. Chem. Soc. 116, 5362.
[42] I. Bertini, F. Briganti, C. Luchinat (1990) Inorg. Chim. Acta 175, 9.
[43] A.L. Macedo, I. Moura, J.J.G. Moura, J. LeGall, B.H. Huynh (1993) Inorg. Chem. 32, 1101.
[44] S.C. Busse, G.N. La Mar, L.P. Yu, J.B. Howard, E.T. Smith, Z.H. Zhou, M.W.W. Adams (1992) Biochemistry 31, 11952.
[45] A. Arseniev, P. Schultze, E. Wörgötter, W. Braun, G. Wagner, M. Vasak, J.H. Kägi, K. Wüthrich (1988) J. Mol. Biol. 201, 637.
[46] P. Schultze, E. Worgotter, W. Braun, G. Wagner, M. Vasak, J.H. Kagi, K. Wuthrich (1988) J. Mol. Biol. 203, 251.
[47] B.A. Messerle, M. Bos, A. Schaffer, M. Vasak, J.H. Kagi, K. Wuthrich (1990) J. Mol. Biol. 214, 781.
[48] A.H. Robbins, D.E. McRee, M. Williamson, S.A. Collett, N.h. Xuong, W.F. Furey, B.C. Wang, C.D. Stout (1991) J. Mol. Biol. 221, 1269.
[49] I. Bertini, C. Luchinat, L. Messori, M. Vasak (1989) J. Am. Chem. Soc. 111, 7296.
[50] I. Bertini, C. Luchinat, L. Messori, M. Vasak (1989) J. Am. Chem. Soc. 111, 7300.
[51] I. Bertini, C. Luchinat, L. Messori, M. Vasak (1993) Eur. J. Biochem. 211, 235.
[52] Cammack, R. and Sykes, A.G. (Eds.) (1992) Advances in Inorganic Chemistry. Iron Sulfur Proteins, Academic Press, San Diego.
[53] K.S. Hagen, A.D. Watson, R.H. Holm (1983) J. Am. Chem. Soc. 105, 3905.
[54] R.H. Holm, W.D. Phillips, B.A. Averill, J.J. Mayerle, T. Herskovitz (1974) J. Am. Chem. Soc. 96, 2109.
[55] I. Bertini, F. Briganti, C. Luchinat, L. Messori, R. Monnanni, A. Scozzafava, G. Vallini (1992) Eur. J. Biochem. 204, 831.
[56] I. Bertini, F. Capozzi, C. Luchinat, M. Piccioli, A.J. Vila (1994) J. Am. Chem. Soc. 116, 651.
[57] J.-M. Mouesca, G.J. Rius, B. Lamotte (1993) J. Am. Chem. Soc. 115, 4714.
[58] P. Middleton, D.P.E. Dickson, C.E. Johnson, J.D. Rush (1980) Eur. J. Biochem. 104, 289.
[59] D.P.E. Dickson, C.E. Johnson, R. Cammack, M.C.W. Evans, D.O. Hall, K.K. Rao (1974) Biochem. J. 139, 105.
[60] I. Bertini, A.P. Campos, C. Luchinat, M. Teixeira (1993) J. Inorg. Biochem. 52, 227.
[61] V. Papaefthymiou, M.M. Millar, E. Münck (1986) Inorg. Chem. 25, 3010.
[62] J. Gloux, P. Gloux, B. Lamotte, J.-M. Mouesca, G.J. Rius (1994) J. Am. Chem. Soc. 116, 1953.
[63] L. Banci, I. Bertini, F. Capozzi, P. Carloni, S. Ciurli, C. Luchinat, M. Piccioli (1993) J. Am. Chem. Soc. 115, 3431.
[64] P. Middleton, D.P.E. Dickson, C.E. Johnson, J.D. Rush (1978) Eur. J. Biochem. 88, 135.
[65] R.N. Mullinger, R. Cammack, K.K. Rao, D.O. Hall, D.P.E. Dickson, C.E. Johnson, J.D. Rush, A. Simonopulos (1975) Biochem. J. 151, 75.
[66] D.P.E. Dickson, C.E. Johnson, P. Middleton, J.D. Rush, R. Cammack, D.O. Hall, R.N. Mullinger, K.K. Rao (1976) J. Phys. Colloq. 37, C6-C6-175.
[67] R.H. Holm, S. Ciurli, J.A. Weigel (1990) In: S.J. Lippard (Ed.), Progress in Inorganic Chemistry. Bioinorganic Chemistry, Vol. 38. John Wiley and Sons, Inc., New York, pp. 1.
[68] J. Meyer, J.-M. Moulis, J. Gaillard, M. Lutz (1992) Adv. Inorg. Chem. 38, 73.
[69] S. Ciurli, P.K. Ross, M.J. Scott, S.-B. Yu, R.H. Holm (1992) J. Am. Chem. Soc. 114, 5415.
[70] I. Bertini, C. Luchinat (1992) Chem. Tracts 4, 269.
[71] A. Roch, R.N. Muller, P. Gillis (1999) J. Chem. Phys. 110, 5403.
[72] J.A. Tainer, E.D. Getzoff, K.M. Beem, J.S. Richardson, D.C. Richardson (1982) J. Mol. Biol. 160, 181.

Chapter 7

Nuclear Overhauser Effect

The nuclear Overhauser effect is something very dear to senior NMR researchers. It is the effect which allows us to know which magnetic nuclei are close to other magnetic nuclei, and information on their distances becomes available. Its understanding used to be a must to move towards 2D spectroscopy. Now times are somewhat changed, but . . .

7.1 INTRODUCTION

We have chosen to dedicate a full chapter to the nuclear Overhauser effect [1,2] (NOE) because its comprehension is of fundamental importance in dealing with nuclear relaxation [3,4]. Especially in paramagnetic compounds, it still represents a most powerful technique to detect dipolar connectivities; finally it allows the comprehension of two-dimensional spectroscopies based on dipolar coupling. The NOE is the fractional variation of the intensity of a signal when another signal is selectively saturated.

The first application to paramagnetic compounds appeared in 1983 on signals with T_1 of the order of 50 ms: they were aliphatic residues in the active cavity of met-myoglobin cyanide which contains low spin iron(III) [5]. Soon it became apparent that the larger the molecular size, the larger the reorientational correlation time and the larger the NOEs. It was possible then to measure NOEs between signals with T_1 shorter than 10 ms with linewidths of several hundred hertz (at half-height) as long as the distances were small, as for geminal protons. This increased the number of successful applications of the NOE technique in the field of metalloproteins. In the meantime, improvements were made in the difference spectra techniques, which permit the recording of the FID when a given signal is saturated and the subtraction of the FID when it is not; this is then repeated a number of times and the differences summed up. The baseline remains flat and small NOEs can be detected even under a large envelope of signals. When no NOE is observed, from the signal to noise ratio of the difference spectrum we can estimate an upper limit for NOE intensities, and therefore a lower limit for the efficiency of the dipolar coupling between the proton of the saturated signal and any other proton nearby.

The NOE is the result of the transfer of spin population (called polarization

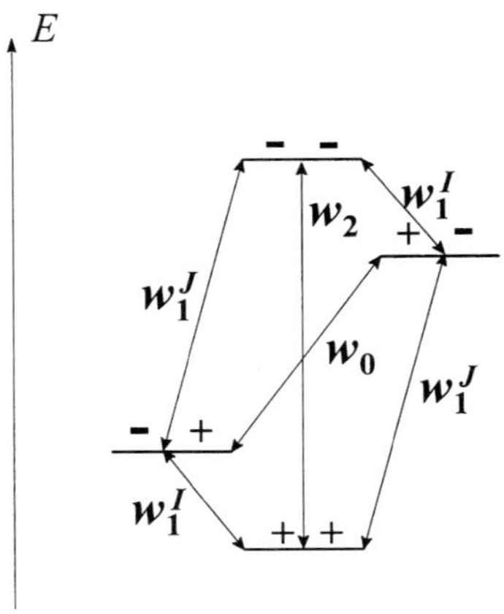

Fig. 7.1. Spin transitions and transition probabilities in a two unlike spin system experiencing dipolar coupling. The + and − refer to the signs of M_I and M_J for the two nuclear spins. The two nuclei are assumed to have positive gyromagnetic ratios.

transfer) between two dipole coupled nuclei I and J, i.e. a change of the $\langle I_z \rangle$ and/or $\langle J_z \rangle$ values due to the coupling. The NOE depends on the square of the magnitude of each nuclear magnetic moment (see later); therefore its application is essentially limited to proton NMR. In protein NMR the ^{15}N–^{1}H NOE in peptide NH groups is measured to learn about the N–H reorientational correlation time.

In order to understand NOEs it is convenient to refer to an energy diagram of the type shown in Fig. 7.1 which shows the energies of two dipole coupled nuclei with $I = 1/2$, $J = 1/2$. In case the gyromagnetic ratios γ_I and γ_J are positive, as in the example, the ground state is that with $M_I = M_J = +1/2$, and the most excited level has $M_I = M_J = -1/2$. At intermediate energies there are the two levels, $M_I = -1/2$, $M_J = +1/2$, and $M_I = +1/2$, $M_J = -1/2$. The dipolar coupling energy averages zero by rotation (Eq. (2.19)). Therefore, the energy levels of Fig. 7.1 do not depend on the extent of the spin–spin dipolar coupling. This figure is similar to Fig. 3.8A, where the dipolar coupling between an electron spin and a nuclear spin was depicted. The NMR spectrum consists of two lines at energies $\hbar\gamma_I B_0$ and $\hbar\gamma_J B_0$. When the equilibrium population of two levels is altered by providing the proper frequency to the system, e.g. the frequency corresponding to the w_1^I transition, the population of all levels will be perturbed. Since the intensity of a signal in NMR depends on the difference in population between the involved levels, selective irradiation of one signal causes a change in the intensity in the other signal. As already defined, the fractional variation of intensity is just the NOE.

In the following sections we will go through the various classical experiments like steady state, truncated and transient NOE, as well as ROE. The presentation has the twofold purpose of sketching (or refreshing) the basic theory to non-specialists, and of underlying the peculiar behavior of each experiment in paramagnetic systems.

7.2 STEADY STATE NOE

The steady state NOE is the fractional variation η_I in the integrated NMR signal intensity of a nuclear spin I when another spin J is saturated for enough time to allow the system to reach a new steady state equilibrium. In order to fully understand the phenomenon it is convenient to refer first to a saturation time long in comparison with the T_1 of the signal on which NOE is going to be measured; for short irradiation times we will then deal with NOE as a function of time $\eta_I(t)$. We now express the intensity of a signal from spin I as the expectation value of I_z, $\langle I_z \rangle$, as already defined (Section 1.6). The larger the difference in population between the M_I levels, the larger $\langle I_z \rangle$. After saturating signal J with a proper soft pulse, we have made the population of the $++$ level equal to that of the $+-$ level, as well as that of the $-+$ level equal to that of the $--$ level (Fig. 7.1). If the saturation lasts for a long enough time (see above), the difference in population between $++$ and $-+$ levels, and between the $+-$ and $--$ levels, reaches a steady state condition and $\langle I_z(t) \rangle$ reaches a new value that is different from the original equilibrium value $\langle I_z(\infty) \rangle$. The NOE $\eta_{I(J)}$ (which indicates that NOE is observed on signal I when J is saturated), will be given by

$$\boxed{\eta_{I(J)} = \frac{\langle I_z(t) \rangle - \langle I_z(\infty) \rangle}{\langle I_z(\infty) \rangle}.} \tag{7.1}$$

It can be shown (see Appendix VII) that the following relationship holds:

$$\langle I_z(t) \rangle = \langle I_z(\infty) \rangle + \left[\sigma_{I(J)}/\rho_{I(J)}\right] \langle J_z(\infty) \rangle. \tag{7.2}$$

In order to define $\sigma_{I(J)}$ and $\rho_{I(J)}$ it is convenient to refer to Fig. 7.1 and to define w_1^I and w_1^J as the transition probabilities between two states involving a single quantum transition either of spin I or J; w_0 is the zero quantum transition and corresponds to the $-+ \rightarrow +-$ transition and vice versa; w_2 corresponds to the $-- \rightarrow ++$ transition and vice versa, which is a double quantum transition. $\rho_{I(J)}$ is defined as

$$\boxed{\rho_{I(J)} = w_0 + 2w_1^I + w_2.} \tag{7.3}$$

It represents the total probability for nucleus I to change its spin component along z in a coupled two spin system.

$\sigma_{I(J)}$ ($= \sigma_{J(I)}$ when I and J have the same spin multiplicity) is called cross-relaxation rate, and is given by

$$\boxed{\sigma_{I(J)} = \sigma_{J(I)} = w_2 - w_0.} \tag{7.4}$$

The transitions corresponding to w_2 and w_0 involve simultaneous changes in spin state of both nuclei. The difference between w_2 and w_0 tells us how the variation in population of J affects the equilibrium of I. In other words, one can think in terms of transfer of spin population from J to I when J is saturated. The larger the cross relaxation, the larger the dipolar coupling. The NOE on I is proportional to the cross relaxation from J and inversely proportional to the capability of I to return to equilibrium once its equilibrium is perturbed through cross relaxation.

The transition probabilities depend on the mean squared interaction energy relative to the mechanism which causes the transition, times the value of the spectral density at the required frequencies (Eq. (3.14)). The square of the dipolar interaction energy is, as usual (see Eq. (1.4) and Appendix V), proportional to $\langle \mu_1 \cdot \mu_2 / r^3 \rangle^2$, where μ_1 and μ_2 are the magnetic moments of the two spins. The actual equations are

$$\boxed{\begin{aligned}\rho_{I(J)} = {} & \left(\frac{\mu_0}{4\pi}\right)^2 \frac{2\hbar^2 \gamma_I^2 \gamma_J^2 J(J+1)}{15 r_{IJ}^6} \\ & \times \left[\frac{\tau_c}{1 + (\omega_I - \omega_J)^2 \tau_c^2} + \frac{3\tau_c}{1 + \omega_I^2 \tau_c^2} + \frac{6\tau_c}{1 + (\omega_I + \omega_J)^2 \tau_c^2}\right]\end{aligned}} \tag{7.5}$$

and

$$\boxed{\begin{aligned}\sigma_{I(J)} = {} & \left(\frac{\mu_0}{4\pi}\right)^2 \frac{2\hbar^2 \gamma_I^2 \gamma_J^2 J(J+1)}{15 r_{IJ}^6} \\ & \times \left[\frac{6\tau_c}{1 + (\omega_I + \omega_J)^2 \tau_c^2} - \frac{\tau_c}{1 + (\omega_I - \omega_J)^2 \tau_c^2}\right].\end{aligned}} \tag{7.6}$$

The correlation time is the reorientational correlation time. This is generally the rotational correlation time unless internal motions of a group in a molecule are faster. In the latter case, pairs of nuclei in the same molecule can have different reorientational correlation times.

Note the analogy between Eq. (7.5) and Eq. (3.16): indeed, $\rho_{I(J)}$ is a longitudinal relaxation time of spin I due to the interaction with spin J upon reciprocal reorientation. The point is that it cannot be measured, except approximately, through a T_1 measurement because T_1 is defined in the absence of links with other nuclear spins, i.e. in the absence of cross relaxation (see Section 2.2).

By combining Eqs. (7.1) and (7.2), we obtain

$$\eta_{I(J)} = \left[\sigma_{I(J)} / \rho_{I(J)}\right] \langle J_z(\infty) \rangle / \langle I_z(\infty) \rangle. \tag{7.7}$$

Since $\langle I_z(\infty)\rangle$ is proportional to $I(I+1)\gamma_I$, and $\langle J_z(\infty)\rangle$ is proportional to $J(J+1)\gamma_J$ (see Eqs. (1.31)), in the general case of heteronuclear spins we have

$$\boxed{\eta_{I(J)} = \left[\sigma_{I(J)}/\rho_{I(J)}\right] J(J+1)\gamma_J / I(I+1)\gamma_I.} \tag{7.8}$$

The relative values of I, J, γ_I, γ_J are important for determining the size and the sign of heteronuclear NOE; obviously, in homonuclear NOE they cancel each other and Eq. (7.7) takes the simple form

$$\eta_{I(J)} = \frac{\sigma_{I(J)}}{\rho_{I(J)}}. \tag{7.9}$$

Cross relaxation can occur in principle also between nuclei coupled by a time-dependent scalar interaction. In this case only w_0 can contribute to it (see Section 3.5). The correlation time for the reorientation is either chemical exchange or the relaxation rate of nucleus J if nucleus I is observed. The correlation time is generally too long to make the experiment successful.

7.2.1 *Steady state NOE in real life*

The first consideration by looking at Eqs. (7.5), (7.6) and (7.7) is that NOE does not contain any structural information within the present scheme. Indeed, both $\rho_{I(J)}$ and $\sigma_{I(J)}$ contain the same r_{IJ} parameter which cancels out in $\eta_{I(J)}$. In practice, the two-spins scheme is never a good description of any real system. The nucleus I, on which NOE is going to be observed, is always coupled to other nuclei. When dealing with hyperfine coupled nuclei, often the major source of relaxation is not the coupling with J but the coupling with the unpaired electrons. The expression for $\eta_{I(J)}$ then is (Appendix VII)

$$\boxed{\eta_{I(J)} = \frac{\sigma_{I(J)}}{\rho_{I(J)} + \rho_{I(\text{other})} + R^I_{1M}} = \frac{\sigma_{I(J)}}{\rho_I}} \tag{7.10}$$

where $\rho_{I(J)}$, as defined above, is the contribution to longitudinal relaxation of nucleus I due to the coupling with nucleus J, $\rho_{I(\text{other})}$ includes all the couplings with other nuclei responsible for relaxation of nucleus I with the exception of the coupling with nucleus J, and R^I_{1M} is the paramagnetic contribution to the longitudinal relaxation rate of nucleus I arising from the various mechanisms described in Chapter 3. The sum of all these relaxation rates is the total rate ρ_I.

When the denominator in Eq. (7.10) is dominated by the paramagnetic contribution to nuclear relaxation ($\rho_I \cong R^I_{1M}$), the following simplified equation holds

$$\eta_{I(J)} = \sigma_{I(J)}/R^I_{1M} = \sigma_{I(J)}T^I_{1M}. \tag{7.11}$$

In Eq. (7.11) the r^{-6} parameter is not canceled out, because $\sigma_{I(J)}$ depends on r_{IJ}^{-6} but R_{1M}^{I} does not. When the nucleus–electron coupling is dipolar in origin, R_{1M}^{I} depends on r^{-6} where r is the distance between the resonating nucleus and the unpaired electron (Eq. (3.19)). This partially holds when R_{1M} is not dominating.

Even if not important in principle, it is important in practice to underline that ρ_I is not a time constant for an exponential process; the process would be exponential only in the absence of cross relaxation. In paramagnetic compounds the total relaxation may be dominated by R_{1M}^{I} and the total time dependence approaches an exponential behavior [6]. Therefore, ρ_I is the measured T_1^{-1} of the nucleus (either from a selective or a non-selective experiment). When R_{1M}^{I} is not dominating, after a selective excitation of I, the initial return to equilibrium can be assumed to be exponential and its rate constant, $T_{1\text{sel}}^{-1}$, can be taken as ρ_I (see Section 2.2).

Of course, the larger the denominator in Eq. (7.10), the smaller the NOE. That is why NOE in paramagnetic complexes has developed so late: the coupling with unpaired electrons makes the relaxation very effective and the NOE very small. In geminal protons belonging to macromolecules ($\tau_r \approx 10^{-8}$ s) with ρ_I of, say, 100 s^{-1} due to paramagnetic contributions, the NOE is about 20% at 500 MHz. If the distance between protons increases to 3 Å the NOE drops to 0.8%. In small complexes τ_r is typically 10–100 times shorter than in the above example and the NOEs have intensities smaller by a similar amount. The problem would be hopeless if we could not gain one to two orders of magnitude in sample concentration. It is therefore important to have a strict control of the temperature (remember that the hyperfine shift is temperature dependent) during the hours of summation of the difference FIDs (see Section 9.3), and to increase the signal to noise ratio until the percent differences can be measured with a precision of two figures by selecting a high enough solute concentration and number of scans.

When measuring steady state NOE, the effect measured on saturating I or on saturating J may be different because ρ_I and ρ_J are in general different. We say that steady state NOE is not symmetric. It follows that larger NOEs, and then more favorable cases, occur when the signal with larger ρ is saturated.

7.2.2 *Selective and non-selective* T_1

It is intuitive that the NOE will be larger the larger the cross relaxation, and smaller the larger ρ_I. It may be appropriate here to discuss in a more quantitative way the response of a dipole-coupled two spin system to selective and non-selective experiments, and the relationship between the results of these experiments and the quantity of interest, ρ_I. The qualitative behavior has been anticipated in Section 3.13. As already described (and see also Section 9.2), when measuring T_1 of a signal one can excite (for instance, invert) all the signals simultaneously (non-selective experiment). If the nuclei are dipole coupled, then they will return to equilibrium also by exchanging polarization through cross relaxation. Alternatively, only one signal is inverted (selective experiment) and its return to equilibrium measured. Cross relaxation will be different from the previous case. In both cases, the return to equilibrium is not exponential in the presence of cross relaxation. If this is the case, the measurement of T_1 is approximate, and even the definition of T_1 may be invalid. This is because the other nuclei are part of the lattice but this lattice does not possess the infinite heat capacity required to have an exponential behavior of the magnetization. In fact, cross relaxation is capable of affecting the spin state populations of the other nuclei.

We start from the expression for $\mathrm{d}\langle I_z\rangle/\mathrm{d}t$ derived in Appendix VII (Eq. (VII.13)) and write the corresponding expression for $\mathrm{d}\langle J_z\rangle/\mathrm{d}t$:

$$\begin{aligned}\frac{\mathrm{d}\langle I_z(t)\rangle}{\mathrm{d}t} &= -\left[\langle I_z(t)\rangle - \langle I_z(\infty)\rangle\right]\rho_I - \left[\langle J_z(t)\rangle - \langle J_z(\infty)\rangle\right]\sigma_{I(J)} \\ \frac{\mathrm{d}\langle J_z(t)\rangle}{\mathrm{d}t} &= -\left[\langle J_z(t)\rangle - \langle J_z(\infty)\rangle\right]\rho_J - \left[\langle I_z(t)\rangle - \langle I_z(\infty)\rangle\right]\sigma_{J(I)}\end{aligned} \tag{7.12}$$

where we have substituted $\rho_{I(J)}$ and $\rho_{J(I)}$ with the corresponding total rates ρ_I and ρ_J (Eq. (7.10)). The two equations, analogous to those encountered in the case of chemical exchange[1] (Section 4.3.4), constitute a system of coupled differential equations which are relevant also for the transient experiments discussed in Sections 7.4 and 7.5, and whose general solutions are

$$\begin{aligned}\langle I_z(t)\rangle &= \langle I_z(\infty)\rangle + C_1\exp(-\lambda_1 t) + C_2\exp(-\lambda_2 t) \\ \langle J_z(t)\rangle &= \langle J_z(\infty)\rangle + \frac{\lambda_1 - \rho_I}{\sigma_{J(I)}}C_1\exp(-\lambda_1 t) + \frac{\lambda_2 - \rho_I}{\sigma_{J(I)}}C_2\exp(-\lambda_2 t)\end{aligned} \tag{7.13}$$

where $\lambda_{1,2} = \rho \pm D$, and

$$\rho = \tfrac{1}{2}(\rho_I + \rho_J) \qquad D = \tfrac{1}{2}\left[(\rho_I - \rho_J)^2 + 4\sigma_{I(J)}\sigma_{J(I)}\right]^{1/2}. \tag{7.14}$$

The explicit expressions for C_1 and C_2 can be derived for many particular cases [3], including the selective and non-selective inversion recovery experiments of interest here. For the selective case

$$C_1 = \frac{2(\lambda_2 - \rho_I)\langle I_z(\infty)\rangle}{\lambda_1 - \lambda_2} \qquad C_2 = -\frac{2(\lambda_1 - \rho_I)\langle I_z(\infty)\rangle}{\lambda_1 - \lambda_2} \tag{7.15}$$

[1] The magnetization, M_z^I, along z is proportional to $\langle I_z\rangle$, as reported in Section 1.6. Therefore Eqs. (7.2), (7.7), (7.12) and (7.13) can be rewritten by substituting $\langle I_z\rangle$ with M_z^I and $\langle J_z\rangle$ with M_z^J in case I and J are nuclei of the same nuclear species.

and for the non-selective experiment

$$\begin{aligned} C_1 &= 2\langle I_z(\infty)\rangle \frac{\lambda_2 - \rho_I}{\lambda_1 - \lambda_2} - 2\langle J_z(\infty)\rangle \frac{\sigma_{J(I)}}{\lambda_1 - \lambda_2} \\ C_2 &= -2\langle I_z(\infty)\rangle \frac{\lambda_1 - \rho_I}{\lambda_1 - \lambda_2} + 2\langle J_z(\infty)\rangle \frac{\sigma_{J(I)}}{\lambda_1 - \lambda_2}. \end{aligned} \tag{7.16}$$

Calculated magnetization recovery profiles are reported in Fig. 3.18. It is clear that in no case is the recovery exponential. However, Eq. (7.13) for $\langle I_z(t)\rangle$ can be approximated to a single exponential, not only when $|\sigma_{J(I)}| \ll |\rho_I - \rho_J|$, as is obvious, but also when, on the contrary, $|\sigma_{J(I)}| \gg |\rho_I - \rho_J|$. In these cases, fitting the recovery to an exponential gives a rate constant close to ρ_I [6]. In contrast, in the non-selective experiment, the recovery is closer to $\rho_I + \sigma_{I(J)}$. In order to estimate ρ_I from T_1-type measurements, the best way is that of performing a selective experiment and to measure the variation of $\langle I_z\rangle$ during an initial short time. Under these circumstances the system maximizes the two contributions to the measured ρ_I (i.e. $\rho_{I(J)}$ and $\rho_{I(\text{other})}$), R^I_{1M} remaining independent of the duration of the measurement. However, in order to extract structural information through, e.g., Eq. (3.19), it is more appropriate to perform non-selective experiments. This is particularly true in the slow motion limit, where $\sigma_{I(J)} \to -\rho_{I(J)}$. Under these conditions, and neglecting $\rho_{I(\text{other})}$, $\rho_I + \sigma_{I(J)} \to R^I_{1M}$ (Eq. (7.10)). This can be intuitively understood because cross relaxation in a non-selective experiment contributes less to signal recovery, especially at the beginning of the experiment, and the latter is dominated by paramagnetic effects (R^I_{1M}). In any case, when ρ_I in paramagnetic systems is very large (e.g. 100–1000 s^{-1}), it means that R^I_{1M} dominates the relaxation processes and the results of selective and non-selective experiments are close to one another.

Cross relaxation is the main cause of deviation from exponentiality, both for diamagnetic and paramagnetic non-selective relaxation recovery. However, a non-selective inversion recovery can be pragmatically interpreted on the basis of a single exponential. The problem arises of whether the extracted R^I_{1M} can be used, according to the Solomon equation, to extract metal–proton distances. Though this problem has been stressed in a negative sense [7], we optimistically advise [8,9] (i) to check exponentiality, (ii) to extract distance parameters, with the only caution regarding geminal protons (having large cross relaxation) whose distances from the metal, however, cannot differ much.

7.2.3 Steady state NOE in paramagnetic compounds

By combining Eqs. (7.5) and (7.6) it appears that there is an NOE dependence on the magnetic field and on the reorientational correlation time. In Fig. 7.2A the NOE dependence on the external magnetic field is shown for τ_r of 10 ns and for two protons at 1.8 Å, by taking $\rho_{I(\text{other})} = 0$ and $R^I_{1M} = 0$, 50 and

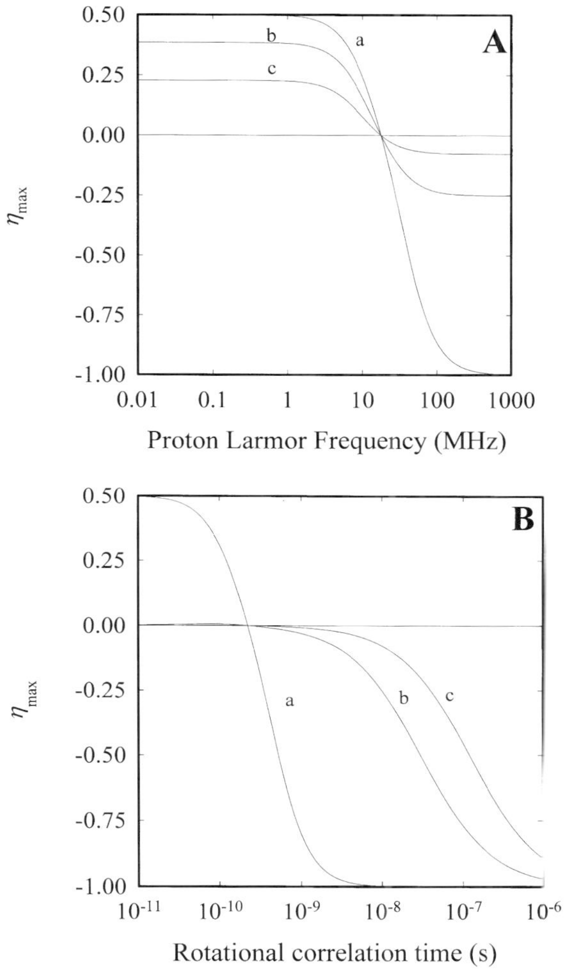

Fig. 7.2. NOE values as a function of (A) magnetic field for $\tau_r = 10$ ns and (B) τ_r at a field of 800 MHz, for two hydrogen nuclei at 1.8 Å distance ($\rho_{I(J)} = 23.6\ s^{-1}$ in the fast motion and 2.36 s^{-1} in the slow motion regime). Curves (a), (b) and (c) refer to a field-independent $R^I_{1M} = 0$, 50 and 200 s^{-1}, respectively.

200 s^{-1}. The dependence on τ_r under the same conditions is shown at 800 MHz (Fig. 7.2B). From Fig. 7.2A it appears that at low $\omega\tau$ values (fast motion limit) and $R^I_{1M} = 0$ the maximal homonuclear NOE is 0.5, whereas at large values of $\omega\tau$ (slow motion limit) the maximal NOE is -1. Of course, these are limiting values in the absence of R^I_{1M} (and $\rho_{I(\mathrm{other})}$): the hyperfine coupled relaxation decreases the above absolute values, but leaves the dependence unaltered. The NOE is larger at larger magnetic fields and/or when rotation is slowed down (Fig. 7.2B). It is important to note that, for very long τ_r values, NOE values close to -1 could

TABLE 7.1
Calculated steady state NOE's (%) for various values of τ_r and R^I_{1M} [a]

τ_r (s)	r_{IJ} (Å)	R^I_{1M} (s^{-1})							
[MW (Da)]		1		2		4		8	
		ν_0 (MHz)		ν_0 (MHz)		ν_0 (MHz)		ν_0 (MHz)	
		200	900	200	900	200	900	200	900
10^{-11}	1.8	7.171	7.078	3.862	3.810	2.009	1.981	1.025	1.010
[25]	3.0	0.388	0.382	0.195	0.192	0.097	0.096	0.049	0.048
$10^{-10.5}$	1.8	17.221	15.509	10.406	9.245	5.808	5.114	3.084	2.700
[80]	3.0	1.197	1.045	0.606	0.528	0.305	0.265	0.153	0.133
10^{-10}	1.8	30.088	14.955	21.697	9.669	13.929	5.665	8.117	3.098
[250]	3.0	3.380	1.229	1.750	0.626	0.891	0.316	0.450	0.159
$10^{-9.5}$	1.8	31.860	−15.959	26.017	−10.409	19.035	−6.139	12.387	−3.373
[800]	3.0	5.700	−1.342	3.063	−0.684	1.591	−0.346	0.811	−0.174
10^{-9}	1.8	−6.365	−57.209	−5.256	−42.124	−3.897	−27.580	−2.569	−16.314
[2500]	3.0	−1.198	−6.878	−0.647	−3.577	−0.337	−1.825	−0.172	−0.922
$10^{-8.5}$	1.8	−68.209	−83.237	−59.727	−71.890	−47.832	−56.490	−34.206	−39.546
[8000]	3.0	−17.482	−19.701	−9.821	−10.941	−5.234	−5.791	−2.706	−2.983
10^{-8}	1.8	−92.235	−94.261	−87.379	−89.239	−79.055	−80.645	−66.403	−67.620
[25000]	3.0	−43.190	−43.842	−27.726	−28.084	−16.156	−16.339	−8.806	−8.897
$10^{-7.5}$	1.8	−97.918	−98.136	−96.139	−96.352	−92.770	−92.970	−86.693	−86.873
[80000]	3.0	−71.060	−71.192	−55.183	−55.274	−38.140	−38.194	−23.576	−23.605
10^{-7}	1.8	−99.383	−99.406	−98.797	−98.819	−97.646	−97.667	−95.421	−95.442
[250000]	3.0	−88.639	−88.658	−79.612	−79.628	−66.140	−66.152	−49.415	−49.423

[a] For each τ_r and R^I_{1M} value two NOEs are calculated for interproton distances of 1.8 and 3.0 Å and for nuclear Larmor frequencies ν_0 of 200 and 900 MHz.

in principle be achieved even in the presence of paramagnetic relaxation; this is because $\sigma_{I(J)}$ increases with τ_r whereas the denominator of Eq. (7.10) increases less because R^I_{1M} does not depend on τ_r but rather on τ_s. Sometimes, with small proteins, it is even convenient to add ethylene glycol to their water solutions in order to increase viscosity and then τ_r. In the literature this has been done for met-myoglobin [10] and for Co_7 thioneins [11].

Steady state NOEs in paramagnetic small complexes under the conditions of fast rotation (MW < 800–2500 Da depending on the magnetic field) are small and sometimes may be below detection. In Table 7.1 the steady state NOE intensities are reported for various τ_r values (and the corresponding molecular weight values estimated from the Stokes–Einstein relation (Eq. (3.8)) and for various R^I_{1M} values. For each pair of τ_r and R^I_{1M} values the NOE intensities are calculated for distances of 1.8 and 3.0 Å and for magnetic fields of 200 MHz and 900 MHz. From inspection of the table it appears that NOE measurements on small complexes may be successful for high concentrations at low magnetic fields, or at large magnetic fields when rotation is slowed down by using viscous solvents or mixtures of solvents. On the contrary, macromolecules seem perfectly suited for this type of experiment since they fall in the slow motion regime.

In Table 7.2 some significant examples of the NOEs observed in paramagnetic

R^I_{1M} (s^{-1})									
16 ν_0 (MHz)		32 ν_0 (MHz)		64 ν_0 (MHz)		128 ν_0 (MHz)		256 ν_0 (MHz)	
200	900	200	900	200	900	200	900	200	900
0.518	0.510	0.260	0.257	0.130	0.129	0.065	0.064	0.033	0.032
0.024	0.024	0.012	0.012	0.006	0.006	0.003	0.003	0.002	0.002
1.591	1.389	0.808	0.705	0.407	0.355	0.205	0.178	0.102	0.089
0.077	0.067	0.038	0.033	0.019	0.017	0.010	0.008	0.005	0.004
4.424	1.625	2.317	0.833	1.186	0.422	0.600	0.212	0.302	0.107
0.226	0.080	0.113	0.040	0.057	0.020	0.028	0.010	0.014	0.005
7.293	−1.774	4.001	−0.911	2.103	−0.461	1.079	−0.232	0.547	−0.117
0.410	−0.087	0.206	−0.044	0.103	−0.022	0.052	−0.011	0.026	−0.005
−1.528	−8.979	−0.844	−4.728	−0.445	−2.428	−0.229	−1.231	−0.116	−0.620
−0.087	−0.463	−0.044	−0.232	−0.022	−0.116	−0.011	−0.058	−0.005	−0.029
−21.791	−24.718	−12.626	−14.125	−6.858	−7.606	−3.583	−3.955	−1.833	−2.018
−1.376	−1.514	−0.694	−0.763	−0.349	−0.383	−0.175	−0.192	−0.087	−0.096
−50.302	−51.111	−33.875	−34.342	−20.491	−20.736	−11.447	−11.569	−6.080	−6.140
−4.611	−4.656	−2.361	−2.384	−1.195	−1.206	−0.601	−0.607	−0.302	−0.304
−76.650	−76.799	−62.232	−62.341	−45.221	−45.289	−29.236	−29.275	−17.128	−17.147
−13.368	−13.382	−7.164	−7.171	−3.715	−3.719	−1.893	−1.895	−0.956	−0.956
−91.263	−91.282	−83.947	−83.964	−72.347	−72.360	−56.682	−56.692	−39.554	−39.559
−32.818	−32.823	−19.631	−19.633	−10.884	−10.885	−5.755	−5.756	−2.963	−2.963

metalloproteins are reported [12]; as expected, as the T_1 values become shorter, the NOE drops to small values, which can be difficult to 'extract' from the noise. However, the NOE increases as the molecular size increases. In Fig. 7.3 the NOEs for a protein of MW 10,000 containing the cluster $Fe_4S_4^{3+}$ are shown when the hyperfine shifted signals due to coordinated β cysteine protons and lying outside the −20 to 20 ppm spectral window are saturated [13]. The NOE difference spectra show many connectivities with other signals which belong to nearby groups, as well as strong connectivities with the respective geminal protons (not shown).

The difference spectra in Fig. 7.3 also show signals (marked with ×) arising from saturation transfer to another species in chemical exchange. Indeed, the experiment for detecting steady state NOE is absolutely identical to that used to detect magnetization transfer in the presence of chemical exchange (Section 4.3.4). Therefore, it is not possible to distinguish, from the experimental point of view, real NOE from chemical exchange effects in the limit of slow motion. Actually, the values of NOE are always a fraction of the signal intensity, which in paramagnetic molecules is often small (of the order of 1%), whereas saturation transfer can change the intensity of the signal really up to 100% even in paramagnetic molecules (see Section 4.3.4).

TABLE 7.2
Some examples of NOE detected on paramagnetically shifted signals in metalloproteins [12]

Group [a]	H–H distance (Å)	NOE [b] (%)	T_1 (ms)	MW	τ_r (ns)
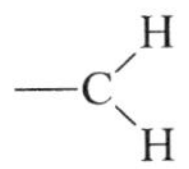	1.8	6	8.4	9000	4
		5	3.6	11000	5
		7	5.6	11000	5
(Asp, Glu, Cys, His)		43	120	16000	9
		7.7	2.4	32000	14
		60	100	36000	21
		50	80	42000	22
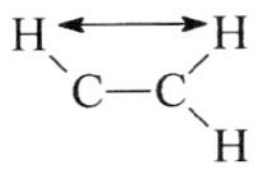	2.2–2.4	4.5	34.0	9000	4
		1.8	5.6	11000	5
(Cys)					
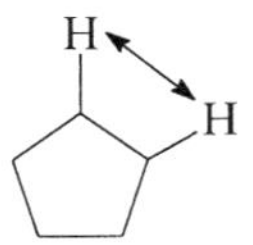	2.4	6	5.4	30000	14
		0.9	1.8	32000	14
		2	4.2	32000	14
(His)					
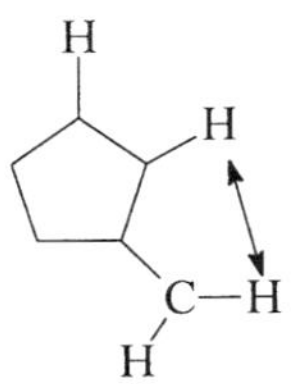	~3.3	0.6	3.5	32000	14
		7–10	–	42000	21
(His)					

[a] The groups bearing the two protons belong to metal ion ligands.
[b] The NOEs in the present systems are all negative; they are reported here without the minus sign.

In diamagnetic compounds we face the problem that after saturation of signal J the Zeeman population of nuclei I may be largely affected, and this may cause further magnetization transfer on other nuclei of the same nuclear species to which nucleus I is coupled. Such coupling indeed accounts for the $\rho_{I(\text{other})}$ term in Eq. (7.10). It follows that a secondary NOE can be observed on signal K not because J is coupled with K but because I is coupled with K. This phenomenon is called *spin diffusion*. It is not relevant in the fast motion limit, because the NOEs are usually small and, therefore, small is the secondary magnetization transfer to nearby nuclei; furthermore, and more importantly, a positive sign of σ means that a decrease in $\langle J_z\rangle(M_z^J)$ causes an *increase* in $\langle I_z\rangle(M_z^I)$ which, in turn, causes a *decrease* in $\langle K_z\rangle(M_z^K)$, and so on. This alternation in sign eventually results in substantial cancellation of the long range effects, because a nucleus very far from

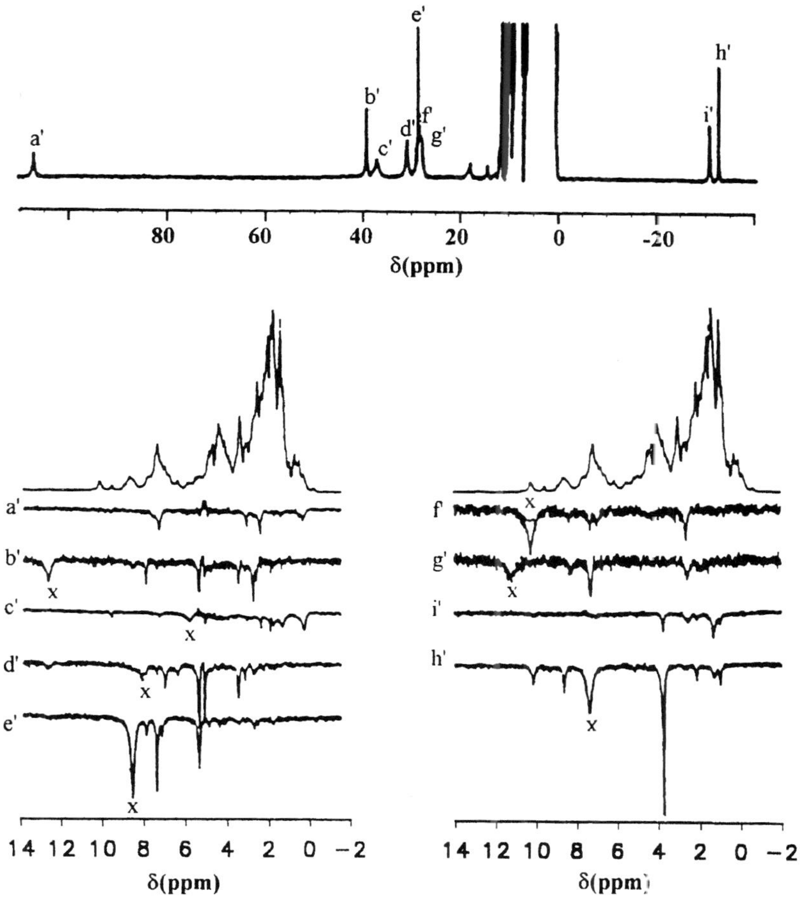

Fig. 7.3. NOE difference spectra obtained upon saturation of the hyperfine shifted signals corresponding to the β-CH_2 protons of the Fe_4S_4-coordinated cysteines in oxidized HiPIP from *C. vinosum* [13]. Signals marked by × arise from saturation transfer to a small amount of reduced species [13].

the saturated nucleus will experience both positive and negative contributions depending on whether the number of intermediate spins on the various cross relaxation pathways is even or odd.

In contrast, sizable spin diffusion may be operative in the slow motion regime. Of course, this phenomenon may terribly complicate the analysis of NOEs, until they become useless. After saturating one signal, many or even all signals of the molecule may be affected in a cascade fashion. However, in paramagnetic macromolecular systems the major source of relaxation for signal I can be the coupling with the unpaired electrons; so it returns to equilibrium prior to appreciably transferring polarization to nucleus K. Therefore, spin diffusion effects are limited.

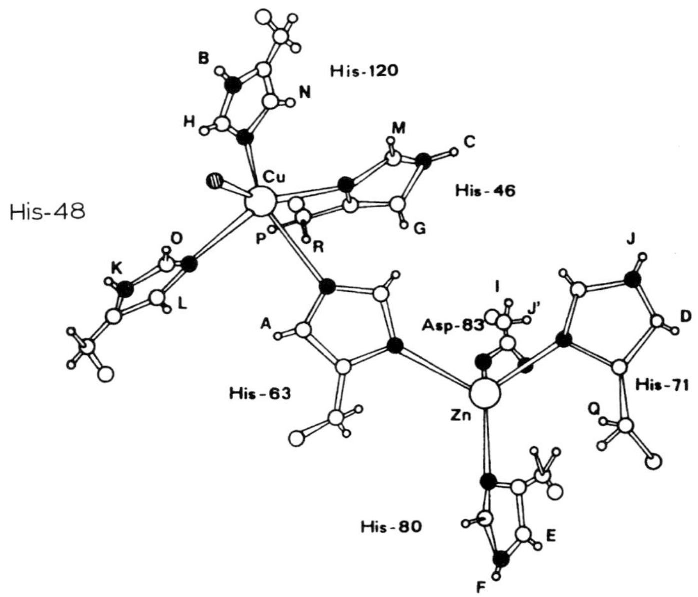

Fig. 7.4. Schematic structure of the active site of Cu_2Zn_2-superoxide dismutase. The labels refer to the proton signal assignment for the Cu_2Co_2 derivative reported in Table 7.3 [15,16].

The counterpart of this simplification is that the NOE is dramatically reduced by the short ρ_I values.

An example is available in the literature where the spin diffusion effects are evaluated in a strongly paramagnetic system [14]. It is the Cu_2Co_2SOD discussed in Section 6.3.3. The scheme is again reported here (Fig. 7.4 with its labeling) for convenience. In Table 7.3 the NOEs at 200 MHz on all the proton ligands are reported when they are saturated one after the other. The T_1 values measured at the same field are also reported [15,16]. The interproton distances are known from the X-ray structure and the value of the reorientational correlation time estimated by the Stokes–Einstein equation (Eq. (3.8)). The NOEs calculated through the two-spin approximation are in satisfactory agreement with the experimental NOEs. If then a matrix is considered where all the ligand protons plus other protons from nearby groups are allowed to cross relax, the newly calculated NOEs are still close to the experimental values [14]. Spin diffusion is quenched by the fast relaxing nature of the system. The ρ_I values are well above 100 s^{-1}. Steady state NOE is then a safe measurement of proton–proton distances. When, however, the ρ_I values are small, spin diffusion may occur even in paramagnetic systems; tricks are developed to detect spin diffusion and to minimize it (see the next two sections).

TABLE 7.3

Non-selective T_1 and NOE values in bovine Cu_2Co_2SOD at 200 MHz [16]. The distances calculated from the NOE values using the two-spin approximation agree with available X-ray distances

Saturated signal	Observed signal	T_1 (obs) (ms)	NOE (%)	r (calc) (Å)	r (X-ray) (Å)
A Hδ2 His 63	L Hδ2 His 48	4.3	0.9 ± 0.1	2.7 ± 0.2	2.7
A Hδ2 His 63	K Hδ1 His 48	8.0	0.6 ± 0.2	3.2 ± 0.3	3.8
A Hδ2 His 63	R Hβ2 His 46	2.4	0.2 ± 0.1	3.3 ± 0.3	3.4
B Hδ1 His 120	H Hε1 His 120	1.8	1.0 ± 0.2	2.3 ± 0.2	2.4
C Hε2 His 46	G Hδ2 His 46	3.5	1.4 ± 0.2	2.4 ± 0.2	2.5
C Hε2 His 46	M Hε1 His 46	2.7	0.9 ± 0.1	2.5 ± 0.2	2.5
G Hδ2 His 46	C Hε2 His 46	4.2	1.7 ± 0.4	2.4 ± 0.2	2.5
M Hε1 His 46	C Hε2 His 46	4.2	2.2 ± 0.2	2.4 ± 0.2	2.5
R Hβ2 His 46	G Hδ2 His 46	3.5	0.6 ± 0.2	2.8 ± 0.3	3.3
Q Hβ1 His 71	D Hδ2 His 71	3.8	1.4 ± 0.2	2.4 ± 0.2	2.8
L Hδ2 His 48	γ_1 CH_3 Val 118	70.0	33.0 ± 6	2.8 ± 0.3	2.8
R Hβ2 His 46	P Hβ1 His 46	1.6	5.4 ± 0.8	1.7 ± 0.2	1.6
P Hβ1 His 46	R Hβ2 His 46	2.4	9.0 ± 1.0	1.7 ± 0.2	1.6
L Hδ2 His 48	P Hβ1 His 46	1.6	1.3 ± 0.3	2.2 ± 0.2	2.3
P Hβ1 His 46	L Hδ2 His 48	4.3	5.5 ± 1.0	2.0 ± 0.2	2.3
L Hδ2 His 48	R Hβ2 His 46	2.4	0.5 ± 0.1	2.6 ± 0.3	3.2
R Hβ2 His 46	L Hδ2 His 48	4.3	2.3 ± 0.2	2.3 ± 0.2	3.2

7.3 TRUNCATED NOE

Up to now steady state NOEs have been considered, i.e. when one signal is saturated for a long time with respect to T_1 of the nucleus on which NOE is going to be measured. Let's consider here what happens when the saturation time is short and variable. The resulting NOE is called truncated NOE [17] because not enough time is left for full polarization transfer. These experiments are of fundamental importance for the measurement of ρ_I, for evaluating cross relaxation, and to avoid or to measure spin diffusion.

In Appendix VII, where the steady state NOE has been derived, the equation for the NOE as a function of the irradiation time is also derived. In the case of homonuclear NOE, it is

$$\eta_{I(J)}(t) = \left[\sigma_{I(J)}/\rho_I\right]\left[1 - \exp(-\rho_I t)\right]. \tag{7.17}$$

It can be easily verified that for t long with respect to ρ_I, the exponential term tends to zero and we are back to the case of steady state NOE (Eq. (7.10)). For t tending to zero, the NOE tends to zero. The dependence of $\eta_{I(J)}(t)$ on t is reported in Fig. 7.5 for the same parameters as in Fig. 7.2. For times short with respect to ρ_I, $1 - \exp(-x) \approx 1 - (1 - x) = x$, and

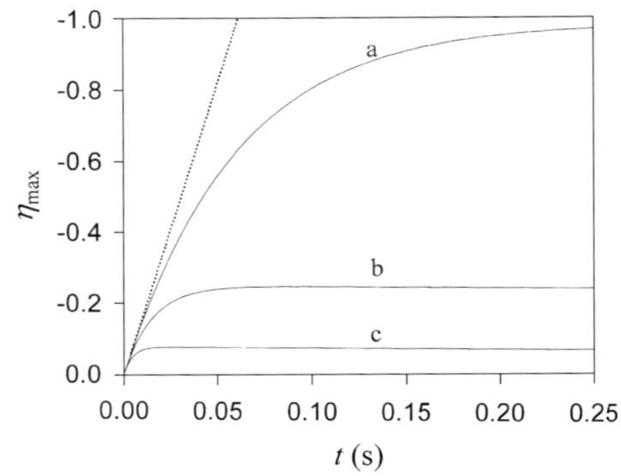

Fig. 7.5. NOE values as a function of irradiation time t, calculated according to Eqs. (7.5), (7.6) and (7.17). Conditions: $\tau_r = 10$ ns, 800 MHz, for two hydrogen nuclei 1.8 Å apart ($\rho_{I(J)} = 2.38$ s^{-1}). $\rho_{I(\text{other})} = 0$, $R^I_{1M} = 0$, 50 and 200 s^{-1} for curves (a), (b) and (c), respectively. The initial slope (dashed line) is the same in all cases and equals $\sigma_{I(J)}$.

$$\eta_{I(J)}(t) = \sigma_{I(J)}t. \tag{7.18}$$

For irradiation times of J short with respect to the relaxation time of I the NOE extent is independent of the relaxation time of the nucleus and provides a direct measurement of $\sigma_{I(J)}$. If the time required to saturate signal J is not negligible compared with t, the response of the system is not linear [18]. The truncated NOE is independent of paramagnetism as it does not depend on ρ_I, which contains the electron spin vector S in the R^I_{1M} term, and only depends on $\sigma_{I(J)}$, which does not contain S. If then the steady state NOE is reached, the value of ρ_I can also be obtained. This is the correct way to measure ρ_I of a nucleus, provided saturation of J can be considered instantaneous. In general, measurements at short t values minimize spin diffusion effects. In fact, in the presence of short saturation times, the transfer of saturation affects mainly the nuclei directly coupled to the one whose signal is saturated. Secondary NOEs have no time to build substantially. As already said, this is more true in paramagnetic systems, the larger the R^I_{1M} contribution to ρ_I.

Measuring the build up of NOE is necessary every time the NOE is provided by a signal buried in an envelope of many other signals. Such a procedure allows one to estimate, besides the shift, ρ_I of such buried signal. From the initial part of the slope the distance between the two nuclei can be obtained. If ρ_I is dominated by R^I_{1M}, the distance from the metal can also be guessed.

7.4 TRANSIENT NOE

The populations of the spin states can be disturbed from their equilibrium condition also by applying strong r.f. pulses at the resonance frequency of one of the spins. Typically a selective 180° pulse is used. Then the decay of the system to equilibrium is observed. This kind of experiment is called transient NOE. This experiment can be used to reduce the effects of spin diffusion. After the 180° pulse the polarization of J is inverted. The system returns to equilibrium with its own ρ_J and through cross relaxation with I (among other nuclear spins). The difference with the steady state NOE is that the time evolution of $\langle J_z \rangle$ now also plays a role. After a delay time τ, a 90° observation pulse is applied and the FID is recorded. The magnitude of the NOE as a function of time is the result of two competing effects. One is the return to equilibrium of nucleus J which causes a decrease in the absolute value of polarization. In fact, when J is at equilibrium, no polarization transfer occurs any more. The larger the difference in spin population with respect to the equilibrium value, the larger is the buildup of NOE. At $t = 0$ the cross relaxation contribution is twice that of a steady state experiment, since in the transient case the polarization is inverted, whereas in the steady state case it is zero. The second competing effect is the buildup of NOE, which is of the type of Eq. (7.17). The resulting transient NOE can be obtained as a particular solution of Eq. (7.13) for $\langle I_z(t) \rangle$ as [3]

$$\boxed{\eta_{I(J)} = -\frac{\sigma_{I(J)}}{D} \exp\left[-(\rho - D)t\right]\left[1 - \exp(-2Dt)\right]} \tag{7.19}$$

where ρ and D have been already defined (Eq. (7.14)). It appears that the transient NOE is described by a buildup function $(1 - \exp(-2Dt))$ multiplied by a decay function $(\exp[-(\rho - D)t])$. The initial slope is $2D$.

In case $\rho_I = \rho_J = \rho$, $D = \sigma_{I(J)}$ and Eq. (7.19) takes the following simpler form:

$$\eta_{I(J)} = \exp\left[-(\rho - \sigma_{I(J)})t\right]\left[1 - \exp(-2\sigma_{I(J)}t)\right]. \tag{7.20}$$

The initial slope in this case is $2\sigma_{I(J)}$.

Note that the transient NOE effect is symmetrical, in the sense that by selective excitation of I or J the same variation in intensity of the other signal is observed, even in the case of different relaxation rates.

The buildup and decay of NOE as a function of time according to Eq. (7.20) is reported in Fig. 7.6. In transient experiments, J is selectively inverted in a time assumed negligible with respect to that necessary for the NOE to build up on I. During the buildup, J also relaxes, and therefore the effect on I will depend on ρ_J, in addition to ρ_I and $\sigma_{I(J)}$. If J is relaxing much faster than I, after selective excitation of J NOE starts to buildup on I but, as J relaxes efficiently, NOE starts to decrease with $\rho \approx \rho_J/2$ according to Eq. (7.14). If, however, the slow relaxing spin I is selectively excited, NOE on the fast relaxing spin J develops, but the effect soon ceases owing to the relaxation of J itself. Therefore, the effect is symmetrical. When nuclear relaxation times are short, as may happen in paramagnetic systems, the transient NOE rapidly drops to zero. This is clearly shown in Fig. 7.6,

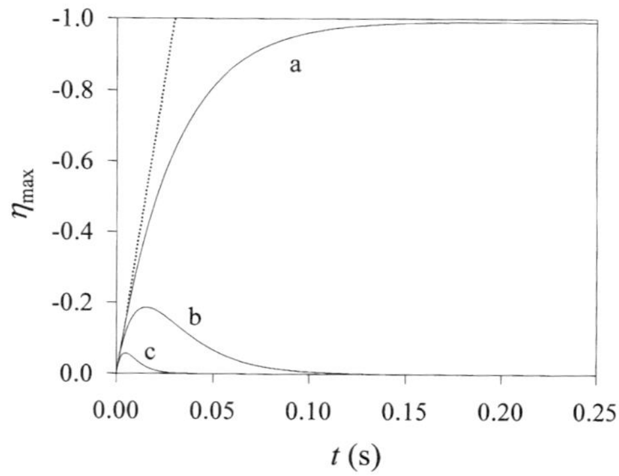

Fig. 7.6. NOE values as a function of time t in transient NOE experiments, calculated according to Eqs. (7.5), (7.6) and (7.20). The parameters are the same as in Fig. 7.5.

where for R^I_{1M} of the order of 200 s^{-1} the transient NOE develops to a small fraction of when R^I_{1M} is 50 s^{-1}, and rapidly disappears with time.

In this type of experiment the NOE buildup tends to disappear with ρ_I. In the steady state case, the saturation time is always long enough to allow spin I to cross-relax with other spins, even if ρ_I is large. In transient experiments, the cross relaxation with spin I is by itself limited in time and, ρ_I being the same, cross relaxation with other spins is drastically limited. In any case, spin diffusion is limited in that region of time in which NOE is growing (Fig. 7.6). Truncated and transient NOEs performed with short NOE buildup times are efficient in quenching spin diffusion.

One important matter is the extent of NOE in the transient and steady state cases. In the case of diamagnetic molecules the former is superior because it reduces spin diffusion and has a maximum value at $t = T_1$ (for the limit case of I and J equally relaxing). In paramagnetic systems the working conditions are much more severe for at least two reasons. One is that the time at which the maximum transient NOE is developed is very short and it may be difficult to set the exact value of the parameter. The second is that a 180° pulse has to occur in a time short with respect to the reciprocal of the average of the longitudinal and transverse relaxation rates of J, $\frac{1}{2}(\rho_{1J} + \rho_{2J})^{-1}$. In order to meet this condition a high power pulse may be needed. Of course, a high power pulse for a short time is hardly selective, and artifacts in the experiment are often encountered. In Fig. 7.7, the calculated time dependence of transient and truncated NOE is reported. A larger NOE is observed in the transient case only in a narrow range of time. In the authors' experience, a steady state or truncated NOE is much safer!

The limiting values of η are 0.385 and -1 in the fast and slow motion limits respectively (Fig. 7.8), as opposed to 0.5 and -1 for steady state NOE. In other words, the steady state NOE is a more sensitive technique for small molecules (0.5 vs. 0.385).

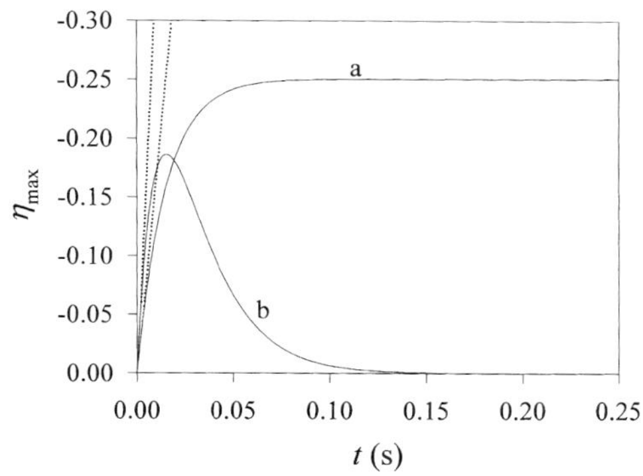

Fig. 7.7. Comparison between truncated or steady state (a) and transient (b) NOE $R^{I}_{1M} = 50$ s^{-1}. Other conditions as in Figs. 7.5 and 7.6. The dotted lines represent the initial slope ($\sigma_{I(J)}$ or $2\sigma_{I(J)}$) in the two cases, respectively.

7.5 NOE IN THE ROTATING FRAME (ROE)

Rotating frame NOE experiments, ROE [19], measure the transfer of transverse polarization, spin locked in the transverse plane, within a pair of spins. The pulse sequence consists of a selective 180° pulse on J, followed by a non-selective 90° pulse. At this point, J is along $-y$ and I is along y (Fig. 7.9). During a time t before acquisition, a spin locking r.f. is applied, in such a way that the B_1 field is parallel to the initial polarization along y in the rotating frame. Polarization transfer can now occur through cross relaxation and an intensity change on I can build up during the time t. This process is similar to that in a transient NOE experiment, with the difference that its extent depends on the rotating frame relaxation rate ρ_ρ and on the rotating frame cross relaxation rate σ_ρ. In analogy with Eqs. (7.5) and (7.6) we can write

$$\rho_{\rho I(J)} = \left(\frac{\mu_0}{4\pi}\right)^2 \frac{\hbar^2\gamma_I^2\gamma_J^2 J(J+1)}{30r_{IJ}^6}\left[\tau_c + \frac{9\tau_c}{1+(\omega_{1I}+\omega_{1J})^2\tau_c^2} + \frac{18\tau_c}{1+\omega_I^2\tau_c^2} + \frac{12\tau_c}{1+(\omega_I+\omega_J)^2\tau_c^2}\right] \tag{7.21}$$

$$\sigma_{\rho I(J)} = \left(\frac{\mu_0}{4\pi}\right)^2 \frac{\hbar^2\gamma_I^2\gamma_J^2 J(J+1)}{30r_{IJ}^6} \times \left[-\tau_c + \frac{9\tau_c}{1+(\omega_{1I}+\omega_{1J})^2\tau_c^2} + \frac{12\tau_c}{1+\omega_I^2\tau_c^2}\right] \tag{7.22}$$

where $\omega_{1I} = \gamma_I B_1$ and $\omega_{1J} = \gamma_J B_1$ are the Larmor frequencies at the spin locking field. It is important to note that, because $\omega_1\tau_c$ is always smaller than unity, σ_ρ is

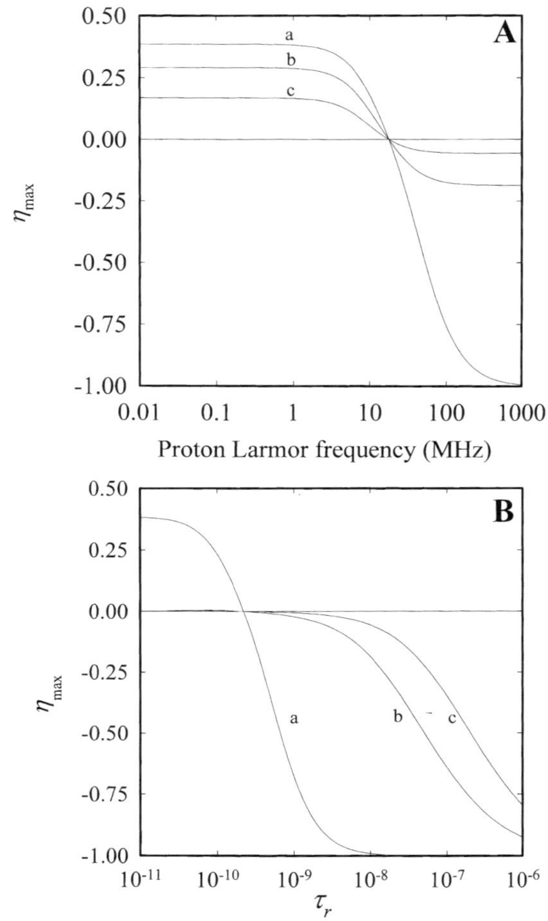

Fig. 7.8. Maximal intensity of transient NOE as a function of (A) magnetic field for $\tau_x = 10$ ns and (B) τ_r at a field of 800 MHz, for two nuclei at 1.8 Å distance ($\rho_{I(J)} = 23.6$ s^{-1} in the fast motion and 2.36 s^{-1} in the slow motion regime). Curves (a), (b) and (c) refer to a field-independent $R^I_{1M} = 0$, 50 and 200 s^{-1}, respectively. Note that in the fast motion regime the maximal NOE for $R^I_{1M} = 0$ is 0.385 vs. 0.5 for steady state NOE (Fig. 7.2).

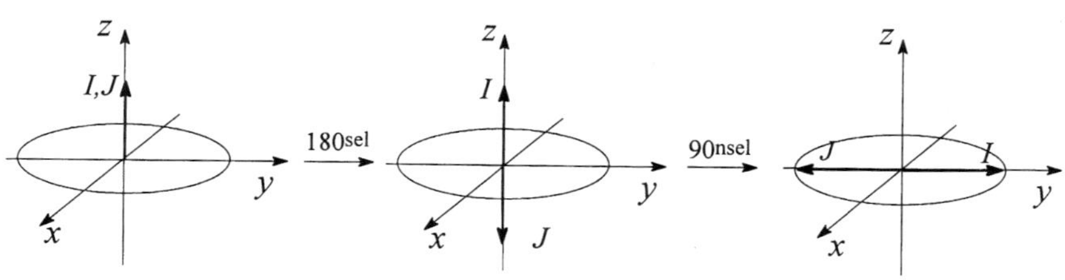

Fig. 7.9. A selective 180° pulse on signal J reverts its magnetization along the z axis while leaving the magnetization of I unaltered. A subsequent non-selective 90° pulse tilts the magnetization of I along y and that of J along $-y$.

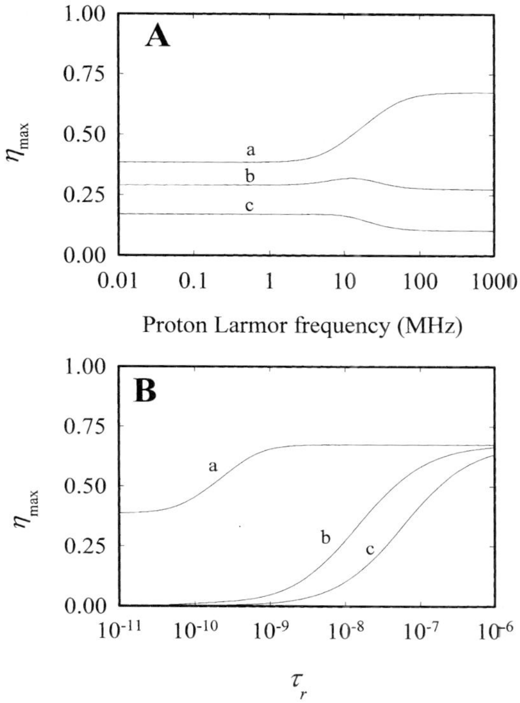

Fig. 7.10. Maximal intensity of the ROE as a function of (A) magnetic field for $\tau_r = 10$ ns and (B) τ_r at a field of 800 MHz, for two hydrogen nuclei at 1.8 Å distance. Curves (a), (b) and (c) refer to a field-independent $R^{I}_{1\rho M}$ (Eq. (3.21)) of 0, 50 and 200 s^{-1}, respectively. For $R^{I}_{1\rho M} = 0$, the fast motion limit is the same as that of transient ROE; the slow motion limit is positive and equal to 0.675.

always positive for every value of ω. Again in analogy with Eq. (7.19):

$$\eta_{I(J)} = \exp\left[-(\rho_\rho - D_\rho)t\right]\left[1 - \exp(-2D_\rho t)\right] \tag{7.23}$$

where, in analogy with Eq. (7.14),

$$\begin{aligned} \rho_\rho &= 1/2(\rho_{\rho I} + \rho_{\rho J}) \\ D_\rho &= 1/2\left[(\rho_{\rho I} - \rho_{\rho J})^2 + 4\sigma_{\rho I(J)}\sigma_{\rho J(I)}\right]^{1/2}. \end{aligned} \tag{7.24}$$

The ROE dependence on the spin lock time has the same profile as that of transient NOE, with the difference that the limiting values are 0.385 and 0.675 at the condition that $\omega_1\tau_c < 1$ (see Fig. 7.10). It appears that the ROE is less convenient than the transient and steady state NOEs in the sense that the expected effect is smaller when all other conditions are the same. Another disadvantage in paramagnetic molecules is that it is difficult to spin lock all the signals in a broad spectral width, whereas it is simple to selectively spin lock a single signal. In order to spin lock two signals, separated by e.g. 1000 Hz, a B_1 field such that $\gamma B_1/2\pi \gg 1000$ Hz is needed. However, a large intensity of the r.f. requires high power irradiation and may cause overheating of the sample. This can be a serious drawback. An advantage of ROE is that the zero value for NOE never occurs. This may constitute a precious help for the measurement of dipolar interactions in small paramagnetic complexes. As in the fast motion limit of NOE, spin diffusion effects in ROE are quenched because of the positive value of σ_ρ. Another advantage of having a positive σ_ρ is that the dipole-coupled nuclei display positive ROEs, whereas signals connected through chemical exchange always provide negative ROEs. This is because, in the presence of chemical exchange, when the J nuclei are aligned along $-y$, the I nuclei receive some $-y$ component and then the signal decreases in intensity. In the ROE experiment in the absence of chemical exchange the I nuclei receive a y component as a result of σ_ρ being always positive.

7.6 REFERENCES

[1] A.W. Overhauser (1953) Phys. Rev. 89, 689.
[2] A.W. Overhauser (1953) Phys. Rev. 92, 411.
[3] J.H. Noggle, R.E. Schirmer (1971) The Nuclear Overhauser Effect. Academic Press, New York.
[4] D. Neuhaus, M. Williamson (1989) The Nuclear Overhauser Effect in Structural and Conformational Analysis. VCH, New York.
[5] R.D. Johnson, S. Ramaprasad, G.N. La Mar (1983) J. Am. Chem. Soc. 105, 7205.
[6] J. Granot (1982) J. Magn. Reson. 49, 257.
[7] G.N. La Mar, J.S. de Ropp (1993) In: L.J. Berliner, J. Reuben (Eds.), Biological Magnetic Resonance, Vol. 12. Plenum Press, New York.
[8] I. Bertini, C. Luchinat, A. Rosato (1996) Progr. Biophys. Mol. Biol. 66, 43.
[9] I. Bertini, A. Donaire, C. Luchinat, A. Rosato (1997) Proteins Struct. Funct. Genet. 29, 348.
[10] L.B. Dugad, G.N. La Mar, S.W. Unger (1990) J. Am. Chem. Soc. 112, 1386.
[11] I. Bertini, C. Luchinat, L. Messori, M. Vasak (1993) Eur. J. Biochem. 211, 235.
[12] L. Banci (1993) In: L.J. Berliner, J. Reuben (Eds.), Biological Magnetic Resonance. Plenum, New York.
[13] I. Bertini, F. Capozzi, S. Ciurli, C. Luchinat, L. Messori, M. Piccioli (1992) J. Am. Chem. Soc. 114, 3332.
[14] L. Banci, I. Bertini, C. Luchinat, M. Piccioli (1990) FEBS Lett. 272, 175.
[15] L. Banci, I. Bertini, C. Luchinat, M.S. Viezzoli (1990) Inorg. Chem. 29, 1438.
[16] I. Bertini, C. Luchinat, M. Piccioli (1994) Progr. NMR Spectrosc. 26, 91.
[17] G. Wagner, K. Wüthrich (1979) J. Magn. Reson. 33, 675.
[18] J.T.J. Lecomte, S.W. Unger, G.N. La Mar (1991) J. Magn. Reson. 94, 112.
[19] A.A. Bothner-By, R.L. Stephens, J. Lee, C.D. Warren, R.W. Jeanloz (1984) J. Am. Chem. Soc. 106, 811.

Chapter 8

Two-Dimensional Spectra and Beyond

In this chapter the effects of paramagnetism on several homocorrelated 2D spectroscopies are discussed, and the conditions to minimize the detectability problems arising from these effects are indicated. Emphasis is given to homocorrelated ^1H–^1H spectroscopies because the adverse effects of paramagnetism are maximal for protons, due to their largest magnetogyric ratio. Once the reader has learned how to optimize these experiments, extension to heterocorrelated 2D or 3D experiments is easy. Finally, those cross correlation effects which are specifically due to the presence of unpaired electrons are discussed.

8.1 INTRODUCTION

Two-dimensional (2D) spectroscopy is used to obtain some kind of correlation between two nuclear spins I and J, for instance through scalar or dipolar connectivities, or to improve resolution in crowded regions of spectra. The parameters to obtain 2D spectra are nowadays well optimized for paramagnetic molecules, and useful information is obtained as long as the conditions dictated by the correlation time for the electron–nucleus interaction are not too severe. Sometimes care has to be taken to avoid that the fast return to thermal equilibrium of nuclei wipes out the effects of the internuclear interactions that are sought through 2D spectroscopy.

2D NMR experiments are characterized by four time periods labeled preparation, evolution, mixing and detection (Fig. 8.1). The preparation period contains at least one pulse which alters the equilibrium population of at least one nucleus and generates some magnetization in the xy plane. The evolution period allows the magnetization to evolve in the xy plane according to the Larmor frequency of the nuclei which have experienced the preparation pulse(s). The evolution period is incremented in successive experiments and constitutes one of the two time domains (t_1) on which the Fourier transform is applied at the end of the 2D experiment. After this time interval a discontinuity is provided by either a pulse or the start of a spin-lock sequence (see later), then the mixing follows. The mixing period t_m which may be lacking or may coincide with a pulse rather than a time period in some 2D experiments, allows for the transfer of magnetization or coherence (see later) between sets of spins which are connected by either dipolar

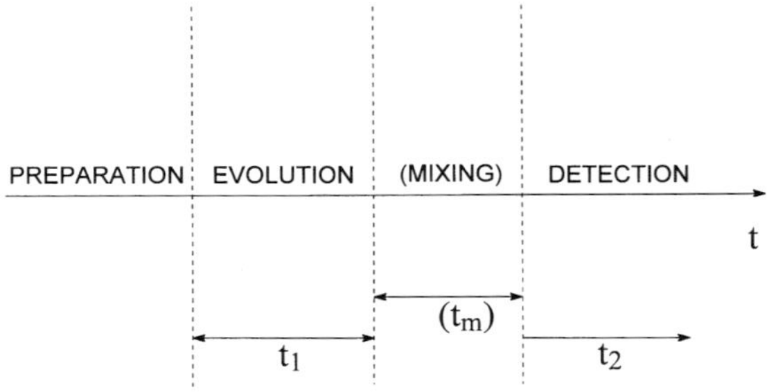

Fig. 8.1. General scheme of the time periods involved in 2D experiments.

or scalar coupling or by chemical exchange. At the end of the mixing period, which usually (but not always) has a constant duration throughout the experiment, the detection pulse(s) provides the xy magnetization which is detected during the detection period as free induction decay. The time axis of the latter constitutes the second time domain (t_2) on which the Fourier transform is applied.

After Fourier transform in both dimensions, a 2D spectrum is obtained in the frequency domain. The spectrum is characterized by two frequency axes, f_1 and f_2, which are related to t_1 and t_2, respectively, and represent the chemical shift scales of the experiment. Cross peaks represent the projections of signals which appear anytime a spin set affects another spin set either through dipolar coupling, scalar J_{IJ} coupling or chemical exchange. The dipolar coupling allows magnetization transfer as has been shown in the NOE experiments (see Chapter 7). The J_{IJ} coupling is new to this book but it is quite familiar to all NMR researchers. It occurs through chemical bonds, i.e. two nuclei interact through the paired electrons of covalent bonds. Its magnitude, for instance for protons separated by three bonds ($^3J_{IJ}$), depends on the dihedral angle H_1–X–Y–H_2 in a way similar to that described in Section 2.4 for the contact hyperfine coupling. Under the effect of J_{IJ} scalar coupling, coherence is transferred to the scalar-coupled partners. The meaning of *coherence transfer* is briefly recalled in Section 8.5. Magnetization transfer and coherence transfer may occur simultaneously depending on the system, and their separation is usually achieved through the so-called phase cycling.

2D experiments are devised in the assumption that the various times involved in the cycle of Fig. 8.1 (with the exception of t_m when present) are small with respect to the nuclear relaxation times. When the latter are short for any reason, e.g. in the case of paramagnetic molecules because of the presence of unpaired electrons, the system of spins may have reached the equilibrium, or almost reached the equilibrium, before the detection pulse. Under these circumstances no memory is left for the state of the spins during the preceding steps. As a consequence, cross peaks may be decreased in intensity until below detectability. It is necessary, therefore, to match all the time intervals with the nuclear relaxation times, in order to detect the maximum possible cross peak intensities. The ideal case is that t_1

and t_2 are as short as possible, compatibly with spectral resolution, and in any case not much longer than T_2. Indeed, in a 2D experiment, coherences decay with time constant T_2 during t_1 and t_2, as it happens during acquisition (t_2) in 1D experiments. As far as the mixing time is concerned, it is again related to nuclear relaxation times in a fashion which will be described experiment by experiment (this consideration does not apply to COSY experiments, for which there is not a formal mixing time). Some gain can be obtained by a more rapid cycling because (1) the spin systems reach equilibrium quite fast, and (2) the overall time required by Fig. 8.1 is much shorter. The first 2D experiment on a system with T_1 in the range between 50 and 100 ms was an EXSY experiment (see later) which appeared in 1984 [1]. In 1985 the first NOESY appeared on a pseudotetrahedral nickel(II) complex [2]; in 1988 the first COSY appeared on a lanthanide complex [3,4]; in 1990 the first TOCSY experiment appeared on a five-coordinated nickel(II) complex [5]. Since then, the reports have been numerous, aiming also at the detection of cross peaks between signals with shorter and shorter T_1 values.

It has been mentioned in Section 7.3, and it was implicit all over Chapter 7, that a finite time is required to achieve selective saturation or inversion of a signal by a soft pulse, during which time polarization starts to be exchanged, causing non-linearity of the response (see also Section 9.3). It should be stressed that this is not the case in all common 2D experiments based on non-selective pulses, which have durations of the order of microseconds instead of milliseconds, as required for selectivity. Selectivity in 2D experiments is intrinsic because of the double frequency labeling along f_1 and f_2.

We are now going to describe some 2D experiments performed on paramagnetic compounds (Fig. 8.2). We should stress that relatively few laboratories are engaged in making 2D spectroscopy suitable for the investigation of this type of compounds. Therefore, whereas the literature is richer and richer in suitable sequences for the various cases in diamagnetic systems, fewer applications are devoted to paramagnetic systems. The following examples will give an overview of the basic achievements which have represented a breakthrough in recent years.

8.2 THE EXSY EXPERIMENT

The 2D exchange spectroscopy (EXSY) is just an extension of the 1D saturation transfer experiment (Section 4.3.4). The simplest pulse sequence used to obtain EXSY spectra is the NOESY sequence reported in Fig. 8.2A. The first 90° pulse, constituting the preparation period of Fig. 8.1, tilts the z-magnetization along, e.g., the x axis in the rotating frame x' (Fig. 8.3); the variable time interval t_1 follows, during which the spins are labeled according to their frequency. This time must be short with respect to T_2, otherwise not enough in-plane magnetization remains to play with. Another 90° pulse is then applied which sends along the z axis the component of the magnetization that is along x' at the end of the

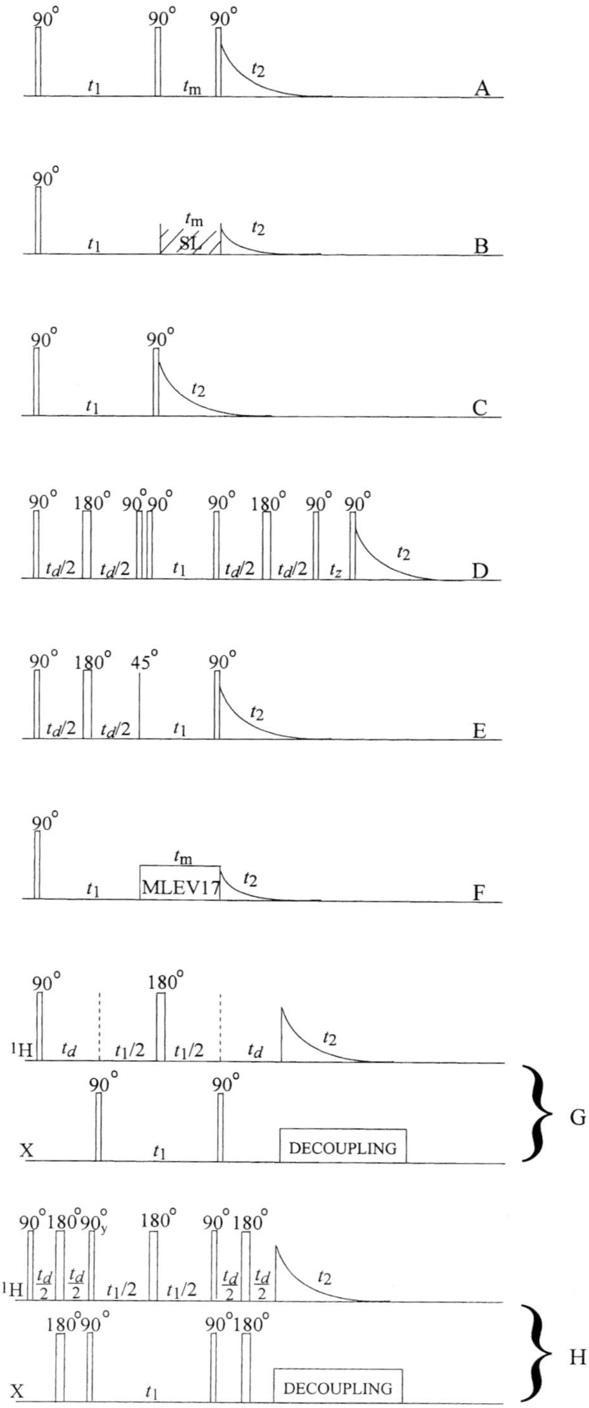
90°
90°
90°
t1
tm
t2
A
90°
tm
t1
SL
t2
B
90°
90°
t1
t2
C
90° 180° 90° 90° 90° 180° 90° 90°
td/2 td/2 t1 td/2 td/2 tz t2
D
90° 180° 45° 90°
td/2 td/2 t1 t2
E
90°
tm
t1
MLEV17
t2
F
90° 180°
1H td t1/2 t1/2 td t2
90° 90°
X t1 DECOUPLING
G
90° 180° 90y 180° 90° 180°
1H td/2 td/2 t1/2 t1/2 td/2 td/2 t2
180° 90° 90° 180°
X t1 DECOUPLING
H

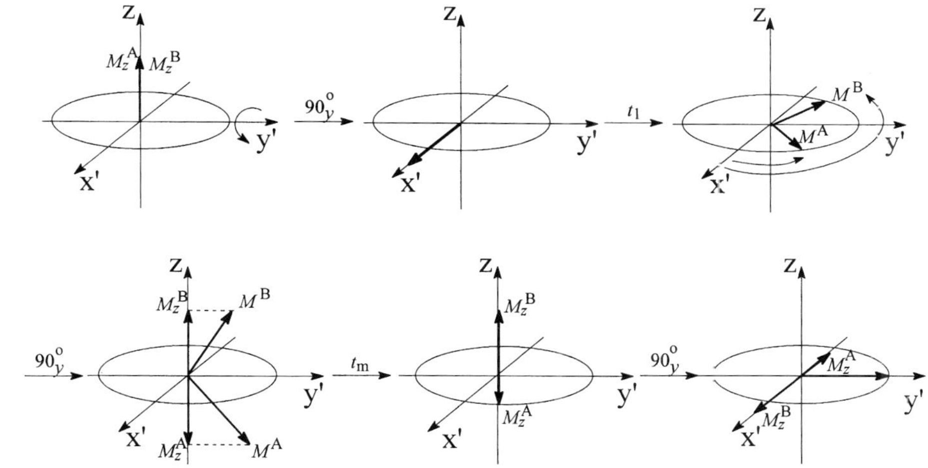

Fig. 8.3. Vector representation of an EXSY (or NOESY) experiment. The first 90° pulse along y' rotates the equilibrium magnetization of the A and B spins from the z axis to the x' axis. During t_1, the two transverse coherences precess in the $x'y'$ plane at their characteristic frequencies, and their x' (and y') components result periodically in and out of phase. The second 90° pulse tilts the two x' components along the z axis. During t_m, magnetization transfer occur, to an extent that depends on how out of phase the two vectors were just before the second 90° pulse (i.e. depending on the value of t_1). The third pulse tilts the two z components, whose intensities have been altered during t_m in a t_1-dependent manner, along x'. Detection can then occur during t_2. In the presence of magnetization transfer, the intensity of (for example) the A signal detected during t_2 is thus modulated not only by its characteristic frequency but also by the frequency of the B signal. A cross peak is thus generated.

evolution period. At this point, during the fixed mixing time t_m, the nuclei of a given chemical species (e.g. A) transfer magnetization to the corresponding nuclei of the other species in chemical equilibrium (e.g. B), to an extent that is modulated by the difference in their z-components. After the time t_m, a 90° detection pulse is applied, which brings back to the xy plane the magnetization of the nuclei of the species in chemical equilibrium. The time t_m ideally should be much longer than the exchange time τ_M and much shorter than the T_1 values of the signals. If the

---

Fig. 8.2. Some of the most common 2D pulse sequences that can be employed using a proper choice of parameters to record 2D spectra of paramagnetic molecules: (A) NOESY, (B) ROESY, (C) COSY, (D) ISECR COSY, (E) zero-quantum (double quantum) COSY, (F) TOCSY, (G) HMQC, (H) HSQC. Sequences (A), (B) and (F) are also used to obtain EXSY spectra. SL indicates a soft spin-lock sequence, while MLEV17 indicates a train of spin-locking hard pulses that optimizes the development of J_{IJ} coupling. In the reverse heteronuclear experiment (G) the upper and lower levels refer to ^{1}H and heteronucleus, respectively. The phase cycles are not indicated. For clarity of discussion, all initial pulses can be thought to be applied along the y' axis, in such a way that the coherence after the first 90° pulse is always along x'.

exchange time is longer than T_1, then the best compromise is to set $t_m \cong T_1$. The maximal information content in EXSY experiments is in the first t_1 and t_2 points. Generally, NMR spectroscopists apply a weighting function to the FID in order to optimize the response. Therefore, in this case we multiply the FID by a weighting function of cos- or cos^2-type in order to give more 'weight' to the first points.

We recall here that the phases of the pulses in Fig. 8.2A sequence, as well as of all other pulse sequences shown in Fig. 8.2 and described later in this chapter, must be properly cycled to achieve selection of the desired connectivities and suppression of artifacts and of other connectivities due to different types of interactions. The criteria to choose the appropriate phase cycling do not depend on the presence of a paramagnetic center in the molecule, and the reader should refer to the many publications on multidimensional NMR for details.

The first EXSY experiment on a paramagnetic system [1] described the chemical exchange between partially reduced species of a cytochrome c containing four hemes (Fig. 8.4). The electron exchange time could be estimated from the relative volumes of the cross peaks. The latter can be determined through numerical integration from the experimental signal intensities over a suitable area of the spectrum. Since the cross peak volume increases with the exchange rate, k_{-1} $(= \tau_M^{-1}$, as defined in Eq. (4.25)), and decreases with $(\rho^{\mathrm{A}} + \rho^{\mathrm{B}})/2$ (A and B being the two species in chemical exchange), in analogy with Eq. (7.19), τ_M can in principle be determined from a single 2D experiment:

$$\begin{aligned} a_{\text{cross}}(t_m) &= \frac{M_{0J}k_{-1}}{2D} \exp[-(\rho - D)t_m][1 - \exp(-2Dt_m)] \\ &= \frac{M_{0J}k_{-1}}{2D} \{\exp[-(\rho - D)t_m] - \exp[-(\rho + D)t_m]\} \end{aligned} \tag{8.1}$$

where M_0 is the intensity of the diagonal peak at zero t_m and

$$D = \tfrac{1}{2}\left[(L^A - L^B)^2 + 4k_{-1}k_1\right]^{1/2} \qquad \rho = \frac{(L^A + L^B)}{2}$$

$$L^A = -\rho^A - k_1 \qquad L^B = -\rho^B - k_{-1}.$$

However, if the aim is only that of measuring τ_M, it is more straightforward to perform a single steady state 1D experiment, as explained in Section 4.3.4. An EXSY experiment is shown in Fig. 8.5, relative to the complex praseodymium diethylenetriaminepentaacetate ($\mathrm{Pr(DTPA)^{2-}}$) [3,4]. The complex undergoes chemical exchange between two conformational isomers. The T_1 values of the signals are around 30 ms.

EXSY experiments have the advantage of displaying cross peaks between many signals belonging to two or more complex species. In Fig. 8.6, the EXSY spectrum of a ferredoxin containing two Fe_4S_4 clusters is shown [6]. Here, the cross peaks connect the fully reduced species containing two $Fe_4S_4^+$ clusters

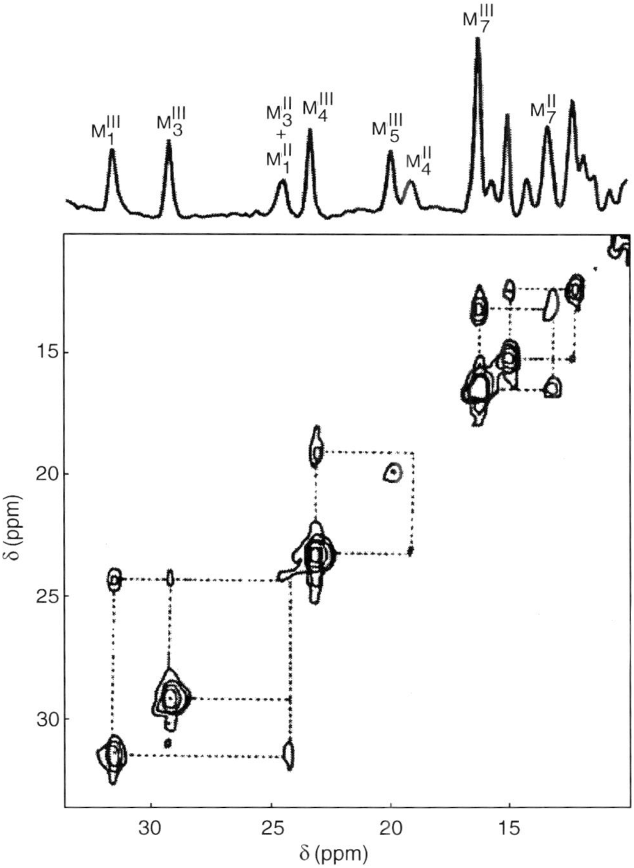

Fig. 8.4. The first EXSY experiment on a paramagnetic system [1]: the 300 MHz spectrum, taken with mixing time of 50 ms, shows species in chemical exchange belonging to two different redox states of a cytochrome c_3, a protein containing four low spin hemes. The signals marked M_1–M_7 represent various heme methyl groups. EXSY cross peaks are observed between M_i of two species containing two (II) or three (III) oxidized hemes, respectively.

with the intermediate species containing two '$Fe_4S_4^{1.5+}$' species and the oxidized species containing two $Fe_4S_4^{2+}$ species. The intermediate species actually contains one oxidized and one reduced cluster, which, however, are in very fast exchange so that, as far as NMR is concerned, only one intermediate species exists. It is noteworthy that we can detect as many as 20 signals of the intermediate species showing two cross peaks connecting them with those of the other two species. One of these patterns is exemplified in Fig. 8.6. Signal (b), arising from the $Fe_4S_4^{1.5+}$ species, is connected to both signal (a), belonging to the fully reduced species, and signal (c), belonging to the fully oxidized species. Signals (a) and (c) are also connected by a 'two-step' exchange cross peak, (a)–(c), whose nature

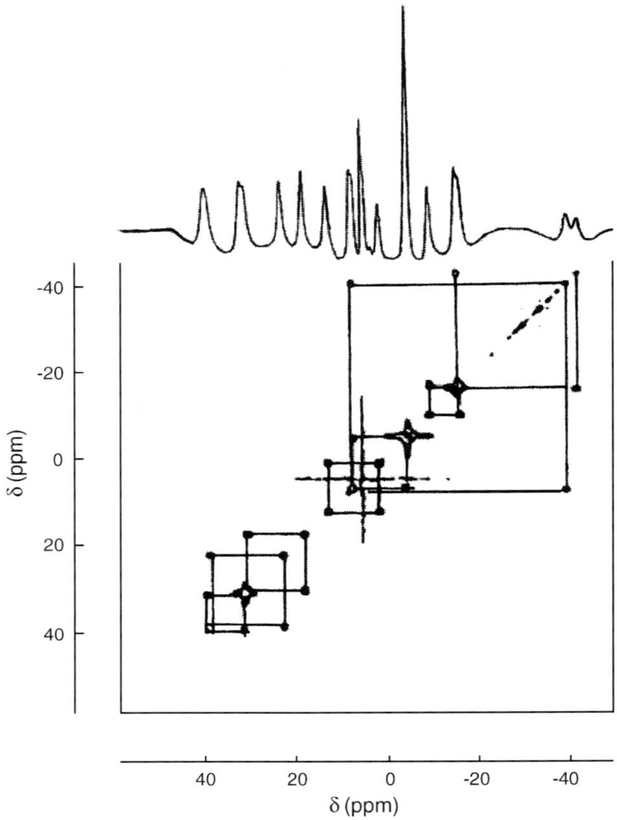

Fig. 8.5. 500 MHz EXSY spectrum of $Pr(DTPA)^{2-}$ recorded with the pulse sequence of Fig. 8.2A [4]. The spectrum demonstrates the presence of an equilibrium between two conformational isomers.

is conceptually related to the spin diffusion mechanism described in Section 7.2.3.

EXSY cross peaks are also obtained in TOCSY experiments (see later) because scalar interactions in the rotating frame are not separable from exchange interactions [7]. An EXSY experiment, performed using a TOCSY sequence (see Section 8.6) is reported in Fig. 8.7 relative to the complex 5Cl–Ni–SAL–MeDPT [5]. This complex, as shown in Fig. 8.8, displays a chemical equilibrium in which the two salicylaldiminate moieties exchange their non-equivalent positions [8]. It is interesting to learn that such complex interconversion occurs with times of the order of the spin-lock time (20 ms) or shorter.

As EXSY cross peak intensities can be as strong as the diagonal peak intensities in favorable cases, EXSY experiments can be performed with relative ease also on nuclei other than protons.

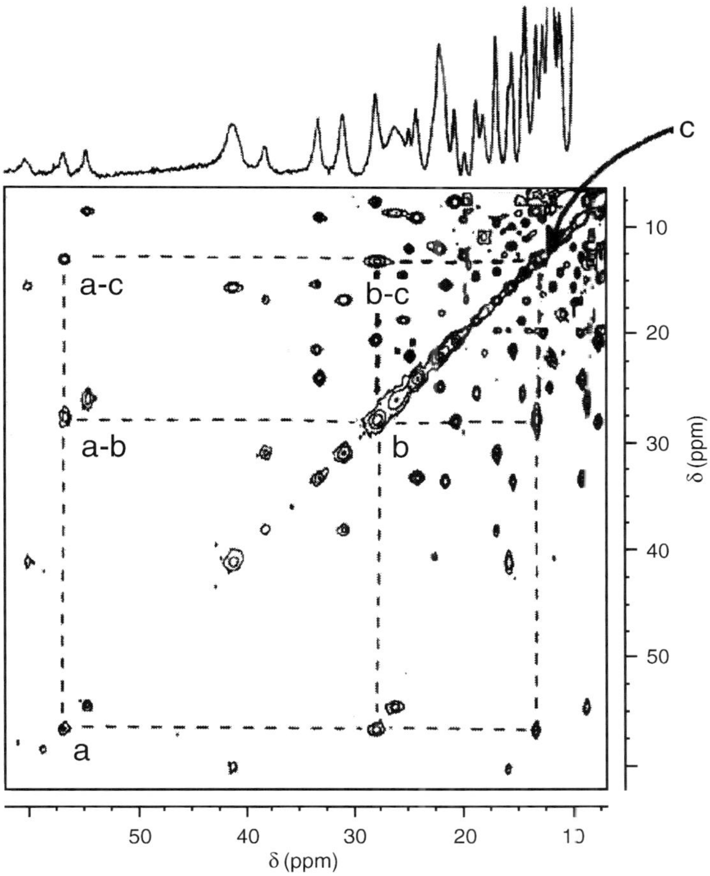

Fig. 8.6. 600 MHz EXSY spectrum of the intermediate reduction product of the $2Fe_4S_4$ ferredoxin from *C. pasteurianum*. Most signals of the intermediate species display cross peaks from both the fully reduced and the fully oxidized species. The sequence used is that in Fig. 8.2A. $t_m = 5$ ms. A pattern belonging to a single signal exchanging among the three species ((a) fully reduced; (b) half reduced; (c) fully oxidized) is highlighted as an example.

EXSY experiments are also sometimes performed with ROESY sequences (see Section 8.4).

8.3 THE NOESY EXPERIMENT

NOESY experiments deal with dipolar interactions between nuclei. Successful experiments are easily planned for nuclei with large magnetic moments like protons, or for heteronuclei when the dipole–dipole interaction is very strong.

The basic pulse sequence for the NOESY experiment is just that illustrated

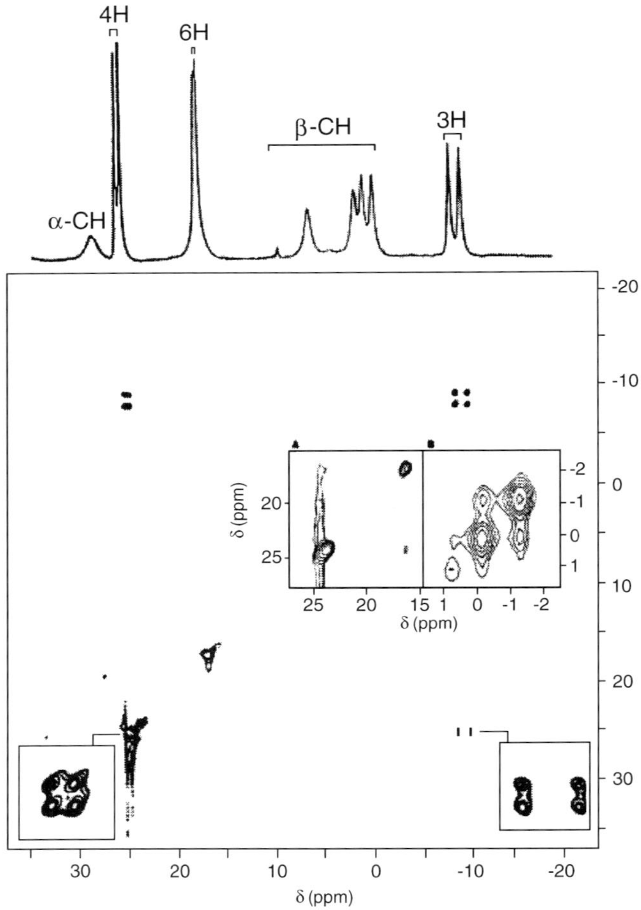

Fig. 8.7. TOCSY spectrum of the complex 5Cl–NiSAL–MeDPT (scheme in Fig. 8.8) showing exchange cross peaks between the two 4H ($T_1 = 5$ ms) and between the two 3H ($T_1 = 37$ ms) salicylaldiminate ring protons. The spin-lock time was 20 ms.

in Fig. 8.2A. In the mixing period t_m magnetization transfer occurs between sets of nuclei I and J which are dipole-coupled. After any given t_1 value and the second 90° pulse, the magnetization of the two sets of spins will be along z with different components. The situation is the same as that illustrated for the EXSY experiment of Fig. 8.3, if one substitutes the labels A and B for the two chemically exchanging sites with I and J for the two sets of nuclei. Just like in a transient NOE experiment, cross relaxation between the two sets of nuclei occurs. In order to predict the intensity of cross peaks, reference should be made to Fig. 7.8 and Eq. (7.19), provided it is considered that positive NOE corresponds to negative NOESY cross peaks and vice versa. This is not surprising if it is remembered

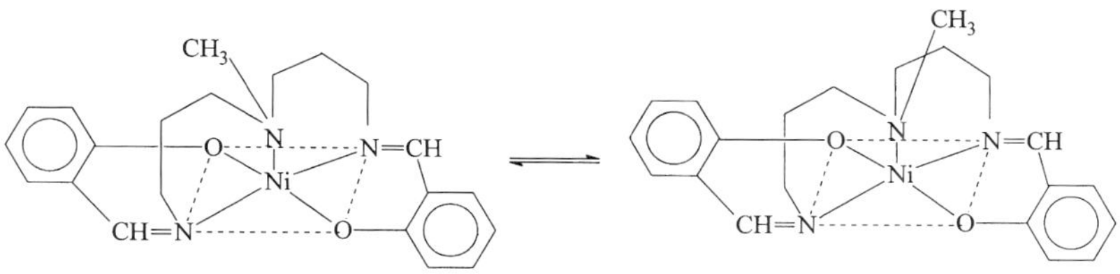

Fig. 8.8. Conformational equilibrium displayed by 5Cl–Ni–SAL–MeDPT. The two halves of the molecule are non-equivalent owing to the CH_3 substituent on the apical nitrogen. The equilibrium involves two equivalent conformations differing in the chirality of the apical nitrogen [8].

that when the NOE is negative it has the same sign of the irradiated peak in the difference spectra. By referring to Fig. 7.8B, and keeping in mind the reversal in sign, when τ_r is small, the NOESY cross peaks are small and negative (i.e. of opposite sign with respect to the diagonal peaks). With increasing τ_r they increase in absolute value and then decrease again until the zero value of the cross peak intensity is obtained. Then, the absolute intensities of positive cross peaks increase. This explains why NOESY experiments are best performed on macromolecules, unless very large concentrations can be reached. The best t_m is of the order of T_1. In the case in which the cross peak occurs between signals with sizably different T_1 values t_m can be optimized according to the following equation:

$$\boxed{t_\mathrm{m}^\mathrm{opt} = \frac{1}{2D}\ln\left(\frac{\rho - D}{\rho + D}\right)} \tag{8.2}$$

where the symbols have the same meaning as in Eq. (7.19). A table is also provided for the reader's convenience (Table 8.1), which reports the best mixing time values $t_\mathrm{m}^\mathrm{opt}$ for each combination of T_1^I and T_1^J (ρ_I^{-1} and ρ_J^{-1}) values for the pair [9]. Note that when the two T_1 values are very different, $t_\mathrm{m}^\mathrm{opt}$ is more than four times longer than the shorter T_1 value. As in EXSY experiments, the maximal information content is in the first t_1 and t_2 points. Again, the most used weighting functions are cos- or $\cos^2$-type.

NOESY experiments may sometimes be a hard task for small molecules, because in the fast motion limit cross relaxation is small and so is the NOE. In Table 8.2 the calculated ^{1}H–^{1}H maximal NOESY intensities are reported for a range of τ_r values and R_{1M} values (taken equal for the two spins). From the τ_r values the corresponding molecular weights are calculated through the Stokes–Einstein equation (Eq. (3.8)) for $\eta = 1$. The mixing time is always equal to ρ_I ($= \rho_{I(J)} + R_{1M}$) to maximize the effect. The NOEs are reported for two magnetic fields and for two proton–proton distances of 1.8 and 3.0 Å. The purpose of this table is to show under which circumstance NOESY spectra of paramagnetic molecules can be attempted. The crossing point between fast and slow motion

TABLE 8.1
Optimal homonuclear NOESY mixing times t_m^{opt} for various T_1 (ρ^{-1}) values (ms) of the two spins I and J, calculated using Eq. (8.2). Cross peaks intensities (%) for a $\sigma_{I(J)}$ value of $-1\ s^{-1}$ are shown in parentheses [9]

T_1^I	T_1^J							
	128	64	32	16	8	4	2	1
128	128 (4.72)	89.0 (3.20)	59.2 (2.02)	38.0 (1.19)	23.7 (0.67)	14.3 (0.36)	8.5 (0.19)	4.9 (0.10)
64	89.0 (3.20)	64.0 (2.36)	44.4 (1.60)	29.6 (1.01)	19.0 (0.59)	11.8 (0.33)	7.2 (0.18)	4.2 (0.09)
32	59.2 (2.02)	44.4 (1.60)	32.0 (1.18)	22.2 (0.80)	14.8 (0.50)	9.5 (0.30)	5.9 (0.17)	3.6 (0.09)
16	38.0 (1.19)	29.6 (1.01)	22.2 (0.80)	16.0 (0.59)	11.1 (0.40)	7.4 (0.25)	4.8 (0.15)	3.0 (0.08)
8	23.7 (0.67)	19.0 (0.59)	14.8 (0.50)	11.1 (0.40)	8.0 (0.29)	5.5 (0.20)	3.7 (0.13)	2.4 (0.07)
4	14.3 (0.36)	11.8 (0.33)	9.5 (0.30)	7.4 (0.25)	5.5 (0.20)	4.0 (0.15)	2.8 (0.10)	1.8 (0.06)
2	8.5 (0.19)	7.2 (0.18)	5.9 (0.17)	4.8 (0.15)	3.7 (0.13)	2.8 (0.10)	2.0 (0.07)	1.4 (0.05)
1	4.9 (0.10)	4.2 (0.09)	3.6 (0.09)	3.0 (0.08)	2.4 (0.07)	1.8 (0.06)	1.4 (0.05)	1.0 (0.04)

regimes occurs at MW $\approx$ 800 at 600 MHz and at MW $\approx$ 2500 at 200 MHz. The maximal intensities in the fast motion regime are always small in absolute value and therefore the observability of cross peaks is predictable but critical. The researcher dealing with low molecular weight compounds may either use low fields or increase τ_r by increasing solvent viscosity and use the highest possible field.

As discussed in Section 7.2.3, spin diffusion is not a problem in small complexes. In macromolecules, the problems are negligible when the nuclear R_{1M} values are much larger than cross relaxation. Such problems become increasingly significant when the R_{1M} values get smaller. Under these circumstances it may be useful to perform the NOESY experiments at different mixing times. At short mixing times (short relative to T_1) only primary NOEs are detected whose intensities are proportional to σ. At increasing mixing time, the build up of these signals can be followed and new cross peaks may appear. In macromolecular systems the best conditions for measuring NOESY cross peaks are met, because τ_r is long (see Table 8.2). In Fig. 8.9 the NOESY spectrum of an Fe_4S_4 protein of MW 6000 is shown where cross peaks are apparent between signals with T_1 values of a few milliseconds [10]. With mixing time of the order of few milliseconds NOESY cross peaks are observed between signals with short T_1, whereas NOESY cross peaks are observed between signals with longer T_1 values if longer mixing times are used. The observation of cross peaks between fast relaxing and slow relaxing

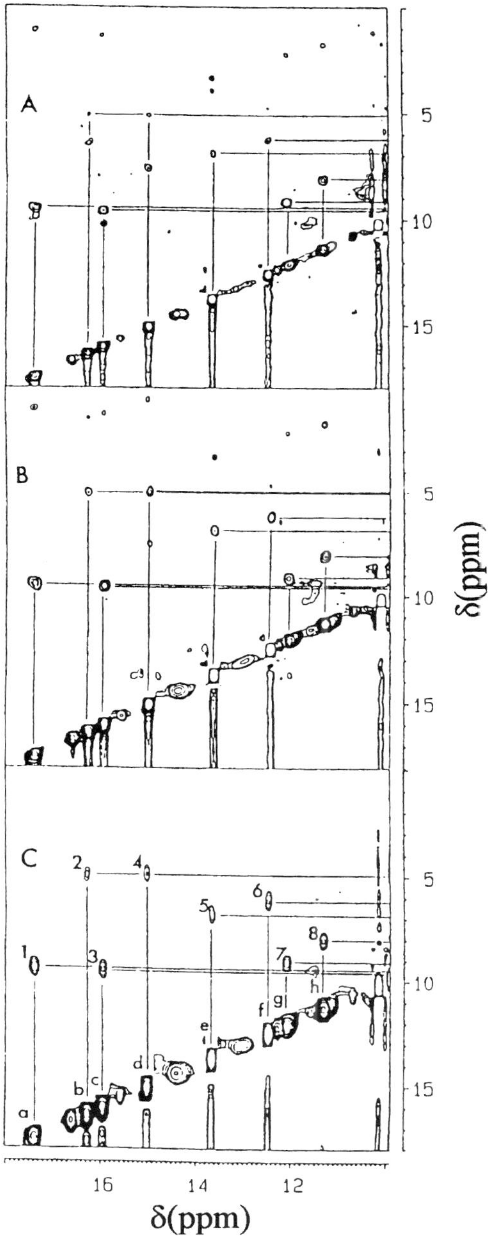

Fig. 8.9. NOESY spectra of the oxidized form of $2(Fe_4S_4)$ ferredoxin from *C. pasteurianum* taken using mixing times of 20 (A), 10 (B) and 5 (C) ms. Signals (a)–(h) belong to β-CH protons of cluster-coordinated cysteines and cross peaks 1–8 connect them to their geminal partners, which lie in the diamagnetic region of the spectrum. The other cross peaks arise from α-CH protons of the cysteines and from other nearby residues. The cross peaks between the fastest relaxing signals (β-CH protons with their geminal partners) are maximal at the shortest mixing time, whereas those between the former and more slowly relaxing signals increase at longer mixing times [10].

TABLE 8.2
Calculated maximal intensities of NOESY cross peaks (%)[a] for various values of τ_r and R^I_{1M} ($= R^J_{1M}$). For each τ_r and R^I_{1M} value two NOE's are calculated for interproton distances of 1.8 and 3.0 Å and for nuclear Larmor frequencies ν_0 of 200 and 600 MHz

τ_r (s) [MW (Da)]	r_{IJ} (Å)	R^I_{1M} (s^{-1}) 1 ν_0 (MHz) 200	1 600	2 ν_0 (MHz) 200	2 600	4 ν_0 (MHz) 200	4 600	8 ν_0 (MHz) 200	8 600
10^{-11}	1.8	−2.640	−2.626	−1.421	−1.413	−0.739	−0.735	−0.377	−0.375
[25]	3.0	−0.143	−0.142	−0.072	−0.071	−0.036	−0.036	−0.018	−0.018
$10^{-10.5}$	1.8	−6.366	−6.089	−3.835	−3.647	−2.138	−2.025	−1.135	−1.072
[80]	3.0	−0.440	−0.416	−0.223	−0.210	−0.112	−0.106	−0.056	−0.053
10^{-10}	1.8	−11.236	−8.248	−8.045	−5.587	−5.141	−3.400	−2.989	−1.909
[250]	3.0	−1.244	−0.772	−0.644	−0.396	−0.328	−0.200	−0.165	−0.101
$10^{-9.5}$	1.8	−11.920	0.890	−9.679	0.616	−7.045	0.381	−4.568	0.216
[800]	3.0	−2.098	0.088	−1.127	0.045	−0.585	0.023	−0.298	0.012
10^{-9}	1.8	2.343	19.647	1.934	14.379	1.434	9.406	0.945	5.581
[2500]	3.0	0.441	2.366	0.238	1.233	0.124	0.630	0.063	0.319
$10^{-8.5}$	1.8	27.084	33.735	23.302	28.378	18.275	21.644	12.831	14.784
[8000]	3.0	6.464	7.238	3.619	4.005	1.926	2.118	0.996	1.091
10^{-8}	1.8	38.951	39.972	36.394	37.299	32.208	32.938	26.264	26.779
[25000]	3.0	16.387	16.633	10.331	10.459	5.969	6.033	3.244	3.275
$10^{-7.5}$	1.8	42.060	42.175	41.073	41.183	39.238	39.339	36.040	36.127
[80,000]	3.0	28.398	28.455	21.347	21.383	14.373	14.393	8.754	8.764
10^{-7}	1.8	42.884	42.896	42.554	42.565	41.909	41.920	40.678	40.689
[250000]	3.0	37.050	37.059	32.481	32.488	26.145	26.150	18.928	18.931

[a] The largest theoretical cross peak intensity for a NOESY experiment is 50%, at variance with a transient NOE experiment (Section 7.4). To predict the maximal values for transient NOE experiments the values of this table should be multiplied by two and the sign should be reversed. Cf. also Table 7.1 for steady state NOE experiments.

signals can be accomplished with mixing times appropriately chosen according to Table 8.1. The difference in T_2, which often parallels the difference in T_1, provides elliptical cross peaks (as shown in Fig. 8.10A). The maximal value of t_1 ($t_1^{\max}$) is best chosen to be equal to or less than one of the two T_2 values, and the maximal value of t_2 ($t_2^{\max}$) to be equal to or less than that of the other. Under these circumstances, one of the two cross peaks on one side of the diagonal is enhanced whereas the other is sacrificed (Fig. 8.10B,C). The difficulty, which is a limit of the technique, is that it is often not possible to see in the same 2D map cross peaks arising from signals with strongly different T_1 and T_2 values. If reference is made to the system of Fig. 8.9 we see that only some cross peaks are evident between β-CH_2 protons of cysteines and between β-CH_2 and β-CH protons. If some other cross peaks are observed between β-CH_2 and other more diamagnetic protons, then cross peaks between the latter and diamagnetic protons are absolutely missing. When we increase the mixing time in order to detect the latter part, the cross peaks with the β-CH_2 are sizably reduced in intensity and will be eventually lost. The procedure used up to now is that of firmly recognizing signals farther from the paramagnetic center to which the β-CH protons are dipolarly connected (mostly

R_{1M}^{I} (s^{-1})									
16 ν_0 (MHz)		32 ν_0 (MHz)		64 ν_0 (MHz)		128 ν_0 (MHz)		256 ν_0 (MHz)	
200	600	200	600	200	600	200	600	200	600
−0.190	−0.189	−0.096	−0.095	−0.048	−0.048	−0.024	−0.024	−0.012	−0.012
−0.009	−0.009	−0.004	−0.004	−0.002	−0.002	−0.001	−0.001	−0.001	−0.001
−0.585	−0.552	−0.297	−0.280	−0.150	−0.141	−0.075	−0.071	−0.038	−0.036
−0.028	−0.027	−0.014	−0.013	−0.007	−0.007	−0.004	−0.003	−0.002	−0.002
−1.628	−1.017	−0.852	−0.526	−0.436	−0.267	−0.221	−0.135	−0.111	−0.068
−0.083	−0.051	−0.042	−0.025	−0.021	−0.013	−0.010	−0.006	−0.005	−0.003
−2.685	0.116	−1.472	0.060	−0.774	0.031	−0.397	0.015	−0.201	0.008
−0.151	0.006	−0.076	0.003	−0.038	0.001	−0.019	0.001	−0.010	0.000
0.562	3.084	0.310	1.629	0.164	0.838	0.084	0.425	0.043	0.214
0.032	0.160	0.016	0.080	0.008	0.040	0.004	0.020	0.002	0.010
8.080	9.111	4.657	5.176	2.525	2.782	1.319	1.446	0.674	0.738
0.506	0.554	0.255	0.279	0.128	0.140	0.064	0.070	0.032	0.035
19.296	19.610	12.702	12.872	7.591	7.677	4.220	4.262	2.238	2.259
1.697	1.712	0.869	0.876	0.440	0.443	0.221	0.223	0.111	0.112
31.041	31.109	24.401	24.446	17.209	17.235	10.909	10.923	6.332	6.339
4.932	4.937	2.638	2.640	1.367	1.368	0.696	0.697	0.352	0.352
38.432	38.442	34.639	34.647	28.998	29.004	21.987	21.991	14.933	14.935
12.291	12.293	7.268	7.269	4.012	4.012	2.118	2.119	1.090	1.090

through 1D NOE), and then try to relate them to known dipolar patterns observed in the diamagnetic region of NOESY spectra recorded with longer t_m, t_1^{max} and t_2^{max}.

From comparison of Table 8.2 with Table 7.1 (or of Eq. (7.20) with Eq. (7.10)), i.e. of transient NOE or NOESY vs. steady state NOE intensities, it appears that the latter are superior under any circumstance. This superiority is striking if the intrinsic asymmetry of the steady state NOE with respect to the symmetry of transient NOE and NOESY experiments (Section 7.4) can be exploited, as in the case of irradiation of fast relaxing nuclei to detect NOE to slow relaxing nuclei. Of course, NOE experiments are advantageous over NOESY experiments if one is looking for dipolar connectivities from only a few specific signals.

Sometimes the shifts of a few signals are so large that it is inconvenient to include them into the spectral window because this will either decrease the resolution or, if resolution is maintained by increasing the data points in the two dimensions, cause cumbersome spectral manipulations. Then, it may be convenient to use a smaller spectral width and allow the excluded signals to appear as folded images, or 'ghost' signals. This can be easily accomplished in the f_1 dimension by increasing the time increment in t_1 between one experiment and the next [11]. It can also be accomplished in the f_2 dimension by likewise increasing the dwell time, provided the filter width is chosen large enough to

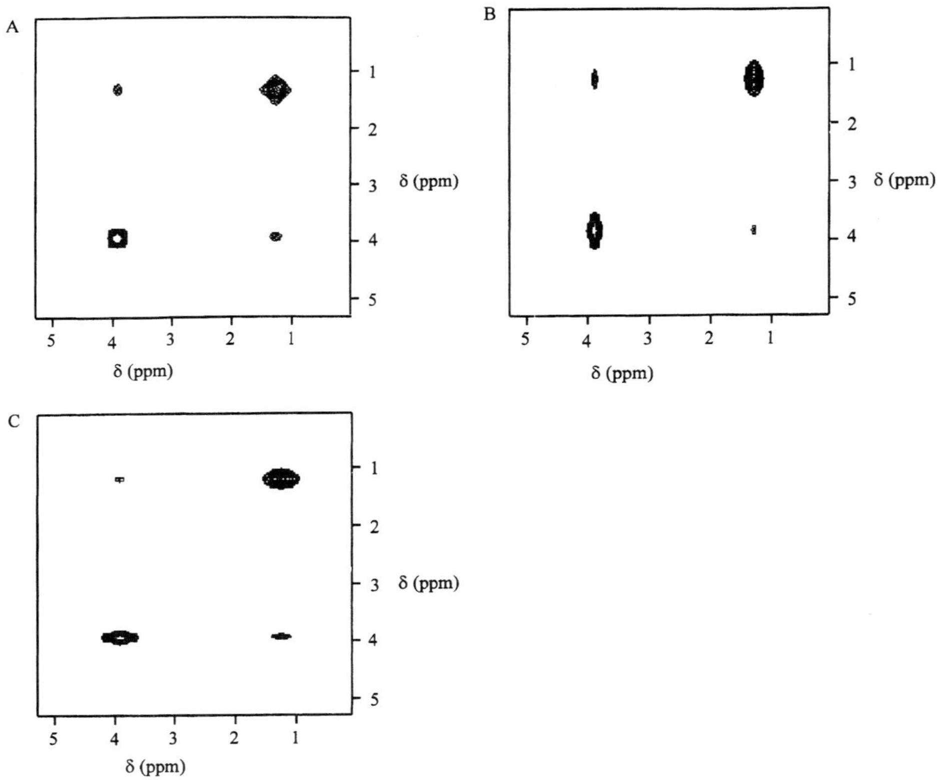

Fig. 8.10. Computer-simulated NOESY cross peaks between signals with different T_2 values. Conditions: $\rho_1^I = \rho_1^J = 40\ \mathrm{s}^{-1}$; $\rho_2^I = 50\ \mathrm{s}^{-1}$; $\rho_2^J = 200\ \mathrm{s}^{-1}$; $\sigma_{IJ} = 5\ \mathrm{s}^{-1}$. (A) $t_1^{\max} = t_2^{\max} = 0.01$ s; (B) $t_1^{\max} = T_2^I$, $t_2^{\max} = T_2^J$; (C) $t_1^{\max} = T_2^J$, $t_2^{\max} = T_2^I$. Note that the upper left cross peak has maximal intensity in case (B), the lower right cross peak has maximal intensity in case (C), whereas in case A the two intensities are equal and intermediate between the larger and the smaller of cases (B) and (C). For simplicity, $t_m = T_1^I = T_1^J$ has been taken. A $\cos^2$ weighting function has been applied.

include all the signals. Ghost signals give ‘ghost’ cross peaks. An example of this is illustrated in Fig. 8.11A, showing the 600 MHz NOESY spectrum of the four-cobalt cluster of Co_7MT, with MT indicating the protein metallothionein (MW $\approx$ 6000) [12].

Sometimes the signals are outside the maximal spectral width allowed by the ADC of the instrument. In this case the folded spectrum is a necessity. If an ADC of the latest generation (see Section 9.1) is not available, another possibility is that of using an ADC with smaller dynamic range (e.g. 12 bit instead of 16 bit) but larger spectral window (e.g. ca. 5–10 MHz instead of ca. 100–200 kHz). Fig. 8.11B shows the 600 MHz NOESY spectrum of the above protein complex [12] covering a spectral width of 250 kHz, recorded with a 12 bit ADC. At least

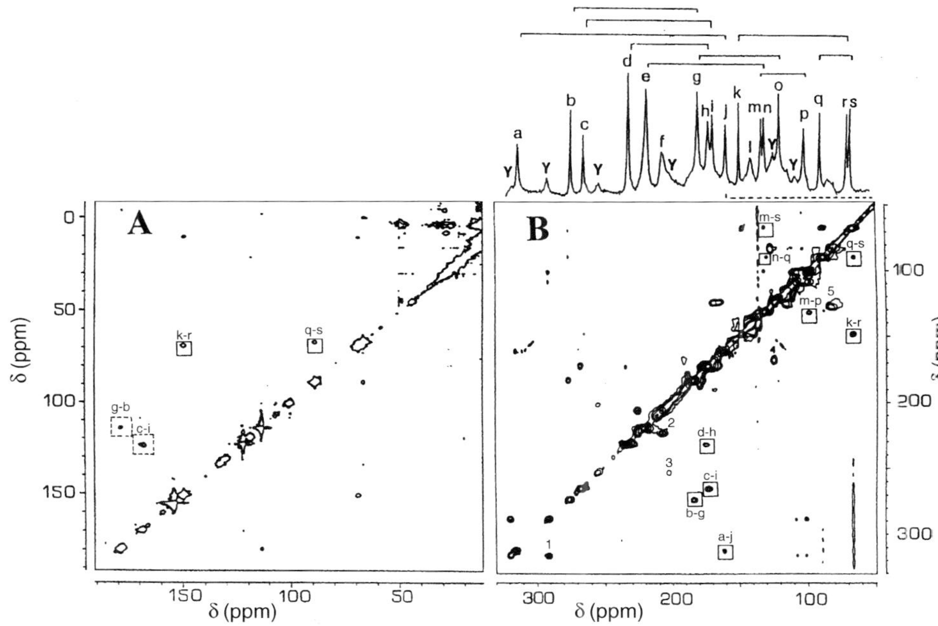

Fig. 8.11. 600 MHz NOESY spectra of the four-cobalt cluster of cobalt-substituted metallothionein. The spectra have been recorded with $t_m = 7$ ms using a specially built probe featuring a 90° pulse of 3.7 μs. (A) Spectrum obtained with a 125 kHz 16 bit ADC by folding the spectrum in both dimensions; cross peaks g–b and c–i are folded. (B) Spectrum recorded using a fast ADC over a 250 kHz spectral width. The nine observed connectivities between geminal protons of cobalt-coordinated cysteine β-CH_2 are summarized in the inset. Cross peaks marked with numbers are EXSY cross peaks to a minor species labeled Y in the inset [12].

nine out of the 11 expected cross peaks between geminal protons of cysteine β-CH_2 (see inset of Fig. 8.11) are observed. Some of the signals giving rise to cross peaks have T_1 values as short as 1 ms.

Of course, these considerations regarding the folded spectra and decreased dynamic range of the ADC hold for any kind of 2D spectroscopy. We have mentioned them here because such problems happen to have been afforded in the NMR literature on paramagnetic compounds in connection with NOESY spectra.

8.4 THE ROESY EXPERIMENT

The 2D ROE or ROESY experiment is an experiment to measure cross-relaxation in the rotating frame (Fig. 8.2B). After an initial 90° pulse and the variable evolution period t_1, a low power or 'soft' spin-lock sequence (SL) is applied for a time t_m during which magnetization transfer in the rotating frame occurs due to cross relaxation. Since scalar connectivities can also develop during spin lock, as in the case of TOCSY experiments (see Section 8.6), ROESY experiments show simultaneous occurrence of the two effects that are difficult to separate. One of the most convenient ways is to apply the SL off resonance, to avoid generation of TOCSY cross peaks [13,14]. Then, acquisition is performed, using weighting functions of cos or $\cos^2$ type. As shown in Fig. 7.10, the ROE is always positive, so that ROESY cross peaks are always negative (i.e. opposite to the diagonal) independently of the magnitude of τ_r. ROESY experiments are thus most useful for relatively small molecules where NOESY cross peak intensities may be close to zero ($\omega_I \tau_r \approx 1$, see Section 7.5). As in NOESY, magnetization transfer occurs through dipolar coupling, and applications are limited to protons. A comparison between a NOESY and a ROESY spectrum of a metalloprotein containing four heme groups is reported in Fig. 8.12 [15].

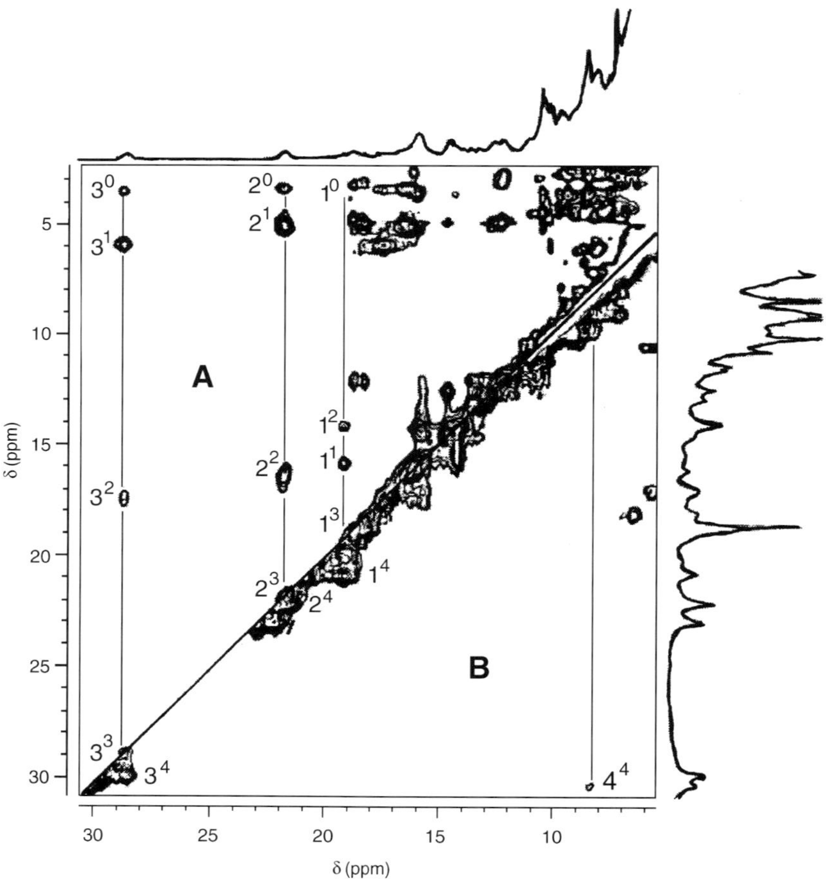

Fig. 8.12. NOESY (A) and ROESY (B) spectrum of a four heme cytochrome. The exchange cross peaks in the two spectra have the same sign and give information on different interactions [15].

TABLE 8.3

Signs of 1D difference spectra or 2D cross peaks arising from dipole–dipole or chemical exchange interactions in fast (FM) and slow (SM) motion regimes [a]

Experiment:	1D NOE		NOESY		1D ROE		ROESY	
Motion regime:	FM	SM	FM	SM	FM	SM	FM	SM
Dipole–dipole	+	–	–	+	+	+	–	–
Chemical exchange	–	–	+	+	–	–	+	+

[a] Signs of 1D experiments are referred to a negative irradiated signal; signs of 2D experiments are referred to a positive diagonal peak.

It should be noted that NOESY and ROESY pulse sequences also provide EXSY spectra, and therefore EXSY cross peaks may appear simultaneously in the 2D NOESY and ROESY spectra. EXSY cross peaks are always positive in both types of experiment, whereas dipolar cross peaks are negative in EXSY spectra independently of molecular weight and in NOESY spectra of small molecules. Therefore, in macromolecules the sign for NOESY and EXSY cross peaks is the same, and the two phenomena cannot be distinguished in NOESY experiments. In contrast, ROESY cross peaks have different sign from EXSY cross peaks and can be distinguished and even plotted selectively in ROESY experiments. These considerations are summarized in Table 8.3 for the reader's convenience.

The problem with spin-locking experiments, which is common to TOCSY (see Section 8.6), is that spin locking of signals over a wide range requires sizable r.f. power, which is sometimes not available or, if available, cannot be delivered to the sample because heat cannot be dissipated fast enough. However, the mixing times used for paramagnetic systems are generally shorter than for diamagnetic systems, so that the total energy (higher power $\times$ shorter time) delivered to the sample may be of the same order of that used in diamagnetic systems. Even so, temperature and lock instabilities may arise. This problem, together with the intrinsic higher sensitivity of NOESY with respect to ROESY makes the former a more common investigation tool. In the problem illustrated in Fig. 8.12 the spectral width is relatively small.

As an example of a system in the fast motion regime, a portion of the ROESY spectrum of a small lanthanide diporphyrin complex is shown in Fig. 8.13, together with portions of the NOESY and TOCSY spectra [16]. The subspectra refer to phenyl rings attached in the meso position to one of the two coordinated porphyrins (Fig. 8.13, inset). The flip rate of the rings is slow on the chemical shift scale but fast on the relaxation time scale. Therefore, separate signals are observed for the exo and endo ortho and meta ring protons. Strong and positive (Table 8.3) EXSY cross peaks are apparent in ROESY and NOESY spectra (as well as in the TOCSY spectrum, see Section 8.6), whereas all dipole–dipole cross peaks are negative in the ROESY and NOESY spectra. In the latter, only one dipole–dipole

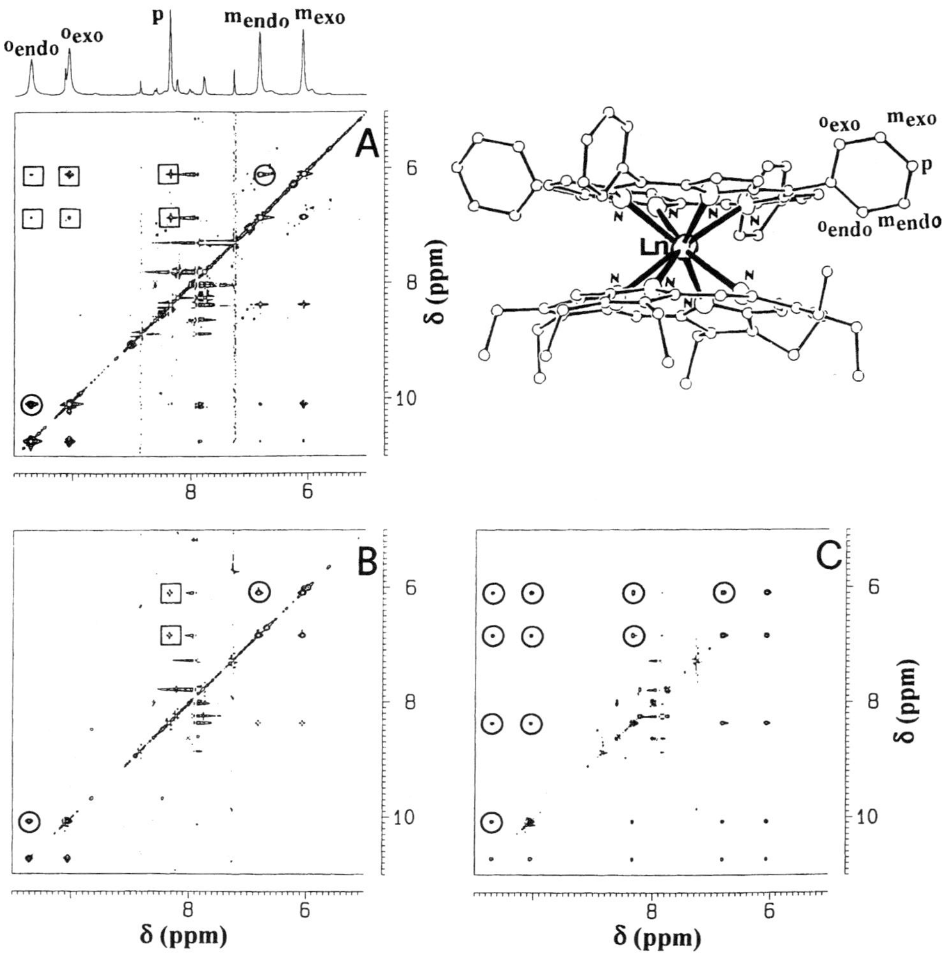

Fig. 8.13. ROESY (A), NOESY (B) and TOCSY (C) cross peaks observed in the phenyl ring 2D pattern of the complex shown in the inset ($Ln = Yb^{3+}$). Negative cross peaks are highlighted by squares, and positive cross peaks by circles [16].

cross peak is observed, on account of the rotational correlation time of the system being close to the null point for the NOE interaction [16] (see also Sections 7.2 and 7.5).

8.5 THE COSY EXPERIMENT

The COSY experiment is the most familiar to 2D NMR spectroscopists. The cross peaks connect protons which are coupled by scalar interactions. Under these

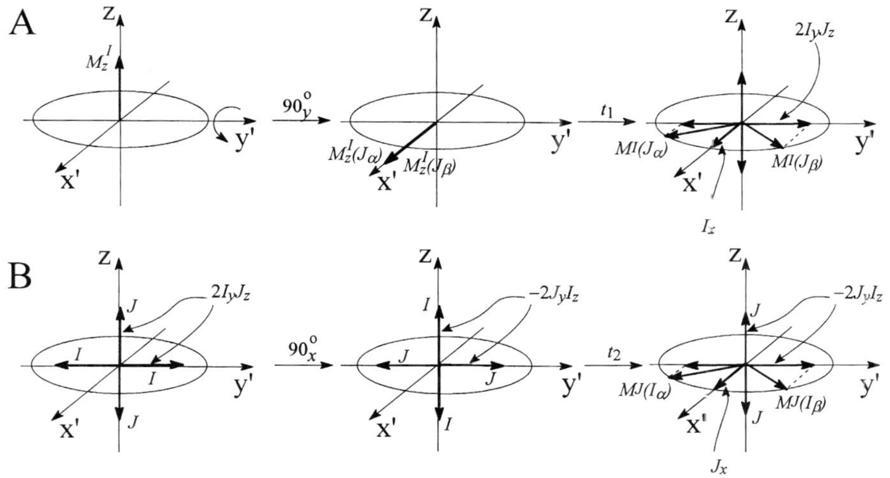

Fig. 8.14. Vector representation of a COSY experiment. The first 90° pulse along y' rotates the equilibrium magnetization of the I and J spins from the z axis to the x' axis. During t_1, the transverse doublet components of, e.g., signal I (assumed to precess at the rotating frame frequency) separate. Their x' projections are in phase, while their y' projections are in antiphase. Antiphase coherence is associated with the two opposite components of the J spin (hence the label $2I_yJ_z$ for the resulting four-vector pattern). The second 90° pulse along x' interchanges the I and J spins, producing a $-2J_xI_z$ antiphase product operator. During t_2 the latter generates a detectable in-phase coherence of the J spin, whose initial intensity had been modulated by the I spin Larmor frequency during t_1, thus originating a cross peak.

conditions each signal of a two-spin system I–J is split into two components, whose separation in frequency is determined by the scalar coupling constant J_{IJ}. The simplest COSY pulse sequence is that reported in Fig. 8.2C. There is an initial 90° pulse followed by the evolution time t_1, during which antiphase coherence of the scalar-coupled spins builds up. The concept of antiphase coherence is recalled in Fig. 8.14. The coherence of the two components of (for example) the I doublet, rotates in the xy plane at different frequency. In a frame rotating at the frequency of the center of the doublet, the two components will rotate in opposite directions with frequency $\pm J_{IJ}/2$. The two in-plane components arise from I spins whose scalar coupled partners have opposite M_z components (α *and* β). The projections of the two I components on the rotating frame axis on which the magnetization was tilted by the first 90° pulse (e.g. the x' axis) oscillate in phase (Fig. 8.14A) while those on the y' axis oscillate in antiphase. The in-phase and antiphase projections are called single quantum in phase and single quantum antiphase coherences. In the product operator language they are indicated as I_x and $2I_yJ_z$ (see Appendix IX). The two are cyclically interconverted into one another during the time t_1.

When the second 90° pulse is applied, the antiphase magnetization of the I and J spins is interchanged (Fig. 8.14B). During t_2, the new antiphase coherence,

$-2J_yI_z$, is again cyclically converted into a single quantum coherence J_x, which is a detectable magnetization of the J spin. The mixing formally occurs with the second 90º, and we have seen that the appropriate coherences must develop during both t_1 and t_2. If it were not for transverse relaxation, the antiphase coherence of the I spin would develop during t_1 proportionally to $\sin(\pi J_{IJ} t_1)$, and the maximal antiphase intensity would be obtained for $t_1 = 1/2J_{IJ}$. This sinusoidal dependence is, however, damped by a term $\exp(-\rho_2^I t_1)$. An analogous behavior is shown by the build up of the single quantum coherence J_x during t_2. Overall, the intensity of the cross peak builds up and decays according to the following relationship:

$$I(t) = \sin(\pi J_{IJ} t_1) \exp(-\rho_2^I t_1) \sin(\pi J_{IJ} t_2) \exp(-\rho_2^J t_2). \tag{8.3}$$

To a good approximation, ρ_2^I is the transverse relaxation rate of spin I and ρ_2^J is the transverse relaxation rate of spin J. The two cross peaks are distinguished according to which of the two T_2^I or T_2^J constants appears in each dimension. From Eq. (8.3), it appears that the maximum information is not contained in the first data points. In fact, considering that Eq. (8.3) represents the evolution of the signal, the derivative of $I(t)$ with respect to t_1 (or t_2) provides the value of t_1 (or t_2) with maximum intensity of the signal:

$$\boxed{t_{1(2)} = \frac{1}{\pi J_{IJ}} \operatorname{atan}\left(\frac{\pi J_{IJ}}{\rho_2^{J(I)}}\right)}$$

For $T_2 \gg 1/(2J_{IJ})$, it is well known that the maximal intensity is obtained when the center of the t_1 and t_2 intervals is at $1/(2J_{IJ})$, i.e. for $t_1^{\max} = t_2^{\max} = 1/J_{IJ}$. For $T_2 \ll 1/(2J_{IJ})$, maximal intensity is obtained for $t_1^{\max} = 2T_2^I$ and $t_2^{\max} = 2T_2^J$. Of course, at these short time values, the sine build up of the interaction is small. Weighting functions of sin, $\sin^2$ or matched-filter-type are used to maximize the signal-to-noise ratio. In any case, since the T_2 values in paramagnetic compounds are small, the cross peak intensity is drastically reduced.

The coherence transfer provides cross peaks which are antiphase for the various J_{IJ}-split components. The antiphase nature of the cross peaks then leads to partial or total cancellation of the cross peaks themselves, especially if they are phased in the absorption mode. This behavior can be simulated (Fig. 8.15) using appropriate treatments of the time evolution of the spin system, for instance using the density matrix formalism [17,18]. It is quite common that signals in paramagnetic systems

---

Fig. 8.15. Calculated shapes of cross peaks in COSY spectra characterized by $T_2 < 1/2J_{IJ}$ (conditions: $J_{IJ} = 8$ Hz; $T_2 = 5$ ms; $t_1^{\max} = t_2^{\max} = 10$ ms). (A) Phase-sensitive spectra, phased in dispersion mode; (B) phase-sensitive spectra, phased in absorption mode; (C) magnitude mode spectra. Sin^2 weighting functions are used.

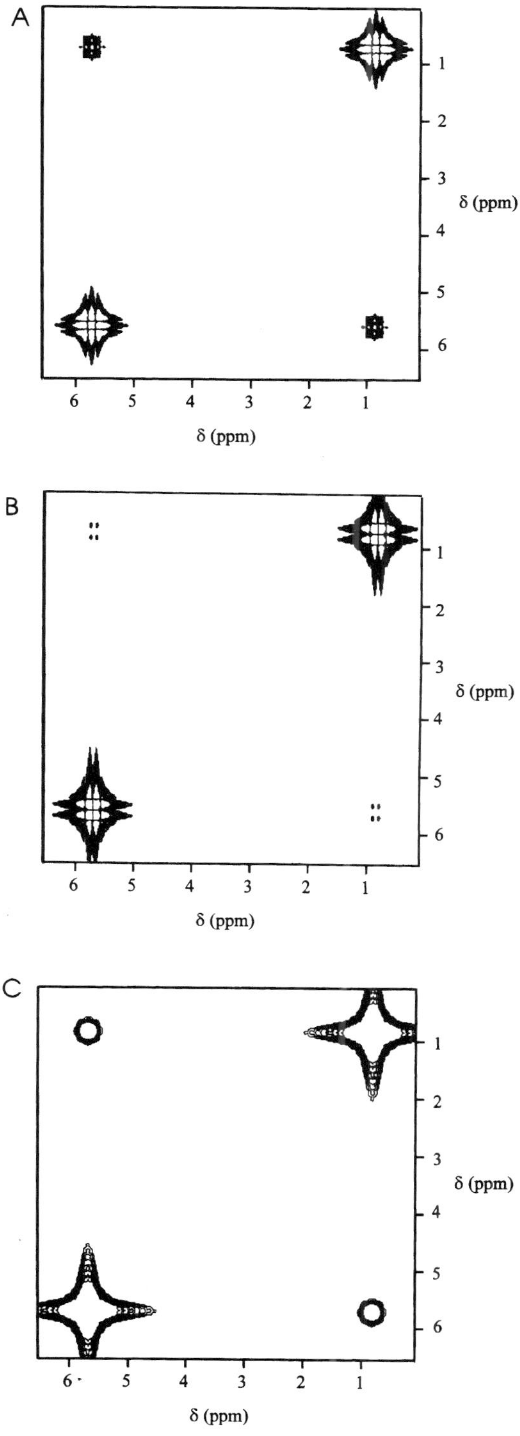
A
δ (ppm)
δ (ppm)
B
δ (ppm)
δ (ppm)
C
δ (ppm)
δ (ppm)

are so broad as to wipe out the J_{IJ} splitting. It has been pointed out that, if the cross peaks are phased in dispersion mode rather than in absorption mode, the loss of intensity is drastically reduced [19]. Accurate phasing may be difficult, however, when the cross peak intensity is small, as happens in paramagnetic compounds. Acquisition and Fourier transform in the so called magnitude mode gives results that are of comparable quality but are absolutely insensitive to phasing. Indeed, in magnitude mode the intensity is given by

$$I = (A^2 + D^2)^{1/2} \tag{8.4}$$

where A and D stand for the pure absorption (or real) and pure dispersion (or imaginary) components of the signal respectively. In Fig. 8.16A part of the magnitude COSY spectrum of the pseudotetrahedral complex NiSAL–i-prop (scheme in the inset) is reported. COSY cross peaks are evident between adjacent protons in the aromatic ring. Although the lines in this case are not very broad, it appears that the phase-sensitive spectrum, routinely phased in absorption mode, has a higher resolution than the magnitude mode spectrum but is of lower quality (Fig. 8.16A,B). A better quality is obtained by phasing the spectrum in dispersion mode (Fig. 8.16C). A good quality spectrum is also obtained with the in-phase cross peaks COSY (ISECR COSY) sequence shown in Fig. 8.2D [20], which provides good resolution and in-phase cross peaks (Fig. 8.16D).

In macromolecular paramagnetic systems, further phenomena may concur which provide COSY cross peaks, when using the sequence of Fig. 8.2C, whose nature is actually dipolar (see Section 8.8).

In several kinds of correlation spectroscopy it is customary to also exploit the evolution of either zero quantum (ZQ) or double quantum (DQ) coherences. As shown in Fig. 7.1, they correspond to the transition $-+ \rightarrow +-$ and $-- \rightarrow ++$, respectively. The typical pulse sequence is described in Fig. 8.2E, where after the first 90° pulse we wait for a time $t_\mathrm{d}/2$, then we send a 180° pulse which has the only role of refocusing the spins. After the same time $t_\mathrm{d}/2$ we send a 45° pulse in order to dislocate the antiphase coherence out of the yz plane in such a way as to create ZQ and DQ coherences (Fig. 8.2E). In the product operator formalism, DQ and ZQ coherences are proportional to $I_x J_y \pm I_y J_x$, respectively. $I_y J_x$ can be visualized as shown in Fig. 8.17. The time t_d should be optimized to either $1/2J_{IJ}$ or T_2. The DQ (ZQ) coherence is let evolve for a time t_1. At this point we can apply a 90° pulse to transform back into an antiphase coherence either the DQ or the ZQ coherence depending on the phase cycling (Fig. 8.17).

Since there is a waste of time in the phase cycling (and some extra loss of signal due to the overall length of the sequence), there is no advantage in using these or other COSY techniques in the place of the simple COSY experiment described in Fig. 8.2C. However, the relaxation of the ZQ transitions may be little affected by the presence of unpaired electrons and under certain conditions may also have some advantages as far as resolution is concerned [21].

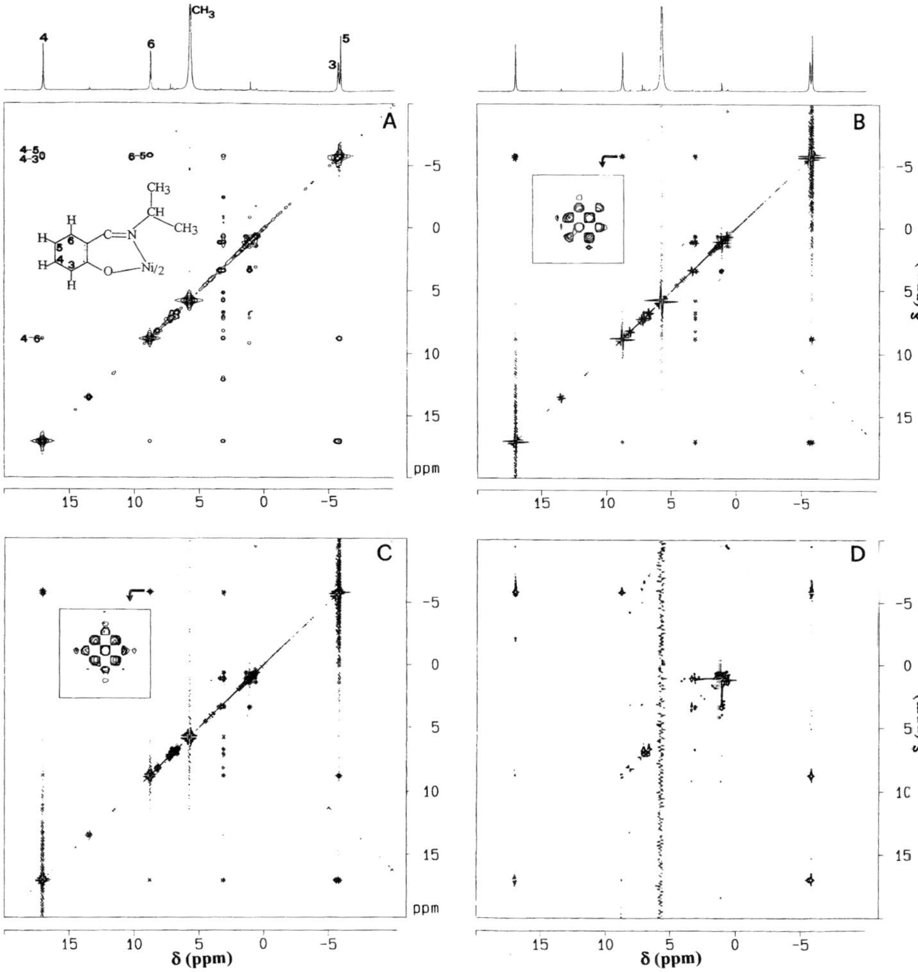

Fig. 8.16. Aromatic ring part of the COSY spectra of the complex shown in the inset. (A) Magnitude mode spectrum; (B) phase-sensitive spectrum, absorption mode; (C) phase-sensitive spectrum, dispersion mode; (D) ISECR [20] spectrum (sequence in Fig. 8.2D). Sin^2 weighting functions have been used for spectra (A)–(C) and a cos^2 weighting function for spectrum (D). Peaks in (A) and (D) are in phase and positive. The positive components of the 6–5 peak in (B) and (C) are shown in the enlargements.

8.6 THE TOCSY EXPERIMENT

TOCSY is a rotating frame experiment designed to detect scalar connectivities over a large range of J_{IJ} values, especially useful for small (ca. 1 Hz) J_{IJ} values. The most common pulse sequence is shown in Fig. 8.2F [22]. The spin lock is achieved by applying a train of relatively high power pulses (the

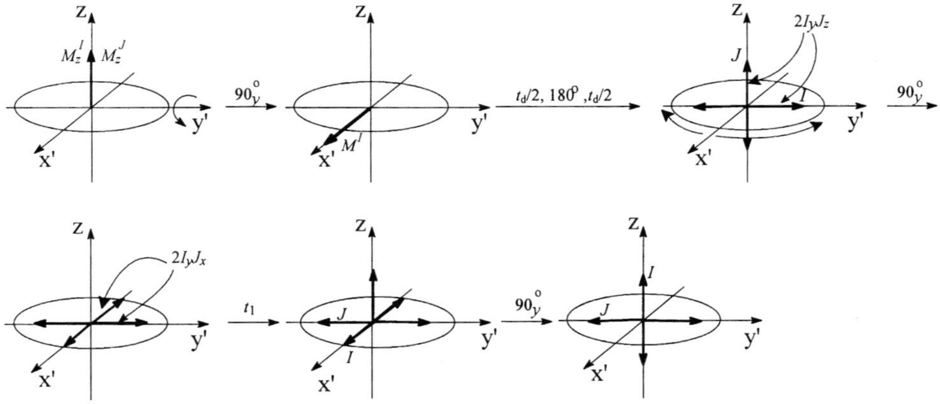

Fig. 8.17. Vector representation of a DQ (ZQ) experiment. The first 90° pulse along y' rotates the equilibrium magnetization of the I (and J) spins from the z axis to the x' axis. After a time $t_d = 1/2J_{IJ}$ (interrupted by a refocusing 180° pulse), the antiphase coherence $2I_yI_z$ (see Appendix IX) is at its maximum. Another pulse along y' (shown as a 90° pulse for clarity, although 45° pulses are more commonly used) then transforms the antiphase coherence into a DQ (ZQ) coherence (the $2I_yJ_x$ component is shown). During t_1 the DQ (ZQ) evolves until a further 90° pulse along y' transforms the $-2I_xJ_y$ component (shown at its maximum for clarity) into a $2J_yI_z$ antiphase coherence. During t_2 the latter generates a detectable in-phase coherence of the J spin, whose initial intensity had been modulated by the DQ (ZQ) Larmor precession frequency during t_1, thus originating a cross peak.

MLEV17 sequence being one of the most used sequences [23]) in such a way as to continuously refocus the chemical shift evolution of the various signals in the xy plane. Analogously to ROESY experiments, the magnetization during the spin-lock (mixing) time disappears with $T_{1\rho}$ (i.e. essentially T_2, see Section 3.4). It follows that coherence transfer in the xy plane, which is built up with a $\sin(\pi J_{IJ} t)$ function, also decreases with time constant $\rho_{1\rho} = (\rho_{1\rho}^I + \rho_{1\rho}^J)/2$:

$$I = \sin(\pi J_{IJ} t_m) \exp(-\rho_{1\rho} t_m). \tag{8.5}$$

A TOCSY experiment on a paramagnetic molecule is reported in Fig. 8.7 for the 5Cl–Ni–SAL–MeDPT complex [5]. Cross peaks between signals with linewidths of the order of 100 Hz were easily detected. In particular, the couplings of each aromatic proton with its neighbors are evident. TOCSY cross peaks between signals with similarly broad lines can also be detected in proteins (see Fig. 8.18) [24].

In TOCSY experiments, the problem of overheating the sample is more serious than in ROESY experiments because of the large irradiation energy required by the spin-lock pulse. Each individual component of the pulse train must have enough power to irradiate the whole spectral window of interest. Spin-lock sequences different from the MLEV17 sequence, that may alleviate the heating problem,

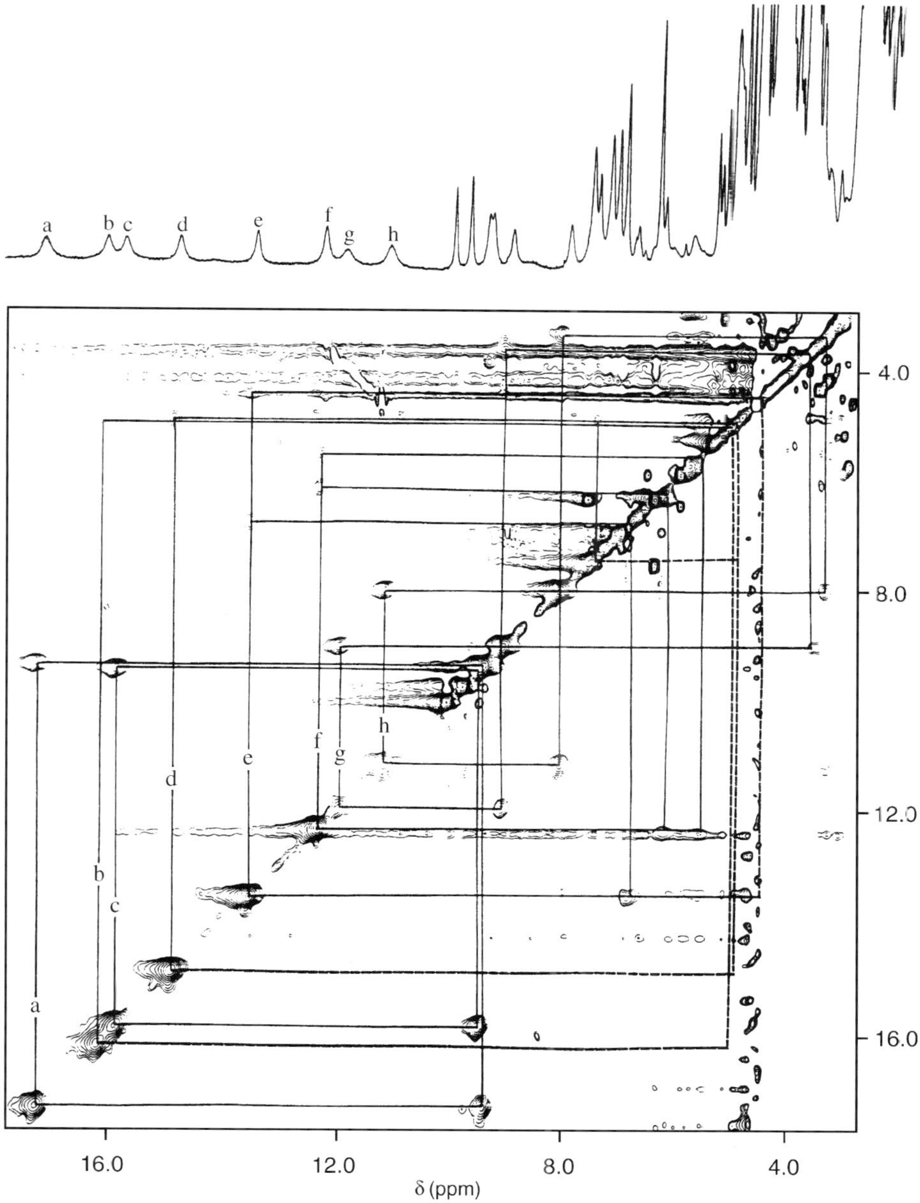

Fig. 8.18. TOCSY spectrum of the oxidized form of ferredoxin from *C. pasteurianum*, showing all the cysteine ligand β-CH_2–β-CH_2 and some β-CH_2–α-CH connections [24].

are discussed in Section 10.4.2. Other modifications that are useful to eliminate unwanted NOESY cross peaks from TOCSY spectra can also be found in Section 9.4.2.

8.7 HETEROCORRELATION SPECTROSCOPY

2D spectra are in principle possible for heteronuclei coupled by either dipolar or scalar interactions. However, the magnetic moments of heteronuclei are sizably smaller than that of the proton, and since cross relaxation depends on the square of the magnetic moment it appears that this is a serious limitation for the observation of NOESY or ROESY cross peaks. However, as already discussed, in scalar-coupled systems the relevant coherences build up with $\sin(\pi J_{IJ} t)$. Since J_{IJ} in directly bound ^{13}C–^{1}H and ^{15}N–^{1}H moieties is of the order of 10^2 Hz, as opposed to about 10 Hz between proton pairs, it is conceivable that scalar correlation experiments are successful. Heterocorrelated spectra have the advantage of allowing one to detect signals of protons attached to carbons or nitrogens when they are within a crowded envelope.

Heterocorrelations can be detected both in direct and reverse modes. In the latter mode, dramatic enhancements of sensitivity can be achieved owing to the larger sensitivity of protons with respect to heteronuclei. In the most common heterocorrelation pulse sequences for reverse detection, called heteronuclear multiple quantum coherence (HMQC) (Fig. 8.2G) [25,26], ^{1}H–^{13}C MQ (multiple quantum) coherence is generated by first applying a 90° pulse on protons and, after a time t_d chosen equal to $1/2J_{IJ}$, by applying a 90° pulse on carbon (Fig. 8.19).

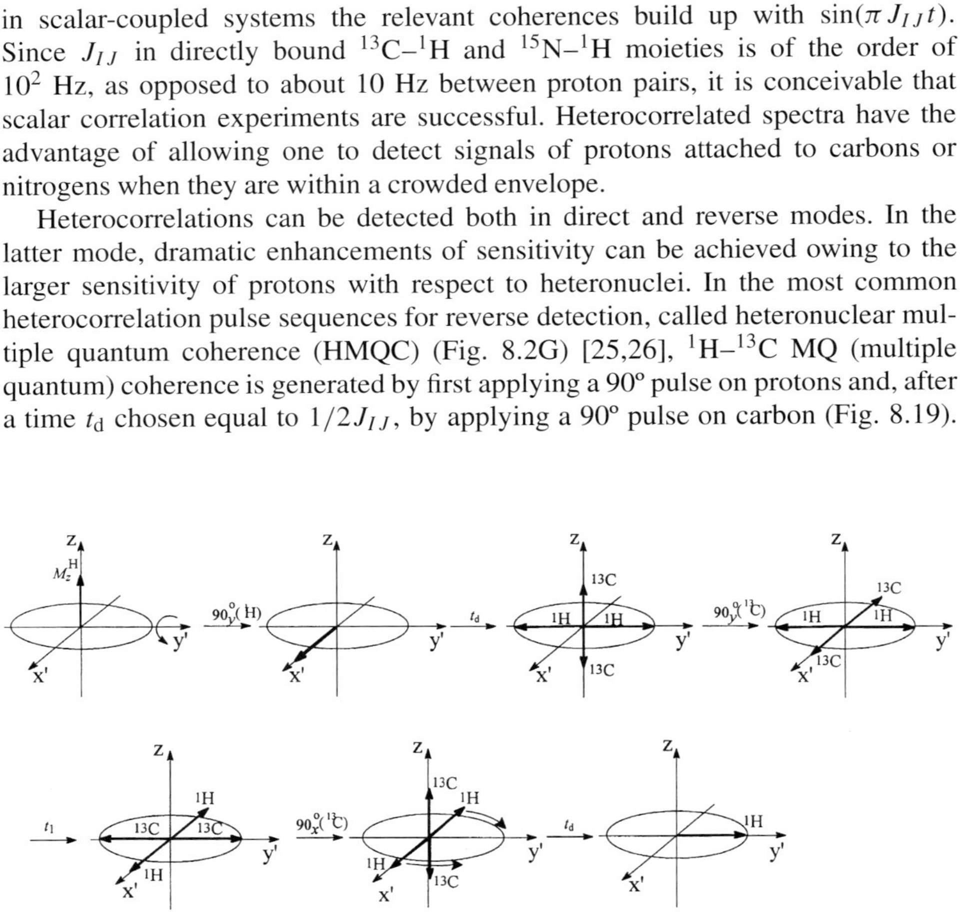

Fig. 8.19. Vector representation of a ^{1}H–^{13}C HMQC experiment. The first 90° pulse along y' rotates the equilibrium magnetization of the proton spin, I^H, from the z axis to the x' axis. After a time $t_d = 1/2J_{HX}$, the antiphase coherence $2I_y^H I_z^C$ (see Appendix IX) is at its maximum. A 90° pulse on carbon along y' then transforms the antiphase coherence into a MQ (multiple quantum) coherence (the $2I_y^H I_x^C$ component is shown). During t_1 the MQ evolves (with a 180° refocusing pulse on proton in the middle), until a further 90° pulse on carbon along x' transforms the $-2I_x^H I_y^C$ component (shown at its maximum for clarity) into a $2I_x^H I_z^C$ antiphase coherence. After the time t_d, in-phase coherence of the proton spin develops. The latter is detected during t_2. Its initial intensity is modulated by the carbon Larmor frequency during t_1 (if proton refocusing has been used), thus originating a proton–carbon cross peak.

At this point the MQ coherence is let evolve for a variable time t_1. A refocusing 180º pulse on proton is applied in the middle of the evolution period, i.e. at $t_1/2$. A further 90º pulse on carbon converts the MQ coherence into an antiphase magnetization (Fig. 8.19), which is let evolve for another time t_d to fully develop the observable single quantum coherence on proton, which is then detected during the acquisition time t_2.

In paramagnetic compounds, in the absence of contact relaxation, heteronuclei are relaxed much less than protons by dipolar coupling with the unpaired electron(s), all other things being equal, because of the γ_I^2 dependence. By looking at the evolution of coherence during the experiment, we can predict the relaxation of the system during the sequence. When the linewidths are larger than J_{IJ}, the first antiphase on proton decays during t_d essentially with ρ_2^I for proton (the heteronucleus being still along the z axis). The MQ decays with the sum of $\rho_2^I + \rho_2^J$, whereas during the second t_d the magnetization decays again with ρ_2^I. In cases where $\rho_2^I \gg \rho_2^J$ (for instance, large Curie relaxation effects on protons, see Section 3.6), direct detection may be less disfavored than reverse detection, because during the two t_d intervals ρ_2^J instead of ρ_2^I is involved.

An example of an HMQC spectrum is reported in Fig. 8.20 [21]. ^{1}H–^{13}C connectivities can be detected for signals for which $T_2 < 1/2J_{IJ}$ by setting $t_d = T_2$.

In principle, HSQC experiments (Fig. 8.2H) should be more effective than HMQC. This is because during the transfer delay t_d the 180º pulses on ^{1}H and ^{13}C eliminate contributions to relaxation arising from field in-homogeneity, diffusion and cross-correlation effects. Furthermore, the antiphase magnetization evolving during t_1 in HSQC experiments roughly decays with the sum of $\rho_1^I + \rho_2^J$ and therefore should relax during t_1 slower than DQ (ZQ). Practically, in our hands we never found substantial difference in the amount of information obtained in HSQC versus HMQC experiments. A possible reason could be that the smaller number of pulses of HMQC makes the latter sequence more robust with respect to experimental missetting. On the other hand it should be stressed that the critical step in this experiment is the polarization transfer from I to J occurring during t_d, which is the same in both experiments.

Usually, HSQC/HMQC spectra are recorded in decoupled mode, i.e. the splitting due to the ^1J coupling between the proton and the heteronucleus, which would give rise to a four-peak pattern in the spectrum, is removed. This is achieved by (1) inserting a 180º pulse on ^{1}H in the middle of the t_1 evolution of the heteronuclear spectrum and (2) broadband decoupling the heteronucleus during proton acquisition in t_2. Alternatively, only one of the four components may be observed by using selective coherence transfer techniques. This latter strategy is used in TROSY experiments [27], and is particularly convenient when the four components of the multiplet have different linewidths, because the sharper one can be selected. In turn, this occurs whenever there are cross-correlation phenomena [28]. One common example of cross-correlation giving rise to different linewidths

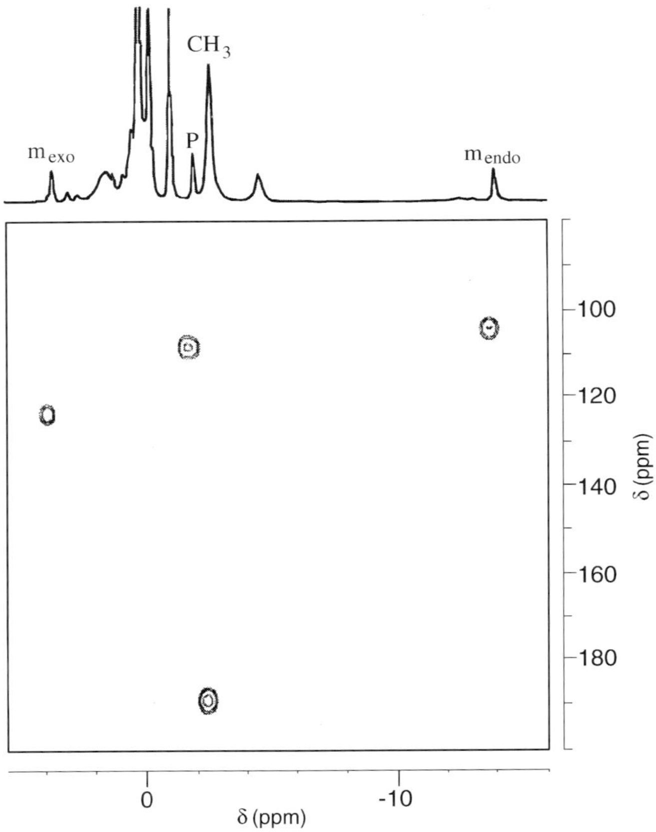

Fig. 8.20. 600 MHz natural abundance ^1H–^{13}C reverse HMQC spectrum of the complex shown in Fig. 8.13 (Ln = Dy^{3+}). The cross peaks involve protons with linewidths of the order of, or larger than, J_{HC} [21].

is in ^1H–^{15}N heterocorrelated spectra of proteins. The nuclei of backbone NH groups experience considerable cross-correlation between the ^1H–^{15}N dipole–dipole coupling and the chemical shift anisotropy (CSA). As both are predictable, it can be also predicted that in a coupled HSQC spectrum the lower-right peak of the multiplet will be always the sharpest, and TROSY experiments that select this component will give the best results in terms of spectral resolution.

In paramagnetic systems, there can be non-negligible contributions to the CSA of NH nuclei from their dipolar coupling with the time-averaged magnetic moment of the electron (see Section 3.6). As a consequence, it cannot be predicted a priori which will be the sharpest component for each NH peak in a heterocorrelated experiment, and TROSY is less useful, unless four different TROSY spectra are acquired by selecting a different component each time [27].

8.8 COHERENCE TRANSFER CAUSED BY DIPOLAR CROSS CORRELATION

When reference was made to the dipolar coupling scheme between two protons of Fig. 7.1, it was implicitly assumed that the two single quantum longitudinal transition probabilities of, for instance, nucleus I, w_1^I, were equal, i.e. that they were independent of the spin state of nucleus J. Similar assumptions are made when dealing with transverse transition probabilities. When the coupling is dipolar in origin, the two transitions are degenerate. Indeed, a signal belonging to a nucleus dipole-coupled to another nucleus actually results from the superposition of two coincident components with equal T_1 and T_2, although one never thinks of it. There are a number of cases of dipole-coupled pairs, however, where the two components, still coincident due to the absence of scalar couplings, have different T_1 and T_2. This phenomenon originates from cross correlation between the dipolar coupling operative within the pair and other interactions experienced by the two nuclei and modulated by the same motion responsible for the modulation of the dipolar coupling [29]. In principle, a third nuclear spin K dipole-coupled to the other two spins I and J is sufficient to differentiate the linewidths. In practice, the effect is small and hardly detectable. A more efficient mechanism is the cross correlation with the chemical shift anisotropy relaxation mechanism.

In paramagnetic systems, a major nuclear relaxation mechanism is Curie relaxation (see Section 3.6). In paramagnetic macromolecules Curie spin relaxation is often the dominant contribution to T_2. Cross correlation between proton–proton dipolar coupling and Curie relaxation may cause the two degenerate signal components of each nucleus to have markedly different linewidths [30]. Under these conditions, the common COSY experiment of Fig. 8.2C yields strong cross peaks between signals coupled by dipolar *but not by scalar* interactions. The origin of the phenomenon can be better understood if one recalls that the components of COSY cross peaks are in antiphase (see, for instance, Fig. 8.16B,C). It is obvious that, if the two components coincide, total cancellation occurs. However, if the two components have different linewidths, they do not cancel and may give rise to very strong cross peaks, as illustrated in Fig. 8.21.

The amount of cross correlation between the dipolar coupling of nuclei A and X and Curie relaxation of the A spin is [31]:

$$\Delta\nu = \frac{2}{15\pi}\left(\frac{\mu_0}{4\pi}\right)^2 \frac{B_0\gamma_A^2\gamma_X\mu_B^2 g_e^2\hbar S(S+1)}{r_r^3 r_{AX}^3 kT} \times \left(4\tau_c + \frac{3\tau_c}{1+\omega_0^2\tau_c^2}\right)\frac{3\cos^2\theta_{S\,AX}-1}{2} \qquad (8.6)$$

where r_{AX} and r_r are the AX internuclear distance and the distance of the A nucleus from the unpaired electron, $\theta_{S\,AX}$ is the angle between the A-electron and the AX vectors, S is the electron spin quantum number and τ_c is the correlation time for molecular tumbling. For lanthanides S and g_e must be replaced with J

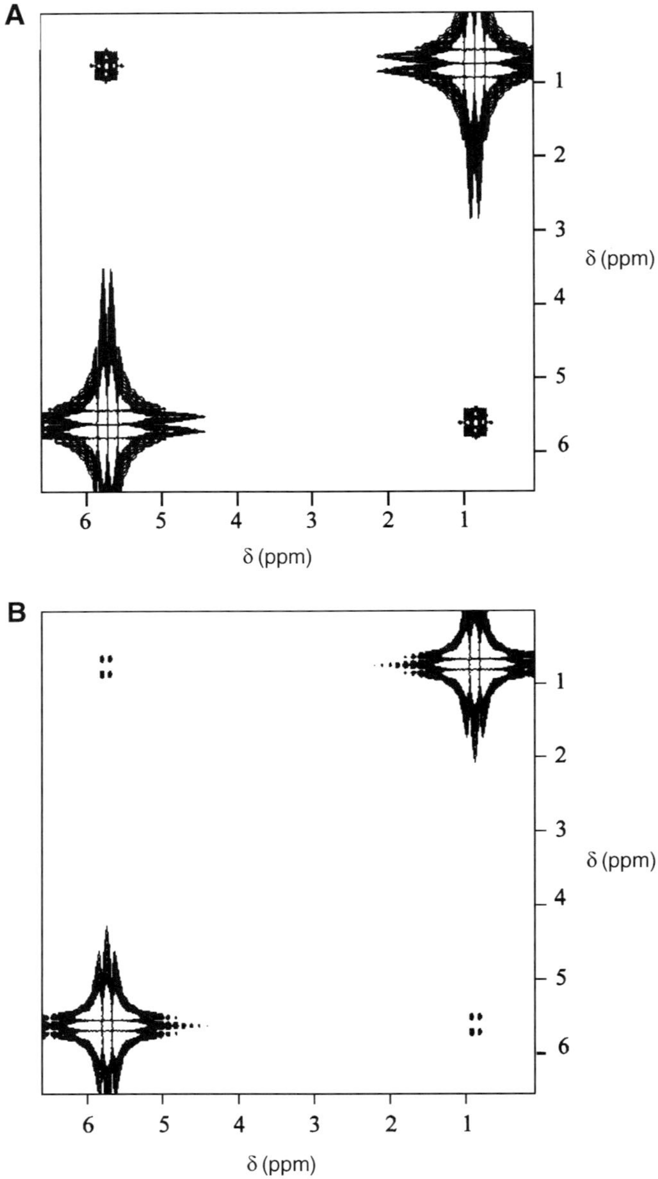

Fig. 8.21. Simulated COSY cross peaks originating from a pair of signals dipole-coupled in the presence of cross relaxation with Curie relaxation and in the absence of scalar coupling. The two degenerate components are in antiphase, but they do not cancel out due to their different linewidths. Note that absorption mode phasing (A) and dispersion mode phasing (B) show opposite patterns with respect to the case of scalar coupling (Fig. 8.15A,B).

and g_J, respectively (see Section 3.6). So, cross correlation can contain structural information. Sequences are available to measure the different transverse relaxation rates of the two components [32]. A sensitivity enhanced TROSY pulse sequence was modified to include a transverse ^1H relaxation period prior to detection. Relaxation rate constants are then calculated from the measured intensity ratios of the individual doublet lines. The dipolar coupling with the time-averaged magnetic moment of the electron can be seen as a contribution to the CSA (Sections 3.6 and 8.7). Therefore, the effect described above cannot be distinguished from the general CSA–dipole–dipole cross correlation. Such cross correlation also provides a contribution to the signal splitting which adds to that of the ^1J coupling of, e.g., the NH group. There is now a debate on its amount and on its dependence on the presence and extent of ZFS [31,33].

8.9 BEYOND 2D SPECTROSCOPY

In a chapter regarding 2D NMR spectroscopy of paramagnetic molecules, the obvious perspective is that of using three-dimensional (3D) NMR for paramagnetic molecules. The demand for 3D spectroscopy is based on a need of increased resolution when macromolecules are concerned. It is possible that for small complexes 3D spectroscopy will never be necessary. However, every time something new has appeared in science, the majority has reacted by saying that the utility was scarce in their own field, and the majority has not always been right. Therefore, we do not commit ourselves.

We have seen in Section 8.3 that one limitation of 2D spectroscopy when dealing with paramagnetic compounds showing a wide range of T_1 (and T_2) values is that the parameters cannot be optimized for the detection of all connectivities in a single experiment. Often, at least two experiments are necessary. An advantage of 3D spectroscopy which is peculiar to paramagnetic systems, and in particular to paramagnetic macromolecules, is that one can cover the whole range of T_1 and T_2 values present in the spectrum by optimizing the spectral parameters in one 2D plane (spectral resolution, maximal evolution time, mixing time, etc.) to the detection of connectivities between fast relaxing signals, and those in another 2D plane to the detection of connectivities with slow relaxing signals. Of course, a clear advantage only appears when fast relaxing signals are many and severely overlapped with slow relaxing signals.

An obvious extension to 3D spectroscopy from 2D spectroscopy is the homonuclear NOESY–NOESY [34]. There are two t_1 variable times and one t_2, which after Fourier transform provide three frequency domains. The 3D NOESY–NOESY spectrum of met-myoglobin cyanide, which contains low spin iron(III) in a heme moiety (see Fig. 5.7), has been successfully measured [35]. In Fig. 8.22 a slice of the 3D spectrum is shown at the 12-CH_3 height. On the diagonal it shows all the dipolar connectivities between 12-CH_3 and other protons; off-diagonal

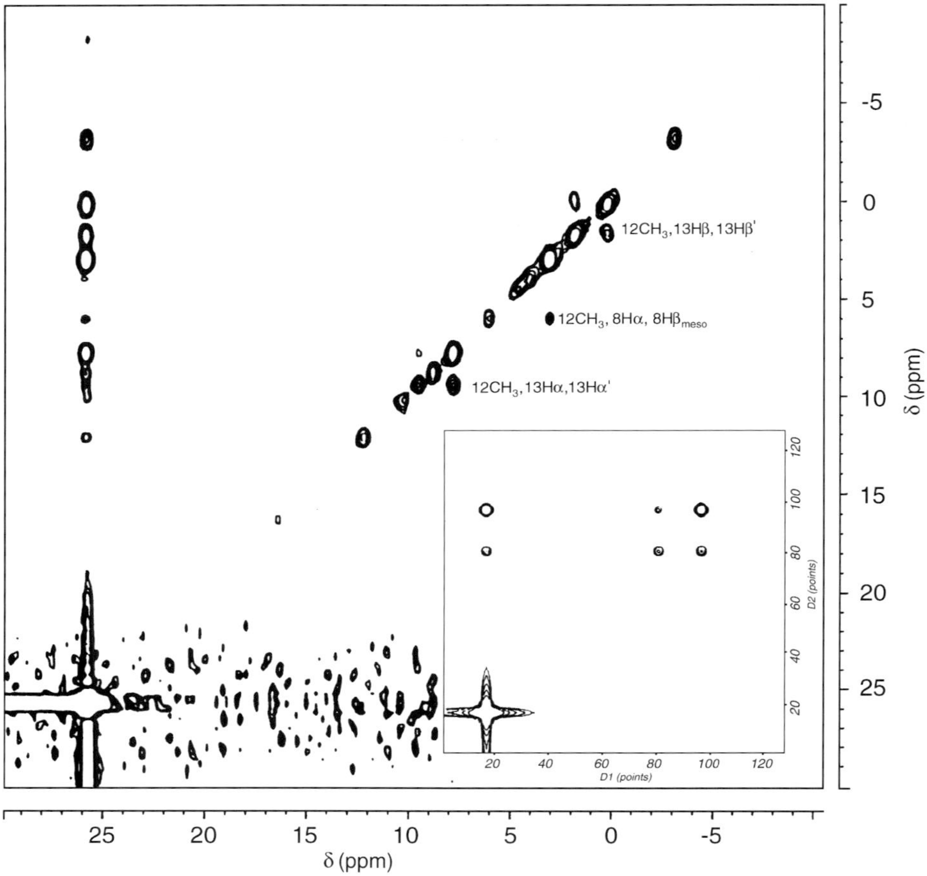

Fig. 8.22. Cross-section of the 600 MHz 3D NOESY–NOESY spectrum of met-myoglobin cyanide. The slice is taken at the 12-CH_3 height. The inset shows the simulated cross-section involving the 12-CH_3, 13-Hα and 13-Hα′ signals [35].

there are dipolar connectivities within the latter set of protons, all belonging to the porphyrin ring.

Results similar to those shown in the slice of Fig. 8.22 can be obtained with the so-called NOE–NOESY sequence [36]. Here a hyperfine shifted signal, e.g. 12-CH_3 of the above compound, is selectively saturated, and then the NOESY pulse sequence is applied. The NOESY difference spectrum obtained by subtracting a NOESY spectrum without presaturation of the 12-CH_3 signal is shown in Fig. 8.23. Here, some more cross peaks are evident with respect to the 3D NOESY–NOESY experiment because secondary NOEs develop much more when the primary NOEs from the 12-CH_3 signal evolve in a steady state experiment like the NOE–NOESY rather than in a transient-type experiment like the NOESY–NOESY. In Fig. 8.23, dipolar connectivity patterns are apparent among protons

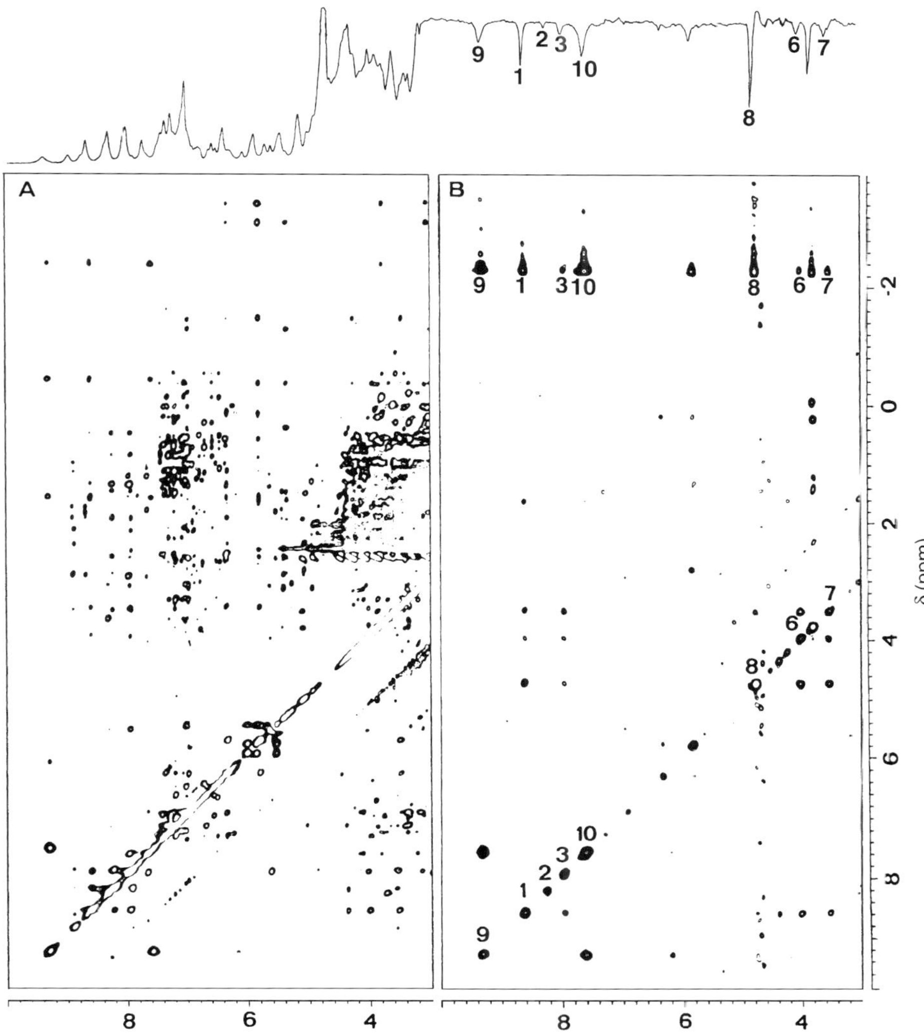

Fig. 8.23. NOESY spectrum (A) and NOE–NOESY spectrum (B) of met-myoglobin cyanide [36]. The latter spectrum is obtained by pre-irradiation of the 12-CH_3 signal.

belonging to Phe-33 (2) Phe-43 (1, 6, 7, 8), Phe-46 (3) and propionate 13 (9, 10). Some of these assignments could only be performed by exploiting the NOE–NOESY experiment.

Other variants are NOE–COSY and NOE–TOCSY. However, these experiments are much less sensitive than NOE–NOESY.

Sometimes it may be useful to suppress signals with long relaxation times in such a way as to detect cross peaks involving one fast relaxing signal under the envelope of diamagnetic signals. In this case a SuperWEFT sequence (180–τ–

90–AQ, see Section 9.1.3) is applied before performing the 2D experiment. By appropriately choosing the τ value, the diamagnetic signals can be brought to have intensities close to zero.

In principle, all the combinations of homonuclear 2D spectroscopies can be performed to originate a 3D spectrum (COSY–COSY, NOESY–COSY, NOESY–TOCSY, etc.). The considerations made in this chapter for the most basic experiments can be easily extended to their combinations. The general guideline should always be that the more complex the pulse sequence is, the more the experimental sensitivity will suffer from fast nuclear relaxation.

The most useful advantage of 3D spectroscopy, however, is that heteronuclei can be included along one dimension. If one of the two 2D sequences is a heterocorrelated sequence, then simultaneous heteronucleus–proton connectivities can be observed. In the case of macromolecules, where there are solubility problems, heteronuclear enrichment is probably required. Heterocorrelated spectroscopies, of the type of those used in diamagnetic systems, have been applied with success, even using natural abundance ^{13}C, on a paramagnetic protein containing the $[Fe_4S_4]^{2+}$ polymetallic center [37,38].

8.10 TRIDIMENSIONAL STRUCTURES OF PARAMAGNETIC PROTEINS IN SOLUTION

NOEs or NOESY cross peaks can be used to define distances between protons. It is well known that the three-bond scalar coupling constants $^3J_{IJ}$ between nuclei I and J obey a Karplus relationship of the type [39,40]

$$^3J_{IJ} = a\cos^2\theta + b\cos\theta + c \tag{8.7}$$

where θ is the I–X–Y–J dihedral angle[1]. When I and J are protons, the problem arises of detecting splitting of COSY cross peak components in the case of hyperfine broadened signals. At this moment the limits of the techniques developed for diamagnetic systems are being tested in paramagnetic molecules. With NOEs and ^3J values, solution structures of proteins have been at hand since the mid-eighties [41–43]. In paramagnetic compounds, the contact shifts may also contain information on the metal–donor–C–H or metal–donor–C–C dihedral angles as discussed in Section 2.4 for cysteines coordinated to paramagnetic metals. Constraints of this type can also be obtained on the orientation of the axial histidine(s) in low spin Fe(III) hemes or high spin Fe(II) heme containing systems (see Sections 5.1.2 and 5.1.3) [44,45]. Pseudocontact shifts (see Section 2.2.2) are essential constraints as well as R_{1M} and R_{2M} (see Section 3.4). On top of

---

[1] Eq. (8.7), which provides the scalar coupling constant due to the interaction between nuclei, is analogous to Eq. (2.29), used to describe the dihedral angle dependence of the contact coupling constant due to the interaction between nuclei and electrons.

them, there are cross correlation effects (see Section 8.8) and the residual dipolar couplings (see Section 8.11).

The solution structure of paramagnetic metalloproteins can now be routinely obtained if the electronic relaxation times are favorable [37,46–48]. In this case, constraints contained in the hyperfine coupling are added to the usual NOE and ^3J constraints [49,50]. It is possible that in small coordination compounds where the number of NOEs is necessarily small, the combination of the various constraints provides a reasonably well-defined average structure in solution [51]. This route has not been pursued much in the literature. Less dramatic than in small coordination compounds, but quite serious, is the problem of DNA/RNA fragments. Early successes were obtained with mononucleotides [52–54]. In oligonucleotides, the signal overlap is severe and the folding is such that the number of NOEs is small. Many J values are needed, which can be obtained only by using fully isotope-labeled samples [55]. If a binding site were available for a paramagnetic metal ion, the induced pseudocontact shifts and nuclear relaxation enhancements could be of significant help.

8.11 THE EFFECTS OF PARTIAL ORIENTATION

Sharp proton NMR lines are associated to short electron relaxation times, which in turn depend on the availability of low lying excited states. Such systems also provide metal ion magnetic anisotropy and pseudocontact shifts (see Chapter 2). Magnetic anisotropy contributes to partial orientation in high magnetic fields (self orientation). Under these circumstances, the dipolar coupling energies average values different from zero (residual dipolar coupling). Pseudocontact shifts are also expected to change under partial orientation. One equation of the hyperfine shift in system partially oriented due to metal ion magnetic anisotropy was proposed (Eq. (2.44)), which however gives only a small correction to the isotropic value [56,57].

Changes in contact shifts may as well be expected as $\langle S_z \rangle$ is not isotropic (see Section 2.2.1) [57], but this research area is not yet explored. Self orientation, instead, is conveniently measured as variation of ^1J ^{15}N–^{1}H or ^2J in ^{1}H–^{15}N–^{1}H systems as a function of the extent of the orientation, i.e. as a function of the magnetic field [58,59] (see Section 2.10):

$$rdc\ (\text{Hz}) = -\frac{1}{4\pi}\frac{B_0^2}{15kT}\frac{\gamma_{\text{N}}\gamma_{\text{H}}\hbar}{2\pi r_{\text{NH}}^3}\left[\Delta\chi_{ax}(3\cos^2\theta - 1) + \frac{3}{2}\Delta\chi_{rh}\sin^2\theta\cos 2\Omega\right] \tag{8.8}$$

where B_0 is the magnetic field, θ is the angle between the ^{15}N–^{1}H vector and the z axis of the χ tensor, Ω is the angle which describes the position of the projection of the ^{15}N–^{1}H vector on the xy plane of the χ tensor, relative to the x axis, and $\Delta\chi_{ax}$ and $\Delta\chi_{rh}$ are defined as in Eq. (2.20). The ^{1}J residual dipolar coupling are meaningful parameters in solution structure determination of macromolecules, as they provide the N–H vector orientation, the magnetic anisotropy values and the χ tensor orientation.

From the ^{1}J residual dipolar coupling the molecular magnetic anisotropy tensor is obtained, which differs from the metal contribution by an extent which depends on the magnetic anisotropy of the diamagnetic part. For example, in cytochrome b_5 the diamagnetic, the paramagnetic and the total susceptibility anisotropy values are $\Delta\chi_{ax} = -0.8$, 2.8, 2.20×10^{-32} m^3, respectively, and $\Delta\chi_{rh} = 0.1$, -1.1, -1.34×10^{-32} m^3, respectively [60]. The corresponding tensors sum up as expected.

8.12 GENERAL REFERENCES

A. Bax (1982) *Two Dimensional Nuclear Magnetic Resonance in Liquids.* Reidel, Dordrecht.

R.R. Ernst, G. Bodenhausen and A. Wokaun (1987) *Principles of Nuclear Magnetic Resonance in One and Two Dimensions.* Oxford University Press, London.

8.13 REFERENCES

[1] H. Santos, D.L. Turner, A.V. Xavier, J. LeGall (1984) J. Magn. Reson. 59, 177.
[2] W. Peters, M. Fuchs, H. Sicius, W. Kuchen (1985) Angew. Chem. 24, 231.
[3] B.G. Jenkins, R.B. Lauffer (1988) Inorg. Chem. 27, 4730.
[4] B.G. Jenkins, R.B. Lauffer (1988) J. Magn. Reson. 80, 328.
[5] C. Luchinat, S. Steuernagel, P. Turano (1990) Inorg. Chem. 29, 4351.
[6] I. Bertini, F. Briganti, C. Luchinat, L. Messori, R. Monnanni, A. Scozzafava, G. Vallini (1992) Eur. J. Biochem. 204, 831.
[7] H. Kessler, H. Oschkinat, C. Griesinger, W. Bermel (1986) J. Magn. Reson. 70, 106.
[8] I. Bertini, L. Sacconi, G.P. Speroni (1972) Inorg. Chem. 11, 1323.
[9] C. Luchinat, M. Piccioli (1995) In: G.N. La Mar GN (Ed.), NMR of Paramagnetic Macromolecules. NATO ASI Series, Kluwer Academic, Dordrecht, pp. 1.
[10] I. Bertini, F. Briganti, C. Luchinat, L. Messori, R. Monnanni, A. Scozzafava, G. Vallini (1991) FEBS Lett. 289, 253.
[11] K. Nagayama, A. Kumar, K. Wüthrich, R.R. Ernst (1980) J. Magn. Reson. 40, 321.
[12] I. Bertini, C. Luchinat, L. Messori, M. Vasak (1993) Eur. J. Biochem. 211, 235.
[13] H. Desvaux, P. Berthault, N. Birlirakis, M. Goldman (1994) J. Magn. Reson. 108, 219.
[14] H. Desvaux, P. Berthault, N. Birlirakis, M. Goldman, M. Piotto (1995) J. Magn. Reson. 113, 47.
[15] C.A. Salgueiro, D.L. Turner, H. Santos, J. LeGall, A.V. Xavier (1992) FEBS Lett. 314, 155.
[16] I. Bertini, A.G. Coutsolelos, A. Dikiy, C. Luchinat, G.A. Spyroulias, A. Troganis (1996) Inorg. Chem. 35, 6308.

[17] A. Bax (1982) Two Dimensional Nuclear Magnetic Resonance in Liquids. Reidel, Dordrecht.
[18] J. Shriver (1992) Concepts Magn. Reson. 4, 1.
[19] A.V. Xavier, D.L. Turner, H. Santos (1993) Methods Enzymol. 227, 1.
[20] S. Talluri, H.A. Scheraga (1990) J. Magn. Reson. 86, 1.
[21] I. Bertini, C. Luchinat, A. Rosato (1996) Chem. Phys. Lett. 250, 495.
[22] L. Braunschweiler, R.R. Ernst (1983) J. Magn. Reson. 53, 521.
[23] A. Bax, D.G. Davis (1985) J. Magn. Reson. 65, 355.
[24] M. Sadek, R.T.C. Brownlee, S.D.B. Scrofani, A.G. Wedd (1993) J. Magn. Reson. 101, 309.
[25] L. Müller (1979) J. Am. Chem. Soc. 101, 4481.
[26] A. Bax, R.H. Griffey, B.L. Hawkins (1983) J. Magn. Reson. 55, 301.
[27] K. Pervushin, R. Riek, G. Wider, K. Wüthrich (1997) Proc. Natl. Acad. Sci. USA 94, 12366.
[28] B. Brutscher (2000) Conc. Magn. Reson. 12, 207.
[29] S. Wimperis, G. Bodenhausen (1989) Mol. Phys. 66, 897.
[30] I. Bertini, C. Luchinat, D. Tarchi (1993) Chem. Phys. Lett. 203, 445.
[31] R. Ghose, J.H. Prestegard (1997) J. Magn. Reson. 128, 138.
[32] J. Boisbouvier, P. Gans, M. Blackledge, B. Brutscher, D. Marion (1999) J. Am. Chem. Soc. 121, 7700.
[33] H. Desvaux, M. Gochin (1999) Mol. Phys. 96, 1317; I. Bertini, J. Kowalewski, C. Luchinat, G. Parigi (2001) J. Magn. Reson., submitted.
[34] R. Boelens, G.W. Vuister, T.M.G. Koning, R. Kaptein (1989) J. Am. Chem. Soc. 111, 8525.
[35] L. Banci, W. Bermel, C. Luchinat, R. Pierattelli, D. Tarchi (1993) Magn. Reson. Chem. 31, S3–S7.
[36] I. Bertini, A. Dikiy, C. Luchinat, M. Piccioli, D. Tarchi (1994) J. Magn. Reson. Ser. B 103, 278.
[37] L. Banci, I. Bertini, L.D. Eltis, I.C. Felli, D.H.W. Kastrau, C. Luchinat, M. Piccioli, R. Pierattelli, M. Smith (1994) Eur. J. Biochem. 225, 715.
[38] I. Bertini, M.M.J. Couture, A. Donaire, L.D. Eltis, I.C. Felli, C. Luchinat, M. Piccioli, A. Rosato (1996) Eur. J. Biochem. 241, 440.
[39] M. Karplus (1959) J. Chem. Phys. 30, 11.
[40] M. Karplus (1963) J. Am. Chem. Soc. 85, 2870.
[41] M.P. Williamson, T.F. Havel, K. Wüthrich (1985) J. Mol. Biol. 135, 295.
[42] K. Wüthrich (1986) NMR of Proteins and Nucleic Acids. Wiley, New York.
[43] G. Wagner, M.H. Frey, D. Neuhaus, E. Worgotter, W. Braun, M. Vasak, J.H. Kagi, K. Wüthrich (1987) EXS 52, 149.
[44] I. Bertini, C. Luchinat, G. Parigi, F.A. Walker (1999) JBIC 4, 515.
[45] I. Bertini, A. Dikiy, C. Luchinat, R. Macinai, M.S. Viezzoli (1998) Inorg. Chem. 37, 4814.
[46] L. Banci, R. Pierattelli (1995) In: G.N. La Mar (Ed.), Nuclear Magnetic Resonance of Paramagnetic Macromolecules. NATO ASI Series, Kluwer Academic, Dordrecht, pp. 281.
[47] I. Bertini, A. Dikiy, D.H.W. Kastrau, C. Luchinat, P. Sompornpisut (1995) Biochemistry 34, 9851.
[48] I. Bertini, A. Donaire, B.A. Feinberg, C. Luchinat, M. Piccioli, H. Yuan (1995) Eur. J. Biochem. 232, 192.
[49] L. Banci, I. Bertini, K.L. Bren, M.A. Cremonini, H.B. Gray, C. Luchinat, P. Turano (1996) JBIC 1, 117.
[50] I. Bertini, A. Donaire, C. Luchinat, A. Rosato (1997) Proteins Struct. Funct. Genet. 29, 348.
[51] L.-J. Ming (2000) In: L. Que Jr. (Ed.), Physical Methods in Bioinorganic Chemistry. University Science Books, Sausalito, CA, pp. 375.

[52] C.D. Barry, A.C.T. North, J.A. Glasel, R.J.P. Williams, A.V. Xavier (1971) Nature 232, 236.
[53] C.D. Barry, J.A. Glasel, R.J.P. Williams, A.V. Xavier (1974) J. Mol. Biol. 84, 471.
[54] C.D. Barry, D.R. Martin, R.J.P. Williams, A.V. Xavier (1974) J. Mol. Biol. 84, 491.
[55] J.P. Marino, H. Schwalbe, C. Anklin, W. Bermel, D.M. Crothers, C. Griesinger (1994) J. Am. Chem. Soc. 116, 6472.
[56] I. Bertini, I.C. Felli, C. Luchinat (1998) J. Magn. Reson. 134, 360.
[57] I. Bertini, C. Luchinat, G. Parigi (2000) Eur. J. Inorg. Chem. 2473.
[58] J.R. Tolman, J.M. Flanagan, M.A. Kennedy, J.H. Prestegard (1995) Proc. Natl. Acad. Sci. USA 92, 9279.
[59] N. Tjandra, A. Bax (1997) Science 278, 1111.
[60] L. Banci, I. Bertini, J.G. Huber, C. Luchinat, A. Rosato (1998) J. Am. Chem. Soc. 120, 12903.

Chapter 9

Hints on Experimental Techniques

This chapter provides some guidelines on how to acquire simple 1D and 2D NMR spectra of paramagnetic molecules. Beginners often loose signals or have problems with the baseline ... or may observe artifacts, and then the whole approach may have serious problems.

9.1 HOW TO RECORD 1D NMR SPECTRA OF PARAMAGNETIC MOLECULES

All over the book we have underlined that paramagnetic molecules have short nuclear relaxation times. This has to be faced when recording the spectra, and proper measures should be taken to minimize the consequences of the fast nuclear relaxation. For example, a broader line has smaller height than a sharp line, if the number of nuclei is the same and if the spectral width is the same. This, in practice, is going to affect the signal to noise (S/N) ratio. The decrease in S/N can be partially avoided by a proper choice of the acquisition time (or of $t_1^{\max}$ and $t_2^{\max}$ in 2D spectra) (see Sections 8.3 and 8.5). As discussed in Chapter 8, the information content of a FID decays with T_2 of the signal of interest, and therefore acquisition times much longer than T_2 should not be used, unless sharp and ill-resolved signals are also looked for in the spectrum (in this case two experiments must be performed: one for the detection of sharp signals and one for broad signals). The reduction in $t_1^{\max}$ does not necessarily decrease the number of sampled data points, as often the large spectral widths needed for paramagnetic compounds require a more frequent data point sampling (shortening of the dwell time). Many modern instruments use oversampling techniques, that allow data points to be always sampled at the highest possible speed, thereby optimizing the S/N for any chosen spectral width. However, oversampling is effective only for spectral widths below 20,000 Hz.

When the spectral width is of hundreds of parts per million, i.e. more than 10^5 Hz on high field instruments, a very short excitation pulse is needed. Of course, high power is needed to reach the r.f. energy corresponding to a 90° pulse in a short time. To best exploit the short relaxation times, it is often convenient to use a full 90° excitation pulse and to recycle fast, because magnetization equilibrium is reached quickly. With suitable power supplies and purpose-built probes, short ^{1}H 90° pulses can be achieved (as short as 2 μs at 800 MHz) [1].

When very large spectral widths are needed, pulses of less than 90º can always be used.

A large spectral width also requires an adequate ADC. For example, an ADC with a minimum dwell time of 2.5 μs covers a spectral width of 400 kHz. Modern ADCs have dwell times of 1 μs or less, so that very large spectral width can be acquired with a dynamic range of 16 bit. We recall that the dynamic range is the maximal number of powers of two that can be used to digitalize the intensity of a signal. The value of the dynamic range is not important in the presence of signals all of similar intensities. It becomes important when a weak signal is in the presence of a strong signal. This is because when the receiver gain is reduced to accommodate the strong signal within the dynamic range of the ADC, the weak signal and even the noise may have an intensity smaller than the digital threshold of the ADC itself. This problem may sometimes be serious in protein ^{1}H NMR at high field, and in general in the presence of undeuterated solvents. For these reasons more and more efficient signal suppression techniques have been developed (see below). In the latest generation instruments, however, it has been shown that under non-extreme signal suppression conditions the receiver gain is already high enough that a further increase — that could be made possible by a further reduction of the strongest signals — amplifies the noise to the same extent as the signal [2]. This is even more true when large spectral windows are used.

Associated with large spectral widths there are serious baseline problems. These arise from different sources, some of which are unavoidable because they are connected with the physics of the experiment. An ideal experiment would require: (1) a pulse of infinitely short length; (2) no dead time between the end of the pulse and the start of the acquisition; (3) immediate linear response of the receiver at the start of the acquisition; (4) a filter of perfectly rectangular shape. In practice, any physical receiver requires a finite time to reach linear response conditions after it is turned on. This time is of the order of microseconds and can be as long as some tens of microseconds. Therefore, if the receiver is turned on immediately after the end of the pulse, and if the dwell time is short because of large spectral width (for example, 2.5 μs) the first several data points may have an altered intensity. After Fourier transform, this altered intensity is translated into a baseline distortion. To reduce this problem, a dead time of the order of the time required by the receiver to achieve linearity is introduced before starting the acquisition. This dead time may also be useful to avoid acoustic ringing from the probe, which may be a serious problem at low Larmor frequencies (typically when observing low γ nuclei, or even protons at low fields). Introducing a dead time is equivalent to loosing the first few data points in the FID. In turn, this causes a first order dephasing of the signals, i.e. a dephasing that increases with increasing offset from the carrier. First order phase correction again introduces a baseline distortion. Reconstruction of the first points of the FID by linear prediction techniques may be useful in this respect.

The finite duration of the pulse in a way causes the same kind of problems, because nuclei resonating far from the carrier frequency appreciably start precessing during the pulse itself and therefore are dephased. The quality of the probe (Q-factor, reflected power, etc.) also influences the quality of the baseline. Finally, the analog filters used to decrease the noise by filtering out the high frequency noise coming from outside the spectral window are not rectangular, and introduce a baseline distortion. This last problem is alleviated by the latest generation instruments performing digital filtering, although filtering is still limited in terms of spectral width. As an extreme limit, there is to consider that when nuclear relaxation is of the order of the dead time, precious information on such nuclei is lost before acquisition. This can be the case in paramagnetic solids, but when this occurs the lines are so broad that we are not in the realm of high resolution NMR any longer.

We are now going to discuss some pulse sequences particularly suitable for ^{1}H NMR of paramagnetic molecules when the signals of protons experiencing hyperfine coupling are present together with an overwhelming number of protons which are virtually diamagnetic. This is the case of paramagnetic metalloproteins and of water solutions of both paramagnetic macromolecules and small molecules displaying a strong solvent signal. These sequences can be classified according to three criteria. The first is based on presaturation of one signal or of the signals within a chosen window. After presaturation the excitation pulse is applied. Presaturation can also be achieved off-resonance without actually offsetting the carrier frequency. The second approach is based on pulse sequences that do not excite the solvent frequency or a narrow window of frequencies. These sequences are most useful when exchangeable protons may be saturated upon saturation of the solvent. The third approach is that of using pulse sequences that take advantage of the different T_1 values in order to suppress signals with long T_1 values. In the absence of chemical exchange phenomena, the choice of the strategy is not philosophical but is simply based on suitability to specific samples, availability of specific spectrometers and, to some extent, on personal taste.

9.1.1 Presaturation sequences

The selective saturation of one signal can be achieved with a soft pulse at a given frequency. Selectivity can be further improved by shaping the pulse, for instance as Gaussian [3], sinc [4] (Fig. 9.1A) or Hermitian [5] functions. Very successful are the compositions of three (G3, Fig. 9.1B) or four (G4) Gaussians (Gaussian cascades) in a single pulse [6]. Shaped pulses are routinely achieved by constructing the desired shape with a relatively large number of short rectangular pulses of variable intensity and phases. A suitable modification of the DANTE [7] sequence allows one to set the carrier frequency at any position with respect to the signal to be presaturated. Such a goal is obtained [8] by subdividing the long presaturation pulse having the low power required to saturate the signal, into a large number of short pulses of the same power with durations corresponding

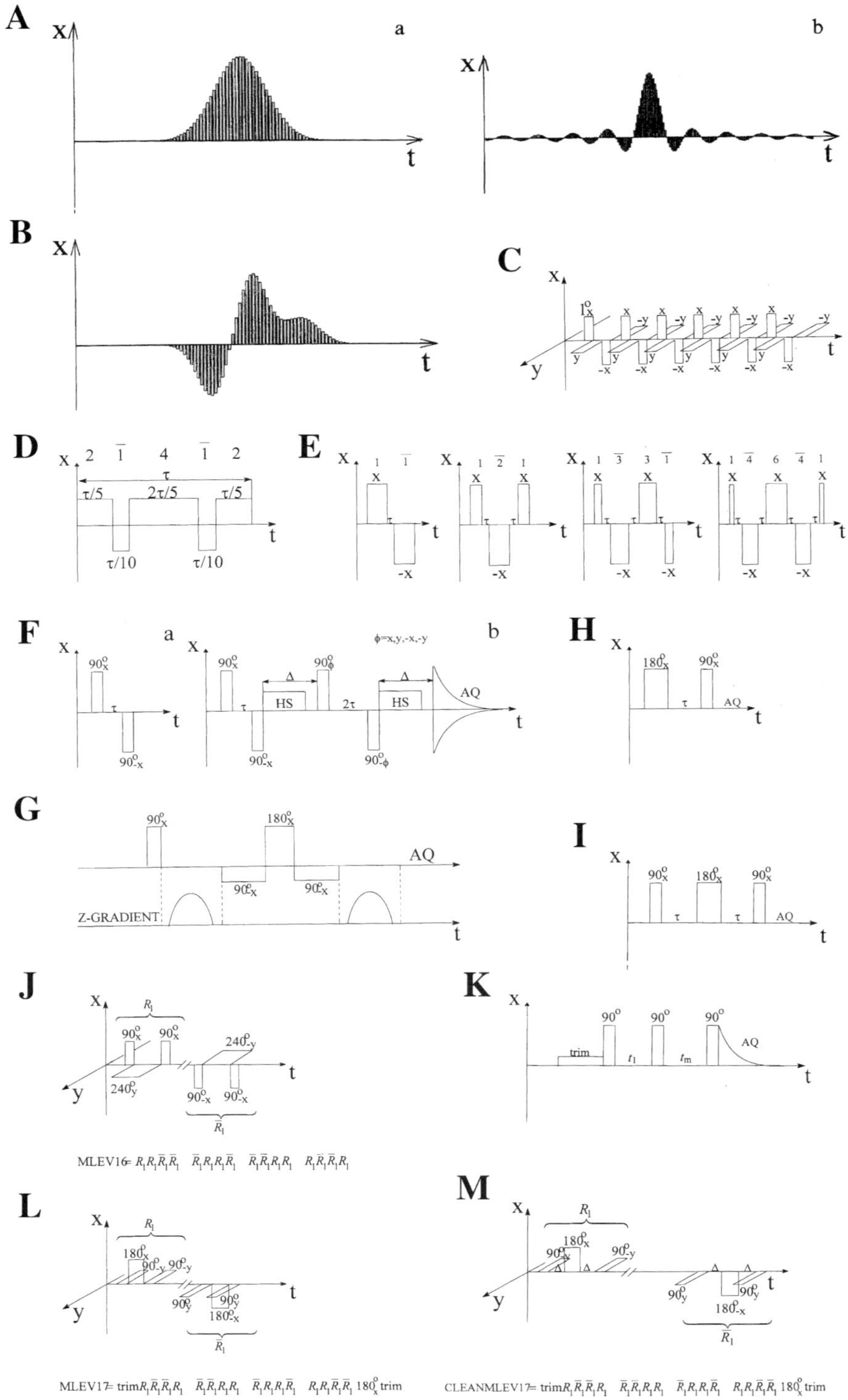
A
a
b
B
C
D
E
F
a
b
φ=x,y,-x,-y
HS
AQ
G
Z-GRADIENT
AQ
H
I
J
K
trim
AQ
L
M

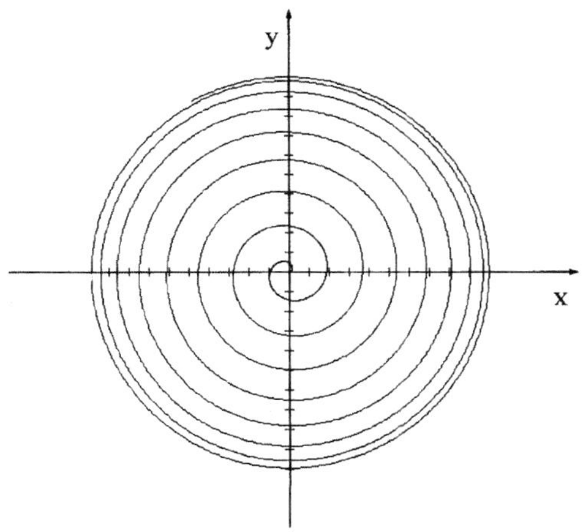

Fig. 9.2. Saturation effect of the modified DANTE sequence on a signal off-resonance with respect to the carrier frequency. The trajectory of the in-plane component of the magnetization is shown. The z magnetization is tilted by the first small angle pulse toward the x axis, and starts precessing toward the y axis. The duration of the pulse corresponds to the duration of a 90° precession, so that at the end of the first pulse the projection lies on the y axis. The phase of the following pulses is rotated in phase with the precession of the signal, in such a way as to 'follow' the spin magnetization in its spiral movement toward the xy plane.

to $1/4$ of the reciprocal of the difference in frequency between the carrier and the signal of interest. The phase of the pulses is incremented by 90° each time, for instance with the scheme $x, y, -x, -y$ (Fig. 9.1C). In this way, after the first pulse the z magnetization of the signal is slightly tilted towards the xy plane and its xy projection has simultaneously precessed by 90°. Therefore, the effect of the second pulse is additive, producing a further tilt toward the xy plane, and so on (Fig. 9.2). Other signals with different carrier frequency end up being periodically out of phase with the pulse phase cycle, and the overall effect is zero. The order in which the phase of the pulses is incremented determines the sign of the offset. For instance, an offset of -2000 Hz is selected by using a time of $1/(4 \times 2000) =$ 0.125 ms and the pulse phases $x, -y, -x, y$.

9.1.2 Selective non-excitation

It is possible to shape an excitation profile with weak pulses in such a way as to have zero excitation at a certain frequency. This was first due to Redfield [9].

Fig. 9.1. (A) Gaussian (a) and sinc (b) excitation profiles. (B) Composite (G3) Gaussian pulse. (C) Train of soft pulses modified after the DANTE sequence to achieve selective off-resonance excitation. (D) Redfield $2\bar{1}4\bar{1}2$ sequence. (E) Binomial $1\bar{1}$, $1\bar{2}1$, $1\bar{3}3\bar{1}$, $1\bar{4}6\bar{4}1$ sequences. (F) JR (a) and compensated JR (or $1\bar{1}1\bar{1}$) (b) sequences. (G) Watergate sequence. (H) Weft (Superweft) sequence. (I) Modeft sequence. (J) MLEV16 sequence. (K) NOESY sequence with trim pulse. (L) MLEV17 sequence with trim pulses. (M) Clean-TOCSY sequence.

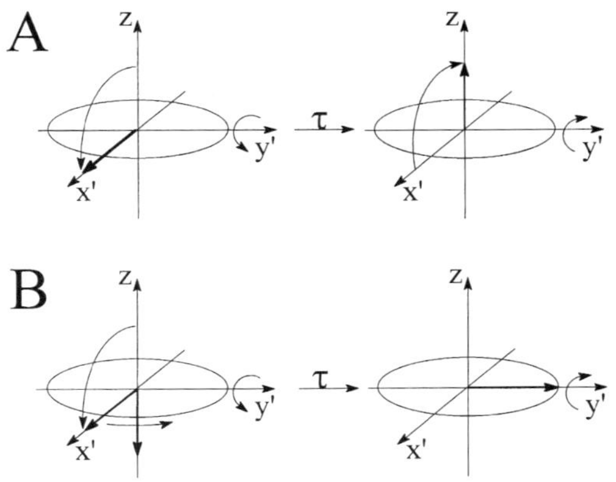

Fig. 9.3. Effect of the JR sequence on a signal on resonance (A) and off-resonance by $1/2\tau$ Hz (B).

The carrier must be placed in the region of the signals to be excited. The $2\bar{1}4\bar{1}2$ Redfield pulse sequence reported in Fig. 9.1D is a suitable example [9]. The total composite pulse duration must be set equal to the reciprocal of the offset of the signal to be non-excited. The carrier can also be located on the unwanted signal, and one of the many binomial pulse sequences (Fig. 9.1E) can be used. Binomial pulses are made of strong rather than weak pulses, interleaved by delays [10,11]. The general concept of the effect of strong pulses can be illustrated by the simple 'Jump and Return' (JR) sequence (Fig. 9.1F), constituted by a 90° pulse followed by a time τ followed by a $-90°$ pulse [12]. If τ is sizably shorter than T_1, the magnetization of the signal at the carrier is first brought along the y axis and then back to the z axis. Signals whose offsets equal $\pm 1/2\tau$ move from the y axis to the $\pm x$ axis after the time τ; therefore, the second $-90°$ pulse leaves them unaltered (see Fig. 9.3). The whole excitation profile is centered at frequencies $\pm 1/2\tau$ and has a width of $1/\tau$. Binomial sequences work similarly, but are constructed in such a way that the total pulse length sums up to a 90° pulse [10,11]. In our experience, the $1\bar{3}3\bar{1}$ sequence has given the best results. A general drawback of all these pulse sequences is that they produce strong baseline distortions in the attempt to phase all signals when they are spread over a large spectral width. A pulse sequence analogous to the JR sequence which minimizes baseline distortions is the $1\bar{1}1\bar{1}$ sequence [13] (Fig. 9.1F).

A very popular method for selective non-excitation employs field gradients along the z axis. After tilting the magnetization of all signals in the xy plane by a non-selective 90° pulse, a field gradient along z applied for a suitable time defocuses all the xy magnetization. If the time of application of the gradient is short enough that no appreciable T_2 relaxation, J modulation and molecular diffusion take place, the defocusing is reversible, and the xy magnetization can be fully recovered by the application of another gradient of opposite sign.

Equivalently, the second gradient can be of the same sign of the first, provided that a 180° pulse is applied in between the two. In the so-called Watergate sequence [14], selective non-excitation is achieved by tailoring the 180° pulse in such a way that all frequencies but the unwanted frequency are irradiated. The simplest scheme employed a $90°_{-x}$(sel)$180°_x$(non-sel)$90°_{-x}$(sel) pulse sequence (Fig. 9.1G). The selectivity increases with the length of the 90° pulses.

The approach discussed in this section is preferable to any approach based on solvent suppression when solvent nuclei are in quasi-slow exchange with signals of interest. Saturation of the solvent nuclei transfers magnetization to the observable signals, decreasing their intensity, whereas non-excitation does not cause any interference.

9.1.3 Selective suppression of signals with long T_1

Whereas the two strategies of solvent suppression techniques described above do not show any particular advantage or disadvantage when applied to paramagnetic systems with respect to diamagnetic systems, the strategies based on differential relaxation properties have obvious intrinsic advantages in paramagnetic systems because of the larger spreading of nuclear relaxation times involved. The simplest approach to take advantage of the fast relaxation times of the signals of interest is that of acquiring the spectrum after a simple 90° pulse with a recycle time short with respect to the T_1 values of the signals to be suppressed, but long with respect to the T_1 values of the paramagnetic signals. Since the latter have also short T_2, short acquisition times are required. The consequence is that the slow relaxing signals are largely saturated, whereas the fast relaxing signals almost fully recover their initial intensity after each cycle. Weft [15] and Superweft [16] pulse sequences (Fig. 9.1H) are based on the 180 τ 90 acquisition pulse sequence. When $\tau = T_1 \ln 2$ the signal with that T_1 has an intensity of zero. If the recycle time is longer than $5T_1$ (i.e. the system has fully recovered at each cycle) zero intensity of the signal of interest is maintained all over the experiment. This is the Weft pulse sequence [15]. During the time τ the fast relaxing signals have essentially recovered their magnetization and are detected. If the recycle time is chosen to be short with respect to the T_1 values of the signals to be suppressed, but long with respect to the T_1 values of the signals of interest, it has been shown that a τ value exists that after a few cycles, i.e. at steady state conditions, effectively zeroes the intensities of the signals with long T_1 [16]. An example of application of Superweft is provided in Fig. 9.4A, where the spectrum of a protein containing a heme with low spin iron(III) in D_2O is reported [17]. This is obtained with a Superweft sequence with $\tau = 250$ ms and recycle time $= 250$ ms. The spectrum shows a number of well-resolved signals outside the diamagnetic region. When the signals of the axially coordinated histidine are looked for, advantage is taken of the fact that they have T_1 values of the order of a few milliseconds, whereas the other hyperfine shifted signals have T_1 values of about 100 ms. Therefore, a Superweft

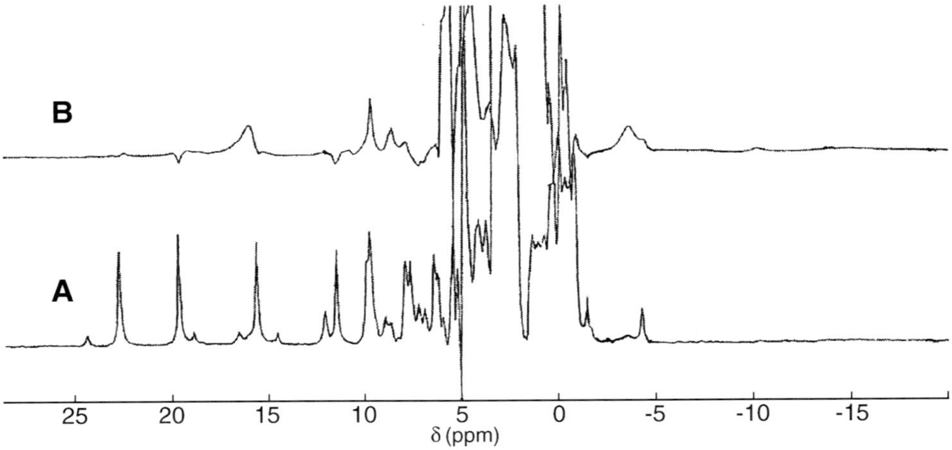

Fig. 9.4. 200 MHz ^{1}H NMR spectra in D_2O of the cyanide adduct of a cytochrome *c* mutant lacking the axial methionine (Ala80cyt *c*) recorded using the Superweft sequence with $\tau = 250$ ms and recycle time of 250 ms (A) and $\tau = 20$ ms and recycle time of 33 ms (B). In the latter spectrum the two signals of the axially coordinated histidine are apparent at 16.1 and −3.4 ppm [17].

with $\tau = 20$ ms and recycle time of 33 ms provides the spectrum of Fig. 9.4B where only the two non-exchangeable histidine ring protons are evident [17].

Weft or Superweft sequences are also widely used in 2D spectroscopies (see Section 9.4); in general, it is sufficient to place the 180°–τ part of the sequence in front of any 2D sequence to achieve the desired signal suppression.

The Modeft pulse sequence [18] drives the slow relaxing signals to equilibrium when the acquisition pulse is delivered. Modeft means modified driven equilibrium Fourier transform [19]. The pulse sequence is 90 τ 180 τ 90 acquisition (Fig. 9.1I). If τ is short with respect to T_1 the protons are driven to equilibrium. Signals with T_1 short with respect to τ recover after each pulse and the last 90° pulse acts as a normal excitation pulse. In Fig. 9.5 ^{1}H Modeft spectra of anion adducts of a cobalt(II) ($S = 3/2$) protein of molecular weight 30,000 are shown [20]. The Modeft sequence is worse than the Superweft sequence in suppressing a single signal like the solvent signal, but performs comparably, and slightly better in some cases, when elimination of more than one slowly relaxing signal is needed. Typically, one might want to record a spectrum of a metalloprotein by the use of the Superweft or the Modeft sequences when the few fast relaxing signals lying underneath the diamagnetic envelope of the many slowly relaxing signals are to be emphasized.

Other strategies to achieve this task are provided by the broadband (BB) saturation sequences (like MLEV16 (Fig. 9.1J) [21], WALTZ16 [22], GARP [23], DIPSI [24], etc.). The power of the BB saturation pulse can be adjusted in such a way as to effectively saturate the slowly relaxing signals and only marginally the

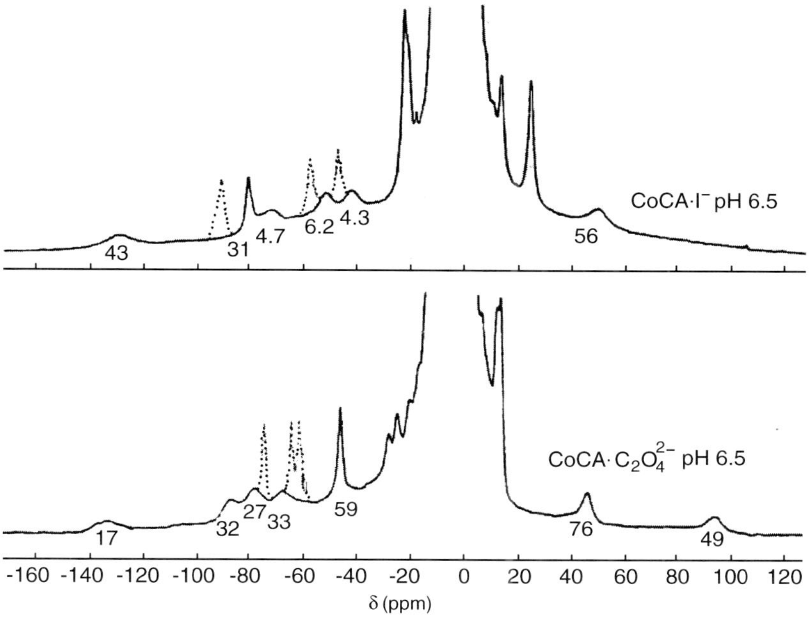

Fig. 9.5. 60 MHz ^{1}H NMR Modeft spectra of cobalt-substituted carbonic anhydrase (MW 30,000) adducts with iodide and oxalate. The T_1 values for some signals obtained with Eq. (9.1) (see later) are indicated. The dashed signals disappear in D_2O [20].

fast relaxing ones. A 90° pulse immediately following the BB saturation sequence then gives a spectrum where the relative intensity of the fast relaxing signals is enhanced. Alternatively, the power of the BB saturation pulse can be high enough to saturate all signals. In this case, the following 90° detection pulse is applied after a time τ, adjusted in such a way that fast relaxing signals are appreciably recovered while slow relaxing signals are not.

While the Weft, Superweft and Modeft sequences are based on differences in T_1, BB saturation techniques are based on the difference in saturability of the signals, which in turn depends on $(T_1T_2)^{1/2}$. Therefore, in cases where $T_2 \ll T_1$ for the paramagnetically affected signals (as can be the case in the presence of strong Curie relaxation), BB techniques may actually give better discrimination. Spin-lock techniques, based on differences in $T_{1\rho}$ ($\cong T_2$), should perform best. However, applications of the latter are scarce, probably owing to the difficulty to efficiently irradiate a large spectral window.

9.1.4 Choice of magnetic field

As the major drawback in the observation of paramagnetic signals is ultimately the width of the lines, it is worth recalling again the various mechanisms causing line broadening. These mechanisms have been illustrated in Sections 3.4–3.6, and their relative importance has been discussed in Section 3.8. The relevant equations describing the three contributions to R_{2M} are Eqs. (3.17) (dipolar), (3.27) (contact) and (3.30) (Curie relaxation). The first two contributions to line broadening decrease with magnetic field (Figs. 3.10 and 3.12), while the third increases with magnetic field (Fig. 3.13). This observation suggests that the experimentalist may be able to use the magnetic field as a parameter to optimize signal detection. This attitude has always been peculiar to paramagnetic NMR, whereas it is now becoming popular also in the field of diamagnetic proteins, thanks to the field dependence of the cross-correlation exploited in TROSY-type experiments (see Section 8.7). It is obvious that, in the absence of adverse effects, the higher the field, the better. In fact, if the linewidth (in hertz) is field-independent, the resolution increases linearly with the field: in other words, the linewidth in parts per million *decreases* with increasing field. However, Curie relaxation increases with the square of magnetic field. Therefore, when Curie relaxation is dominant, an increase in field will actually cause a decrease in resolution. In practice, Curie relaxation is dominant when its correlation time τ_r is much larger than τ_s, as in this case the latter is the dominant correlation time for dipolar and contact mechanisms. Furthermore, the importance of Curie relaxation increases with the electron spin quantum number S (or J for lanthanides), because dipolar and contact relaxation mechanisms depend on $S(S+1)$ (or $J(J+1)$), whereas Curie relaxation depends on the square of these quantities. Fig. 9.6 illustrates

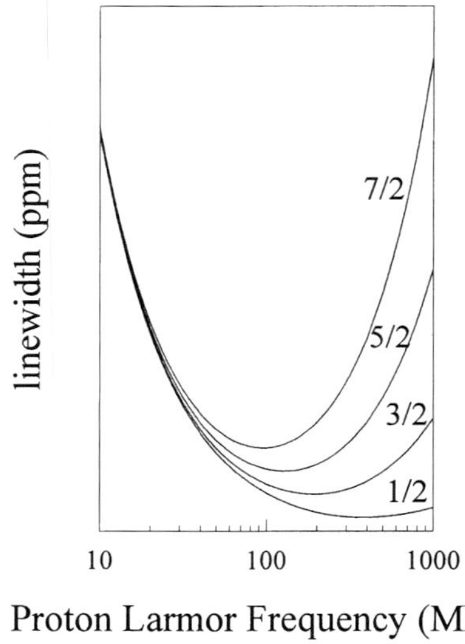

Fig. 9.6. Calculated field dependence of the linewidth (ppm) for a proton interacting with a paramagnetic metal ion in a macromolecule with $\tau_r = 10^{-8}$ s, $\tau_s = 10^{-12}$ s, and $S = 1/2, 3/2, 5/2$ (or $J = 7/2$ for a lanthanide).

the field dependence of the linewidth (in parts per million) for a middle-sized macromolecular system ($\tau_r = 10^{-8}$ s) containing a fast relaxing paramagnetic metal ion ($\tau_s = 10^{-12}$ s) with S (or J) equal to ½, 3/2, 5/2 and 7/2. It appears that the optimal magnetic field decreases from about 400 MHz to 200 MHz, 120 MHz and 90 MHz. Note that these conditions are relatively common, and far from being extreme. In fact, τ_r values may be easily longer than 10^{-8} s (a τ_r of 10^{-8} s corresponds to a 30,000 MW protein); τ_s values may be shorter, and J values may be larger (for example in lanthanides, where τ_s values close to 10^{-13} s have been reported and J values can be as large as 8).

9.2 MEASUREMENTS OF T_1 AND T_2

As far as T_1 is concerned, we should remember that longitudinal relaxation can be multiexponential. Therefore, different information can be obtained, depending on whether a given signal is selectively excited or it is excited together with other signals. We usually refer to T_1 as an experimental T_1 value obtained by assuming that the magnetization recovery is exponential. This is very often a better approximation in paramagnetic systems than in diamagnetic ones, because the unpaired electron acts as a sink for nuclear spin energy, thereby making the whole magnetization recovery closer to a single exponential (Sections 1.7.4 and 7.2.2). When a single signal is excited, we refer to selective T_1, which is thus distinguishable from a non-selective T_1 obtained when more signals are simultaneously excited.

Non-selective T_1 can be measured with the usual inversion recovery 180 τ 90 pulse sequences [25]. It should be noted that, if the frequency range covered is large, the 180 or even the 90 pulse may not be the same all over the spectral region of interest. It may then be convenient to select ranges of excitation. Sometimes it may be convenient to measure T_1 with the Modeft pulse sequence 90 τ 180 τ 90 (see Section 1.3): by making τ shorter and shorter we eventually have zero intensity also of the fast relaxing signals. The signal intensity as a function of τ is given by [18]

$$M_z(\tau) = M_z(\infty)\left[1 - 2\exp(-R_1\tau) + (1-a)\exp(-2R_1\tau)\right] \tag{9.1}$$

where $R_1 = T_1^{-1}$ and a is a parameter that compensates for misadjustments of the first 90° pulse.

Selective T_1 values are generally measured with the 180 τ 90 pulse sequence using a soft 180° pulse. When the nuclear relaxation times are short, it may become impossible to invert the magnetization and at the same time maintain the required selectivity with the soft pulse. When such difficulties arise, a good compromise is to use a soft pulse that can at least saturate the signal. Then, the sequence becomes equivalent to a 90 τ 90 pulse sequence.

2D experiments can also be modified to allow for the measurements of T_1. A simple way to measure T_1 is to introduce a 180°–τ module in front of any homonuclear 2D sequence, and recording several 2D spectra with different τ values [26–29]. In this way, the initial magnetization value of each nucleus at the beginning of the 2D experiment will be a function of τ, and so will be the intensities of the corresponding diagonal peak and of all its cross peaks in the same row. In heterocorrelated experiments, if the 180°–τ module is used at the proton frequency, then the T_1 of the protons is obtained [28]. To obtain the T_1 of the heteronucleus, whose evolution is usually in the indirect dimension, the τ value is inserted within an HSQC-type sequence (see Section 8.7). A refocused Inept provides single quantum (SQ) coherence of nitrogen spins. After t_1 evolution, a 90° pulse rotates nitrogen magnetization on the z axis. The latter relaxes during the τ value and the resulting magnetization is then converted to proton SQ coherence through a inverse refocused Inept [30,31].

Fast relaxing nuclei are characterized by a sizable linewidth. The measurement of the linewidth represents the easiest and most straightforward way of measuring T_2^{-1} from the relationship

$$T_2^{-1} = \pi \Delta\nu \tag{9.2}$$

where $\Delta\nu$ is the linewidth at half peak height. Such procedure is absolutely adequate in paramagnetic systems when the lines are relatively large, since the relative contribution of field inhomogeneity is in any case small. Of course, when the linewidth becomes narrow, as in a diamagnetic system, the usual Carr–Purcell–Meiboom–Gill (CPMG) technique [32] should be used. CPMG modules can also be used in homonuclear and heteronuclear 2D experiments [33,34].

In the case of T_1 measurements we have mentioned that cross relaxation provides multiexponential magnetization recovery (Sections 1.7.4 and 7.2.2). A far less known analogy may occur in the linewidths, as already discussed (Section 8.8) when two protons are dipole–dipole coupled and cross correlation occurs between Curie relaxation and proton–proton dipolar relaxation. In this case, we are in the presence of two overlapping signal components with different linewidths, i.e. of biexponentiality in T_2 [35]. Pulse sequences are available to remove the effects of cross correlation [36]. Such effects are common in paramagnetic metalloproteins where Curie relaxation is usually relevant (in principle, such cross correlation effects can be operative also in the case of T_1, although only to the extent that Curie relaxation on T_1 is effective).

9.3 MEASUREMENTS OF NOE

The nuclear Overhauser effect was predicted in 1953 [37], experimentally demonstrated in 1955 [38], and widely used since then to obtain structural and conformational information in diamagnetic small molecules. As it appears from

Section 7.2, NOEs are small and positive in small molecules. Cross relaxation is small and spin diffusion effects are also small. When the rotational correlation time increases, the sign of the NOE becomes negative, the absolute value increases, and cross relaxation effects become larger. As a result of the latter effects, the so-called spin diffusion takes place, which makes less straightforward the interpretation of the relationship between NOE and internuclear distances. In diamagnetic macromolecules, nuclear Overhauser effects are almost exclusively measured through 2D or 3D experiments, and distance information often extracted with the help of proper algorithms. In small molecules, 1D NOE is still used to obtain accurate selected internuclear distances when the signals are well separated.

The 1D NOE technique has had a strong revival when applied to paramagnetic molecules. Here the signals to be irradiated can be well separated (hyperfine shifted), nuclear relaxation is more affected by the coupling with the unpaired electrons than by the coupling with other protons and, if the molecule is large, large effects can be measured without the drawback of having all the signals in a narrow spectral region. If one is at the extreme limit of detectability of dipolar connectivities (as is often the case in fast relaxing and relatively diluted samples), 1D NOE is the most suitable technique, as discussed in Chapter 7.

As discussed in Sections 1.1 and 1.3, it is difficult to saturate in a selective way a fast relaxing signal. Selectivity here is an absolute requirement because the wings of the excitation profile of a soft pulse may well partially saturate other signals, with the consequence of appearance of artifacts which vary from case to case but may lead to gross misinterpretations. The use of suitably shaped pulses designed to be more selective than a rectangular pulse is advisable. Even with a perfectly selective irradiation, other signals can be partly saturated because their tails may extend down to where the irradiation is applied. To further remove the residual effects of non-perfect selectivity and of tail excitation, the most common procedure is that of alternating experiments with the saturation pulse applied on-resonance and off-resonance, and subtracting the latter from the former. Even so, off-resonance effects on signals very close to the irradiated one cannot be readily discriminated from true NOEs that in macromolecules are negative. Fig. 9.7A provides a clear example [39]. Upon irradiation of signal *g*, signal *f* gives NOE while signal *d* is affected by off-resonance effects. This could be demonstrated by performing an irradiation 'profile' (Fig. 9.7B): the effect of *f* is maximal when the irradiation frequency is on *g*, whereas the effect on *d* steadily increases with increasing irradiation frequency [39].

In our experience, the best experimental scheme is to choose two off-resonance positions symmetrical with respect to the irradiated signal and typically offset by one to two times the linewidth of the irradiated signal (Fig. 9.8).

It often happens that there is at least one signal too close to the signal to be irradiated to allow for the setting of the two off-resonance frequencies in the optimal position. In fact, if another signal is close to one of the two off-resonance frequencies, in the difference spectrum this signal appears as strongly positive.

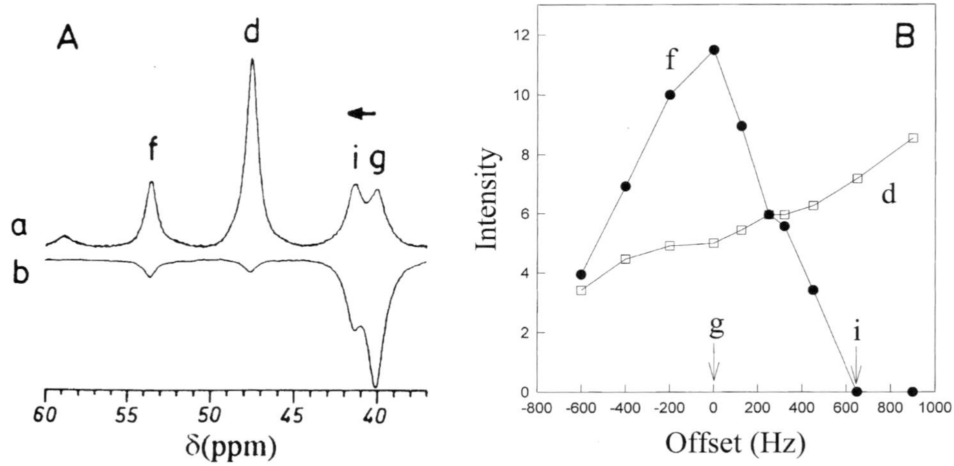

Fig. 9.7. 1D NOE difference experiments on met-aquo myoglobin. (A) Reference spectrum (a) and difference spectrum (b) observed upon saturation of peak *g*. (B) Intensity of negative *f* and *d* signals in (b) as a function of the irradiation frequency. Signal *f* is maximal when the irradiation frequency is on *g*, signal *d* shows a steady increase with increasing frequency [39].

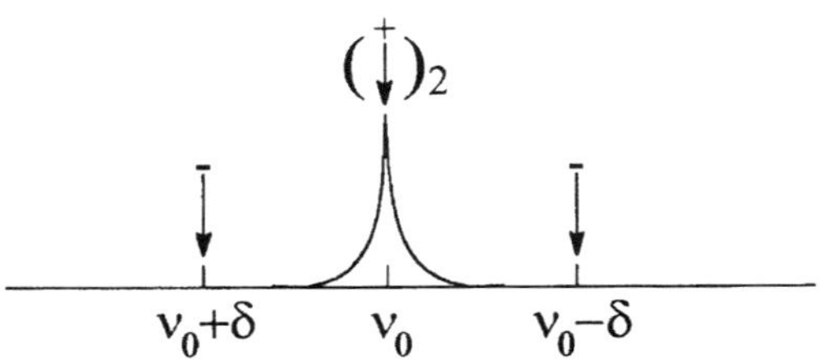

Fig. 9.8. Irradiation scheme used in NOE experiments to minimize off-resonance effects. The r.f. is alternately placed at ν_0, $\nu_0 + \delta$, $\nu_0 - \delta$. The spectra obtained upon irradiation at $\nu_0 \pm \delta$ are subtracted from those obtained upon irradiation at ν_0.

If this signal gives NOEs with the same signal(s) which experience NOE from the irradiated signal, the two NOEs are opposite in sign and the unwanted one may reduce or cancel the one which is looked for. A general strategy to eliminate this effect is to place the off-resonance offset beyond the unwanted signal. If the offset is far enough, the latter signal is now negative (for macromolecules), since the off-resonance from the central irradiation position is dominant. It is easy to show that an off-resonance position always exists for which the intensity of the unwanted signal is rigorously zero. The other offset is, of course, always symmetrical with respect to the irradiated signal. The optimal position can be determined empirically with a small number of scans before starting the real experiment. In Fig. 9.9, the spectrum of the oxidized high potential iron–sulfur protein from *E. halophila* iso I is reported [40], together with the NOE difference

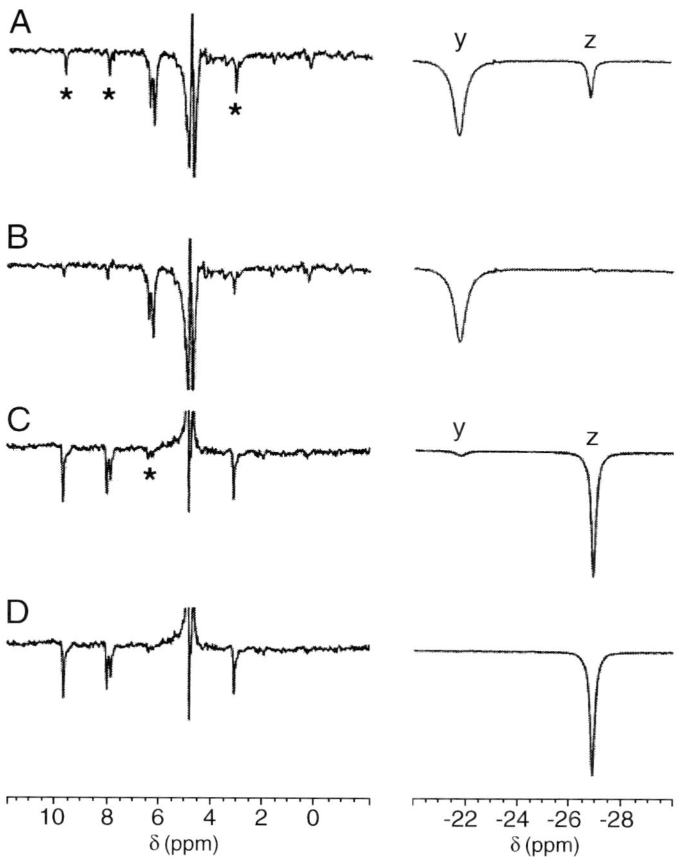

Fig. 9.9. 600 MHz 1D NOE difference spectra of the oxidized form of the high potential Fe_4S_4 protein from *E. halophila* iso I [40]. Spectra (A) and (C) are obtained by irradiating each of the two geminal β-CH_2 protons (y and z) of the cluster-ligated Cys-39. Spectrum (B) is obtained by subtracting from spectrum (A) spectrum (C) multiplied by a factor such that signal z is canceled; spectrum (D) is obtained by subtracting from spectrum (C) spectrum (A) multiplied by a factor such that signal y is canceled. Note the reduction of intensity of the signals marked with asterisks.

spectra obtained upon saturation of signals y and z, using the above irradiation scheme. Signals y and z happen to be geminal β-CH_2 protons of the same cysteine residue and to have rather long T_1 values. Therefore, despite the optimization of the off-resonance frequencies, a sizable real NOE is present from one signal to another in both cases. To further eliminate the secondary NOEs from the geminal signal, double difference NOE spectra were obtained by subtracting from one NOE difference spectrum the other one, appropriately scaled.

Approaches of this kind are also particularly important when the NOE to be observed is on a signal close to the irradiated one. In the case of the two β-CH_2 protons of an Asp residue bound to a high spin cobalt(II) in the protein superoxide

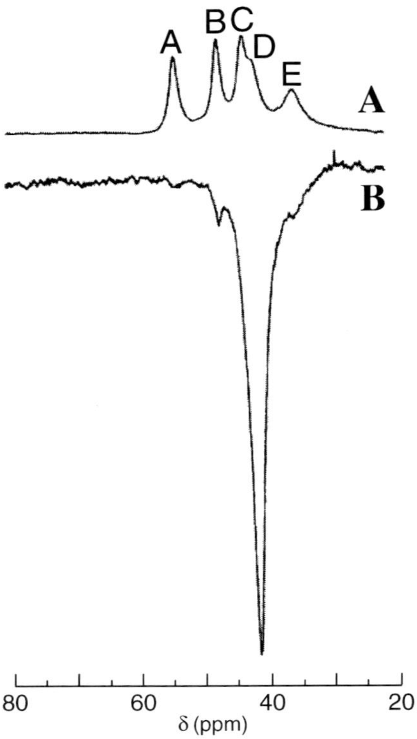

Fig. 9.10. (A) 200 MHz ^{1}H NMR spectrum of copper-depleted, cobalt-substituted Cu,Zn superoxide dismutase; (B) difference between the difference NOE spectra obtained by irradiating signals *D* and *C*, respectively. The spectra demonstrate that the NOE on *E* arises from *D* and not from *C*. The off-resonance position was placed symmetrical with respect to *E* in both cases [41].

dismutase, the two signals *D* and *E* are very broad (Fig. 9.10A). The NOE between the two signals was observed by irradiating signal *D* and by recording the difference spectrum with one off-resonance position symmetrical with respect to the position of signal *E* [41]. In principle, in this way any intensity detected at the *E* position is due to a real NOE. In practice, a strong distortion of the baseline at the *E* position is present because of the tails of signal *C*, experiencing off-resonance from *D*. Another experiment is performed by irradiating this third signal with the off-resonance position still symmetrical with respect to *E*, and by further subtracting the two NOE difference spectra one from the other. The double difference NOE spectrum is shown in Fig. 9.10B.

It is often necessary to obtain a profile of NOE intensities vs. the irradiation time, called NOE buildup. A series of truncated NOEs are thus needed (Section 7.3). The only caveat with respect to buildup experiments in a diamagnetic case is that there is a lower limit in the irradiation time because even a truncated NOE needs saturation of the irradiated signals. Since 'instantaneous' saturations are not possible, but short saturation times are needed to measure the first points of the buildup, there is the risk that at the beginning of the buildup the irradiated signal is not 100% saturated [39]. In practice, buildup measurements are feasible when the irradiated signal relaxes faster than the responding signals, but hardly when the reverse is true. Even in the most favorable case, the intensities of the very first points may not be accurate; as a consequence, it is always a risk to use buildup experiments to discriminate between primary and secondary NOEs. However, in paramagnetic systems, secondary NOEs are usually small and, in general, recognizable from other indirect evidence (see Chapter 7).

As far as the transient NOE is concerned, the problem of the selective 180° pulse has already been addressed. The non-optimal selectivity leads to dramatic artifacts [42]. Therefore, transient NOE is not the technique of choice in paramagnetic systems [43].

9.4 2D SPECTRA

9.4.1 NOESY

When setting a NOESY experiment on a paramagnetic molecule the mixing time t_m should be of the order of the T_1 of the signals between which cross peaks are looked for. When the T_1 values are very different, and this is common in paramagnetic molecules, especially if there are protons close to the paramagnetic center and protons far from it, the best mixing time should be calculated case by case from Eq. (8.2). For the reader's convenience, Table 8.1 is provided, with the relative T_1 values spanning two orders of magnitude. Note that, when the T_1 values are very different, the best mixing time, which is always intermediate between the two T_1 values, can be as large as five times the shorter T_1 (Section 8.3).

A comment is due to the comparison of the intensity of a steady state NOE with that of a transient NOE in the case of signals with different T_1 values. This is because the intensity of a NOESY cross peak is related to the intensity of a transient NOE (Sections 7.4 and 8.3) and because in paramagnetic systems we may be at the lower limit of cross peak detectability. Therefore, a steady state NOE obtained by saturating the fast relaxing signal is always advisable to extend the detectability of its dipolar connectivities as much as possible. Table 9.1 shows the advantage of steady state vs. transient NOE for a range of ratios of T_1 values. It appears that steady state NOE is always superior to transient NOE, except when the T_1 of the irradiated signal is longer than that of the responding signal by more than a factor of two. In paramagnetic metalloproteins, the faster relaxing signals are often outside the diamagnetic envelope and are separated one from the other, so that they can be irradiated quite easily. The slower relaxing signals are often in the diamagnetic envelope. It is therefore a good practice to measure the steady state NOE by irradiating each of them prior to performing 2D experiments.

TABLE 9.1
Comparison between steady state (SS) and maximal transient (TR) NOE intensities (%) for various T_1 (ρ^{-1}) values (ms) of the two spins I and J, calculated using Eqs. (7.10) and (7.19) (the negative signs are omitted)

T_1^I		T_1^J							
		128	64	32	16	8	4	2	1
128	SS	12.80	6.40	3.20	1.60	0.80	0.40	0.20	0.10
	TR	9.44	6.40	4.03	2.38	1.33	0.72	0.37	0.19
64	SS	12.80	6.40	3.20	1.60	0.80	0.40	0.20	0.10
	TR	6.41	4.71	3.20	2.02	1.19	0.67	0.36	0.19
32	SS	12.80	6.40	3.20	1.60	0.80	0.40	0.20	0.10
	TR	4.03	3.20	2.35	1.60	1.01	0.59	0.33	0.18
16	SS	12.80	6.40	3.20	1.60	0.80	0.40	0.20	0.10
	TR	2.38	2.02	1.60	1.18	0.80	0.50	0.30	0.17
8	SS	12.80	6.40	3.20	1.60	0.80	0.40	0.20	0.10
	TR	1.33	1.19	1.01	0.80	0.59	0.40	0.25	0.15
4	SS	12.80	6.40	3.20	1.60	0.80	0.40	0.20	0.10
	TR	0.72	0.67	0.59	0.50	0.40	0.29	0.20	0.13
2	SS	12.80	6.40	3.20	1.60	0.80	0.40	0.20	0.10
	TR	0.37	0.36	0.33	0.30	0.25	0.20	0.15	0.10
1	SS	12.80	6.40	3.20	1.60	0.80	0.40	0.20	0.10
	TR	0.19	0.19	0.18	0.17	0.15	0.13	0.10	0.07

The calculations are performed for a $\sigma_{I(J)}$ value of $-1\ \text{s}^{-1}$. Steady state NOE is always superior to transient NOE, except when T_1^I of the irradiated signal is more than twice the T_1^J of the responding signal (upper right part of the table).

In summary, when performing a NOESY spectrum of a paramagnetic molecule containing both fast and slow relaxing nuclei, it is convenient: (1) to record a NOESY spectrum with a mixing time which matches the short relaxation times in order to detect connectivities between fast relaxing signals; (2) to record a NOESY spectrum with a mixing time which matches the relaxation times of the slow relaxing nuclei in order to detect connectivities between slow relaxing signals; (3) to record a NOESY with a mixing time taken from Table 8.1 depending on the T_1 of the fast and slow relaxing nuclei in order to detect connectivities between fast and slow relaxing signals; (4) to measure the steady state NOEs by saturating the fast relaxing signals in order to be sure to detect connectivities with slow relaxing nuclei as much as possible.

A further parameter in measuring NOESY spectra is the recycle time. In small molecules it should be ideally five times the longest T_1 of the signals for which connectivities are looked for. The $5T_1$ is needed to let the nuclei of interest recover their magnetization. In macromolecules such time is better set approximately equal to the longest T_1 among the signals of interest: the loss in cross peak intensity is overcome by the gain in number of cycles within the same experimental time. When recording a NOESY spectrum with a short recycle time the transverse magnetization of the slow relaxing signals has no time to disappear before the next scan; this gives rise to various artifacts in the 2D spectrum. For this reason, a CW pulse of the duration of some milliseconds, called trim pulse, can be used to destroy transverse coherence before the following experiment. A convenient pulse sequence is shown in Fig. 9.1K. The intensity of the pulse should be set according to the spectral region over which the trim is needed. Typically, trim pulses of the order of 1 ms are used for paramagnetic systems.

In a NOE–NOESY experiment (Section 8.9) the connectivities between a fast relaxing signal and a set of slow relaxing signals are obtained simultaneously with those within the above set of slow relaxing signals. Therefore the NOESY parameters should be set according to the slow relaxing signals.

NOESY experiments aimed at identifying dipolar connectivities between bound water molecules and protein protons have been developed over the last decade [44]. Among them, the ePHOGSY experiment [45] has been successfully applied on paramagnetic metalloproteins [46,47].

9.4.2 *COSY and spin-lock experiments*

A typical COSY sequence is 90 τ 90 acquisition. If two signals A and B are coupled by scalar interactions, a cross peak is expected. The shape of the cross peak depends on several factors. The first one regards whether the signal linewidths are smaller or larger with respect to J. In the former case, the cross peak consists of four components arranged on a square and separated by J. The four components have antiphase structure, i.e. in a phase-sensitive experiment phased in absorption mode there are two positive and two negative peaks. The routine phase-sensitive experiment introduces sensitivity to the phase, which allows the operator to choose absorption or dispersion mode for the cross peaks. There is no advantage in choosing the dispersion mode because a line in dispersion mode is broader than in absorption mode. In magnitude mode the phase is lost and all four peaks are positive.

In the case in which T_2^{-1} is larger than J, one would think that the antiphase character of the cross peaks in the TPPI mode cancels partially or totally the cross peaks. Indeed, much of the intensity of the signal is lost. The loss can, however, be reduced if the cross peaks are phased in dispersion mode [48] or if the experiments are performed in magnitude mode [36] (Section 8.5). The occurrence of coherence transfer phenomena may give rise to COSY cross peaks even in the absence of scalar coupling (Section 8.8) [35]. As long as this is kept in mind, COSY cross peaks between very broad lines can be looked for, and interpreted as dipolar correlation cross peaks.

TOCSY spectra are critical to perform on paramagnetic systems when the required spectral window is large, because of the need to irradiate the spectral window efficiently with each component of a long train of many pulses (like for instance the MLEV17 sequence [49], Fig. 9.1L). Overheating of the sample can occur, and in extreme cases the probe coil may not stand the pulse power for the required time. Other spin lock sequences constructed from WALTZ-type [22] or DIPSY-type [24] sequences may allow some reduction of the power needed, the spectral window being the same.

Another improvement with respect to these problems is represented by the so-called Clean-TOCSY experiment [50]. Clean-TOCSY was originally devised to avoid the build up of ROE during a TOCSY experiment. In small molecules ROE cross peaks in a TOCSY spectrum may be mistaken for scalar TOCSY cross peaks, while in large molecules, where ROESY and TOCSY have opposite signs, the presence of ROESY may accidentally cancel TOCSY cross peaks. In Clean-TOCSY, short delays are interleaved between the pulses of the MLEV17 sequence (Fig. 9.1M) in such a way that the ROE effect is lost. This has also the side-effect that, for the same overall spin lock time t_m, less energy (power $\times$ time) is delivered to the sample. Therefore, the length of each pulse can be, for instance, halved, and its power doubled without delivering to the sample more energy than in the original TOCSY experiment. In principle, even full high power pulses can be used, provided that enough delay is interleaved between them.

9.4.3 Heteronuclear correlation experiments

A crucial aspect related to the successful use of heteronuclear HSQC and HMQC experiments (see Section 8.7 and Figs. 8.2G and 8.2H) in paramagnetic systems is the efficiency of polarization transfer between proton and the attached heteronucleus. Polarization transfer allows to obtain transverse magnetization of the heteronucleus with a larger sensitivity, close to that of proton. In fact, the sensitivity gain due to polarization transfer with respect to the direct excitation of the heteronucleus depends on γ_p/γ_X, where γ_p is the proton magnetogyric ratio and γ_X is the magnetogyric ratio of spin X. In the absence of relaxation the maximum of polarization transfer occurs for $t_d = 1/(2J_{HX})$. When paramagnetic contributions enhance R_2 proton relaxation, the efficiency of polarization transfer decreases and the maximum transfer occurs for shorter t_d [51]. This is analogous to what already described for the COSY experiments (see Section 8.5). At increasing R_2 proton relaxation, the loss of efficiency of polarization transfer can be larger than the sensitivity gain due to polarization transfer. This will make the use of HSQC or HMQC sequences unfavorable with respect to direct excitation of the heteronucleus.

Indeed, paramagnetic broadening is proportional to the square of the nuclear magnetogyric ratio γ. Therefore, the inverse detection of heteronucleus (i.e. the transfer of magnetization from the X nucleus to the proton followed by proton detection) may prevent the observation of X signal because of proton T_2 relaxation. In fact, the relative values of γ for ^1H, ^{13}C and ^{15}N nuclei are $1:1/4:1/10$, and thus the relative contribution to overall relaxation arising from the hyperfine interaction is $1:1/16:1/100$, respectively. Therefore, to identify

^{13}C or ^{15}N resonances that escaped detection through inverse detection technique, direct detection of ^{13}C or ^{15}N can be used [52].

9.4.4 3D experiments

As the variety of 3D and higher dimensionality experiments increases with the number of dimensions, it is inappropriate to describe any of them in detail here. However, it may be useful to summarize a few guidelines on how to select the best high-dimensionality experiments for a paramagnetic system. (i) Among different sequences that provide the same — or almost the same — type of information, always select the simplest in terms of number of pulses and the shortest in terms of duration. This suggestion is obviously dictated by the need to minimize loss of magnetization/coherence due to paramagnetic relaxation. (ii) In macromolecules, always select the sequences that minimize the duration of in-plane evolution of spins, as T_2 is usually shorter than T_1 especially when paramagnetic relaxation is dominated by Curie relaxation. (iii) If possible, let the longer in-plane evolution be that of heteronuclei rather than that of protons, as relaxation depends on the square of the magnetogyric ratio and is therefore 16 times and 100 times less for carbons and nitrogens (respectively) than it is for protons (see Section 4.3). (iv) Whenever possible, substitute a 3D experiment with a few 2D experiments . . .

9.5 SUGGESTIONS FOR SPECTRAL ASSIGNMENT

At the end of this book it may be useful to summarize a possible strategy for spectral assignments. Some hints are so simple that they have never been mentioned previously, whereas other hints are implicit in the various parts of the book.

When analyzing a proton NMR spectrum one should look at the intensities of the signals in a spectrum acquired with recycle time much longer than the T_1 values of the signals. This helps in deciding whether the signal belongs, for instance, to a methyl or to a set of magnetically equivalent nuclei. Another obvious criterion is to compare the spectra where a proton in a given position is substituted by another nucleus.

The first criterion really related to the content of this book is the analysis of T_1 and T_2. As the dominant contribution to nuclear relaxation is dipolar in nature, T_1^{-1} and linewidths will decrease as we move farther from the paramagnetic center. Even the contact contribution to relaxation often decreases with the number of chemical bonds from the paramagnetic center. A caveat, however, should be given. Spin density transfer causes ligand-centered relaxation. Significant spin density on a π orbital of an sp^2 carbon may relax an attached proton more than the paramagnetic center itself, owing to the different distances and to the sixth power dependence on distance.

With relaxation data at hand, and with some experience, a tentative assignment can be proposed. This is how NMR spectroscopists of paramagnetic molecules operated from the 1960s to the 1980s. The first order J splitting is only rarely observed in paramagnetic molecules, and therefore is not a useful tool for the assignment.

At this point NOE and NOESY experiments are needed. In this way we learn which proton is close to which other proton, and the picture becomes sharper. Sometimes, when the correlation times governing cross relaxation are unfavorable, ROE or ROESY experiments can be a valid alternative. Both NOE and ROE types of experiment also provide information on the presence of chemical equilibria when the interconversion rate is of the order of T_1. It is also clear that NOE and ROE types of experiment may not always provide a unique picture, because they ignore chemical bonds. COSY and TOCSY experiments provide the information on which proton is chemically bound to which other proton.

The last consideration regards the pattern of spin densities, which should match with expectations based on the possible delocalization mechanisms illustrated in Chapter 2.

If high sample concentrations can be reached, ^{1}H–^{13}C heterocorrelated experiments can be performed and further information can be obtained to study spin density patterns. If then the sample can be enriched with ^{13}C or ^{15}N, experiments are available to determine heteronuclear J coupling involving ^{1}H, ^{13}C, ^{15}N, ^{31}P, etc. The J values contain structural information, and therefore can be used for assignment. Owing to the small γ values, heteronuclei can be conveniently investigated in paramagnetic molecules, although their full potentiality has still to be further investigated.

9.6 NUCLEAR MAGNETIC RELAXATION DISPERSION (NMRD)

The information content of nuclear longitudinal relaxation measurements in both paramagnetic and diamagnetic systems can be greatly increased by performing such measurements as a function of the magnetic field. For paramagnetic species, the reason is apparent from the functional form of the equations discussed in Chapter 3 and from the relevant experimental data, reported in Chapter 5. The field dependence of a relaxation rate is called relaxation dispersion, and is abbreviated as NMRD. In principle, NMRD would be helpful for any chemical system, but practical limitations, as will be shown, restrict its use, with a few exceptions, to water protons.

9.6.1 Changing the magnetic field

The basic instrument for performing routine NMRD experiments must contain: (1) a variable field magnet (usually an electromagnet); (2) a frequency synthesizer, and a broadband transmitter and receiver; and (3) one (or more) broadband tunable probes covering a large frequency range. A variable field external lock or, at least, a reasonably good flux stabilizer for the magnetic field are also desirable although not strictly necessary (see below). Once these requirements are met, the experiment can be done by first measuring the water proton T_1 at the regular proton frequency — 100 MHz for a 2.35 T magnet, for example — with a usual pulse sequence — such as that used in an inversion-recovery experiment, for example. Then the electromagnet current is reduced to bring the field to a lower value (for instance, 1.88 T, corresponding to 80 MHz resonance frequency), the synthesizer is set to 80 MHz, the transmitter and receiver are retuned accordingly (changing the probehead and the matching/tuning circuitries if necessary), the pulse durations are readjusted, and the T_1 measurement is repeated. Using this procedure, it is possible to go as low as is permitted by the lowest frequency probe (typically 2 to 4 MHz); but sensitivity problems will ultimately arise. Of course the set of data may be implemented by measurements at 200, 400, 500, 600, 700, 800, 900 MHz, each on a different commercial instrument, if available. Except at 100 MHz, the standard lock system is useless, be it an external ^{1}H lock or an internal ^{2}H lock, since both operate at a fixed frequency and once the field has been lowered the lock signal remains far off-resonance. This problem may be partially overcome by choosing magnetic field values at which an easily observable nucleus resonates at the same frequency at which ^{2}H resonates when the field is set at 2.35 T; in this way the ^{2}H lock channel may be used. Alternatively, by using an additional frequency synthesizer and an especially built tunable external lock probe, the proton signal of an external lock can be followed over the entire field range.

If none of these instrumental systems is used the field will be unlocked, and care should be taken to keep it as stable as possible. Long-term stability is not usually required, as the water signal can be observed with a single pulse, but short-term field variations during the sampling of the FID (typically from 0.1 to 10 s) will be reflected in the signal lineshape. Since a T_1 experiment requires reliable signal intensities to follow the recovery of magnetization on the z axis, it is better to use the areas of the peaks rather then their height; alternatively, the initial intensity of the FID can be used directly in the diode-detection mode, instead of the phase-detection mode. In the diode-detection mode, several transients can be summed up to average out the noise without the worry of field fluctuations, which would alter the shape of the FID in the phase-detection mode. Since the field homogeneity of the magnet is in general optimized only for the usual high field working conditions, and for the dramatic decrease in sensitivity at low fields, the experiments are usually limited to single and intense resonance lines, which in general are solvent lines.

9.6.2 Field-cycling relaxometry

Besides the complications inherent with the type of measurements just discussed, and their being rather tedious, the main disadvantage of performing NMRD by changing the operating magnetic field is the rather high lower limit of practically accessible fields. This problem has been overcome by using a completely different approach, developed in the 1950s by Redfield (see, for example, Ref. [53]), pursued later to a high degree of refinement and reliability by Koenig and Brown [54,55] and recently marketed in a version derived from a design by F. Noack. According to this approach a solenoid without any iron core to generate a magnetic field is used. In this way the field can be turned on and off, or changed from one value to any other, rapidly; typical rise or fall times for the field are from 1 to 0.1 millisecond per MHz. Depending on the design of the solenoid, the highest magnetic field obtainable ranges between 0.5 and 1.2 T (from 20 to 50 MHz for protons).

T_1 is defined as the time constant for the first-order process in which the Boltzmann equilibrium of the macroscopic magnetization along the z axis is reached from *any* nonequilibrium situation. Thus, the onset of the equilibrium when either a sample is placed into a magnetic field or after its magnetization has been reversed, for instance, follows the same rate law. Now, suppose a sample is taken from a strong magnetic field and rapidly put into a weak one: the magnetization will decay according to Fig. 9.11A. The time constant required to bring it to the equilibrium value at that field is the same as that for a sample that started at zero field (Fig. 9.11B). T_1 will then be the time constant for the field at which the magnetization evolves towards equilibrium: that is, the weak field (or measuring field).

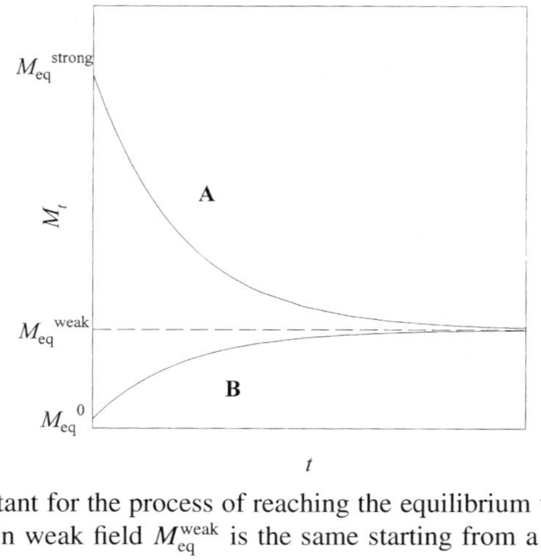

Fig. 9.11. The time constant for the process of reaching the equilibrium value of the macroscopic magnetization at a certain weak field M_{eq}^{weak} is the same starting from a higher field (A) or from zero field (B).

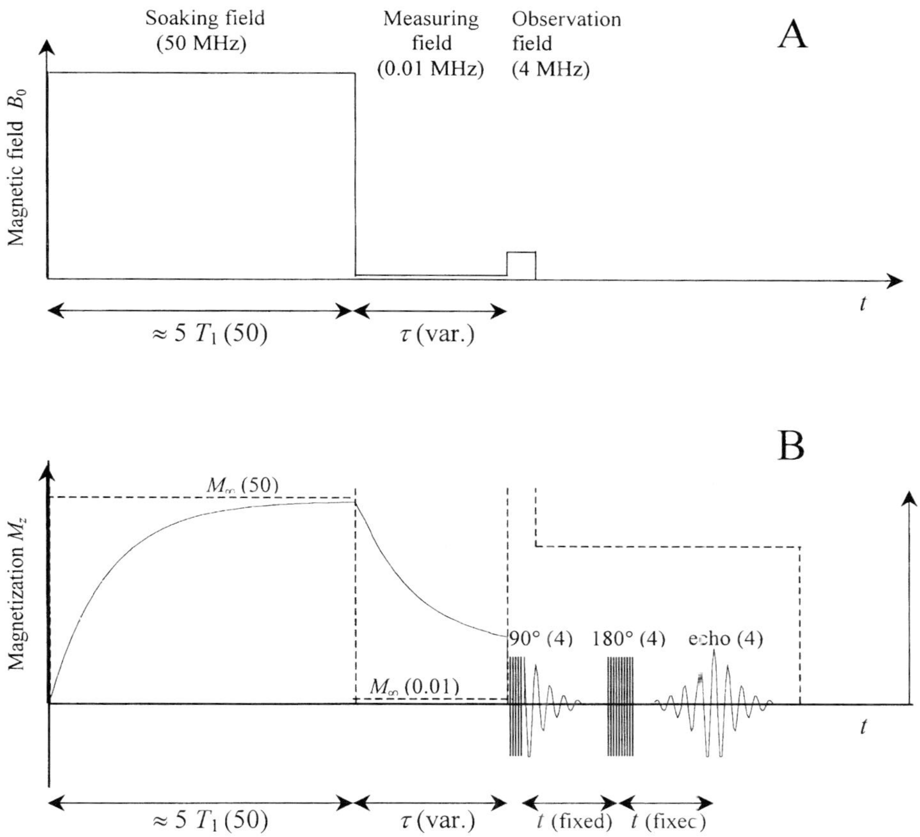

Fig. 9.12. Pulsed field-water proton T_1 measurements. Sequence for measuring T_1 at fields lower than 22.5 MHz.

If data points could be taken along the curve in Fig. 9.11A, this would measure the T_1 at the weak field by sampling magnetization values of the order of magnitude of those obtained in the high field magnet. This can, in fact, be accomplished by using the previously described solenoid, and a transmitter and receiver system tuned at a single frequency value (4 MHz in the following example). The procedure is as follows (Fig. 9.12):

(1) The magnetic field is turned on and set to a value corresponding to, say, 50 MHz (Fig. 9.12A). It is maintained at this value for a time of the order of 5 $T_{1(50)}$ (where $T_{1(50)}$ is the T_1 at 50 MHz) to allow the magnetization on the z axis to build up substantially (Fig. 9.12B).

(2) The field is then lowered to a value (as low as 0.01 MHz, as in this example) corresponding to the desired Larmor frequency and kept at this value for a time τ (Fig. 9.12A), during which M_z decays towards its new equilibrium value, M_∞ (Fig. 9.12B).

(3) The field is then raised to a value corresponding to a fixed observation frequency (4 MHz here, Fig. 9.12A). As soon as the field is set to 4 MHz a 90° or a 90°–t–180° pulse sequence is applied (with t of the order of milliseconds) and the receiver turned on (enlargement in Fig. 9.12B). The height of the echo that appears is proportional to the value of M_z at the time τ if the raise time of the magnetic field from 0.01 to 4 MHz and the time t are negligible compared to $T_{1(4)}$.

(4) The field is then turned off to let the magnetization dissipate (and the system cool down), and the whole procedure is repeated using a different value of τ. T_1 can then be calculated for that particular magnetic field from the heights of the echoes, as a function of τ.

In summary, T_1 at 0.01 MHz is measured at 4 MHz by taking advantage of the sensibility obtainable at 50 MHz. Of course, the same kind of measurement can be made for any field value, provided it is not too close to 50 MHz, as the

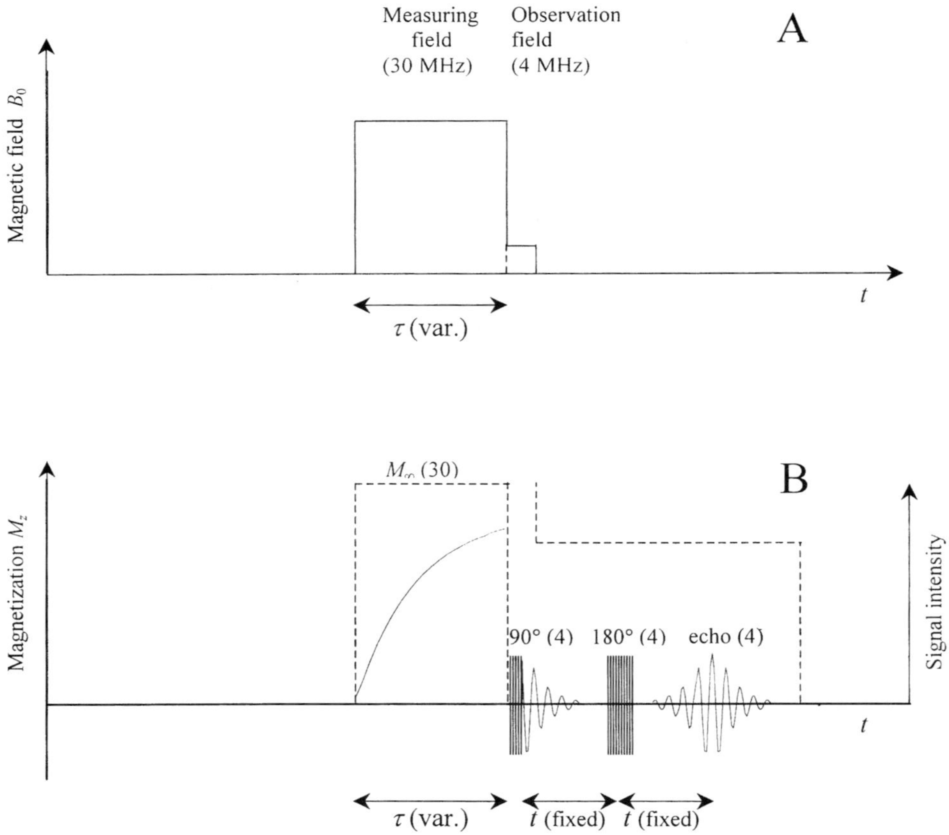

Fig. 9.13. Pulsed field-water proton T_1 measurements. Sequence for measuring T_1 at fields higher than 22.5 MHz.

observed variation in $\boldsymbol{M}$ would then be too small. It should be remembered that the observation field is always 4 MHz, so only one transmitter and receiver system is needed, and there is no need to retune the coil, either.

When measuring relaxation rates at fields higher than 25 MHz, there is a larger change in magnetization when starting from zero magnetic field and going up as compared to starting from 50 MHz and going down. Therefore, for fields higher than 25 MHz the procedure is simplified as follows (Fig. 9.13): the magnetic field is turned on and set to the value of interest (for instance, 30 MHz, Fig. 9.13A) for a time τ, during which magnetization starts building up according to $T_{1(30)}$ (Fig. 9.13B); the field is then switched to the 4 MHz observation value and steps 3 and 4 are executed. Again, everything is repeated for different values of τ.

9.7 REFERENCES

[1] C. Luchinat, M. Piccioli, R. Pierattelli, F. Engelke, T. Marquardsen, R. Ruin (2001) J. Magn. Reson., in press
[2] M. Piccioli (1996) J. Magn. Reson. Ser. B 110, 202.
[3] C. Bauer, R. Freeman, T. Frenkiel, J. Keeler, A.J. Shaka (1984) J. Magn. Reson. 58, 442.
[4] A.J. Temps, Jr., C.F. Brewer (1984) J. Magn. Reson. 56, 355.
[5] W.S. Warren (1984) J. Chem. Phys. 81, 5437.
[6] L. Emsley, G. Bodenhausen (1990) Chem. Phys. Lett. 165, 469.
[7] G.A. Morris, R. Freeman (1978) J. Magn. Reson. 29, 433.
[8] L.E. Kay, D. Marion, A. Bax (1989) J. Magn. Reson. 84, 72.
[9] A.G. Redfield, S.D. Kunz, E.K. Ralph (1975) J. Magn. Reson. 19, 114.
[10] D.L. Turner (1983) J. Magn. Reson. 54, 146.
[11] P.J. Hore (1983) J. Magn. Reson. 55, 283.
[12] P. Plateau, M. Guéron (1982) J. Am. Chem. Soc. 104, 7310.
[13] V. Sklenar, A. Bax (1987) J. Magn. Reson. 74, 469.
[14] M. Piotto, V. Saudek, V. Sklenar (1992) J. Biomol. NMR 2, 661.
[15] S.L. Patt, B.D. Sykes (1972) J. Chem. Phys. 56, 3182.
[16] T. Inubushi, E.D. Becker (1983) J. Magn. Reson. 51, 128.
[17] K.L. Bren, H.B. Gray, L. Banci, I. Bertini, P. Turano (1995) J. Am. Chem. Soc. 117, 8067.
[18] J. Hochmann, H. Kellerhals (1980) J. Magn. Reson. 38, 23.
[19] E.D. Becker, J.A. Ferretti, T.C. Farrar (1969) J. Am. Chem. Soc. 91, 7784.
[20] I. Bertini, G. Canti, C. Luchinat, F. Mani (1981) J. Am. Chem. Soc. 103, 7784.
[21] M.H. Levitt, R. Freeman, T. Frenkiel (1982) J. Magn. Reson. 47, 328.
[22] A.J. Shaka, J. Keeler, R. Freeman (1983) J. Magn. Reson. 53, 313.
[23] A.J. Shaka, P.B. Barker, R. Freeman (1985) J. Magn. Reson. 64, 547.
[24] A.J. Shaka, C.J. Lee, A. Pines (1988) J. Magn. Reson. 77, 274.
[25] R.L. Vold, J.S. Waugh, M.P. Klein, D.E. Phelps (1968) J. Chem. Phys. 48, 3831.
[26] A. Arseniev, A.G. Sobol, V.F. Bystrov (1986) J. Magn. Reson. 70, 427.
[27] I. Bertini, A. Donaire, I.C. Felli, C. Luchinat, A. Rosato (1996) Magn. Reson. Chem. 34, 948.
[28] I. Bertini, M.M.J. Couture, A. Donaire, L.D. Eltis, I.C. Felli, C. Luchinat, M. Piccioli, A. Rosato (1996) Eur. J. Biochem. 241, 440.
[29] J.G. Huber, J.-M. Moulis, J. Gaillard (1996) Biochemistry 35, 12705.
[30] L.E. Kay, D.A. Torchia, A. Bax (1989) Biochemistry 28, 8972.

[31] J.W. Peng, G. Wagner (1992) J. Magn. Reson. 98, 308.
[32] S. Meiboom, D. Gill (1960) Rev. Sci. Instr. 29, 688.
[33] L.E. Kay, L.K. Nicholson, F. Delaglio, A. Bax, D.A. Torchia (1992) J. Magn. Reson. 97, 359.
[34] D.M. Korzhnev, E.V. Tischenko, A.S. Arseniev (2000) J. Biomol. NMR 17, 231.
[35] I. Bertini, C. Luchinat, D. Tarchi (1993) Chem. Phys. Lett. 203, 445.
[36] I. Bertini, C. Luchinat, M. Piccioli, D. Tarchi (1994) Concepts Magn. Reson. 6, 307.
[37] A.W. Overhauser (1953) Phys. Rev. 89, 689.
[38] I. Solomon (1955) Phys. Rev. 99, 559.
[39] J.T.J. Lecomte, S.W. Unger, G.N. La Mar (1991) J. Magn. Reson. 94, 112.
[40] L. Banci, I. Bertini, F. Capozzi, P. Carloni, S. Ciurli, C. Luchinat, M. Piccioli (1993) J. Am. Chem. Soc. 115, 3431.
[41] L. Banci, I. Bertini, C. Luchinat, M.S. Viezzoli (1990) Inorg. Chem. 29, 1438.
[42] M. Paci, A. Desideri, M. Sette, M. Falconi, G. Rotilio (1990) FEBS Lett. 261, 231.
[43] L. Banci, I. Bertini, C. Luchinat, M. Piccioli (1990) FEBS Lett. 272, 175.
[44] G. Otting (1997) Progr. NMR Spectrosc. 31, 259.
[45] C. Dalvit, U. Hommel (1995) J. Magn. Reson. Ser. B 109, 334.
[46] I. Bertini, C. Dalvit, J.G. Huber, C. Luchinat, M. Piccioli (1997) FEBS Lett. 415, 45.
[47] I. Bertini, J.G. Huber, C. Luchinat, M. Piccioli (2000) J. Magn. Reson. 147, 1.
[48] A.V. Xavier, D.L. Turner, H. Santos (1993) Methods Enzymol. 227, 1.
[49] A. Bax, D.G. Davis (1985) J. Magn. Reson. 65, 355.
[50] C. Griesinger, G. Otting, K. Wüthrich, R.R. Ernst (1988) J. Am. Chem. Soc. 110, 7870.
[51] I. Bertini, C. Luchinat, R. Macinai, M. Piccioli, A. Scozzafava, M.S. Viezzoli (1994) J. Magn. Reson. Ser. B 104, 95.
[52] I. Bertini, Y.-M. Lee, C. Luchinat, M. Piccioli, L. Poggi (2001) Chem. Bio. Chem., in press.
[53] A.G. Anderson, A.G. Redfield (1959) Phys. Rev. 116, 583.
[54] K. Hallenga, S.H. Koenig (1976) Biochemistry 15, 4255.
[55] R.D. Brown III, C. Brewer, S.H. Koenig (1977) Biochemistry 16, 3883.

Appendices

Appendix I

NMR Properties of Nuclei

Isotope	Spin (I)	Natural abundance (%)	Relative sensitivity [a]	g_I	γ_I [b]	NMR frequency (MHz at 2.3488 T)	Q (10^{-28} m^2)
^{1}H	1/2	99.98	1.00	5.5856912	2.6751987×10^{8}	100.000	–
^{2}H	1	1.56×10^{-2}	9.65×10^{-3}	0.8574376	4.106593×10^{7}	15.351	2.875×10^{-3}
^{3}H	1/2	0	1.21	5.957920	2.853473×10^{8}	106.664	–
^{3}He	1/2	1.3×10^{-4}	0.442	−4.255248	-2.037999×10^{8}	76.181	–
^{6}Li	1	7.42	8.50×10^{-3}	0.8220467	3.937092×10^{7}	14.717	-6.44×10^{-4}
^{7}Li	3/2	92.58	0.294	2.170949	1.039750×10^{8}	38.866	-3.66×10^{-2}
^{9}Be	3/2	100	1.39×10^{-2}	−0.7852	-3.7606×10^{7}	14.057	5.3×10^{-2}
^{10}B	3	19.58	1.99×10^{-2}	0.600217	2.87466×10^{7}	10.746	8.472×10^{-3}
^{11}B	3/2	80.42	0.165	1.792425	8.58460×10^{7}	32.090	4.065×10^{-2}
^{13}C	1/2	1.108	1.59×10^{-2}	1.404822	6.72822×10^{7}	25.150	–
^{14}N	1	99.63	1.01×10^{-3}	0.4037607	1.933763×10^{7}	7.228	1.56×10^{-2}
^{15}N	1/2	0.37	1.04×10^{-3}	−0.5663784	-2.712600×10^{7}	10.140	–
^{17}O	5/2	3.7×10^{-2}	2.91×10^{-2}	−0.75752	-3.62803×10^{7}	13.562	-2.578×10^{-2}
^{19}F	1/2	100	0.834	5.257732	2.518127×10^{8}	94.129	–
^{21}Ne	3/2	0.257	2.46×10^{-3}	−0.441197	-2.11306×10^{7}	7.899	1.029×10^{-1}
^{23}Na	3/2	100	9.27×10^{-2}	1.478347	7.080361×10^{7}	26.467	1.01×10^{-1}
^{25}Mg	5/2	10.13	2.68×10^{-3}	−0.34218	-1.63883×10^{7}	6.126	0.22
^{27}Al	5/2	100	0.207	1.456602	6.976216×10^{7}	26.077	0.140
^{29}Si	1/2	4.7	7.86×10^{-3}	−1.11058	-5.3190×10^{7}	19.883	–
^{31}P	1/2	100	6.65×10^{-2}	2.26320	1.08393×10^{8}	40.518	–
^{33}S	3/2	0.76	2.27×10^{-3}	0.429214	2.05567×10^{7}	7.684	-6.4×10^{-2}
^{35}Cl	3/2	75.53	4.72×10^{-3}	0.5479157	2.624176×10^{7}	9.809	-8.249×10^{-2}
^{37}Cl	3/2	24.47	2.72×10^{-3}	0.4560820	2.184349×10^{7}	8.165	-6.493×10^{-2}
^{39}K	3/2	93.1	5.10×10^{-4}	0.2609772	1.249918×10^{7}	4.672	4.9×10^{-2}

^{41}K	3/2	6.88	8.43×10^{-5}	0.1432466	6.860621×10^{6}	2.565	6.0×10^{-2}
^{43}Ca	7/2	0.145	6.42×10^{-3}	−0.37636	-1.80253×10^{7}	6.738	–
^{45}Sc	7/2	100	0.302	1.358995	6.508741×10^{7}	24.330	−0.22
^{47}Ti	5/2	7.28	2.10×10^{-3}	−0.31539	-1.51053×10^{7}	5.646	0.29
^{49}Ti	7/2	5.51	3.78×10^{-3}	−0.31548	-1.51094×10^{7}	5.648	0.24
^{50}V	6	0.24	5.58×10^{-2}	0.557908	2.67203×10^{7}	9.988	7×10^{-2}
^{51}V	7/2	99.76	0.384	1.47183	7.0491×10^{7}	26.350	-5.2×10^{-2}
^{53}Cr	3/2	9.55	9.08×10^{-4}	−0.31636	-1.51517×10^{7}	5.664	2.2×10^{-2}
^{55}Mn	5/2	100.00	0.176	1.3813	6.6155×10^{7}	24.729	0.40
^{57}Fe	1/2	2.19	3.42×10^{-5}	0.1812459	8.680549×10^{6}	3.245	–
^{59}Co	7/2	100.00	0.278	1.322	6.332×10^{7}	23.67	0.404
^{61}Ni	3/2	1.19	3.59×10^{-3}	−0.50001	-2.39475×10^{7}	8.952	0.162
^{63}Cu	3/2	69.09	9.34×10^{-2}	1.4822	7.0988×10^{7}	26.536	−0.209
^{65}Cu	3/2	30.91	0.115	1.5878	7.6046×10^{7}	28.426	−0.195
^{67}Zn	5/2	4.11	2.87×10^{-3}	0.350191	1.677198×10^{7}	6.269	0.150
^{69}Ga	3/2	60.4	6.97×10^{-2}	1.34439	6.43881×10^{7}	24.069	0.168
^{71}Ga	3/2	39.60	0.143	1.70818	8.18112×10^{7}	30.581	0.106
^{73}Ge	9/2	7.76	1.41×10^{-3}	−0.1954371	-9.360221×10^{6}	3.499	−0.173
^{75}As	3/2	100.00	2.54×10^{-2}	0.95965	4.5961×10^{7}	17.180	0.29
^{77}Se	1/2	7.58	7.03×10^{-3}	1.07012	5.1252×10^{7}	19.158	–
^{79}Br	3/2	50.54	7.94×10^{-2}	1.404266	6.72556×10^{7}	25.140	0.293
^{81}Br	3/2	49.46	9.95×10^{-2}	1.513707	7.24971×10^{7}	27.100	0.27
^{83}Kr	9/2	11.55	1.90×10^{-3}	−0.215704	-1.033089×10^{7}	3.862	0.270
^{85}Rb	5/2	72.15	1.06×10^{-2}	0.54121	2.59207×10^{7}	9.689	0.274
^{87}Rb	3/2	27.85	0.177	1.83416	8.78449×10^{7}	32.837	0.132
^{87}Sr	9/2	7.02	2.71×10^{-3}	−0.242849	-1.16309×10^{7}	4.348	0.15
^{89}Y	1/2	100.00	1.19×10^{-4}	−0.2748306	-1.316268×10^{7}	4.920	–
^{91}Zr	5/2	11.23	9.49×10^{-3}	−0.52145	-2.49741×10^{7}	9.335	–
^{93}Nb	9/2	100.00	0.488	1.37122	6.5673×10^{7}	24.549	−0.36
^{95}Mo	5/2	15.72	3.27×10^{-3}	−0.3657	-1.7514×10^{7}	6.547	-1.9×10^{-2}
^{97}Mo	5/2	9.46	3.49×10^{-3}	−0.3734	-1.7884×10^{7}	6.785	−0.102
^{99}Ru	5/2	12.72	1.13×10^{-3}	−0.2565	-1.2286×10^{7}	4.592	7.6×10^{-2}

[a] NMR sensitivity at constant field and equal number of nuclei, relative to ^{1}H nuclei, given by $4/3\,(|g_I|/|g_H|)^3 I(I+1)$.
[b] γ_I is defined by: $\gamma_I = g_I \mu_N / \hbar$, where the nuclear magneton, μ_N is 5.050824×10^{-27} J T^{-1}, and $\hbar$ is $1.0545877 \times 10^{-34}$ J s rad^{-1}.

Appendix I (continued)

Isotope	Spin (I)	Natural abundance (%)	Relative sensitivity[a]	g_I	γ_I[b]	NMR frequency (MHz at 2.3488 T)	Q (10^{-28} m^2)
^{101}Ru	5/2	17.07	1.59×10^{-3}	−0.2875	-1.3770×10^{7}	5.147	0.44
^{103}Rh	1/2	100.00	3.17×10^{-5}	−0.17680	-8.468×10^{6}	3.165	–
^{105}Pd	5/2	22.23	1.13×10^{-3}	−0.2568	-1.230×10^{7}	4.60	0.8
^{107}Ag	1/2	51.82	6.72×10^{-5}	−0.227140	-1.08786×10^{7}	4.066	–
^{109}Ag	1/2	48.18	1.02×10^{-4}	−0.2613810	-1.251852×10^{7}	4.679	–
^{111}Cd	1/2	12.75	9.66×10^{-3}	−1.1897712	-5.698264×10^{7}	21.300	–
^{113}Cd	1/2	12.26	1.11×10^{-2}	−1.2446010	-5.960865×10^{7}	22.282	–
^{113}In	9/2	4.28	0.351	1.22864	5.8844×10^{7}	21.996	0.846
^{115}In	9/2	95.72	0.353	1.23129	5.8971×10^{7}	2.044	0.861
^{115}Sn	1/2	0.35	3.56×10^{-2}	−1.83766	-8.8012×10^{7}	32.899	–
^{117}Sn	1/2	7.61	4.60×10^{-2}	−2.00208	-9.5887×10^{7}	35.843	–
^{119}Sn	1/2	8.58	5.27×10^{-2}	−2.09456	-1.00316×10^{8}	37.499	–
^{121}Sb	5/2	57.25	0.163	1.34536	6.4434×10^{7}	24.086	−0.20
^{123}Sb	7/2	42.75	4.66×10^{-2}	0.72851	3.4891×10^{7}	13.043	−0.26
^{123}Te	1/2	0.87	1.84×10^{-2}	−1.47358	-7.0575×10^{7}	26.381	–
^{125}Te	1/2	6.99	3.22×10^{-2}	−1.77656	-8.5086×10^{7}	31.806	–
^{127}I	5/2	100.00	9.54×10^{-2}	1.12531	5.38953×10^{7}	20.146	−0.789
^{129}Xe	1/2	26.44	2.16×10^{-2}	−1.555952	-7.45204×10^{7}	27.856	–
^{131}Xe	3/2	21.18	2.82×10^{-3}	0.461241	2.209056×10^{7}	8.258	−0.120
^{133}Cs	7/2	100.00	4.84×10^{-2}	0.7377209	3.533224×10^{7}	13.207	-3×10^{-3}
^{135}Ba	3/2	6.59	5.00×10^{-3}	0.558629	2.675484×10^{7}	10.001	0.18
^{137}Ba	3/2	11.32	7.00×10^{-3}	0.624910	2.992930×10^{7}	11.188	0.28
^{138}La	5	0.089	9.41×10^{-2}	0.74278	3.5575×10^{7}	13.298	0.51
^{139}La	7/2	99.911	6.06×10^{-2}	0.79520	3.8085×10^{7}	14.236	0.22
^{141}Pr	5/2	100.00	0.303	1.6544	7.924×10^{7}	29.62	-5.89×10^{-2}
^{143}Nd	7/2	12.17	3.39×10^{-3}	−0.3043	-1.457×10^{7}	5.45	−0.484
^{145}Nd	7/2	8.30	7.93×10^{-4}	−0.1874	-8.98×10^{6}	3.36	−0.253
^{147}Sm	7/2	14.97	1.52×10^{-3}	−0.23280	-1.1150×10^{7}	4.168	−0.18
^{149}Sm	7/2	13.83	8.52×10^{-4}	−0.19191	-9.192×10^{6}	3.436	5.2×10^{-2}
^{151}Eu	5/2	47.82	0.179	1.3887	6.6509×10^{7}	24.861	1.15

^{153}Eu	5/2	52.18	1.54×10^{-2}	0.6132	2.9368×10^{7}	10.978	2.94
^{155}Gd	3/2	14.73	1.48×10^{-4}	−0.1727	-8.273×10^{6}	3.092	1.59
^{157}Gd	3/2	15.68	3.34×10^{-4}	−0.2265	-1.0850×10^{7}	4.056	2.03
^{159}Tb	3/2	100.00	6.94×10^{-2}	1.343	6.431×10^{7}	24.04	1.18
^{161}Dy	5/2	18.88	4.75×10^{-4}	−0.1922	-9.205×10^{6}	3.441	2.44
^{163}Dy	5/2	24.97	1.30×10^{-3}	0.2690	1.2885×10^{7}	4.817	2.57
^{165}Ho	7/2	100.00	0.204	1.1923	5.710×10^{7}	21.35	2.73
^{167}Er	7/2	22.94	5.11×10^{-4}	−0.16186	-7.752×10^{6}	2.898	2.827
^{169}Tm	1/2	100.00	5.70×10^{-4}	−0.4632	-2.218×10^{7}	8.293	–
^{171}Yb	1/2	14.31	5.52×10^{-3}	0.98734	4.7287×10^{7}	17.676	–
^{173}Yb	5/2	16.13	1.35×10^{-3}	−0.271956	-1.30250×10^{7}	4.869	2.8
^{175}Lu	7/2	97.41	3.13×10^{-2}	0.63791	3.0552×10^{7}	11.421	5.68
^{176}Lu	7	2.59	4.05×10^{-2}	0.456	2.18×10^{7}	8.16	8
^{177}Hf	7/2	18.5	1.40×10^{-3}	0.22671	1.0858×10^{7}	4.059	4.5
^{179}Hf	9/2	13.75	5.47×10^{-4}	−0.14242	-6.821×10^{6}	2.550	5.1
^{181}Ta	7/2	99.99	3.74×10^{-2}	0.6771	3.243×10^{7}	12.12	3.9
^{183}W	1/2	14.40	7.50×10^{-5}	0.2355694	1.128231×10^{7}	4.217	–
^{185}Re	5/2	37.07	0.139	1.2748	6.1057×10^{7}	22.823	2.36
^{187}Re	5/2	62.93	0.143	1.2879	6.1681×10^{7}	23.057	2.24
^{187}Os	1/2	1.64	1.24×10^{-5}	0.12930	6.193×10^{6}	2.315	–
^{189}Os	3/2	16.10	2.44×10^{-3}	0.439955	2.107112×10^{7}	7.876	0.91
^{191}Ir	3/2	37.30	2.65×10^{-5}	0.0974	4.665×10^{6}	1.744	0.78
^{193}Ir	3/2	62.70	3.42×10^{-5}	0.1061	5.080×10^{6}	1.899	0.70
^{195}Pt	1/2	33.80	1.04×10^{-2}	1.21898	5.8382×10^{7}	21.823	–
^{197}Au	3/2	100.00	2.76×10^{-5}	0.098772	4.73056×10^{6}	1.768	0.594
^{199}Hg	1/2	16.84	5.94×10^{-3}	1.0117702	4.845750×10^{7}	18.114	–
^{201}Hg	3/2	13.22	1.49×10^{-3}	−0.373483	-1.788753×10^{7}	6.686	0.455
^{203}Tl	1/2	29.50	0.196	3.244514	1.553920×10^{8}	58.086	–
^{205}Tl	1/2	70.50	0.202	3.2764268	1.5692047×10^{8}	58.657	–
^{207}Pb	1/2	22.60	9.06×10^{-3}	1.16438	5.57666×10^{7}	20.846	–
^{209}Bi	9/2	100.00	0.144	0.91347	4.3749×10^{7}	16.354	−0.46
^{235}U	7/2	0.72	1.2×10^{-4}	−0.10	-4.8×10^{6}	1.8	4.55

[a] NMR sensitivity at constant field and equal number of nuclei, relative to ^{1}H nuclei, given by $4/3\,(|g_I|/|g_H|)^3 I(I+1)$.

[b] γ_I is defined by: $\gamma_I = g_I \mu_N/\hbar$, where the nuclear magneton, μ_N is 5.050824×10^{-27} J T^{-1}, and $\hbar$ is $1.0545877 \times 10^{-34}$ J s rad^{-1}.

Dipolar Coupling Between Two Spins

The Hamiltonian for the dipolar coupling between two magnetic moments associated with the spins I and J is given by an equation similar to Eq. (1.1)

$$H = -\frac{\mu_0}{4\pi}\frac{\hbar^2\gamma_I\gamma_J}{r^3}[3(\boldsymbol{I}\cdot\boldsymbol{r})(\boldsymbol{J}\cdot\boldsymbol{r}) - \boldsymbol{I}\cdot\boldsymbol{J}] \tag{II.1}$$

where $\boldsymbol{r}$ is a unit vector oriented along the line of length r connecting $\boldsymbol{I}$ and $\boldsymbol{J}$. In the most general case, the $\boldsymbol{I}$ and $\boldsymbol{J}$ vectors are not parallel, and each of them will form different angles with $\boldsymbol{r}$ and with the applied magnetic field, $\boldsymbol{B_0}$. If the Zeeman interaction is the dominant among all interactions involving the spins, the quantization axis of both spins will be along $\boldsymbol{B_0}$ (see Fig. II.1). The first spin $\boldsymbol{I}$ creates a dipole $\boldsymbol{\mu_1}$ which precesses around the field $\boldsymbol{B_0}$ and thus has a static component along the field and a rotating component in the plane perpendicular to the field. The static component of $\boldsymbol{\mu_1}$ produces at the site of the dipole $\boldsymbol{\mu_2}$, related to the second spin $\boldsymbol{J}$, a small static field that depends on the relative position of the spins. For large $\boldsymbol{B_0}$, only the component parallel or antiparallel to $\boldsymbol{B_0}$ significantly changes the net static field. From Eq. (II.1), when $\theta_I = \theta_J = \gamma$, it results that the dipolar interaction energy depends on the interspin distance r, on the eigenstates of the I and J spins, and on the angle γ between the interspin vector $\boldsymbol{r}$ and the

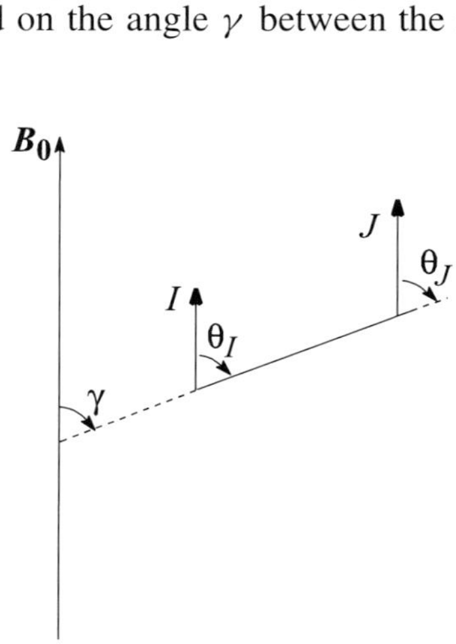

Fig. II.1. Reciprocal orientation of a nuclear spin I and an electron spin J, in the presence of an external magnetic field.

external magnetic field (along which both I and J are quantized)

$$E \propto 3\cos^2\gamma(t) - 1. \tag{II.2}$$

In general, dipolar relaxation occurs whenever these quantities are modulated by time-dependent phenomena. In solids r and γ are fixed for a single I–J pair. In liquids, γ varies upon rotation of the molecule bearing the I–J pair, and r may vary because of chemical exchange.

It is interesting to note that on rapid (with respect to the anisotropy of the interaction energy in frequency units) rotation, the dipolar interaction energy averages to zero. In contrast, the square of this quantity (which determines the extent of the contribution to relaxation) does not average to zero. The average values obtained by integration of the functions results in

$$\frac{1}{2}\int_{-1}^{1}(3\cos^2\gamma - 1)\,\mathrm{d}(\cos\gamma) = \frac{1}{2}(x^3 - x)\Big|_{-1}^{1} = 0 \tag{II.3}$$

$$\begin{aligned}\frac{1}{2}\int_{-1}^{1}(3\cos^2\gamma - 1)^2\,\mathrm{d}(\cos\gamma) &= \frac{1}{2}\int_{-1}^{1}(9x^4 + 1 - 6x^2)\,\mathrm{d}x \\ &= \frac{1}{2}\left(\frac{9}{5}x^5 + x - \frac{6}{3}x^3\right)\Big|_{-1}^{1} = \frac{4}{5}.\end{aligned} \tag{II.4}$$

Derivation of the Equations for Contact Shift and Relaxation in a Simple Case

The spin Hamiltonian for two contact coupled $S = 1/2$, $I = 1/2$ spins is given by

$$\begin{aligned} H &= g_e \mu_B B_0 S_z - \hbar \gamma_I B_0 I_z + A \boldsymbol{I} \cdot \boldsymbol{S} \\ &= \tfrac{1}{2} Z_e - \tfrac{1}{2} Z_N + A\left[S_z I_z + \tfrac{1}{2}(S_+ I_- + S_- I_+)\right] \end{aligned} \qquad \text{(III.1)}$$

where Z_e and Z_N are the electron and nuclear Zeeman energies, and recalling the definition of ladder operators

$$\begin{aligned} I_+ &= I_x + i I_y \qquad & I_- &= I_x - i I_y \\ S_+ &= S_x + i S_y \qquad & S_- &= S_x - i S_y \end{aligned} \qquad \text{(III.2)}$$

from which we obtain

$$\begin{aligned} I_x &= \frac{1}{2}(I_+ + I_-) \qquad & I_y &= \frac{1}{2i}(I_+ - I_-) \\ S_x &= \frac{1}{2}(S_+ + S_-) \qquad & S_y &= \frac{1}{2i}(S_+ - S_-). \end{aligned} \qquad \text{(III.3)}$$

The energy matrix is

M_S, M_I	$\lvert\frac{1}{2}, \frac{1}{2}\rangle$	$\lvert\frac{1}{2}, -\frac{1}{2}\rangle$	$\lvert -\frac{1}{2}, \frac{1}{2}\rangle$	$\lvert -\frac{1}{2}, -\frac{1}{2}\rangle$
$\langle\frac{1}{2}, \frac{1}{2}\rvert$	$\frac{1}{2}Z_e - \frac{1}{2}Z_N + \frac{1}{4}A$	0	0	0
$\langle\frac{1}{2}, -\frac{1}{2}\rvert$	0	$\frac{1}{2}Z_e + \frac{1}{2}Z_N - \frac{1}{4}A$	$\frac{1}{2}A$	0
$\langle -\frac{1}{2}, \frac{1}{2}\rvert$	0	$\frac{1}{2}A$	$-\frac{1}{2}Z_e - \frac{1}{2}Z_N - \frac{1}{4}A$	0
$\langle -\frac{1}{2}, -\frac{1}{2}\rvert$	0	0	0	$-\frac{1}{2}Z_e + \frac{1}{2}Z_N + \frac{1}{4}A$

Diagonalization of the above matrix yields the following energies and eigenfunctions:

$$E_1 = \tfrac{1}{2}Z_e - \tfrac{1}{2}Z_N + \tfrac{1}{4}A \qquad \Psi_1 = |\tfrac{1}{2}, \tfrac{1}{2}\rangle$$

$$E_2 = -\tfrac{1}{4}A + \tfrac{1}{2}R \qquad \Psi_2 = c_1|\tfrac{1}{2}, -\tfrac{1}{2}\rangle + c_2|-\tfrac{1}{2}, \tfrac{1}{2}\rangle$$

$$E_3 = -\tfrac{1}{4}A - \tfrac{1}{2}R \qquad \Psi_3 = -c_2|\tfrac{1}{2}, -\tfrac{1}{2}\rangle + c_1|-\tfrac{1}{2}, \tfrac{1}{2}\rangle$$

$$E_4 = -\tfrac{1}{2}Z_e + \tfrac{1}{2}Z_N + \tfrac{1}{4}A \qquad \Psi_4 = |-\tfrac{1}{2}, -\tfrac{1}{2}\rangle$$

where

$$R = (A^2 + (Z_e + Z_N)^2)^{1/2}$$

$$c_1 = \left[\frac{1}{2}\left(1 + \frac{Z_e + Z_N}{R}\right)\right]^{1/2} \qquad c_2 = \left[\frac{1}{2}\left(1 - \frac{Z_e + Z_N}{R}\right)\right]^{1/2}$$

Nuclear transition energies are given by $E_2 - E_1$ and $E_4 - E_3$. By referring for simplicity to the high field limit, i.e. when $Z_e \gg A$, we have

$$E_2 - E_1 = Z_N - \tfrac{1}{2}A \qquad E_4 - E_3 = Z_N + \tfrac{1}{2}A$$

The first transition is between states characterized by positive Z_e, the second between states with negative Z_e. Note that the nuclear transitions are separated by A and independent of S. However, the nucleus is only able to experience an average additional field from the magnetic moment of the electron, owing to its fast relaxation between the two electronic states. The average transition energy is thus given by

$$\Delta E_{\text{av}} = P_+(Z_N - \tfrac{1}{2}A) + P_-(Z_N + \tfrac{1}{2}A) \tag{III.4}$$

where P_+ and P_- are the Boltzmann population of the electronic levels, as defined in Eq. (1.25). Eq. (III.4) can thus be rewritten as

$$\begin{aligned}\Delta E_{\text{av}} = {} & \frac{\exp[-(Z_e/2kT)]}{\exp[-(Z_e/2kT)] + \exp(Z_e/2kT)}(Z_N - \tfrac{1}{2}A) \\ & + \frac{\exp(Z_e/2kT)}{\exp[-(Z_e/2kT)] + \exp(Z_e/2kT)}(Z_N + \tfrac{1}{2}A)\end{aligned} \tag{III.5}$$

and, in the limit $Z_e \ll 2kT$:

$$\begin{aligned}\Delta E_{\text{av}} &= \tfrac{1}{2}(1 - Z_e/2kT)(Z_N - \tfrac{1}{2}A) + \tfrac{1}{2}(1 + Z_e/2kT)(Z_N + \tfrac{1}{2}A) \\ &= Z_N + \tfrac{1}{2}A\frac{Z_e}{2kT} = \hbar\gamma_I B_0 + A\frac{g_e\mu_B B_0}{4kT}\end{aligned} \tag{III.6}$$

The contribution to the nuclear energy due to coupling with the unpaired electron, relative to the nuclear Zeeman energy equals the contact shift in ppm and is thus given by

$$\delta^{\text{con}} = A\frac{g_e\mu_B}{4\hbar\gamma_I kT} \tag{III.7}$$

which can be generalized for $S \neq 1/2$ to obtain Eq. (2.5):

$$\delta^{\text{con}} = \frac{A}{\hbar}\frac{g_e\mu_B S(S+1)}{3\gamma_I kT} \tag{III.8}$$

which is the simplest equation for the contact shift. Eq. (III.8) can also be obtained in a simpler way by recalling that for large magnetic fields the quantization axis of both the nuclear and the electron spins is along the external magnetic field. Therefore, the energy of the contact interaction from Hamiltonian (2.4) can be written as

$$E^{\text{con}} = AI_z\langle S_z\rangle = -AI_z\frac{g_e\mu_B S(S+1)B_0}{3kT} \tag{III.9}$$

where we have used the definition of $\langle S_z\rangle$ given in Eq. (1.31). The contact shift is again obtained by dividing the contact coupling energy by the nuclear Zeeman energy:

$$\delta^{\text{con}} = \frac{E_{\text{con}}}{\hbar\gamma_I I_z B_0} = \frac{A}{\hbar}\frac{g_e\mu_B S(S+1)}{3\gamma_I kT} \tag{III.10}$$

Contact interactions also give rise to relaxation. The perturbing Hamiltonian for contact interaction, H' (last term in Eq. (III.1)), is analogous to the first term of the perturbing Hamiltonian of the dipolar interaction (see Appendix V, Eq. (V.10)) except that the part containing the ladder operators is multiplied by $1/2$ instead of $-1/4$. The transition probability w_0 (see Fig. 3.8) is provided by (see Appendix V)

$$w_0 = \frac{1}{\hbar^2}\left|\int_{-t}^{t}\langle +-|H'(0)|-+\rangle\langle -+|H'(t')|+-\rangle e^{-i(\omega_I-\omega_S)t'}\,dt'\right|. \tag{III.11}$$

The only part of H' extracting $|+-\rangle$ from $|-+\rangle$ is that containing I_+S_-; at variance with the dipolar case, $w_1^I = w_1^S = 0$ and $w_2 = 0$. Therefore, in this case, since $T_1^{-1} = w_0 + 2w_1^I + w_2$ (see Eq. (7.3) and Appendix VII), we have $T_1^{-1} = w_0$.

By integration of Eq. (III.11), we obtain

$$T_1^{-1} = \frac{A^2}{2\hbar^2}\frac{\tau_c}{1+(\omega_I-\omega_S)^2\tau_c^2}. \tag{III.12}$$

The corresponding equation for T_2^{-1} can be obtained analogously:

$$T_2^{-1} = \frac{A^2}{4\hbar^2}\left(\frac{\tau_c}{1+(\omega_I-\omega_S)^2\tau_c^2} - \tau_c\right). \tag{III.13}$$

Derivation of the Pseudocontact Shift in the Case of Axial Symmetry

We refer to the generic geometric arrangement of the electron and nuclear magnetic moments depicted in Fig. 2.6. The z axis of the cartesian coordinate frame, defined by the external magnetic field, is indicated by the unitary vector $\boldsymbol{\kappa}$. The orientation of the molecular z axis is indicated by the unitary vector $\boldsymbol{\lambda}$. The metal–nucleus vector is indicated by $\boldsymbol{r}$. The angle between $\boldsymbol{\kappa}$ and $\boldsymbol{\lambda}$ is called α, the angle between $\boldsymbol{\lambda}$ and $\boldsymbol{r}$ is called θ, and the angle between $\boldsymbol{\kappa}$ and $\boldsymbol{r}$ is called γ. We also define a direction $\boldsymbol{\nu}$ perpendicular to $\boldsymbol{\lambda}$ and lying in the $\boldsymbol{\kappa\lambda}$ plane, and a direction $\boldsymbol{\iota}$ perpendicular to the $\boldsymbol{\kappa\lambda}$ plane. When performing rotational averaging, we will be dealing with rotations of the molecule about its principal axis $\boldsymbol{\lambda}$; upon such rotation, the vector $\boldsymbol{r}$ will define a cone around $\boldsymbol{\lambda}$, and it will be useful to define an angle Ω to locate $\boldsymbol{r}$ along the surface of the cone. Ω is zero when $\boldsymbol{r}$ lies in the $\boldsymbol{\kappa\lambda}$ plane.

Our aim is that of evaluating the energy of the dipolar interaction between the nuclear and electron magnetic moments from the classical expression (see Eq. (1.1)):

$$E^{\text{dip}} = -\left(\frac{\mu_0}{4\pi}\right)\frac{1}{r^3}\left[3\frac{(\langle\boldsymbol{\mu}\rangle\cdot\boldsymbol{r})(\boldsymbol{\mu}_{I\kappa}\cdot\boldsymbol{r})}{r^2} - \langle\boldsymbol{\mu}\rangle\cdot\boldsymbol{\mu}_{I\kappa}\right] \tag{IV.1}$$

where $\langle\boldsymbol{\mu}\rangle$ is the average induced magnetic moment of the electron, and $\boldsymbol{\mu}_{I\kappa}$ is the projection of the nuclear magnetic moment along $\boldsymbol{B}_0$. By referring to Fig. 2.6, $\boldsymbol{\mu}_{I\kappa} = \hbar\gamma_I I_z\boldsymbol{\kappa}$. From Eq. (1.27), we can write

$$\langle\mu\rangle = \frac{\chi_M}{N_\text{A}}\frac{B_0}{\mu_0} = \chi\frac{B_0}{\mu_0} \tag{IV.2}$$

where χ is defined as the molecular magnetic susceptibility. We also know that χ is a tensorial quantity (cf. Eq. (1.40)). By analogy, the vector $\langle\boldsymbol{\mu}\rangle$ can be obtained by taking the vector projection of the $\boldsymbol{\chi}$ tensor along $\boldsymbol{\kappa}$ and rewriting Eq. (IV.2) as

$$\langle\boldsymbol{\mu}\rangle = \frac{B_0}{\mu_0}\boldsymbol{\chi}\cdot\boldsymbol{\kappa}. \tag{IV.3}$$

By recalling that $\boldsymbol{\lambda}\cdot\boldsymbol{\kappa} = \cos\alpha$, $\boldsymbol{\nu}\cdot\boldsymbol{\kappa} = -\sin\alpha$, and $\boldsymbol{\iota}\cdot\boldsymbol{\kappa} = 0$, we obtain

$$\langle\boldsymbol{\mu}\rangle = \frac{B_0}{\mu_0}(\chi_\parallel\cos\alpha\boldsymbol{\lambda} - \chi_\perp\sin\alpha\boldsymbol{\nu}) \tag{IV.4}$$

Note that $\langle \boldsymbol{\mu} \rangle$ is slightly misaligned with respect to the direction of $\boldsymbol{B_0}$, $\boldsymbol{\kappa}$; it would be coincident with the $\boldsymbol{\kappa}$ direction only if $\chi_\parallel = \chi_\perp$, i.e. in the absence of magnetic susceptibility anisotropy.

By substituting Eq. (IV.4) in Eq. (IV.1), and recalling also that

$$\boldsymbol{\lambda} \cdot \boldsymbol{r} = r\cos\theta \qquad \boldsymbol{\nu} \cdot \boldsymbol{r} = -r\sin\theta\cos\Omega$$

$$\boldsymbol{\kappa} \cdot \boldsymbol{r} = r\cos\gamma = r(\cos\alpha\cos\theta + \sin\alpha\sin\theta\cos\Omega)$$

we obtain

$$\begin{aligned} E^{\text{dip}} = & -\frac{B_0}{4\pi r^3}\hbar\gamma_I I_z\big[3(\chi_\parallel\cos\alpha\cos\theta + \chi_\perp\sin\alpha\sin\theta\cos\Omega) \\ & \times(\cos\alpha\cos\theta + \sin\alpha\sin\theta\cos\Omega) - (\chi_\parallel\cos^2\alpha + \chi_\perp\sin^2\alpha)\big] \end{aligned} \tag{IV.5}$$

which can be rearranged as

$$\begin{aligned} E^{\text{dip}} = & -\frac{B_0}{4\pi r^3}\hbar\gamma_I I_z\big[\chi_\parallel\cos^2\alpha(3\cos^2\theta - 1) + \chi_\perp\sin^2\alpha(3\sin^2\theta\cos^2\Omega - 1) \\ & + \tfrac{3}{4}(\chi_\parallel + \chi_\perp)\sin 2\alpha\sin 2\theta\cos\Omega\big]. \end{aligned} \tag{IV.6}$$

Eq. (IV.6) is the general formula for the dipolar interaction energy when the principal axis of $\boldsymbol{\chi}$ is in a generic $\boldsymbol{\lambda}$ direction. The shift is then obtained by calculating the energy difference between two states differing by $\Delta M_I = \pm 1$:

$$\begin{aligned} \Delta E^{\text{dip}} = & -\frac{B_0}{4\pi r^3}\hbar\gamma_I\big[\chi_\parallel\cos^2\alpha(3\cos^2\theta - 1) + \chi_\perp\sin^2\alpha(3\sin^2\theta\cos^2\Omega - 1) \\ & + \tfrac{3}{4}(\chi_\parallel + \chi_\perp)\sin 2\alpha\sin 2\theta\cos\Omega\big] \end{aligned} \tag{IV.7}$$

and by dividing the result by the nuclear Zeeman energy $\hbar\gamma_I B_0$

$$\begin{aligned} \frac{\Delta E^{\text{dip}}}{\hbar\gamma_I B_0} = \delta^{\text{dip}} = & \frac{1}{4\pi r^3}\big[\chi_\parallel\cos^2\alpha(3\cos^2\theta - 1) + \chi_\perp\sin^2\alpha(3\sin^2\theta\cos^2\Omega - 1) \\ & + \tfrac{3}{4}(\chi_\parallel + \chi_\perp)\sin 2\alpha\sin 2\theta\cos\Omega\big]. \end{aligned} \tag{IV.8}$$

Eq. (IV.8) gives the dipolar shift in the solid state. Note that when χ is isotropic it reduces to

$$\begin{aligned} \delta^{\text{dip}} & = \frac{1}{4\pi r^3}\chi\big[3(\cos\alpha\cos\theta + \sin\alpha\sin\theta\cos\Omega)^2 - 1\big] \\ & = \frac{1}{4\pi r^3}\chi(3\cos^2\gamma - 1) \end{aligned} \tag{IV.9}$$

which gives the dipolar shift in magnetically isotropic solids and which averages zero in solution (cf. Eq. (2.19) and Appendix II).

To derive the pseudocontact shift in solution δ^{pc} we must now take the rotational average of Eq. (IV.8). By recalling that

$$\frac{1}{2\pi}\int_0^{2\pi} \cos\Omega \, \mathrm{d}\Omega = 0 \qquad \frac{1}{2\pi}\int_0^{2\pi} \cos^2\Omega \, \mathrm{d}\Omega = \frac{1}{2}$$

integration of Eq. (IV.8) in $\mathrm{d}\Omega$ gives

$$\delta = \frac{1}{4\pi r^3}[\chi_{\parallel} \cos^2\alpha(3\cos^2\theta - 1) + \chi_{\perp} \sin^2\alpha(\tfrac{3}{2}\sin^2\theta - 1)]. \tag{IV.10}$$

We then need integrating over the solid angle α ($\mathrm{d}\cos\alpha = \sin\alpha\,\mathrm{d}\alpha$), and recalling that

$$\frac{1}{2}\int_0^{\pi} \cos^2\alpha \sin\alpha\,\mathrm{d}\alpha = \frac{1}{3} \qquad \frac{1}{2}\int_0^{\pi} (1 - \cos^2\alpha)\sin\alpha\,\mathrm{d}\alpha = \frac{2}{3}$$

we obtain

$$\begin{aligned} \delta^{\mathrm{dip}} &= \frac{1}{4\pi r^3}\left[\chi_{\parallel}\tfrac{1}{3}(3\cos^2\theta - 1) + \chi_{\perp}\tfrac{1}{3}(3 - 3\cos^2\theta - 2)\right] \\ &= \frac{1}{12\pi r^3}(\chi_{\parallel} - \chi_{\perp})(3\cos^2\theta - 1) \end{aligned} \tag{IV.11}$$

which is Eq. (2.18). The derivation of the general equation for a non axial case (Eq. (2.20)) can be obtained through analogous reasoning.

Eq. (2.20), written in a form containing direction cosines (Eq. (2.21)), can be also intuitively derived from Eq. (2.15), i.e. from the shift experienced by the nucleus when the magnetic field is along the χ_{zz} direction. According to Eq. (2.15), the dipolar shift in this situation is given by:

$$\delta_{zz}^{\mathrm{dip}} = \frac{1}{4\pi}\frac{1}{r^3}\chi_{zz}(3\cos^2\gamma - 1) = \frac{1}{4\pi}\frac{1}{r^3}\chi_{zz}(3n^2 - 1) \tag{IV.12}$$

and analogous relationships hold for the shift experienced by the molecule when oriented with the x or y axis along the magnetic field, with l^2 and m^2 substituting n^2, respectively. l, m and n are the direction cosines of the metal–nucleus vector with respect to the x, y and z principal directions of the χ tensor, respectively. As the dipolar shift is a tensorial property, its average in solution is just the average of the three values of the shift along the three principal directions:

$$\begin{aligned} \delta^{\mathrm{pc}} &= \frac{1}{4\pi}\frac{1}{r^3}\frac{\left[\chi_{xx}(3l^2 - 1) + \chi_{yy}(3m^2 - 1) + \chi_{zz}(3n^2 - 1)\right]}{3} \\ &= \frac{1}{4\pi}\frac{1}{r^3}\left[\chi_{xx}l^2 + \chi_{yy}m^2 + \chi_{zz}n^2 - \overline{\chi}\right] \end{aligned} \tag{IV.13}$$

which can be reduced to Eq. (2.21) by simple algebra.

Appendix V

Relaxation by Dipolar Interaction Between Two Spins

A. Transition probabilities along the z direction

A system constituted by two spins, I and S, is considered. The Hamiltonian is given by

$$H = -\hbar\gamma_I \boldsymbol{B_0} \cdot \boldsymbol{I}_z - \hbar\gamma_S \boldsymbol{B_0} \cdot \boldsymbol{S}_z + H' = H_0 + H' \tag{V.1}$$

where the first two terms are the Zeeman terms for each spin and are time-independent, and H' is the interaction Hamiltonian, which is time-dependent. For the dipolar interaction, H' is given by the first term of the equation written in note 2 of Section 3.4; the interaction depends, besides other variables, on the magnitude of r and orientation of $\boldsymbol{r}$, the vector connecting the atoms bearing the I and S spins. H' is a time-dependent Hamiltonian, whose modulation induces spin relaxation. In the present case we shall assume, without loss of generality, that the only time-dependent part in H' is $\boldsymbol{r}$, i.e., the reciprocal orientation of the two spins with respect to the external magnetic field, while r is constant. In this way H' can only be modulated by rotation and not, for instance, by chemical exchange which would make r time-dependent. Even in this case the treatment is analogous and the final result identical. If $|m_i\rangle$ and $|m_j\rangle$ are two eigenstates of the unperturbed Hamiltonian, H_0, with energies E_i and E_j, the transition probability per unit time between these two states, w_{ij}, is, in the first order, given by the equation written in note 3 of Section 3.4, which can be rewritten as

$$w_{ij} = \frac{1}{\hbar^2}\left|\int_{-t}^{t} \langle m_j|H'(0)|m_i\rangle\langle m_i|H'(t')|m_j\rangle \mathrm{e}^{-i\omega_{ij}t'}\,\mathrm{d}t'\right| \tag{V.2}$$

where $\omega_{ij} = (E_i - E_j)/\hbar$. This integral can be calculated if the perturbing Hamiltonian $H'(t')$ fluctuates randomly; when the fluctuations are rapid, w_{ij} is independent of time.

In the coupled system I–S there are four possible eigenstates according to the values of m_{Iz} (first term in the ket) and m_{Sz} (second term in the ket), as shown in Fig. 3.8A. The transition probabilities in this system are w_1^I, w_1^S, w_0 and w_2.

The perturbing Hamiltonian H' is given by

$$H' = k\big[\boldsymbol{I} \cdot \boldsymbol{S} - 3(\boldsymbol{I} \cdot \boldsymbol{r})(\boldsymbol{S} \cdot \boldsymbol{r})\big] \tag{V.3}$$

where $k = (\mu/4\pi)\hbar\gamma_I\gamma_S/r^3$ will be omitted in the following calculations ($\boldsymbol{r}$ is the unit vector oriented along the line of length r connecting I and S). By developing the vector products we obtain

$$I_x S_x + I_y S_y + I_z S_z - 3(I_x l + I_y m + I_z n)(S_x l + S_y m + S_z n) \tag{V.4}$$

where l, m and n are the direction cosines of $\boldsymbol{r}$. Eq. (V.4) can be rearranged into

$$\begin{aligned} &I_x S_x(1 - 3l^2) + I_y S_y(1 - 3m^2) + I_z S_z(1 - 3n^2) - 3(I_x S_y + I_y S_x)lm \\ &\quad - 3(I_x S_z + I_z S_x)ln - 3(I_y S_z + I_z S_y)mn. \end{aligned} \tag{V.5}$$

We shall now further rearrange Eq. (V.5) so as to distinguish terms which leave the total z component, $M_z = m_{Iz} + m_{Sz}$, unchanged, terms for which $\Delta M_z = \pm 1$, and terms for which $\Delta M_z = \pm 2$. This distinction is important because these terms will contribute to w_0, w_1^I and w_1^S, and w_2, respectively. The direction cosines can be expressed by the polar angles, according to the relationships

$$\begin{aligned} l &= \sin\theta\cos\varphi \\ m &= \sin\theta\sin\varphi \\ n &= \cos\theta. \end{aligned} \tag{V.6}$$

Using Eqs. (III.2), (III.3) and (V.6), Eq. (V.5) can be rearranged to

$$\begin{aligned} &I_z S_z(1 - 3\cos^2\theta) + \tfrac{1}{4}(I_+S_+ + I_+S_- + I_-S_+ + I_-S_-)(1 - 3\sin^2\theta\cos^2\varphi) \\ &\quad - \tfrac{1}{4}(I_+S_+ - I_+S_- - I_-S_+ + I_-S_-)(1 - 3\sin^2\theta\sin^2\varphi) \\ &\quad + \tfrac{3}{2}i(I_+S_+ - I_-S_-)\sin^2\theta\sin\varphi\cos\varphi \\ &\quad - \tfrac{3}{2}(I_+S_z + I_-S_z + I_zS_+ + I_zS_-)\sin\theta\cos\theta\cos\varphi \\ &\quad + \tfrac{3}{2}i(I_+S_z - I_-S_z + I_zS_+ - I_zS_-)\sin\theta\cos\theta\sin\varphi \end{aligned} \tag{V.7}$$

By recalling that $\cos^2\alpha + \sin^2\alpha = 1$, $\cos^2\alpha - \sin^2\alpha = \cos 2\alpha$, $\sin\alpha\cos\alpha = ½\sin 2\alpha$, and $\cos\alpha \pm i\sin\alpha = e^{\pm i\alpha}$, Eq. (V.7) can be further rearranged to

$$\begin{aligned} &I_z S_z(1 - 3\cos^2\theta) - \tfrac{1}{4}(I_+S_- + I_-S_+)(1 - 3\sin^2\theta) - I_+S_+(\tfrac{3}{4}\sin^2\theta \mathrm{e}^{-2i\varphi}) \\ &\quad - I_-S_-(\tfrac{3}{4}\sin^2\theta \mathrm{e}^{2i\varphi}) - (I_+S_z + I_zS_+)(\tfrac{3}{2}\sin\theta\cos\theta \mathrm{e}^{-i\varphi}) \\ &\quad - (I_-S_z + I_zS_-)(\tfrac{3}{2}\sin\theta\cos\theta \mathrm{e}^{i\varphi}). \end{aligned} \tag{V.8}$$

By defining now

$$
\begin{aligned}
F_0(t) &= k(1 - 3\cos^2\theta(t)) \\
F_1(t) &= -\tfrac{3}{2}k\sin\theta(t)\cos\theta(t)\mathrm{e}^{-i\varphi(t)} \\
F_2(t) &= -\tfrac{3}{4}k\sin^2\theta(t)\mathrm{e}^{-2i\varphi(t)}
\end{aligned} \tag{V.9}
$$

with $k = (\mu_0/4\pi)\hbar\gamma_I\gamma_S/r^3$, and recalling that the time dependence is indeed contained in θ and φ, the final dipolar Hamiltonian is

$$
\begin{aligned}
H' = {} & [I_zS_z - \tfrac{1}{4}(I_+S_- + I_-S_+)]F_0 + [I_+S_z + I_zS_+]F_1 + \\
& [I_-S_z + I_zS_-]F_1^* + I_+S_+F_2 + I_-S_-F_2^*.
\end{aligned} \tag{V.10}
$$

Note that the F_i functions are just the spherical harmonics of order i.

We can now express the transition probabilities. Let us start from

$$
w_0 = \frac{1}{\hbar^2}\left|\int_{-t}^{t} \langle + - |H'(0)| - +\rangle\langle - + |H'(t')| + -\rangle \mathrm{e}^{-i(\omega_I-\omega_S)t'}\,\mathrm{d}t'\right|. \tag{V.11}
$$

The only part of H' extracting $| + -\rangle$ from $| - +\rangle$ is that containing I_+S_-, and therefore

$$
w_0 = \frac{1}{\hbar^2}\left|\int_{-t}^{t} \tfrac{1}{16}\langle F_0(0)F_0^*(t')\rangle \mathrm{e}^{-i(\omega_I-\omega_S)t'}\,\mathrm{d}t'\right|. \tag{V.12}
$$

Analogously,

$$
w_1^I = \frac{1}{\hbar^2}\left|\int_{-t}^{t} \tfrac{1}{4}\langle F_1(0)F_1^*(t')\rangle \mathrm{e}^{-i\omega_I t'}\,\mathrm{d}t'\right| \tag{V.13}
$$

$$
w_1^S = \frac{1}{\hbar^2}\left|\int_{-t}^{t} \tfrac{1}{4}\langle F_1(0)F_1^*(t')\rangle \mathrm{e}^{-i\omega_S t'}\,\mathrm{d}t'\right| \tag{V.14}
$$

$$
w_2 = \frac{1}{\hbar^2}\left|\int_{-t}^{t} \langle F_2(0)F_2^*(t')\rangle \mathrm{e}^{-i(\omega_I+\omega_S)t'}\,\mathrm{d}t'\right|. \tag{V.15}
$$

We shall recall that the time dependence in the perturbing Hamiltonian is now confined to the F functions, which are only affected by rotation. If we assume that the rotation is random we can write the expectation value of the products $F_i(0)F_i(t)$ as an exponential function (see note 3 of Section 3.4)

$$\langle F_i(0)F_i^*(t)\rangle = \langle |F_i(0)|^2\rangle e^{-t/\tau_c} \tag{V.16}$$

so that we can extract $\langle |F_i|^2\rangle$ from the integrals in Eqs. (V.12)–(V.15)[1]. In turn, the $\langle |F_i|^2\rangle$ can be easily computed; we show here the calculations for $\langle |F_0|^2\rangle$:

$$\begin{aligned}\langle |F_0|^2\rangle &= k^2\langle |1-3\cos^2\theta|^2\rangle \\ &= \tfrac{1}{2}k^2\int_{-1}^{1}(1-6\cos^2\theta+9\cos^4\theta)\,\mathrm{d}(\cos\theta) = \tfrac{4}{5}k^2.\end{aligned} \tag{V.17}$$

The other averages are obtained in a similar manner:

$$\langle |F_1|^2\rangle = \langle |F_2|^2\rangle = \tfrac{3}{10}k^2. \tag{V.18}$$

Before performing the integration we should distinguish between the two cases of I and S being like or unlike spins. In the case of like spins $T_1^{-1} = 2(w_1 + w_2)$, and we have integrals of the type

$$\int_{-t}^{t}\langle F_\alpha(0)F_\alpha^*(t')\rangle e^{-i\alpha\omega t'}\,\mathrm{d}t' \tag{V.19}$$

where $\omega = \omega_I = \omega_S$ and $\alpha = 1$ in the case of w_1^I (or w_1^S) and $\alpha = 2$ in the case of w_2. These integrals must be evaluated for $t \gg \tau_c$. Under these conditions it has been shown that Eqs. (V.14) and (V.15) become

$$\begin{aligned} w_1 &= \frac{\tau_c}{2\hbar^2}\langle |F_1|^2\rangle\frac{1}{1+\omega^2\tau_c^2} \\ w_2 &= \frac{2\tau_c}{\hbar^2}\langle |F_2|^2\rangle\frac{1}{1+4\omega^2\tau_c^2}\end{aligned} \tag{V.20}$$

and thus

$$T_1^{-1} = \frac{3}{10}\left(\frac{\mu_0}{4\pi}\right)^2\frac{\gamma^4}{r^6}\left(\frac{\tau_c}{1+\omega^2\tau_c^2}+\frac{4\tau_c}{1+4\omega^2\tau_c^2}\right). \tag{V.21}$$

In the case of unlike spins, in general, relaxation does not follow an exponential behavior. However, a special case of primary interest in this book is when S is an electron spin. In this case $\langle S_z(t)\rangle = \langle S_z(0)\rangle$ during the entire relaxation process of

[1] This holds when $\tau_c = \tau_r$; otherwise, if $\tau_c < \tau_r$, the average should be performed on the integral value and the F function cannot be extracted from the integral. However, the final results are unchanged if H_0 is isotropic.

I. T_1^{-1} is given by $w_0 + 2w_1^I + w_2$ (see Eq. (7.3) and Appendix VII). In analogy to Eq. (V.20) w_0 is given by

$$w_0 = \frac{1}{16}\frac{2\tau_c}{\hbar^2}\langle|F_0|^2\rangle\frac{1}{1+(\omega_S-\omega_I)^2\tau_c^2}. \tag{V.22}$$

Thus, by analogy with the like-spins case,

$$T_1^{-1} = \frac{1}{10}\left(\frac{\mu_0}{4\pi}\right)^2\frac{\hbar^2\gamma_I^2\gamma_S^2}{r^6}\left(\frac{3\tau_c}{1+\omega_I^2\tau_c^2}+\frac{\tau_c}{1+(\omega_I-\omega_S)^2\tau_c^2}\right.$$
$$\left.+\frac{6\tau_c}{1+(\omega_I+\omega_S)^2\tau_c^2}\right) \tag{V.23}$$

which, given that $\omega_S \gg \omega_I$, can be approximated by

$$T_1^{-1} = \frac{1}{10}\left(\frac{\mu_0}{4\pi}\right)^2\frac{\hbar^2\gamma_I^2\gamma_S^2}{r^6}\left(\frac{7\tau_c}{1+\omega_S^2\tau_c^2}+\frac{3\tau_c}{1+\omega_I^2\tau_c^2}\right). \tag{V.24}$$

B. Transition probabilities in the xy plane

In calculating longitudinal relaxation, we started from four eigenstates $|--\rangle$, $|-+\rangle$, $|+-\rangle$, $|++\rangle$. For transverse relaxation we are interested in the eigenstates of I_x (or I_y) and S_x (or S_y). We can define the states $|\alpha\rangle$ and $|\beta\rangle$ so that

$$I_x|\alpha\rangle = \tfrac{1}{2}|\alpha > \qquad S_x|\alpha\rangle = \tfrac{1}{2}|\alpha\rangle$$
$$I_x|\beta\rangle = \tfrac{1}{2}|\beta\rangle \qquad S_x|\beta\rangle = \tfrac{1}{2}|\beta\rangle.$$

However, here $|\alpha\alpha\rangle$ $|\alpha\beta\rangle$ $|\beta\alpha\rangle$ $|\beta\beta\rangle$ are not eigenstates of the energy. Nevertheless, they are still orthogonal and can always be expanded in eigenvectors of the energy. For example, in the present case of spin 1/2 we have:

$$|\alpha\rangle = \tfrac{1}{\sqrt{2}}(|+\rangle + |-\rangle)$$
$$|\beta\rangle = \tfrac{1}{\sqrt{2}}(|+\rangle - |-\rangle).$$

In general, two orthogonal states $|a\rangle$ and $|b\rangle$ are expanded

$$|a\rangle = \sum_i a_i|m_i\rangle \qquad |b\rangle = \sum_i b_i|m_i\rangle$$

and, therefore, the transition probability between the states $|a\rangle$ and $|b\rangle$ is

$$u_{ab} = \frac{1}{\hbar^2}\left|\int_{-t}^{t}\sum_{ij}\langle m_j|H'(0)|m_i\rangle\langle m_i|H'(t')|m_j\rangle a_i b_j \mathrm{e}^{-i\omega_{ij}t'}\,\mathrm{d}t'\right|.$$

This equation is analogous to Eq. (V.2), and through a procedure analogous to that already reported, the equations for T_2^{-1} in the like-spin case (Eq. (3.12)) and unlike-spin case (Eq. (3.17)) are obtained.

Appendix VI

Calculation of $\langle S_z \rangle$: Curie's Law

The expectation values of physical operators O for which spin states are eigenfunctions are given by

$$\langle O \rangle = \frac{\sum_i \langle i|O|i\rangle \mathrm{e}^{-E_i/kT}}{\sum_i \mathrm{e}^{-E_i/kT}}. \tag{VI.1}$$

In the case of the operator S_z, Eq. (VI.1) becomes

$$\langle S_z \rangle = \frac{\sum_{S,M_S} \langle S, M_S|S_z|S, M_S\rangle \mathrm{e}^{-E_{S,M_S}/kT}}{\sum_{S,M_S} \mathrm{e}^{-E_{S,M_S}/kT}} \tag{VI.2}$$

where the eigenfunctions $|S, M_S\rangle$ are the spin functions of the S multiplet without any splitting at zero magnetic field and E_{S,M_S} are the energies. The latter are non degenerate in the presence of a static magnetic field and equal to $g_e\mu_B B_0 M_S$. Now, for current values of magnetic fields, $E_{S,M_S}/kT \ll 1$; therefore, $\mathrm{e}^{E_{S,M_S}/kT} \approx 1 - E_{S,M_S}/kT$, and Eq. (VI.2) can be written as

$$\langle S_z \rangle = \frac{\sum_{S,M_S} \langle S, M_S|S_z|S, M_S\rangle (1 - E_{S,M_S}/kT)}{\sum_{S,M_S} (1 - E_{S,M_S}/kT)}. \tag{VI.3}$$

The matrix elements in the numerator yield the M_S values, i.e. numbers ranging from $-S$ to S. The terms in the sum will thus be of the type

$$-S - \frac{g_e\mu_B B_0}{kT}S^2, -(S-1) - \frac{g_e\mu_B B_0}{kT}(S-1)^2, \ldots, S - \frac{g_e\mu_B B_0}{kT}S^2 \tag{VI.4}$$

and it can be seen that the sum of $-S$, $-(S-1)$, … , S is zero. The numerator becomes

$$-\frac{g_e\mu_B B_0}{kT} \sum_{M_S=-S}^{S} M_S^2. \tag{VI.5}$$

In the denominator the terms containing the energy cancel, and we are left with $\sum_{M_S} 1 = 2S + 1$. Recalling that $\sum M_S^2 = (2S+1)S(S+1)/3$, Eq. (VI.3) becomes

$$\langle S_z \rangle = -\frac{g_e \mu_B B_0}{3kT} S(S+1) \tag{VI.6}$$

which gives the magnetic field and temperature dependence of $\langle S_z \rangle$, known as Curie's law.

Appendix VII

Derivation of the Equations Related to NOE

A necessary step to derive Eq. (7.2) and other equations in Chapter 7 is the obtainment of a suitable expression for the time dependence of $\langle I_z \rangle$ in a dipole-coupled two spin system. By referring to Fig. 7.1, $\langle I_z \rangle$ and $\langle J_z \rangle$ are given by

$$\begin{aligned} \langle I_z \rangle &= K(P_{++} + P_{+-} - P_{-+} - P_{--}) \\ \langle J_z \rangle &= K(P_{++} + P_{-+} - P_{+-} - P_{--}). \end{aligned} \tag{VII.1}$$

The rate of variation of $\langle I_z \rangle$ with time, $\mathrm{d}\langle I_z(t) \rangle / \mathrm{d}t$, is proportional to the rate of population increase of levels $++$ and $+-$ minus the rate of population increase of levels $-+$ and $--$:

$$\frac{\mathrm{d}\langle I_z(t) \rangle}{\mathrm{d}t} = K\left(\frac{\mathrm{d}P_{++}}{\mathrm{d}t} + \frac{\mathrm{d}P_{+-}}{\mathrm{d}t} - \frac{\mathrm{d}P_{-+}}{\mathrm{d}t} - \frac{\mathrm{d}P_{--}}{\mathrm{d}t}\right). \tag{VII.2}$$

Among the various transitions shown in Fig. 7.1, we can neglect those involving only changes in spin state of J. For instance, the first of the four terms in the right-hand side of Eq. (VII.2) is:

$$\frac{\mathrm{d}P_{++}}{\mathrm{d}t} = -(w_1^I + w_2)P_{++} + w_1^I P_{-+} + w_2 P_{--} + c. \tag{VII.3}$$

The constant c can be evaluated by setting $\mathrm{d}P_{++}/\mathrm{d}t = 0$:

$$c = (w_1^I + w_2)P_{++}(\infty) - w_1^I P_{-+}(\infty) - w_2 P_{--}(\infty) \tag{VII.4}$$

and Eq. (VII.3) becomes

$$\frac{\mathrm{d}P_{++}}{\mathrm{d}t} = -(w_1^I + w_2)\Delta P_{++} + w_1^I \Delta P_{-+} + w_2 \Delta P_{--} \tag{VII.5}$$

where

$$P - P(\infty) = \Delta P. \tag{VII.6}$$

Analogous equations can be written for the other three terms in Eq. (VII.2). By grouping all the terms together, Eq. (VII.2) becomes

$$\frac{\mathrm{d}\langle I_z(t)\rangle}{\mathrm{d}t} = K\big[-2\Delta P_{++}(w_1^I + w_2) - 2\Delta P_{+-}(w_1^I + w_0) + 2\Delta P_{-+}(w_1^I + w_0) + 2\Delta P_{--}(w_1^I + w_2)\big] \quad \text{(VII.7)}$$

which can be rearranged to

$$\frac{\mathrm{d}\langle I_z(t)\rangle}{\mathrm{d}t} = 2K\big[(\Delta P_{--} - \Delta P_{++})(w_1^I + w_2) + (\Delta P_{-+} - \Delta P_{+-})(w_1^I + w_0)\big] \quad \text{(VII.8)}$$

The factor 2 derives from the fact that a single transition decreases the population of the starting state and at the same time increases the population of the target state, thereby changing the population difference by twice as much.

However, from Eq. (VII.1) we can write

$$\begin{aligned}\langle I_z\rangle + \langle J_z\rangle &= -2K(P_{--} - P_{++})\\ \langle I_z\rangle - \langle J_z\rangle &= -2K(P_{-+} - P_{+-})\end{aligned} \quad \text{(VII.9)}$$

and, using Eq. (VII.6)

$$\begin{aligned}\langle I_z(t)\rangle - \langle I_z(\infty)\rangle + \langle J_z(t)\rangle - \langle J_z(\infty)\rangle &= -2K(\Delta P_{--} - \Delta P_{++})\\ \langle I_z(t)\rangle - \langle I_z(\infty)\rangle - \langle J_z(t)\rangle + \langle J_z(\infty)\rangle &= -2K(\Delta P_{-+} - \Delta P_{+-})\end{aligned} \quad \text{(VII.10)}$$

Substituting Eq. (VII.10) into Eq. (VII.8) gives

$$\frac{\mathrm{d}\langle I_z(t)\rangle}{\mathrm{d}t} = -\big[\langle I_z(t)\rangle - \langle I_z(\infty)\rangle + \langle J_z(t)\rangle - \langle J_z(\infty)\rangle\big](w_1^I + w_2) - \big[\langle I_z(t)\rangle - \langle I_z(\infty)\rangle - \langle J_z(t)\rangle + \langle J_z(\infty)\rangle\big](w_1^I + w_0) \quad \text{(VII.11)}$$

that finally becomes

$$\frac{\mathrm{d}\langle I_z(t)\rangle}{\mathrm{d}t} = -\big[\langle I_z(t)\rangle - \langle I_z(\infty)\rangle\big](w_0 + 2w_1^I + w_2) - \big[\langle J_z(t)\rangle - \langle J_z(\infty)\rangle\big](w_2 - w_0) \quad \text{(VII.12)}$$

Eq. (VII.12) is the starting point to derive not only the equations relevant for the NOE phenomenon (Chapter 7) but also Eq. (3.15) and the following ones (Section 3.4). A somewhat different form of Eq. (VII.12) has already been encountered when dealing with transfer of magnetization between two sites in chemical exchange (Section 4.3.4).

According to the definitions given in Eqs. (7.3) and (7.4), Eq. (IV.12) can be rewritten as

$$\frac{\mathrm{d}\langle I_z(t)\rangle}{\mathrm{d}t} = -\big[\langle I_z(t)\rangle - \langle I_z(\infty)\rangle\big]\rho_{I(J)} - \big[\langle J_z(t)\rangle - \langle J_z(\infty)\rangle\big]\sigma_{I(J)}. \quad \text{(VII.13)}$$

Eq. (7.2), relevant for steady state NOE, can easily be obtained from Eq. (VII.13) by setting $\mathrm{d}\langle I_z(t)\rangle/\mathrm{d}t = 0$ and $\langle J_z(t)\rangle = 0$:

$$0 = -\big[\langle I_z(t)\rangle - \langle I_z(\infty)\rangle\big]\rho_{I(J)} + \langle J_z(\infty)\rangle\sigma_{I(J)} \quad \text{(VII.14)}$$

which rearranges to

$$\langle I_z(t)\rangle = \langle I_z(\infty)\rangle + \big[\sigma_{I(J)}/\rho_{I(J)}\big]\langle J_z(\infty)\rangle. \quad \text{(7.2)}$$

For the homonuclear case, generalized by substituting $\rho_{I(J)}$ with the total relaxation rate ρ_I in Eq. (VII.13) and the following, we obtain

$$\frac{\langle I_z(t)\rangle - \langle I_z(\infty)\rangle}{\langle I_z(\infty)\rangle} = \frac{\sigma_{I(J)}}{\rho_I}. \quad \text{(7.10)}$$

The equation for truncated NOE (Section 7.3) can also be derived from Eq. (VII.13), again generalized by substituting $\rho_{I(J)}$ with the total relaxation rate ρ_I, by setting $\langle J_z(t)\rangle = 0$ (instantaneous saturation). Integration then gives

$$\langle I_z(t)\rangle = A\exp(-\rho_I t) + B. \quad \text{(VII.15)}$$

To evaluate the constants A and B we start by deriving again Eq. (VII.15):

$$\frac{\mathrm{d}\langle I_z(t)\rangle}{\mathrm{d}t} = -A\rho_I\exp(-\rho_I t). \quad \text{(VII.16)}$$

From Eq. (VII.15) we can then write

$$A\exp(-\rho_I t) = \langle I_z(t)\rangle - B \quad \text{(VII.17)}$$

and, by substituting Eq. (VII.17) into Eq. (VII.16), we obtain

$$\frac{\mathrm{d}\langle I_z(t)\rangle}{\mathrm{d}t} = -\rho_I\langle I_z(t)\rangle + \rho_I B. \quad \text{(VII.18)}$$

However, from Eq. (VII.13) under instantaneous saturation of J we can write

$$\frac{\mathrm{d}\langle I_z(t)\rangle}{\mathrm{d}t} = -\big[\langle I_z(t)\rangle - \langle I_z(\infty)\rangle\big]\rho_I + \langle J_z(\infty)\rangle\sigma_{I(J)} \quad \text{(VII.19)}$$

and, by equating the right-hand sides of Eqs. (VII.18) and (VII.19)

$$B = \langle I_z(\infty)\rangle + [\sigma_{I(J)}/\rho_I]\langle J_z(\infty)\rangle. \quad \text{(VII.20)}$$

By substituting Eq. (VII.20) back into Eq. (VII.15) we obtain

$$\langle I_z(t)\rangle = A\exp(-\rho_I t) + \langle I_z(\infty)\rangle + [\sigma_{I(J)}/\rho_I]\langle J_z(\infty)\rangle \tag{VII.21}$$

at $t = 0$, $\langle I_z(0)\rangle = \langle I_z(\infty)\rangle$, and therefore

$$A = -[\sigma_{I(J)}/\rho_I]\langle J_z(\infty)\rangle \tag{VII.22}$$

from which Eq. (VII.21) becomes

$$\langle I_z(t)\rangle = \langle I_z(\infty)\rangle + (\sigma_{I(J)}/\rho_I)\langle J_z(\infty)\rangle[1 - \exp(-\rho_I t)] \tag{VII.23}$$

by setting $\langle J_z(\infty)\rangle = \langle I_z(\infty)\rangle$ and rearranging we obtain the equation for truncated NOE:

$$\frac{\langle I_z(t)\rangle - \langle I_z(\infty)\rangle}{\langle I_z(\infty)\rangle} = \eta_{I(J)}(t) = [\sigma_{I(J)}/\rho_I][1 - \exp(-\rho_I t)]. \tag{7.17}$$

Magnetically Coupled Dimers in the High-Temperature Limit

A. Contact shift

We perform here a sample calculation to show that Eq. (6.9)

$$\delta^{\text{con}} = A_1 \frac{g_e \mu_B}{\hbar \gamma_I 3kT} \frac{\sum_i C_{i1} S_i'(S_i' + 1)(2S_i' + 1) \exp(-E_i/kT)}{\sum_i (2S_i' + 1) \exp(-E_i/kT)} \tag{6.9}$$

reduces to Eq. (VIII.1)

$$\delta^{\text{con}} = A_1 \frac{1}{\hbar \gamma_I B_0} \langle \hat{S}_{1z} \rangle = -A_1 \frac{g_e \mu_B}{\hbar \gamma_I 3kT} S_1'(S_1' + 1) \tag{VIII.1}$$

in the high temperature limit. If $E_i/kT \ll 1$ the exponential terms in the numerator and denominator are all ca. 1 and Eq. (6.9) becomes

$$\delta^{\text{con}} = A_1 \frac{g_e \mu_B}{\hbar \gamma_I 3kT} \frac{\sum_i C_{i1} S_i'(S_i' + 1)(2S_i' + 1)}{\sum_i (2S_i' + 1)}. \tag{VIII.2}$$

Let us take $S_1 = 2$ and $S_2 = 5/2$, as in the case of reduced Fe_2S_2 (see Section 6.3.4). From Table 6.1 it is

$S' = 1/2$	$S' = 3/2$	$S' = 5/2$	$S' = 7/2$	$S' = 9/2$
$C_{11} = -4/3$	$C_{12} = 2/15$	$C_{13} = 12/35$	$C_{14} = 26/63$	$C_{15} = 4/9$
$C_{21} = 7/3$	$C_{22} = 13/15$	$C_{23} = 23/35$	$C_{24} = 37/63$	$C_{25} = 5/9$

and the contribution from metal 1 ($S_1 = 2$) is

$$\frac{-4/3 \cdot 3/4 \cdot 2}{2+4+6+8+10} + \frac{2/15 \cdot 15/4 \cdot 4}{2+4+6+8+10} + \frac{12/35 \cdot 35/4 \cdot 6}{2+4+6+8+10}$$
$$+ \frac{26/63 \cdot 63/4 \cdot 8}{2+4+6+8+10} + \frac{4/9 \cdot 99/4 \cdot 10}{2+4+6+8+10}$$
$$= \frac{1}{30}(-2+2+18+52+110)$$
$$= \frac{180}{30} = 6 = S_1(S_1+1).$$

As expected, this contribution is the same as that of the isolated ion.

B. Nuclear relaxation

In the high temperature limit and assuming a single correlation time, Eqs. (6.13) and (6.13b) can be simplified to

$$R_{1M_1} = \frac{2}{15}\left(\frac{\mu_0}{4\pi}\right)^2 \frac{\gamma_I^2 g_e^2 \mu_B^2}{r^6} \times \frac{\sum_i \left[C_{i1}^2 S_i'(S_i'+1)(2S_i'+1)\left(\frac{7\tau_{c2}}{1+\omega_S^2\tau_{c2}^2} + \frac{3\tau_{c1}}{1+\omega_I^2\tau_{c1}^2}\right)\right]}{\sum_i [(2S_i'+1)]} \tag{VIII.3}$$

$$R_{1M_1} = \frac{2}{3}\frac{A_1^2}{\hbar^2} \times \frac{\sum_i \left[C_{i1}^2 S_i'(S_i'+1)(2S_i'+1)\left(\frac{\tau_{c2}}{1+\omega_S^2\tau_{c2}^2}\right)\right]}{\sum_i [(2S_i'+1)]}. \tag{VIII.4}$$

These equations are related to the equations for dipolar and contact relaxation in the case of isolated metal ions (Eq. (3.19) and Eq. (3.26), respectively) by a coefficient. By comparison of Eq. (3.19) with Eq. (VIII.3) and Eq. (3.26) with Eq. (VIII.4), this coefficient is found to be [1]

$$X_1 = \frac{1}{S_1(S_1+1)} \frac{\sum_i [C_{i1}^2 S_i'(S_i'+1)(2S_i'+1)]}{\sum_i [(2S_i'+1)]}.$$

Therefore, in the high temperature limit, the equations for nuclear relaxation are related to those for the isolated systems only by a multiplicative coefficient [1].

REFERENCES

[1] C. Owens, R.S. Drago, C. Luchinat, I. Bertini, L. Banci (1986) J. Am. Chem. Soc. 108, 3298.

Appendix IX

Product Operators: Basic Tools [1]

I. BERTINI, C. LUCHINAT and A. ROSATO

The evolution of magnetization during NMR experiments can be followed by means of the so-called product operators formalism. This approach has the advantage of being simple, and of being pictorially representable.

The basic idea of this approach is that of representing the magnetization of one spin through a combination of spin angular momentum operators. For instance, magnetization of a spin I can be represented through a linear combination of the three operators I_x, I_y, I_z. In the presence of an external magnetic field, the nuclear spin will be aligned along the direction of the field, say z; its magnetization is then represented by (is proportional to) I_z. A generic α degrees pulse applied along one of the in-plane axes will produce a rotation of the magnetization vector around that axis of α degrees. This means that after a 90° pulse applied along the y axis, magnetization will be along the x axis (Fig. IX.1), and therefore will be represented by I_x [2].

If the nuclear spin has a Larmor frequency ω, then its time evolution under the effect of chemical shift is given by the following rules:

$$\begin{aligned} I_z &\longrightarrow I_z \\ I_x &\longrightarrow I_x \cos \omega t + I_y \sin \omega t \\ I_y &\longrightarrow I_y \cos \omega t - I_x \sin \omega t \end{aligned} \qquad \text{(IX.1)}$$

This kind of representation for a single nuclear spin is absolutely equivalent to the classic vector model.

Things get a little bit more complicated when one has to deal with a two spin (I, K) system. In this case, the spin system is described by a combination of the following spin operators:

$$\begin{aligned} &I_x, I_y, I_z \\ &K_x, K_y, K_z \\ &I_xK_x, I_yK_x, I_zK_x, I_xK_y, I_yK_y, I_zK_y, I_xK_z, I_yK_z, I_zK_z \end{aligned}$$

[1] This appendix will deal only with spin ½ nuclei. However, extensions of this approach have been developed for other nuclear spins.

[2] The convention used here is that a pulse applied along the y axis rotates spin magnetization from the z axis to the positive side of the x axis.

Operator name	Matrix form
I	$\begin{bmatrix} 1 & 0 & 0 & 0 \\ 0 & 1 & 0 & 0 \\ 0 & 0 & 1 & 0 \\ 0 & 0 & 0 & 1 \end{bmatrix}$
$I_x\ \frac{1}{2}$	$\begin{bmatrix} 0 & 0 & 1 & 0 \\ 0 & 0 & 0 & 1 \\ 1 & 0 & 0 & 0 \\ 0 & 1 & 0 & 0 \end{bmatrix}$
$I_y\ \frac{i}{2}$	$\begin{bmatrix} 0 & 0 & -1 & 0 \\ 0 & 0 & 0 & -1 \\ 1 & 0 & 0 & 0 \\ 0 & 1 & 0 & 0 \end{bmatrix}$
$I_z\ \frac{1}{2}$	$\begin{bmatrix} 1 & 0 & 0 & 0 \\ 0 & 1 & 0 & 0 \\ 0 & 0 & -1 & 0 \\ 0 & 0 & 0 & -1 \end{bmatrix}$
$S_x\ \frac{1}{2}$	$\begin{bmatrix} 0 & 1 & 0 & 0 \\ 1 & 0 & 0 & 0 \\ 0 & 0 & 0 & 1 \\ 0 & 0 & 1 & 0 \end{bmatrix}$
$S_y\ \frac{i}{2}$	$\begin{bmatrix} 0 & -1 & 0 & 0 \\ 1 & 0 & 0 & 0 \\ 0 & 0 & 0 & -1 \\ 0 & 0 & 1 & 0 \end{bmatrix}$
$S_z\ \frac{1}{2}$	$\begin{bmatrix} 1 & 0 & 0 & 0 \\ 0 & -1 & 0 & 0 \\ 0 & 0 & 1 & 0 \\ 0 & 0 & 0 & -1 \end{bmatrix}$
$I_zS_z\ \frac{1}{4}$	$\begin{bmatrix} 0 & 0 & 1 & 0 \\ 0 & 0 & 0 & -1 \\ 1 & 0 & 0 & 0 \\ 0 & -1 & 0 & 0 \end{bmatrix}$

I spins   *S* spins

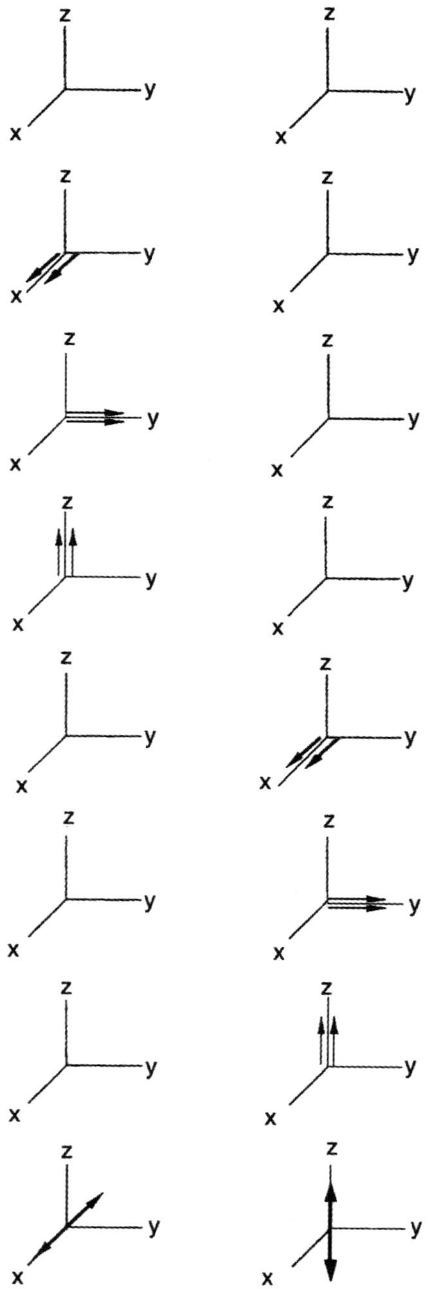

Fig. IX.1. Product operators for a two-spin system. The density matrix form is shown along with the vector representation (adapted from [1]).

Operator name	Matrix form	*I* spins	*S* spins
$I_yS_x\,\frac{i}{2}$	$\begin{bmatrix} 0 & 0 & -1 & 0 \\ 0 & 0 & 0 & 1 \\ 1 & 0 & 0 & 0 \\ 0 & -1 & 0 & 0 \end{bmatrix}$	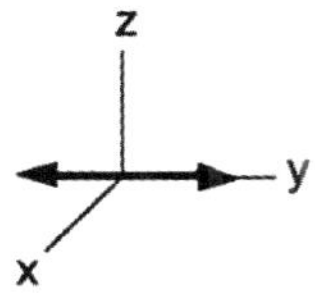	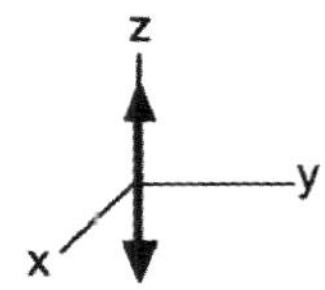
$I_zS_z\,\frac{1}{4}$	$\begin{bmatrix} 1 & 0 & 0 & 0 \\ 0 & -1 & 0 & 0 \\ 0 & 0 & -1 & 0 \\ 0 & 0 & 0 & 1 \end{bmatrix}$	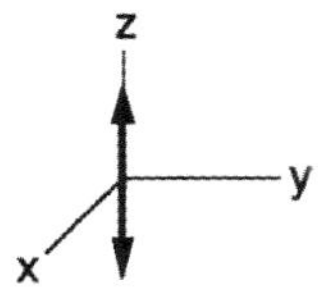	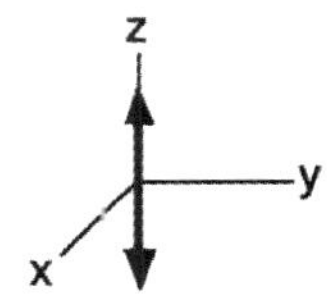
$I_zS_x\,\frac{1}{4}$	$\begin{bmatrix} 0 & 1 & 0 & 0 \\ 1 & 0 & 0 & 0 \\ 0 & 0 & 0 & -1 \\ 0 & 0 & -1 & 0 \end{bmatrix}$	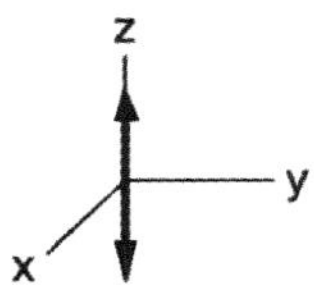	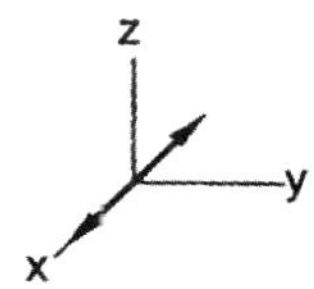
$I_zS_y\,\frac{i}{4}$	$\begin{bmatrix} 0 & -1 & 0 & 0 \\ 1 & 0 & 0 & 0 \\ 0 & 0 & 0 & -1 \\ 0 & 0 & -1 & 0 \end{bmatrix}$	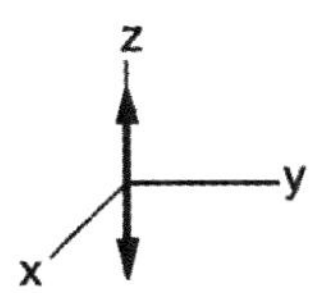	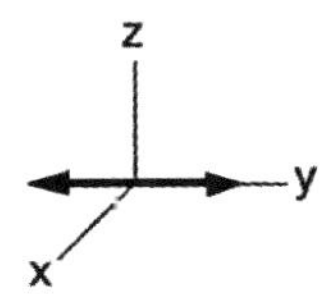
$I_zS_x\,\frac{1}{4}$	$\begin{bmatrix} 0 & 0 & 0 & 1 \\ 0 & 0 & 1 & 0 \\ 0 & 1 & 0 & 0 \\ 1 & 0 & 0 & 0 \end{bmatrix}$	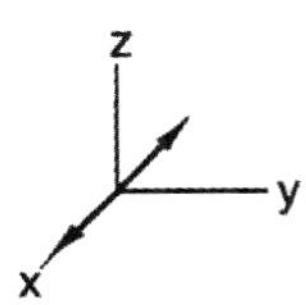	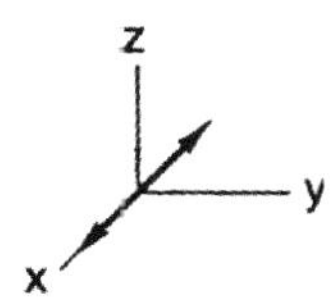
$I_yS_y\,\frac{1}{4}$	$\begin{bmatrix} 0 & 0 & 0 & -1 \\ 0 & 0 & 1 & 0 \\ 0 & 1 & 0 & 0 \\ -1 & 0 & 0 & 0 \end{bmatrix}$	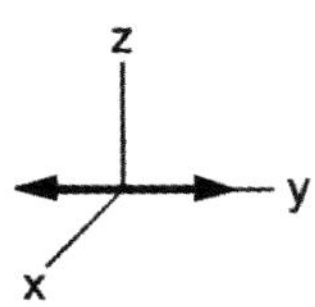	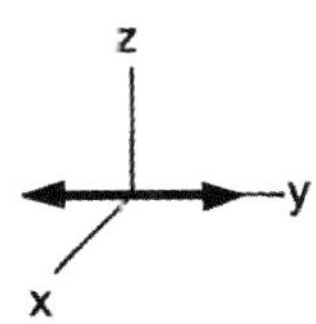
$I_zS_y\,\frac{i}{4}$	$\begin{bmatrix} 0 & 0 & 0 & -1 \\ 0 & 0 & 1 & 0 \\ 0 & -1 & 0 & 0 \\ 1 & 0 & 0 & 0 \end{bmatrix}$	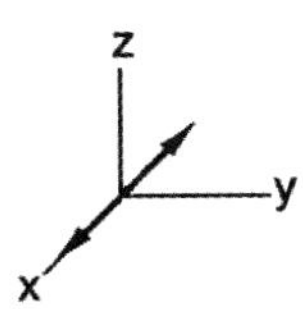	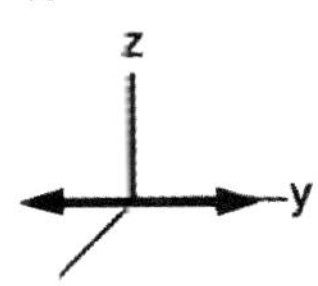
$I_yS_x\,\frac{i}{4}$	$\begin{bmatrix} 0 & 0 & 0 & -1 \\ 0 & 0 & -1 & 0 \\ 0 & 1 & 0 & 0 \\ 1 & 0 & 0 & 0 \end{bmatrix}$	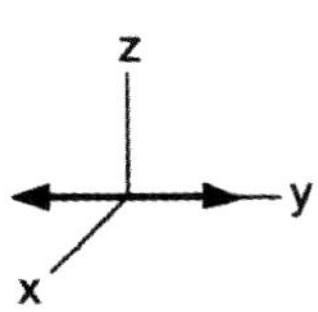	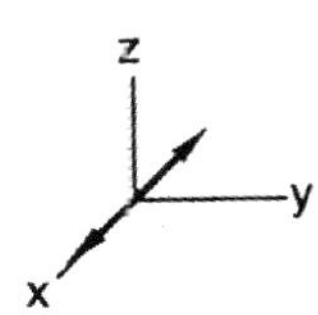

Fig. IX.2.

Notice that the latter nine operators are formed by all the possible products (hence the name of the formalism), two at time, of the three spin angular momentum operators of spin I and the three spin angular momentum operators of spin K. Vector representations of all these operators are shown in Fig. IX.2.

In the presence of an external magnetic field, the two nuclear spins will be aligned along z and their magnetization will be represented by the operators I_z and K_z. If I and K are two spins of the same nuclear species, after a non-selective pulse applied along the y axis, their magnetization will be represented by I_x and K_x respectively. Evolution under the effect of chemical shift is still given by Eqs. (IX.1). Spin I will precess with frequency ω_I and spin K will precess with frequency ω_K. So far, each spin behaves as it would do if the other spin were not present. However, if the two spins are coupled, the evolution of one spin will affect the evolution of the other spin as well. Let us restrict ourselves, for the sake of simplicity, to the case of scalar coupling in solution, the coupling constant between I and K being J. The effect of scalar coupling on the time evolution of one spin (say spin I) can be computed from the following equations:

$$\begin{aligned}
I_z &\longrightarrow I_z \\
I_x &\longrightarrow I_x \cos \pi J t + 2I_y K_z \sin \pi J t \\
I_y &\longrightarrow I_y \cos \pi J t - 2I_x K_z \sin \pi J t \\
2I_y K_z &\longrightarrow 2I_y K_z \cos \pi J t - I_x \sin \pi J t \\
2I_x K_z &\longrightarrow 2I_x K_z \cos \pi J t + I_y \sin \pi J t
\end{aligned} \qquad \text{(IX.2)}$$

The same holds for spin K; the effect is still described by Eqs. (IX.2), in which I and K are swapped.

The effects of chemical shift and scalar coupling on the time evolution of one spin can be accounted for independently, regardless of the order (the two effects are said to commute). For example:

$$I_x$$

$$\swarrow \qquad \searrow$$

$$I_x \cos \omega_I t + I_y \sin \omega_I t \qquad\qquad I_x \cos \pi J t + 2I_y K_z \sin \pi J t$$

$$\Downarrow \qquad\qquad \Downarrow$$

$$\begin{aligned}&\cos \omega_I t\,[I_x \cos \pi J t + 2I_y K_z \sin \pi J t] + \\ &+ \sin \omega_I t[I_y \cos \pi J t - 2I_x K_z \sin \pi J t]\end{aligned} \qquad \begin{aligned}&\cos \pi J t[I_x \cos \omega_I t + I_y \sin \omega_I t] + \\ &+ \sin \pi J t[2I_y K_z \cos \omega_I t - 2I_x K_z \sin \omega_I t]\end{aligned}$$

$$\searrow \qquad \swarrow$$

$$\begin{aligned}&I_x \cos \omega_I t \cos \pi J t + 2I_y K_z \cos \omega_I t \sin \pi J t + \\ &+ I_y \sin \omega_I t \cos \pi J t - 2I_x K_z \sin \omega_I t \sin \pi J t\end{aligned}$$

In other words, in a term which is the product of two operators, each operator evolves under the effect of chemical shift and scalar coupling independently. A further example is

$$\begin{aligned} I_x K_y \xrightarrow{\omega_I,\omega_K} & I_x \cos\omega_I t[K_y \cos\omega_K t - K_x \sin\omega_K t] \\ & + I_x \sin\omega_I t[K_y \cos\omega_K t - K_x \sin\omega_K t] \\ = & I_x K_y \cos\omega_I t \cos\omega_K t - I_x K_x \cos\omega_I t \sin\omega_K t \\ & + I_x K_y \sin\omega_I t \cos\omega_K t - I_x K_x \sin\omega_I t \sin\omega_K t \end{aligned}$$

Unfortunately, there is an exception: a term which is product of two transverse operators does not evolve under scalar coupling between the two spins. That is, scalar coupling between I and K does not affect terms like $I_x K_x$, $I_x K_y$, etc. Anyway, one should keep in mind that such terms do evolve under the effect of scalar coupling to a third nucleus. As an example, suppose that spin I is coupled to spin S with a constant J_{IS}:

$$I_x K_x \xrightarrow{\pi J_{IS} t} I_x K_x \cos\pi J_{IS} t + 2I_y K_x S_z \sin\pi J_{IS} t$$

Note that the operator K_x just acts as a multiplicative constant.

In products of two operators, each one evolves independently also under the effect of pulses. For instance:

$$I_z K_z \xrightarrow{90_y{}^\circ} I_x K_x$$

$$I_x K_y \xrightarrow{90^\circ{}_y} -I_z K_y$$

As can be understood by looking at Fig. IX.2, only single spin operators correspond to directly detectable magnetization. So, in common NMR experiments, only terms like I_x, I_y, K_x, K_y correspond to an observable, whereas all other terms ($I_x K_x$, $I_x K_y$, $I_y K_z$, etc.) are not directly observable.

By means of the few basic ideas exposed so far it is possible to follow a COSY experiment. Say the two coupled nuclei are I and K, the coupling constant between the two being J. We can follow the evolution of one spin: the evolution of the other can be easily obtained by swapping the I and K operators. After the first $90^\circ{}_y$ pulse, z magnetization is converted into x magnetization. Then, evolution during t_1 follows

$$\begin{aligned} I_x \longrightarrow & I_x \cos\omega_I t_1 \cos\pi J t_1 + 2I_y K_z \cos\omega_I t_1 \sin\pi J t_1 \\ & + I_y \sin\omega_I t_1 \cos\pi J t_1 - 2I_x K_z \sin\omega_I t_1 \sin\pi J t_1 \end{aligned}$$

After the second 90° pulse is applied along the x axis, the system is transformed into

$$I_x \cos\omega_I t_1 \cos\pi J t_1 - 2I_z K_y \cos\omega_I t_1 \sin\pi J t_1$$
$$+ I_z \sin\omega_I t_1 \cos\pi J t_1 + 2I_x K_y \sin\omega_I t_1 \sin\pi J t_1$$

At this point, signal detection occurs. As I_z and $2I_x K_y$ do not yield observable magnetization, the third and fourth terms are of no further interest and will be dropped. During t_2, the evolution of the first two terms is as follows (only relevant terms are shown):

$$I_x \cos\omega_I t_1 \cos\pi J t_1 \longrightarrow I_x \cos\omega_I t_1 \cos\pi J t_1 \cos\omega_I t_2 \cos\pi J t_2$$
$$+ I_y \cos\omega_I t_1 \cos\pi J t_1 \sin\omega_I t_2 \cos\pi J t_2 + \ldots$$

$$- 2I_z K_y \cos\omega_I t_1 \sin\pi J t_1 \longrightarrow K_x \cos\omega_I t_1 \sin\pi J t_1 \cos\omega_K t_2 \sin\pi J t_2$$
$$+ K_y \cos\omega_I t_1 \sin\pi J t_1 \sin\omega_K t_2 \sin\pi J t_2 + \ldots$$

The terms in the first row give rise to a peak at frequency ω_I in the first dimension and ω_I in the second dimension: that is, they give rise to the diagonal peak of spin I. The peak arising from the terms in the second row has frequency ω_K in the first dimension and ω_I in the second dimension, thus yielding a cross-peak between I and K. If one carries out the same calculation starting from K, its diagonal peak and the cross-peak between I and K at $(\omega_I; \omega_K)$ are obtained.

Product operators can thus be used to predict the behavior of an NMR experiment. The calculations are relatively simple to perform. Computer programs are available that also take into account the effects of phase-cycling to select the desired terms and reject unwanted ones. A drawback of the product operators approach is that, in its simplest version, it does not take into account the effect of relaxation. This is a must when dealing with paramagnetic substances. Exponential decay terms can be introduced to multiply each term and take relaxation into account. The method then becomes more cumbersome, and the effect of relaxation is introduced in a phenomenological way. A more detailed approach is that of using the concept of Redfield density matrix [1,2].

REFERENCES

[1] J. Shriver (1992) Concepts Magn. Reson. 4, 1.
[2] A. Bax (1982) Two Dimensional Nuclear Magnetic Resonance in Liquids. Reidel, Dordrecht.

Appendix X

Reference Tables

Physical and mathematical constants

Quantity	Symbol	Value	
		SI	CGS emu
Physical constants			
Permeability of vacuum	μ_0	$4\pi \times 10^{-7}$ kg m s^{-2} A^{-2}	1
Speed of light in vacuo	c	2.9979×10^{8} m s^{-1}	2.9979×10^{10} cm s^{-1}
Elementary charge (absolute value of the electron charge)	e	1.6022×10^{-19} A s	1.6022×10^{-20} abcoulombs
Plank constant	h	6.6261×10^{-34} J s	6.6261×10^{-27} erg s
	$\hbar = h/2\pi$	1.0546×10^{-34} J s rad^{-1}	1.0546×10^{-27} erg s rad^{-1}
Avogadro constant	N_A	6.0221×10^{23} mol^{-1}	
Electron rest mass	m_e	0.9109×10^{-30} kg	0.9109×10^{-27} g
Proton rest mass	m_p	1.6726×10^{-27} kg	1.6726×10^{-24} g
Free-spin electron g factor [a]	g_e	2.0023	
Electron Bohr magneton	μ_B	9.2740×10^{-24} J T^{-1}	9.2740×10^{-21} erg G^{-1}
Nuclear Bohr magneton	μ_N	5.0508×10^{-27} J T^{-1}	5.0508×10^{-24} erg G^{-1}
Free-electron magnetic moment	$\mu_e = -\frac{g_e \mu_B}{2}$	-9.2848×10^{-24} J T^{-1}	-9.2848×10^{-21} erg G^{-1}
Proton magnetic moment	$\mu_p = \frac{g_p \mu_N}{2}$	1.4106×10^{-26} J T^{-1}	1.4106×10^{-23} erg G^{-1}
Electron-to-proton magnetic moments ratio	$\lvert\mu_e/\mu_p\rvert$	658.21	
Free-electron magnetogyric ratio [a]	γ_e	-1.7609×10^{11} rad s^{-1} T^{-1}	-1.7609×10^{7} rad s^{-1} G^{-1}
Proton magnetogyric ratio	γ_p	2.6752×10^{8} rad s^{-1} T^{-1}	2.6752×10^{4} rad s^{-1} G^{-1}
Boltzmann constant	k	1.3807×10^{-23} J K^{-1}	1.3807×10^{-16} erg K^{-1}
Hyperfine coupling constant of the hydrogen atom	$(a/h)_{\mathrm{H}}$	1.4204×10^{9} Hz	
	$(a/\hbar)_{\mathrm{H}}$	8.9247×10^{9} rad s^{-1}	
Mathematical constants			
	e	2.7183	
	π	3.1416	

[a] The electron spin possesses a negative γ and therefore a negative g_e. However, it is common practice to use a positive g_e and to revert the sign of all equations involving g_e (see Eq. (1.10)).

Conversion factors

	To convert from	To	Multiply by
Length	Angstroms (Å)	meters (m)	1×10^{-10}
Pressure	atmospheres (atm)	pascals (Pa) (kg m^{-1} s^{-2})	1.01325×10^{5}
Mass	atomic mass unit (amu)	kilograms (kg)	1.6606×10^{-27}
Energy	calories (cal)	joules (J)	4.1840
	electron volts (eV)	joules (J)	1.6022×10^{-19}
	kilowatt-hours (kWh)	joules (J)	3.6×10^{6}
Viscosity	centipoises	kg m^{-1} s^{-1}	1×10^{-3}
Angles	degrees (deg)	radians (rad)	0.017453
Volume	liters (l)	cubic meters (m^3)	1×10^{-3}
Frequency	radians/seconds (rad s^{-1})	cycles/seconds (cps) (Hz)	0.15915
	wavenumber (cm^{-1})	energy (J)	1.9865×10^{-23}
		frequency (s^{-1}) (Hz)	2.9979×10^{10}
		frequency (rad s^{-1})	1.8837×10^{11}
	frequency (s^{-1}) (Hz)	energy (J)	6.6262×10^{-34}
	temperature (K)	energy (J)	1.3807×10^{-23}
		wavenumber (cm^{-1})	0.69467

Some physical quantities and their SI units

Electric charge (quantity of electricity)	Q	C	A s
Electric current	i	A	A
Electric potential	V	V	kg m^2 s^{-3} A^{-1} (= J A^{-1} s^{-1})
Energy	E	J	kg m^2 s^{-2}
Frequency	ν	Hz	s^{-1}
Frequency (angular velocity)	ω	rad s^{-1}	rad s^{-1}
Magnetic field strength	H	A m^{-1}	A m^{-1}
Magnetic induction (flux density)	B	T	kg A^{-1} s^{-2}
Magnetic moment	μ	J T^{-1}	A m^2
Magnetic susceptibility	χ	m^3	m^3
Magnetization	M	J T^{-1} m^{-3}	A m^{-1}
Magnetogyric ratio	Γ	rad s^{-1} T^{-1}	rad A s kg^{-1}
Molar magnetic susceptibility	χ_M	m^3 mol^{-1}	m^3 mol^{-1}
Power	P	W	kg m^2 s^{-3} (=J s^{-1})
Viscosity (dynamic)	η	kg m^{-1} s^{-1}	kg m^{-1} s^{-1}

Subject Index